THE HUBBARD MODEL

THE HUBBARD MODEL

A Reprint Volume

edited by

Arianna Montorsi
Dipartimento di Fisica
Politecnico di Torino

World Scientific
Singapore • New Jersey • London • Hong Kong

Published by

World Scientific Publishing Co. Pte. Ltd.
P O Box 128, Farrer Road, Singapore 9128
USA office: Suite 1B, 1060 Main Street, River Edge, NJ 07661
UK office: 73 Lynton Mead, Totteridge, London N20 8DH

The editor and publisher would like to thank the authors and the following publishers of the various journals for their assistance and permission to reproduce the selected reprints found in this volume:

American Association for the Advancement of Science (*Science*);
American Institute of Physics (*Sov. Phys. JETP*);
American Physical Society (*Phys. Rev., Phys. Rev. Lett.*);
Elsevier Science Publishers (*Physica B, Phys. Lett.*);
European Physical Society (*Europhys. Lett.*);
The Physical Society of Japan (*J. Phys. Soc. Jpn.*);
Progress of Theoretical Physics (*Prog. Theor. Phys.*);
The Royal Society (*Proc. Roy. Soc.*);
Springer-Verlag (*Z. Phys. B*).

THE HUBBARD MODEL — A Reprint Volume

ISBN 981-02-0585-6
ISBN 981-02-0586-4 (pbk)

Printed in Singapore by JBW Printers and Binders Pte. Ltd.

PREFACE

This book contains a collection of reprints on the Hubbard model. The major contributions to the subject since its origin are included, with the aim of providing all scientists working on the model and its applications with easy access to the relevant literature.

The book is divided into five parts. The first one is concerned with the physical origin and motivations of the model, whereas the remaining four sections are intended to present a coherent scenario of the different approaches to the model solution. The conceptual difficulty of the model is such that the exact and rigorous statistical mechanics results on it are relatively few (chapter 2). On the other hand, many relevant approximate results have been obtained in the strong coupling limit (chapter 3), and in other physically relevant limits (chapter 4). Furthermore, numerical Quantum Monte Carlo and exact diagonalization studies have also been shown to provide useful information (chapter 5).

Each section is preceded by commentary notes which introduce the papers presented (denoted by [#chapter.#order]), and add some references to papers not contained in the volume (quoted by [r#chapter.#order]).

CONTENTS

THE HUBBARD MODEL

Chapter 1

THE HUBBARD HAMILTONIAN IN PHYSICS

The model referred to as the Hubbard model appeared in the literature for the first time in 1963, in two subsequent independent papers — the first by Gutzwiller (see [1.1], eq. (11)), and the second by Hubbard ([1.2], eq. (10)) — as an attempt to describe in a simplified way the effect of correlations for d-electrons in transition metals. Indeed in the latter case the feature that electrons form a (narrow) delocalized d-band is in competition with the quasi-atomic behavior originated from correlations, which would make plausible an atomic description of the problem. The model hamiltonian H consists of two contributions,

$$H = \sum_{\mathbf{i},\mathbf{j}} \sum_{\sigma} t_{\mathbf{i},\mathbf{j}} \left(a^{\dagger}_{\mathbf{i},\sigma} a_{\mathbf{j},\sigma} + \text{h.c.} \right) + U \sum_{\mathbf{i}} n_{\mathbf{i},\uparrow} n_{\mathbf{i},\downarrow} \quad , \tag{1.1}$$

a kinetic term describing the motion of electrons between neighbouring sites (the hopping integral $t_{\mathbf{i},\mathbf{j}}$ is usually restricted to nearest-neighbours, and is assumed translationally invariant, namely $t_{\mathbf{i},\mathbf{j}} = -t$, $t > 0$), and an on-site term, which approximates the interactions among electrons, whose strength is given by the parameter U. $U > 0$ corresponds to repulsive Coulomb interaction, whereas $U < 0$ could eventually describe an effective attractive interaction mediated by the ions. $\mathbf{i},\mathbf{j}$ label the sites of a D-dimensional lattice Λ, $\sigma = \uparrow, \downarrow$ denotes the spin, and $a^{\dagger}_{\mathbf{i},\sigma}, a_{\mathbf{j},\sigma}$ are the electrons creation and annihilation operators, with $n_{\mathbf{i},\sigma} \doteq a^{\dagger}_{\mathbf{i},\sigma} a_{\mathbf{i},\sigma}$.

Hamiltonian (1.1) is expected to be capable of describing the main collective features of the above materials, namely itinerant magnetism and metal-insulator (Mott) transition. Indeed, for $U = 0$, H reduces to a system of non-interacting moving electrons, while for $t = 0$ (atomic limit) the electrons are fully localized, and at half-filling the ground state contains exactly one electron per site, i.e. the system is insulating. The latter feature still holds for finite t and $U = \infty$, and the corresponding system has been shown to be an antiferromagnetic insulator[r1.1]. A first question is then for which correlation strength one has the Mott transition, and under which conditions the system exhibits (anti-) ferromagnetic long-range order. Besides, one would like to know how these features depend on temperature, as well as on the filling of the band. The latter can be controlled, as usual, through the chemical potential μ, by adding to H a term $-\mu\Sigma_{\mathbf{i},\sigma}n_{\mathbf{i},\sigma}$, and fixing then μ so that the expectation value of the electron number operator per site is equal to 2δ. At half-filling and $T = 0$, $\mu = U/2$. Notice that the case $\delta > \frac{1}{2}$ can be discussed similarly to the case $\delta < \frac{1}{2}$ by considering holes instead of electrons, and changing the sign of t[3.6].

A considerable amount of work has been devoted to the solution of the Hubbard model since its introduction in physics. Nevertheless, exact results are still very rare, and their validity is mainly confined to the one-dimensional case, in which the metal-insulator transition is absent at any $T \neq 0$, according to the Mermin-Wagner theorem on the absence of long-range order in one- (and two-) dimensional systems. Even at $T = 0$ the exact solution exhibits no metal-insulator transition for any $U > 0$ and δ (ch. 2), and the magnetic properties of the model turn out to be peculiar of the dimension (ch. 3), in particular that of having always a total average spin equal to zero[r1.2]. Many diverse approaches have been proposed in order to gain information about the behavior of such an oversimplified system, and a first account of these attempts, and of the established results obtained up to the late seventies, is given in [1.3].

Successively, the interest for the Hubbard model, and in general for the field of strongly correlated electron systems, has been revived by the discovery of heavy fermions and high-T_c superconductors[1.4]. It is in fact believed that in the latter materials the relevance of the interplay between magnetic behavior and Mott transition near half-filling is crucial, in that on the one hand most of the new ceramic superconductors are good Mott insulators, and on the other hand, especially in the Cu–O_2 two-dimensional planes in which superconductivity takes place, they can exhibit strong antiferromagnetic correlations, and the antiferromagnetic phase is remarkably close to the superconducting one. Actually, a recent result of Yang[r1.3],[1.5] shows that it is possible to construct metastable eigenstates of the Hubbard hamiltonian which exhibit off-diagonal long-range order — essential for superconductivity —, and these can always be labelled in terms of two quantum numbers, one of which is related to the existence of superconductivity in the state, and the other to its magnetic properties. It is well established[3.3] that superconductivity can occur for $U < 0$, but it is not clear whether this is still true for repulsive interaction.

The important physical applications related to superconductivity have stimulated the achievement of new, significant results, concerning the properties of the model, especially for the two-dimensional case, even if nowadays, there is still no general agreement about their meaning within the framework of high-T_c materials.

[r1.1] P.W. Anderson, *Phys. Rev.* **115**, 2 (1959).
[r1.2] E. Lieb and D.C. Mattis, *Phys. Rev.* **125**, 164 (1962).
[r1.3] C.N. Yang, *Phys. Rev. Lett.* **63**, 2144 (1989).

VOLUME 10, NUMBER 5 PHYSICAL REVIEW LETTERS 1 MARCH 1963

EFFECT OF CORRELATION ON THE FERROMAGNETISM OF TRANSITION METALS

Martin C. Gutzwiller
Research Laboratory Zurich, International Business Machines Corporation, Rüschlikon ZH, Switzerland
(Received 27 September 1962)

The purpose of this Letter is to present a new approach to the problem of ferromagnetism in a metal. A correlated wave function for the electrons in the 3*d* band is proposed as approximation to the ground state. The expectation value of the energy is evaluated by diagram techniques. The simplest example of a face-centered cubic structure (whose density-of-states curve is parabolic at the bottom and has a peak at the top) is discussed. Under these assumptions the arguments show that the ferromagnetic state is lower if the band is nearly full, whereas the nonmagnetic state has the lower energy if the band is nearly empty.

The main attempt so far to explain ferromagnetism in metals is based on the collective electron theory of ferromagnetism,[1] in which both the magnetic and the nonmagnetic ground states are assumed to be antisymmetrized products of Bloch functions. The expectation value of the energy in the nonmagnetic state contains a large term which is due to the repulsion of two electrons of opposite spin at the same lattice site. Slater,[2] in particular, pointed out that this term should be reduced by considering correlated wave functions before the effects of exchange are discussed. The collective electron theory fails especially in the limit of large spacing between the lattice sites, a situation comparable to that of the relatively tight 3*d* levels in the transition metals.

The present model is an attempt to deal with

this particular difficulty. The correlated wave function Ψ is obtained from the antisymmetrized product Φ of Bloch functions by simply eliminating those parts in Ψ in which two electrons of opposite spin happen to be at the same lattice site.

Consider a lattice of L sites which are numbered by an index g. To each site belongs, for a given spin, only one orbital $\varphi(x-g)$ of the Wannier type

$$\int \varphi^*(x-g_1)\varphi(x-g_2)dx = \delta_{g_1g_2}. \tag{1}$$

The restriction to Wannier functions is made in this Letter in order to bring out more clearly the way in which the correlation between electrons of opposite spin is achieved. The generalization to nonorthogonal orbits is feasible, and the present theory is viewed as an approximation which is valid for small overlap between two orbits $\varphi(x-g_1)$ and $\varphi(x-g_2)$ belonging to different sites in the lattice. Bloch waves $\psi_k(x)$ are constructed by forming

$$\psi_k(x) = L^{-1/2}\sum_g \exp(ikg)\varphi(x-g). \tag{2}$$

Each of these wave functions is to be multiplied with a spin function, indicated by an arrow, $\uparrow$ or $\downarrow$, as index. Corresponding to $\varphi(x-g)$, there is a Fermion creation operator a_g^+, in the usual manner; and corresponding to $\psi_k(x)$, there is a Fermion creation operator a_k^+, with the relation (2) now becoming

$$a_k^+ = L^{-1/2}\sum_g \exp(ikg)a_g^+. \tag{3}$$

The wave function Φ is represented by

$$\Phi = \prod_{\{k\}} a_{k\uparrow}^+ \prod_{\{\kappa\}} a_{\kappa\downarrow}^+ \Phi_0, \tag{4}$$

where Φ_0 is the vacuum. The sets $\{k\}$ and $\{\kappa\}$ extend over the appropriate regions in reciprocal space. These regions are, in general, the inside of some Fermi surface, and it may well be that the Fermi surface S for the set $\{k\}$ with spin up does not coincide with the Fermi surface Σ for the set $\{\kappa\}$ with spin down. The total number N of electrons described by Φ does not have to be in any simple relation to the number L of lattice sites. Indeed, if one takes the liberty of grouping the 4s electrons in Ni with the effective periodic potential, one is left with a fractional value of N/L for the 3d band. If there are more electrons than lattice sites, the whole formalism can be written in terms of holes instead of electrons, so that one may always assume $N \leq L$.

The correlated wave function Ψ is defined by

$$\Psi = \prod_g [1 - (1-\eta)N_{g\uparrow}N_{g\downarrow}]\Phi, \tag{5}$$

where $N_g = a_g^+ a_g$. As $L \to \infty$, the function Ψ becomes an eigenfunction of the operator $W = \sum_g N_{g\uparrow}N_{g\downarrow}$. For a given antisymmetrized product Φ, the value of η is to be considered as a function of $\langle W\rangle$.

By a straightforward computation, the norm of Ψ is shown to be

$$(\Psi|\Psi) = 1 + \sum_{m=1}^{L} \frac{(\eta^2-1)^m}{m!} \sum_{g_1\cdots g_m} \begin{vmatrix} W(g_1-g_1) & \cdots & W(g_1-g_m) \\ \vdots & & \vdots \\ W(g_m-g_1) & \cdots & W(g_m-g_m) \end{vmatrix} \cdot \begin{vmatrix} \omega(g_1-g_1) & \cdots & \omega(g_1-g_m) \\ \vdots & & \vdots \\ \omega(g_m-g_1) & \cdots & \omega(g_m-g_m) \end{vmatrix}, \tag{6}$$

where the kernels $W(g)$ and $\omega(g)$ are given by

$$W(g) = \frac{1}{L}\sum_{k \subset S} \exp(ikg), \quad \omega(g) = \frac{1}{L}\sum_{\kappa \subset \Sigma} \exp(i\kappa g). \tag{7}$$

Similar but somewhat more complicated expressions can be obtained for $(\Psi|a_f^+ a_h|\Psi)$, etc. Of particular interest is the expectation value

$$\langle W\rangle = \tfrac{1}{2}\eta \cdot \partial \ln(\Psi|\Psi)/\partial\eta. \tag{8}$$

Expressions such as (6) cannot be evaluated directly. However, a fairly simple diagrammatic analysis can be used. Each of the $m!$ permutations which occur in expanding the determinants of (6) can be viewed as the result of performing the appropriate cycle permutations. Each cycle is represented by an oriented polygon whose vertices are labeled by the numbers of the cycle in their correct order. There are two kinds of polygons, one for spin-up particles and one for spin-down particles. Corresponding vertices, i.e., vertices which are labeled by the same number, are connected by a dashed line to complete the diagram. The contribution of each diagram to $(\Psi|\Psi)$ is easily found.

The subdivision of a diagram into linked clusters, the assignment of a symmetry number to each linked cluster, and the counting of equivalent diagrams is done in the usual manner. The result is

$$(\Psi|\Psi)=\exp\left[\sum_{\nu=1}^{L}\frac{(\eta^2-1)^\nu}{\nu!}\sum\Lambda_\nu\right], \tag{9}$$

where Λ_ν stands for the contribution of a particular linked cluster of order ν. The contribution Λ_ν of each linked cluster is proportional to L. Moreover, Λ_ν is proportional to $(N/L)^{\nu+1}$. For instance, $\langle W\rangle=\eta^2N_\uparrow N_\downarrow/L$ from the term $\nu=1$ in (9). All expectation values of interest can, therefore, be computed with good accuracy for $N\ll L$. (In the 50-50 alloy of Ni and Cu, which is still ferromagnetic, one has $N/L\lesssim 1/10$.)

The occupation probability $N_{k\uparrow}=\langle a_{k\uparrow}{}^+a_{k\uparrow}\rangle$ is given to second order in N/L by

$$N_{k\uparrow}=\begin{cases}1-(1-\eta)^2n_\downarrow & \text{for } k\subset S,\\ (1-\eta)^2n_\uparrow n_\downarrow & \text{for } k\not\subset S,\end{cases} \tag{10}$$

where $n_\uparrow=N_\uparrow/L$ and $n_\downarrow=N_\downarrow/L$. The average number of particles inside the Fermi surface is reduced by a factor $[1-(1-\eta)^2n_\downarrow]$. In the special case $\eta=0$, it turns out that $N_{k\uparrow}=1-n_\downarrow$ for $k\subset S$ quite independently of the magnitude of $n_\uparrow$ and $n_\downarrow$. For $N=L$ and $\eta=0$ one has, therefore, $N_{k\uparrow}=N_{k\downarrow}=\frac{1}{2}$ over the whole Brillouin zone, independently of the choice of Φ, in agreement with the Heisenberg model of ferromagnetism.

It is instructive to apply these results to the following example. The Hamiltonian is given by

$$H=\sum_k E_k(N_{k\uparrow}+N_{k\downarrow})+C\sum_g N_{g\uparrow}N_{g\downarrow}. \tag{11}$$

The first part of H arises from solving Schrödinger's equation for a single electron in the effective periodic potential of the lattice. The second part describes the repulsion between two electrons of opposite spin which happen to be in the same orbit around a particular lattice site. All those interactions whose matrix elements involve integrals over orbits at different sites, and this includes the interaction which is associated with the exchange integral, are grouped with the effective periodic potential.

If the crystal potential has a nonvanishing matrix element ϵ only between nearest neighbors, the function E_k is given for an fcc lattice by

$$E_k=-4\epsilon(\cos ak_1\cos ak_2+\cos ak_2\cos ak_3+\cos ak_3\cos ak_1), \tag{12}$$

with respect to cubic axes. For $\epsilon>0$ the bottom of the band is at $E=-12\epsilon$, where the density of states dn/dE as a function of energy E vanishes like $(E+12\epsilon)^{1/2}$. Near the top of the band at $E=4\epsilon$ the density of states has a peak like $-\log(4\epsilon-E)^{1/2}$. The expectation value $\langle H\rangle_u$ can be obtained for Ψ in the nonmagnetic state by minimizing $\langle H\rangle$ with respect to η which is then found to be $(1+C/8\epsilon)^{-1}$ for small values of $n=N/L$. The resulting value of $\bar{H}_u=\langle H\rangle_u/N$ is compared to $\bar{H}_f=\langle H\rangle_f/N$ for the magnetic state. The calculations are easily worked out if n is only a few percent.

The following results and typical figures are obtained. If the band contains only few electrons, then the nonmagnetic state Ψ has the lower $\langle H\rangle$, e.g., for $n=0.01$ and $\eta=0$, one has $\bar{H}_f=1.07$ above the bottom of the band, and $\Delta E=\bar{H}_f-\bar{H}_u=0.26\,\epsilon$. However, if the band is nearly full, then the magnetic state has a lower $\langle H\rangle$ than Ψ, provided C is sufficiently larger than the bandwidth $16\,\epsilon$. For instance, for $n=0.01$ and $\eta=0$, one has $\bar{H}_f=-0.022\,\epsilon$ below the top of the band, and $\Delta E=-0.007\,\epsilon$, whereas for $n=0.01$ and $\eta=0.2$ (corresponding to C equals twice the bandwidth) one still has $\bar{H}_f=-0.022\,\epsilon$, but $\Delta E=-0.0004\,\epsilon$.

The wave function (5) is only an approximation for the ground state, and, therefore, it is not certain that for a nearly full band the Hamiltonian (11) with the band structure (12) leads to ferromagnetism, while it does not for a nearly empty band. It seems rather that the exact ground state of the Hamiltonian (11) is never ferromagnetic. But it may be conjectured on the basis of the above figures that the exact nonmagnetic ground state of the Hamiltonian (11) lies much closer to the ferromagnetic ground state of the same Hamiltonian if the band is nearly full than if it is nearly empty. Also, this difference in behavior is certainly due to the different behavior of the density-of-states curve at the top and at the bottom of the band. The occurrence of ferromagnetism would then depend on the interaction of the electrons which is associated with the two-center Coulomb integrals, in particular, the exchange integral. But in view of the above correlation through the one-center Coulomb integral, the top of the band structure (12) has gained a decisive advantage for achieving ferromagnetism over the bottom. Indeed, the expectation value for the terms associated with the two-center Coulomb integrals is the same for the nearly empty and the nearly full bands, positive for the nonmagnetic state, zero for the ferromagnetic state.

The author is especially indebted to Professor R. Brout and Dr. D. Mattis, Dr. S. Nettel, and Dr. H. Thomas for many stimulating discussions.

[1]See E. P. Wohlfahrt, Rev. Modern Phys. 25, 211 (1953).

[2]J. C. Slater, Rev. Modern Phys. 25, 199 (1953).

Electron correlations in narrow energy bands

By J. Hubbard

Theoretical Physics Division, A.E.R.E., Harwell, Didcot, Berks

(*Communicated by B. H. Flowers, F.R.S.—Received* 23 *April* 1963)

It is pointed out that one of the main effects of correlation phenomena in d- and f-bands is to give rise to behaviour characteristic of the atomic or Heitler–London model. To investigate this situation a simple, approximate model for the interaction of electrons in narrow energy bands is introduced. The results of applying the Hartree–Fock approximation to this model are examined. Using a Green function technique an approximate solution of the correlation problem for this model is obtained. This solution has the property of reducing to the exact atomic solution in the appropriate limit and to the ordinary uncorrelated band picture in the opposite limit. The condition for ferromagnetism of this solution is discussed. To clarify the physical meaning of the solution a two-electron example is examined.

1. Introduction

In recent years much attention has been given to the theory of correlation effects in the free electron gas (Bohm & Pines 1953; Gell-Mann & Brueckner 1957; Sawada, Brueckner, Fukuda & Brout 1957; Hubbard 1957, 1958; Pines & Nozières 1958). Apart from the intrinsic interest of this problem, the free electron gas serves as a model for the conduction bands of metals and alloys. Transition and rare-earth metals have in addition to their conduction bands partly filled d- or f-bands which give rise to the characteristic properties of these metals. Correlation phenomena are of great importance in determining the properties of these narrow energy bands, indeed more important than the corresponding effects in conduction bands. Unfortunately, however, the free-electron gas does not provide a good model for these bands. Rather, one requires a theory of correlations which takes into account adequately the atomistic nature of the solid. Indeed, in the case of the f-electrons of rare earth metals it is probable that for most purposes a purely atomic (sometimes referred to as a Heitler–London or localized) model will prove satisfactory. The same cannot be said, however, of the d-electrons of transition metals. It is with one approach to a theory of correlation effects in the d-bands of transition metals that this paper is concerned.

A theory of correlation effects in narrow energy bands is inevitably of a somewhat different nature from a theory of correlation effects in the free electron gas. The electron charge density in a d-band is concentrated near the nuclei of the solid and sparse between the atoms, making it possible to speak with some meaning of an electron being 'on' a particular atom. This circumstance gives rise to the possibility of an atomic description of the d-band despite its considerable bandwidth. It is, in fact, found experimentally that the d-electrons of transition metals exhibit behaviour characteristic of both the ordinary band model and the atomic model. For example, the occurrence of spin-wave phenomena in ferromagnetic metals and the strong temperature dependence of the susceptibilities of some transition metals represent properties which can be understood on the basis of an atomic model, while the large d-electron contribution to the low temperature specific heat and the occurrence in ferromagnets of magnetic moments per atom which are far from integral numbers of Bohr magnetons are properties which are easily explained by band theory. As will be tried to make plausible below, it is correlation effects in narrow bands which lead to the atomic behaviour and it is only by taking correlation effects into account that one can understand how d-electrons exhibit both kinds of behaviour simultaneously. Thus a theory of correlations in d-bands will be mainly concerned with understanding this situation in greater detail and determining the balance between bandlike and atomic-like behaviour.

In its most naïve form the atomic theory would picture a transition metal as a collection of (say singly charged) ions immersed in the conduction electron gas and interacting with each other in much the same way as the corresponding ions in salts. If, as is generally supposed, the number of d-electrons per atom is non-integral this simple picture is untenable. However, it is possible to substitute for it a less restrictive model which nevertheless guarantees most of the characteristic properties of the atomic model. It is sufficient to assert that, despite the band motion of the d-electrons, the electrons on any atom are strongly correlated with each other but only weakly with electrons on other atoms; such intra-atomic correlations are inevitably of such a type as to make the metal behave to some extent according to the predictions of the atomic model.

It may be that this situation can be made clear by considering one or two examples. Consider first a partly filled d-band of non-interacting electrons. In such a system the spin of an atom (that is the total spin of all the electrons on that atom) is a quantity which fluctuates randomly in magnitude and direction, the characteristic time of fluctuation being of the order of the d-electron hopping time, i.e. the time ($\sim \hbar/\Delta$, $\Delta = d$-electron bandwidth) in which a d-electron hops from one atom to another in performing its band motion. In this situation it is reasonable to think of the spin being associated with each of the moving d-electrons.

Let us now inquire what effect one might expect the electron interaction will have in this situation. As a guide one may note that Hund's first rule for atoms indicates that the intra-atomic interactions are of such a nature as to aline the electron spins on an atom, so one may expect a similar effect in a metal. Suppose now that the electrons have their spins quantized in what will be called the up and down directions and that at some instant a given atom has its total spin in the up direction. Then the intra-atomic interactions are, according to Hund's rule, of such a nature that this atom tends to attract electrons with spin up and repel those with spin down. In this way the property of an atom on having total spin at some instant tends to be self-perpetuating. If these intra-atomic forces are strong enough to produce appreciable correlations, then it follows that the state of total spin up on an atom may persist for a period long compared with the d-electron hopping time. This persistence of the atomic spin state is not due to the same up-spin electrons being localized on the atom. The actual electrons on the atom are always changing as a result of their band motion, but the electron motions are correlated in such a way as to keep a preponderance of up-spin electrons on the atom. In these circumstances (i.e. if the correlations are strong enough) one can think of the spin as being associated with the atom rather than with the electrons and the possibility of an atomic or Heisenberg model emerges.

This example illustrates the possibilities of the situation. Although one may still suppose the electrons to move rapidly from atom to atom as assumed in the band model, their motion may be correlated in such a manner as to give properties characteristic of the atomic theory. In this way one may understand how the electrons can exhibit both types of behaviour simultaneously. The degree of atomic behaviour exhibited depends upon the strength of the correlations.

A second example which has been studied by various authors (Slater 1937; Herring 1952; Thompson 1960; Edwards 1962; Kubo, Izuyama & Kim 1962) is the theory of spin-waves in the band model of ferromagnetic metals. These authors show that the spin-wave can be regarded as a collective motion which appears when the electron interactions are taken into account. More precisely, the spin-wave appears as a bound state of an electron of one spin with a hole of opposite spin, the relative motion of the electron and hole being such that they spend most of their time on the same atom. Now, an electron of one spin and a hole of opposite spin on the same atom look just like a reversed spin on that atom, the motion of the bound electron-hole pair resembling a motion of the reversed spin from atom to atom, which is just the Heisenberg model picture of a spin-wave. Thus again the atomic picture emerges as a consequence of correlation effects, this time the correlation between an electron and a hole.

Yet another important example concerns the fluctuation in the number of electrons on a given atom. It is, of course, one of the more obvious features of the atomic model that it assumes that there are the same number of electrons on each atom. But one can show that for uncorrelated electrons belonging to a band containing ν states per atom that the probability of finding n electrons on a given atom is given by the binomial distribution

$$\frac{\nu!}{n!\,(\nu-n)!}\left(\frac{s}{\nu}\right)^n\left(1-\frac{s}{\nu}\right)^{\nu-n},$$

where s is the mean number of electrons per atom. Thus n fluctuates about its mean value s, the root-mean-square fluctuation being $\sqrt{\{s(1-s/\nu)\}}$ and the frequency of fluctuation of the order of an electron hopping time. Now one general effect of electrostatic interactions is a tendency to even out the electron charge distribution, opposing the build-up of an excess of charge in one place and a deficiency in another. Thus the correlations produced by the interaction will be of such a nature as to reduce the fluctuation in the electron number on each atom. It is this type of correlation which is most important in the hypothetical case of a collection of atoms arranged on a lattice but widely separated from each other. Formally ordinary band theory is applicable to such a situation, but the correlation effects of the type discussed above are dominant and make the system behave like a set of isolated neutral atoms, which is clearly the correct description physically.

It is clear from the above discussion that an important requirement of a theory of correlations in narrow energy bands is that it have the property of reducing to the atomic solution in the appropriate limit, i.e. when applied to a hypothetical system of atoms on a lattice but widely separated from each other and interacting only weakly. It is one of the purposes of this paper to describe a very simple, approximate theory having this property. Although one has always in mind the case of d-electrons, the theory to be described is concerned with the case of an s-band having two states per atom (up and down spin states). The advantage of this particular case is its comparative mathematical simplicity. One may expect that some important aspects of the real (d-electron) case will be missed in a study of the s-band case but may nevertheless hope to obtain some results of general application.

It might seem that in view of the fact that no adequate theory of correlations in free electron gases at metallic densities exists at the present time that it is over-ambitious to attempt a study of the formally more difficult case of band electrons. However, it turns out that in the case of narrow energy bands one can take account of the atomicity of the electron distribution to introduce a very simple approximate representation of the electron interactions. This approximate interaction is, in fact, mathematically much simpler to handle than the Coulomb interaction itself. This possibility has been well known for many years and has been applied to the spin-wave problem by the authors mentioned in that connexion above, but does not seem to have been exploited hitherto in connexion with the general correlation problem. In § 2 this approximate interaction and the adequacy of the approximation involved is discussed.

For the sake of comparison with the results of the theory of correlations developed later, in § 3 the application of the Hartree–Fock approximation to the simplified interaction is considered and in particular the condition for ferromagnetism predicted by Hartree–Fock theory is examined.

In §§ 5 and 6 the approximate correlation theory for an s-band mentioned above is developed. To this end a Green function technique of the type described by Zubarev (1960) is used; to establish the notation the basic definitions and equations of this technique are briefly reviewed in § 4. In § 5 it is shown how, using this technique, an exact solution can be obtained in the atomic (zero bandwidth) limit. In § 6 the same method is applied to the general (finite bandwidth) case to obtain the approximate solution. In § 7 the nature and some of the properties of this solution are discussed.

In § 8 we examine a 2-electron problem which has been studied previously in a related context (Slater, Statz & Koster 1953) and which throws some light upon the physical interpretation of the solution obtained in the preceding sections.

Finally in § 9 the condition for ferromagnetism predicted by the new calculation is discussed. It is found to be considerably more restrictive than the corresponding criterion derived from Hartree–Fock theory, and, in fact, can only be satisfied in rather special circumstances.

2. An approximate representation of electron interactions

In this section the approximate model of electron interactions in narrow energy bands used in later calculations is described. As pointed out in the introduction, for reasons of mathematical simplicity the case of an s-band will be considered. However, when discussing below the validity of the various approximations which have gone into the derivation of the model we shall assume we are dealing with $3d$-transition metal electrons since this is the case of real interest.

Consider a hypothetical partly-filled narrow s-band containing n electrons per atom. The Bloch functions of the band will be denoted by $\psi_{\mathbf{k}}$ and the corresponding energy by $\epsilon_{\mathbf{k}}$ where $\mathbf{k}$ is the wave vector. These wave functions and energies are assumed to have been calculated in some appropriate Hartree–Fock potential representing the average interaction of the s-band electrons with the electrons of other bands and the n electrons per atom of the s-band itself. This Hartree–Fock potential will be assumed to be spin independent so one has the same energies and wave-functions for both spins.

Now let $c_{\mathbf{k}\sigma}$, $c^+_{\mathbf{k}\sigma}$ be the destruction and creation operators for electrons in the Bloch state $(\mathbf{k}, \sigma)$, where $\sigma = \pm 1$ is the spin label. Then the dynamics of the electrons of the band may be described approximately by the Hamiltonian

$$
\begin{aligned}
H = & \sum_{\mathbf{k}\sigma} \epsilon_{\mathbf{k}} c^{\dagger}_{\mathbf{k}\sigma} c_{\mathbf{k}\sigma} \\
& + \tfrac{1}{2} \sum_{\mathbf{k}_1 \mathbf{k}_2 \mathbf{k}_1' \mathbf{k}_2'} \sum_{\sigma_1 \sigma_2} (\mathbf{k}_1 \mathbf{k}_2 |\, 1/r \,| \mathbf{k}_1' \mathbf{k}_2') c^{\dagger}_{\mathbf{k}_1 \sigma_1} c^{\dagger}_{\mathbf{k}_2 \sigma_2} c_{\mathbf{k}_2' \sigma_2} c_{\mathbf{k}_1' \sigma_1} \\
& - \sum_{\mathbf{k}\mathbf{k}'} \sum_{\sigma} \{2(\mathbf{k}\mathbf{k}' |\, 1/r \,| \mathbf{k}\mathbf{k}') - (\mathbf{k}\mathbf{k}' |\, 1/r \,| \mathbf{k}'\mathbf{k})\} \nu_{\mathbf{k}'} c^{+}_{\mathbf{k}\sigma} c_{\mathbf{k}\sigma},
\end{aligned} \tag{1}
$$

where the $\mathbf{k}$ sums run over the first Brillouin zone (all sums over momenta in this paper are to be understood in this way) and where

$$
(\mathbf{k}_1 \mathbf{k}_2 |\, 1/r \,| \mathbf{k}_1' \mathbf{k}_2') = e^2 \int \frac{\psi^*_{\mathbf{k}_1}(\mathbf{x})\, \psi_{\mathbf{k}_1'}(\mathbf{x})\, \psi^*_{\mathbf{k}_2}(\mathbf{x}')\, \psi_{\mathbf{k}_2'}(\mathbf{x}')}{|\mathbf{x} - \mathbf{x}'|} d\mathbf{x}\, d\mathbf{x}'. \tag{2}
$$

The first term of H represents the ordinary band energies of the electrons, the second their interaction energy. The last term subtracts the potential energy of the electrons in that part of the Hartree–Fock field arising from the electrons of the s-band itself. This term has to be subtracted off to avoid counting the interactions of the electrons of the band twice, once explicitly in the Hamiltonian and also implicitly through the Hartree–Fock field determining the $\epsilon_{\mathbf{k}}$. The $\nu_{\mathbf{k}}$ are the assumed occupation numbers of the states of the band in the Hartree–Fock calculation; it has been assumed that up and down spin states are occupied equally.

It is convenient to transform the Hamiltonian of (1) by introducing the Wannier functions

$$
\phi(\mathbf{x}) = N^{-\frac{1}{2}} \sum_{\mathbf{k}} \psi_{\mathbf{k}}(\mathbf{x}), \tag{3}
$$

where N is the number of atoms. One can then write

$$
\psi_{\mathbf{k}}(\mathbf{x}) = N^{-\frac{1}{2}} \sum_{i} e^{i\mathbf{k}.\mathbf{R}_i} \phi(\mathbf{x} - \mathbf{R}_i), \tag{4}
$$

where the sum runs over all the atomic positions $\mathbf{R}_i$. Introducing the creation and destruction operators $c^{\dagger}_{i\sigma}$ and $c_{i\sigma}$ for an electron of spin σ in the orbital state $\phi(\mathbf{x} - \mathbf{R}_i)$, one can also write

$$
c_{\mathbf{k}\sigma} = N^{-\frac{1}{2}} \sum_{i} e^{i\mathbf{k}.\mathbf{R}_i} c_{i\sigma}, \quad c^{\dagger}_{\mathbf{k}\sigma} = N^{-\frac{1}{2}} \sum_{i} e^{-i\mathbf{k}.\mathbf{R}_i} c^{\dagger}_{i\sigma}. \tag{5}
$$

These results can now be used to rewrite the Hamiltonian of (1) as

$$H = \sum_{i,j}\sum_{\sigma} T_{ij} c_{i\sigma}^{\dagger} c_{j\sigma} + \tfrac{1}{2} \sum_{ijkl} \sum_{\sigma\sigma'} (ij\,|1/r|\,kl)\, c_{i\sigma}^{\dagger} c_{j\sigma'}^{\dagger} c_{l\sigma'} c_{k\sigma}$$

$$- \sum_{ijkl} \sum_{\sigma} \{2(ij\,|1/r|\,kl) - (ij|\,1/r\,|lk)\}\, \nu_{jl} c_{i\sigma}^{\dagger} c_{k\sigma}^{\dagger}, \tag{6}$$

where

$$T_{ij} = N^{-1} \sum_{\mathbf{k}} \epsilon_{\mathbf{k}}\, e^{i\mathbf{k}.(\mathbf{R}_i - \mathbf{R}_j)}, \tag{7}$$

$$(ij|\,1/r\,|kl) = e^2 \int \frac{\phi^*(\mathbf{x}-\mathbf{R}_i)\,\phi(\mathbf{x}-\mathbf{R}_k)\,\phi^*(\mathbf{x}'-\mathbf{R}_j)\,\phi(\mathbf{x}'-\mathbf{R}_l)}{|\mathbf{x}-\mathbf{x}'|}\, d\mathbf{x}\, d\mathbf{x}' \tag{8}$$

and

$$\nu_{jl} = N^{-1} \sum_{\mathbf{k}} \nu_{\mathbf{k}}\, e^{i\mathbf{k}.(\mathbf{R}_j - \mathbf{R}_l)}. \tag{9}$$

It is now possible to make the essential simplifying approximation. Since one is dealing with a narrow energy band the Wannier functions ϕ will closely resemble atomic s-functions. Furthermore, if the bandwidth is to be small these s-functions must form an atomic shell which has a radius small compared with the inter-atomic spacing. From (8) it may be seen that in this circumstance the integral $(ii|\,1/r\,|ii) = I$ will be much greater in magnitude than any of the other integrals (8), suggesting that a possible approximation is to neglect all the integrals (8) apart from I. If this approximation, the validity of which is discussed in greater detail below, is made, then the Hamiltonian of (6) becomes

$$H = \sum_{i,j}\sum_{\sigma} T_{ij} c_{i\sigma}^{\dagger} c_{j\sigma} + \tfrac{1}{2} I \sum_{i,\sigma} n_{i\sigma} n_{i,-\sigma} - I \sum_{i,\sigma} \nu_{ii} n_{i\sigma}, \tag{10}$$

where $n_{i\sigma} = c_{i\sigma}^{\dagger} c_{i\sigma}$. From (9), $\nu_{ii} = N^{-1} \sum_{\mathbf{k}} \nu_{\mathbf{k}} = \tfrac{1}{2} n$, so the last term of (10) reduces to $-\tfrac{1}{2} I n \sum_{i,\sigma} n_{i\sigma} = -\tfrac{1}{2} I N n^2 =$ constant and may be dropped. Equation (10) gives the approximate Hamiltonian used in the later sections of this paper.

Obviously many approximations, explicit and implicit, have gone into the derivation of the simplified Hamiltonian of (10). We will next try to assess the validity of these approximations when applied to the case of transition metal $3d$-electrons.

The most obvious approximation has been the neglect of all the interaction terms in (6) other than the $(ii|\,1/r\,|ii)$ term. For the sake of comparison one may note that I has the order of magnitude 20 eV for $3d$-electrons in transition metals. The largest of the neglected terms are those of the type $(ij|\,1/r\,|ij)$ where i and j are nearest neighbours. From (9) these integrals can be estimated to have the order of magnitude $(2/R)\,\mathrm{Ry} \sim 6\,\mathrm{eV}$ ($R =$ interatomic spacing in Bohr units). Actually this figure should be reduced appreciably to allow for the screening of the interactions of electrons on different atoms by the conduction electron gas. This screening effect may be allowed for approximately by multiplying the above estimate by a factor $e^{-\kappa R}$ where κ is an appropriate screening constant. In the case of $3d$ transition metals $e^{-\kappa R} \sim \tfrac{1}{3} - \tfrac{1}{2}$, reducing the $(ij|\,1/r\,|ij)$ term to the order of magnitude 2 to 3 eV. For the case in which i and j are now nearest neighbours

$$(ij|\,1/r\,|ij) \sim \frac{2e^{-\kappa|\mathbf{R}_i - \mathbf{R}_j|}}{|\mathbf{R}_i - \mathbf{R}_j|}\,\mathrm{Ry}$$

which falls off rapidly with increasing $|\mathbf{R}_i - \mathbf{R}_j|$ on account of the exponential factor. Thus the term

$$\tfrac{1}{2} \sum_{i,j} \sum_{\sigma,\sigma'} (ij|\, 1/r \,|ij)\, n_{i\sigma} n_{j\sigma'}, \tag{11}$$

in (6) is quite appreciable, but can, perhaps be neglected compared to I as a first approximation.

The next biggest terms neglected are those of the types:

$$(ii|\, 1/r \,|ij) \sim q\,\mathrm{Ry} \sim \tfrac{1}{2}\mathrm{eV},$$
$$(ij|\, 1/r \,|ik) \sim \tfrac{1}{4}q\,\mathrm{Ry} \sim \tfrac{1}{10}\mathrm{eV},$$
$$(ii|\, 1/r \,|jj) \sim (ij|\, 1/r \,|ji) \sim q^2\,\mathrm{Ry} \sim \tfrac{1}{40}\mathrm{eV},$$

where i, j and k are all nearest neighbours and $q \sim \frac{1}{20}$ is the overlap charge (in units of e) between two $3d$-electrons on nearest neighbour atoms. All the other interaction terms in (6) which have been neglected are smaller still than these which one sees are already small compared to those of (11).

A different type of approximation that has been made is to assume that only the interactions of importance are those between the $3d$-electrons (actually between the electrons of the s-band in the equations above), the interactions with electrons of other bands being represented only through the Hartree–Fock field. One question concerning this point is raised at once by the fact that in estimating the order of magnitude of the terms of (11) allowance was made for the screening effect of the conduction electron gas on the interactions. It might therefore be inquired whether there is not a similar screening effect reducing the magnitude of I. There is, in fact, such an effect. Because the speed at which d-electrons move from atom to atom is slow compared with the velocity of a typical conduction electron the latter can correlate efficiently with the d-electrons and screen their fields. Thus, if a given atom has an extra d-electron its negative charge will repel conduction electrons producing a correlation hole about that atom in the conduction electron gas. The presence of this correlation hole reduces the electrostatic potential at the atom (and therefore at each of its d-electrons) by about 5 V, which is equivalent to reducing I by 5 eV. This reduction is appreciable but does not change the order of magnitude of I.

It might also be thought that I will be reduced by the screening of the interactions of the d-electrons by the core electrons and by the d-electrons themselves. This is not expected to be a big effect, however, because the kinetic energies of the orbital motion of the d-electrons are large compared to I. In fact, one may estimate the reduction in I due to this effect by noticing that a similar effect will occur in free atoms. In the case of free atoms it has been found that these effects make the $F^2(3d, 3d)$ and $F^4(3d, 3d)$ parameters (using the notation of Condon & Shortley 1935) determined from experiment about 10 to 20 % smaller than those calculated from Hartree–Fock wave functions (see Watson 1960) so one may expect a reduction in I of a similar order of magnitude.

It would seem from the above discussion, although it may be more realistic to use in the Hamiltonian of (10) an 'effective' I ($\sim$ 10 eV) rather than that given by the integral (8), the approximations involved in (10) are otherwise not so poor as to make it an unreasonable starting-point for a theory of correlations when suitably generalized from the s-band to the d-band case. It may, perhaps, be hoped that the terms omitted in going from (6) to (10) may be treated as perturbations on solutions obtained from (10).

3. The Hartree–Fock approximation

For the sake of comparison with the results of the correlation theory developed in later sections it is convenient now to investigate the results obtained by applying the Hartree–Fock approximation to the Hamiltonian of (10). Actually, we shall not make an exhaustive study of all possible Hartree–Fock solutions, but will restrict attention to a particularly simple class of solutions which may represent non-magnetic or ferromagnetic states but not more complicated spin arrangements. A similar restriction applies also to the correlated solutions investigated in later sections.

As is well known, one may obtain the effective Hartree–Fock Hamiltonian by 'linearizing' the interaction terms in the true Hamiltonian. In the case of the Hamiltonian of (10) this amounts to simply replacing the term $n_{i\sigma} n_{i,-\sigma}$ by $n_{i\sigma}\langle n_{i,-\sigma}\rangle + n_{i,-\sigma}\langle n_{i\sigma}\rangle$ where $\langle n_{i\sigma}\rangle$ is the average of the expectation of $n_{i\sigma}$ over a canonical ensemble at some temperature Θ. Dropping the last term of (10) which has been shown to be a constant, the Hartree–Fock Hamiltonian is found to be

$$H_{hf} = \sum_{i,j} \sum_{\sigma} T_{ij} c^{\dagger}_{i\sigma} c_{j\sigma} + I \sum_{i,\sigma} n_{i\sigma} \langle n_{i,-\sigma}\rangle. \tag{12}$$

Attention will now be restricted to the class of solutions for which

$$\langle n_{i\sigma}\rangle = n_{\sigma} \quad \text{for all } i. \tag{13}$$

Then (12) becomes

$$H_{hf} = \sum_{i,j} \sum_{\sigma} T_{ij} c^{\dagger}_{i\sigma} c_{j\sigma} + I \sum_{i\sigma} n_{-\sigma} c^{\dagger}_{i\sigma} c_{i\sigma}, \tag{14}$$

or, transforming back to the operators $c^{\dagger}_{\mathbf{k}\sigma}$, $c_{\mathbf{k}\sigma}$

$$H_{hf} = \sum_{\mathbf{k}} \sum_{\sigma} \{\epsilon_{\mathbf{k}} + I n_{-\sigma}\} c^{\dagger}_{\mathbf{k}\sigma} c_{\mathbf{k}\sigma}, \tag{15}$$

which is simply the Hamiltonian for a collection of non-interacting electrons with a slightly modified band structure, the energy of the $(\mathbf{k}, \sigma)$ state now being $\epsilon_{\mathbf{k}} + I n_{-\sigma}$. It follows at once that if $P(E)$ is the density of states per atom corresponding to the band structure $\epsilon_{\mathbf{k}}$, then the densities of states $\rho_{\sigma}(E)$, where $\sigma = \pm 1$, for the electrons described by the Hamiltonian of (15) are

$$\rho_{\sigma}(E) = P(E - I n_{-\sigma}) = P(E - In + I n_{\sigma}), \tag{16}$$

where the last step follows from

$$n_{\uparrow} + n_{\downarrow} = n. \tag{17}$$

If μ is the chemical potential of the electrons, then at the absolute zero of temperature one will have

$$n_{\sigma} = \int_{-\infty}^{\mu} \rho_{\sigma}(E)\,\mathrm{d}E = \int_{-\infty}^{\mu} P(E - In + I n_{\sigma})\,\mathrm{d}E. \tag{18}$$

The pair of equations (18) must now be solved together with (17) for $n_{\uparrow}$, $n_{\downarrow}$ and μ.

One possible solution of (18) is that for which

$$n_\uparrow = n_\downarrow = \tfrac{1}{2}n, \tag{19}$$

which represents a non-magnetic state of the system: μ is determined by

$$\tfrac{1}{2}n = \int_{-\infty}^{\mu} P(E-\tfrac{1}{2}In)\,\mathrm{d}E. \tag{20}$$

If I is sufficiently large it may also be possible to find ferromagnetic solutions for which $n_\uparrow \neq n_\downarrow$. In this case equation (18) must have two distinct solutions which are such that they can satisfy (17). The condition that ferromagnetism is just possible can now easily be seen to be the condition that (19) and (20) give a double solution of (18). But this condition can at once be found from (18) to be

$$1 = IP(\mu-\tfrac{1}{2}In). \tag{21}$$

Thus, if for any E the condition $IP(E) > 1$ is satisfied, then for some n and μ determined by (20) and (21) Hartree–Fock theory predicts that the system will become ferromagnetic. It will be found that when correlation effects are taken into account one obtains a somewhat more restrictive condition for ferromagnetism.

4. A Green function technique

In the next two sections an approximate solution of the correlation problem for the Hamiltonian of (10) is derived. The method of calculation is based upon the Green function technique described by Zubarev (1960). In order to establish the notation, the principal definitions and basic equations of this technique are briefly reviewed in this section.

Let X be any operator. Then define

$$\langle X\rangle = Z^{-1}\,\mathrm{tr}\{X\,\mathrm{e}^{-\beta(H-\mu N)}\}, \quad Z = \mathrm{tr}\{\mathrm{e}^{-\beta(H-\mu N)}\}, \tag{22}$$

where H is the Hamiltonian and N the total number operator, $\beta = 1/\kappa\Theta$, κ = Boltzmann's constant, Θ = absolute temperature and μ = chemical potential of the electrons.

Now let $A(t) = \mathrm{e}^{\mathrm{i}Ht}A(0)\,\mathrm{e}^{-\mathrm{i}Ht}$ (in units in which $\hbar = 1$) and $B(t')$ be two operators. Then retarded (+) and advanced (−) Green functions may be defined by

$$\langle\langle A(t); B(t')\rangle\rangle^{(\pm)} = \mp\,\mathrm{i}\theta\{\pm(t-t')\}\langle[A(t), B(t')]_\eta\rangle, \tag{23}$$

where $[A, B]_\eta = AB-\eta BA$, $\eta = \pm 1$ (whichever is the more convenient), and $\theta(x)$ is the step function $\theta(x) = 1$ if $x > 0$, $= 0$ otherwise. These Green functions can be shown to satisfy the equation of motion

$$\mathrm{i}\frac{\mathrm{d}}{\mathrm{d}t}\langle\langle A(t); B(t')\rangle\rangle^{(\pm)} = \delta(t-t')\langle[A(t), B(t)]_\eta\rangle + \langle\langle[A(t), H]; B(t')\rangle\rangle^{(\pm)}. \tag{24}$$

Since $\langle\langle A(t); B(t')\rangle\rangle^{(\pm)}$ are functions of $t-t'$ only, one can define for real E the Fourier transforms

$$\langle\langle A; B\rangle\rangle_E^{(\pm)} = \frac{1}{2\pi}\int_{-\infty}^{\infty}\langle\langle A(t); B(0)\rangle\rangle^{(\pm)}\,\mathrm{e}^{\mathrm{i}Et}\,\mathrm{d}t. \tag{25}$$

In the case of the retarded (+) function the integral (25) converges also for complex E provided $\mathscr{I}E > 0$, so $\langle\langle A; B\rangle\rangle_E^{(+)}$ can be defined and is a regular function of E in the upper half of the complex E-plane. Similarly, $\langle\langle A; B\rangle\rangle_E^{(-)}$ is a regular function in the lower half of the complex E-plane. One may now define

$$\langle\langle A; B\rangle\rangle_E = \langle\langle A; B\rangle\rangle_E^{(+)} \quad \text{if} \quad \mathscr{I}E > 0, \qquad \langle\langle A; B\rangle\rangle_E^{(-)} \quad \text{if} \quad \mathscr{I}E < 0, \tag{26}$$

which will be a function regular throughout the whole complex E-plane except on the real axis. From (24) it can be shown that $\langle\langle A; B\rangle\rangle_E$ satisfies

$$E\langle\langle A; B\rangle\rangle_E = 1/2\pi \langle [A, B]_\eta\rangle + \langle\langle [A, H]; B\rangle\rangle_E. \tag{27}$$

It can be shown (Zubarev 1960) that

$$\langle B(t')\,A(t)\rangle = \mathrm{i} \lim_{\epsilon\to 0+} \int_{-\infty}^{\infty} [\langle\langle A; B\rangle\rangle_{E+i\epsilon} - \langle\langle A; B\rangle\rangle_{E-i\epsilon}] \frac{\mathrm{e}^{-\mathrm{i}E(t-t')}}{\mathrm{e}^{\beta(E-\mu)} - \eta}\,\mathrm{d}E. \tag{28}$$

Equations (27) and (28) together with the method of approximation described by Zubarev (1960) form the essential basis of calculations with these Green functions.

5. The exact solution in the atomic limit

In this section the application of the technique described in the preceding section to the Hamiltonian (10) in the limiting case of zero bandwidth is discussed. This limit corresponds to the situation in which the wave functions on different atoms have only a negligible overlap, in which case one knows that the atomic theory gives the exact solution. It will be shown that in this case the Green function technique also leads to an exact solution. Of course, for the Hamiltonian (10) the results of these calculations are rather trivial, but they do serve to reveal the essential trick required to make more elaborate theories go over into the exact atomic solution in the appropriate limit.

All effort will be concentrated on obtaining an expression for the Green's function.

$$G_{jk}^{\sigma}(E) = \langle\langle c_{j\sigma}; c_{k\sigma}^{\dagger}\rangle\rangle_E \quad (\eta = -1), \tag{29}$$

since, as is well known, a knowledge of this Green function enables one to calculate pseudo-particle energies, the Fermi energy, free energy, etc. For example, substituting (29) into (28), putting $j = k$, $t - t' = 0$ and summing on j, one obtains for the mean number n_σ of electrons per atom of spin σ the expression

$$\begin{aligned} n_\sigma &= N^{-1} \sum_j \langle c_{j\sigma}^{\dagger} c_{j\sigma}\rangle \\ &= \frac{i}{N} \lim_{\epsilon\to 0+} \sum_j \int [G_{jj}^{\sigma}(E+i\epsilon) - G_{jj}^{\sigma}(E-i\epsilon)] \frac{\mathrm{d}E}{\mathrm{e}^{\beta(E-\mu)}+1}, \end{aligned} \tag{30}$$

from which one may infer that

$$\rho_\sigma(E) = \frac{i}{N} \lim_{\epsilon\to 0+} \sum_j [G_{jj}^{\sigma}(E+i\epsilon) - G_{jj}^{\sigma}(E-i\epsilon)] \tag{31}$$

gives the density of (pseudo-particle) states per atom of spin σ.

Define

$$T_0 = N^{-1} \sum_{\mathbf{k}} \epsilon_{\mathbf{k}}; \tag{32}$$

T_0 is the mean band energy. In the limit of zero bandwidth $\epsilon_{\mathbf{k}} = T_0$ for all $\mathbf{k}$, whence it follows that $T_{ij} = T_0 \delta_{ij}$ and the Hamiltonian (10) becomes

$$H = T_0 \sum_{i,\sigma} n_{i\sigma} + \tfrac{1}{2} I \sum_{i,\sigma} n_{i\sigma} n_{i,-\sigma}, \tag{33}$$

from which it follows that

$$[c_{i\sigma}, H] = T_0 c_{i\sigma} + I c_{i\sigma} n_{i,-\sigma}, \tag{34}$$

so equation (27) gives

$$E G^{\sigma}_{ij}(E) = \frac{1}{2\pi} \delta_{ij} + T_0 G^{\sigma}_{ij}(E) + I \Gamma^{\sigma}_{ij}(E), \tag{35}$$

where
$$\Gamma^{\sigma}_{ij}(E) = \langle\langle n_{i\,-\sigma} c_{i\sigma}; c^{\dagger}_{j\sigma} \rangle\rangle_E \quad (\eta = -1). \tag{36}$$

Now, $[n_{i,-\sigma} c_{i\sigma}, c^{\dagger}_{j\sigma}]_+ = \delta_{ij} n_{i,-\sigma}$; also from (33) $[n_{i\sigma}, H] = 0$, so (27) gives for $\Gamma^{\sigma}_{ij}(E)$ the equation

$$E \Gamma^{\sigma}_{ij}(E) = \frac{1}{2\pi} \delta_{ij} \langle n_{i,-\sigma} \rangle + T_0 \Gamma^{\sigma}_{ij} + I \langle\langle n^2_{i,-\sigma} c_{i\sigma}; c^{\dagger}_{j\sigma} \rangle\rangle_E. \tag{37}$$

At this point one may notice that since $n^2_{i\sigma} = n_{i\sigma}$, the last term of (37) is just $I \Gamma^{\sigma}_{ij}$, so (37) can be solved at once to give

$$\Gamma^{\sigma}_{ij}(E) = \frac{1}{2\pi} \delta_{ij} \frac{\langle n_{i,-\sigma} \rangle}{E - T_0 - I}. \tag{38}$$

The usual infinite sequence of equations of the type (27) involving higher and higher order Green functions has been avoided here by the simple observation that the last term of (37) can be expressed in terms of Γ^{σ}_{ij}. It is this possibility that enables one to obtain an exact solution in the present case and solutions reducing to the correct solution in the zero bandwidth limit in the general case discussed in the next section.

It follows from the definition (22) and the symmetry of the problem that $\langle n_{i\sigma} \rangle$ is independent of i and σ, so one has at once $\langle n_{i\sigma} \rangle = \tfrac{1}{2} n$. Using this result and substituting (38) into (35) one obtains

$$G^{\sigma}_{ij}(E) = \frac{1}{2\pi} \delta_{ij} \left\{ \frac{1 - \tfrac{1}{2} n}{E - T_0} + \frac{\tfrac{1}{2} n}{E - T_0 - I} \right\}, \tag{39}$$

which gives at once from (31)

$$\rho_\sigma(E) = (1 - \tfrac{1}{2} n) \delta(E - T_0) + \tfrac{1}{2} n \delta(E - T_0 - I). \tag{40}$$

Thus the calculation shows that the system behaves as though it has two energy levels T_0 and $T_0 + I$ containing $1 - \tfrac{1}{2} n$ and $\tfrac{1}{2} n$ states per atom respectively. Thus as electrons are added to the band, initially the Fermi-energy will be fixed at $\mu = T_0$ whilst the lower level fills up. The lower level will become full when $\tfrac{1}{2} n = 1 - \tfrac{1}{2} n$, i.e. $n = 1$, and the chemical potential μ then jumps to $\mu = T_0 + I$ whilst the remaining electrons are added. This is just the correct result. As electrons are added they will (at $\Theta = 0$) distribute themselves on different atoms giving $\mu = T_0$ until when $n = 1$ this is no longer possible and any further electrons added have to go on atoms which have already one electron so μ jumps to $T_0 + I$. A similar discussion can be given for other properties such as specific heat, behaviour in a magnetic field, at $\Theta = 0$ and at finite temperatures; in all cases the Green function solution yields the correct result for a collection of isolated atoms.

6. An approximate solution of the correlation problem

In this section the correlation problem for the Hamiltonian (10) will be studied in the finite bandwidth case by the same technique used in the last section. In the present calculation, however, certain additional terms appear which have to be treated approximately to obtain a solution.

Returning to the Hamiltonian (10), one finds

$$[c_{i\sigma}, H] = \sum_j T_{ij} c_{j\sigma} + I n_{i,-\sigma} c_{i\sigma}, \tag{41}$$

$$[n_{i\sigma}, H] = \sum_j T_{ij} (c^\dagger_{i\sigma} c_{j\sigma} - c^\dagger_{j\sigma} c_{i\sigma}), \tag{42}$$

so the equation for G^σ_{ij} defined by (29) becomes

$$E G^\sigma_{ij}(E) = \frac{1}{2\pi}\delta_{ij} + \sum_k T_{ik} G^\sigma_{kj} + I \Gamma^\sigma_{ij}, \tag{43}$$

where Γ^σ_{ij} is again defined by (36), but now satisfies the equation

$$\begin{aligned} E\Gamma^\sigma_{ij}(E) = {} & \frac{1}{2\pi}\delta_{ij}\langle n_{i,-\sigma}\rangle + T_0 \Gamma^\sigma_{ij} + I\Gamma^\sigma_{ij} \\ & + \sum_{k\neq i} T_{ik} \langle\langle n_{i,-\sigma} C_{k\sigma}; c^\dagger_{j\sigma}\rangle\rangle_E \\ & + \sum_{k\neq i} T_{ik} \{\langle\langle c^\dagger_{i,-\sigma} c_{k,-\sigma} c_{i\sigma}; c^\dagger_{j\sigma}\rangle\rangle_E - \langle\langle c^\dagger_{k,-\sigma} c_{i,-\sigma} c_{i\sigma}; c^\dagger_{j\sigma}\rangle\rangle_E\}. \end{aligned} \tag{44}$$

The term $I\Gamma^\sigma_{ij}$ has been obtained by using $n^2_{i,-\sigma} = n_{i,-\sigma}$, while the term $T_0\Gamma^\sigma_{i,j}$ has been obtained by separating out the $i = k$ part of the fourth term. The first three terms are identical with those of (37) while the latter pair of terms vanish in the zero band-width limit. Thus, whatever approximations are made in the last pair of terms of (44), one will obtain a theory that goes over into the exact solution in the zero bandwidth limit.

In order to break off the sequence of Green function equations an approximate expression will be substituted for the last pair of terms in (44). These approximations are obtained by the methods indicated by Zubarev, and are given by

$$\langle\langle n_{i,-\sigma} c_{k\sigma}; c^\dagger_{j\sigma}\rangle\rangle_E \simeq \langle n_{i,-\sigma}\rangle G^\sigma_{kj}(E); \tag{45}$$

$$\langle\langle c^\dagger_{i,-\sigma} c_{k,-\sigma} c_{i\sigma}; c^\dagger_{j\sigma}\rangle\rangle_E \simeq \langle c^\dagger_{i,-\sigma} c_{k-\sigma}\rangle G^\sigma_{ij}(E); \tag{46}$$

$$\langle\langle c^\dagger_{k-\sigma} c_{i,-\sigma} c_{i\sigma}; c^\dagger_{j\sigma}\rangle\rangle_E \simeq \langle c^\dagger_{k-\sigma} C_{i,-\sigma}\rangle G^\sigma_{ij}(E). \tag{47}$$

By making these approximations one obtains what is practically the crudest theory possible consistent with the condition that it reduces to the correct zero bandwidth limit. One shortcoming of the theory which arises from these approximations is pointed out in § 8.

Other important physical effects neglected as a consequence of these approximations are associated with collective motions of the spin-wave type (see authors cited in the Introduction) and zero-sound type (Landau 1957).

With the approximations (46), (47), the last term of (44) vanishes as a consequence of translational symmetry, since

$$\sum_{k\neq i} T_{ik}\langle c^\dagger_{i,-\sigma}c_{k,-\sigma}\rangle = N^{-1}\sum_{i,k}{}' T_{ik}\langle c_{i,-\sigma}c_{k,-\sigma}\rangle$$
$$= N^{-1}\sum_{i,k}{}' T_{ik}\langle c^\dagger_{k,-\sigma}c_{i,-\sigma}\rangle = \sum_{k\neq i} T_{ik}\langle c^\dagger_{k,-\sigma}c_{i,-\sigma}\rangle,$$

where $T_{ik} = T_{ki}$ which follows from $\epsilon_{\mathbf{k}} = \epsilon_{-\mathbf{k}}$ has been used, and $\sum'_{i,k}$ means the double sum with the term $i = k$ omitted.

In (45) we will now put

$$\langle n_{i\sigma}\rangle = n_\sigma, \tag{48}$$

which follows from translational symmetry. Strictly speaking it follows from the symmetry of the problem that n_σ is independent of σ and therefore equal to $\frac{1}{2}n$. However, it would seem intuitively evident that when solutions of the equations with $n_\uparrow \neq n_\downarrow$ exist, these solutions are connected with the possible ferromagnetism of the system (provided they have lower energy than the non-magnetic solution). One can, perhaps, rationalize this situation by imagining that a minute magnetic field is applied to the system; this field destroys the symmetry between up and down spin, but is so small that it can be neglected in the calculations. To discuss this situation in detail would go far beyond the scope and intent of the present paper. It may be remarked that a similar situation exists in the zero bandwidth case discussed in the preceding section, but that nothing new would be found there by considering solutions for which $n_\uparrow \neq n_\downarrow$. One might also inquire whether meaning can be assigned to solutions (forbidden by symmetry) for which $\langle n_{i\sigma}\rangle$ is not independent of i or even for which quantities like $\langle c^\dagger_{i\sigma}c_{j,-\sigma}\rangle$ do not vanish. The answer would seem to be that these solutions correspond to the possibilities of antiferromagnetism, spiral spin arrangements, etc., but only solutions falling under (46) will be investigated here.

Substituting the approximations (45) to (47) into (44) one obtains

$$E\Gamma^\sigma_{ij} = n_{-\sigma}\frac{\delta_{ij}}{2\pi} + (T_0+I)\,\Gamma^\sigma_{ij} + n_{-\sigma}\sum_{k\neq i} T_{ik}G^\sigma_{kj}, \tag{49}$$

whence

$$\Gamma^\sigma_{ij}(E) = \frac{n_{-\sigma}}{E-T_0-I}\left\{\frac{\delta_{ij}}{2\pi} + \sum_{k\neq i} T_{ik}G^\sigma_{kj}\right\}, \tag{50}$$

which when substituted into (43) gives

$$EG^\sigma_{ij}(E) = T_0G^\sigma_{ij} + \left\{1+\frac{In_{-\sigma}}{E-T_0-I}\right\}\left\{\frac{\delta_{ij}}{2\pi} + \sum_{k\neq i} T_{ik}G^\sigma_{kj}\right\}. \tag{51}$$

This equation may be solved by Fourier transformation. Writing

$$G^\sigma_{ij}(E) = N^{-1}\sum_{\mathbf{q}} G^\sigma(\mathbf{q},E)\exp[i\mathbf{q}.(\mathbf{R}_i-\mathbf{R}_j)] \tag{52}$$

and using (7), one obtains from (51)

$$EG^\sigma(\mathbf{q},E) = T_0G^\sigma(\mathbf{q},E) + \left\{1+\frac{In_{-\sigma}}{E-T_0-I}\right\}\left\{\frac{1}{2\pi N} + (\epsilon_{\mathbf{q}}-T_0)\,G^\sigma(\mathbf{q},E)\right\}, \tag{53}$$

whence
$$G^\sigma(\mathbf{q}, E) = \frac{1}{2\pi N} \frac{E - T_0 - I(1-n_{-\sigma})}{(E-\epsilon_\mathbf{q})(E - T_0 - I) + n_{-\sigma} I (T_0 - \epsilon_\mathbf{q})}, \tag{54}$$

which gives the approximate solution to the correlation problem which has been sought. The properties of this solution are discussed in the next section.

7. Properties of the approximate solution

The general nature of the solution given by (54) will next be investigated. The expression (54) for $G^\sigma(E)$ is a rational function of E and may be resolved into partial fractions according to

$$G^\sigma(\mathbf{q}, E) = \frac{1}{2\pi N} \frac{1}{E^{(1)}_{\mathbf{q}\sigma} - E^{(2)}_{\mathbf{q}\sigma}} \left\{ \frac{E^{(1)}_{\mathbf{q}\sigma} - T_0 - I(1-n_{-\sigma})}{E - E^{(1)}_{\mathbf{q}\sigma}} - \frac{E^{(2)}_{\mathbf{q}\sigma} - T_0 - I(1-n_{-\sigma})}{E - E^{(2)}_{\mathbf{q}\sigma}} \right\}, \tag{55}$$

where $E^{(1)}_{\mathbf{q}\sigma} < E^{(2)}_{\mathbf{q}\sigma}$ are the two roots of

$$(E - \epsilon_\mathbf{q})(E - T_0 - I) + n_{-\sigma} I (T_0 - \epsilon_\mathbf{q}) = 0. \tag{56}$$

It can be shown that $E^{(1)}_{\mathbf{q}\sigma} < T_0 + I(1-n_{-\sigma}) < E^{(2)}_{\mathbf{q}\sigma}$, so (56) has the form

$$G^\sigma(\mathbf{q}, E) = \frac{1}{2\pi N} \left\{ \frac{A^{(1)}_{\mathbf{q}\sigma}}{E - E^{(1)}_{\mathbf{q}\sigma}} + \frac{A^{(2)}_{\mathbf{q}\sigma}}{E - E^{(2)}_{\mathbf{q}\sigma}} \right\}, \tag{57}$$

with $A^{(1)}_{\mathbf{q}\sigma}, A^{(2)}_{\mathbf{q}\sigma} > 0$. If one had $A^{(1)}_{\mathbf{q}\sigma} = A^{(2)}_{\mathbf{q}\sigma} = 1$, then the expression (57) would be the Green function appropriate to a band structure having two bands with the dispersion laws $E = E^{(1)}_{\mathbf{q}\sigma}$ and $E = E^{(2)}_{\mathbf{q}\sigma}$. The effect of the factors $A^{(i)}_{\mathbf{q}\sigma}$ cannot be given any very simple interpretation beyond saying that they reduce the density of states in each band in such a way that the total number of states per atom in both bands together is just 1 and not 2 as it would be if $A^{(1)}_{\mathbf{q}\sigma} = A^{(2)}_{\mathbf{q}\sigma} = 1$. One may see this directly by noticing that from (31) and (52) one has

$$\rho_\sigma(E) = i \lim_{\epsilon \to 0+} \sum_\mathbf{q} \{G^\sigma(\mathbf{q}, E + i\epsilon) - G^\sigma(\mathbf{q}, E - i\epsilon)\}, \tag{58}$$

which gives on substitution of the expression (57) for $G^\sigma(\mathbf{q}, E)$

$$\rho_\sigma(E) = N^{-1} \sum_\mathbf{q} \{A^{(1)}_{\mathbf{q}\sigma} \delta[E - E^{(1)}_{\mathbf{q}\sigma}] + A^{(2)}_{\mathbf{q}\sigma} \delta[E - E^{(2)}_{\mathbf{q}\sigma}]\}, \tag{59}$$

and finally noting that
$$A^{(1)}_{\mathbf{q}\sigma} + A^{(2)}_{\mathbf{q}\sigma} = 1. \tag{60}$$

The general form of the band structure $E^{(1)}_{\mathbf{q}\sigma}, E^{(2)}_{\mathbf{q}\sigma}$ given by (56) is sketched in figure 1. In the limit $I \to 0$, the lower $E^{(1)}_{\mathbf{q}\sigma}$ curve goes over into APX and at the same time $A^{(1)}_{\mathbf{q}\sigma} \to 1$ along AP and $A^{(1)}_{\mathbf{q}\sigma} \to 0$ along PX. Similarly, the upper curve goes over into BPY, $A^{(2)}_{\mathbf{q}\sigma} \to 1$ along PY. Thus as the interaction is switched off the two portions AP and PY combine to make up the unperturbed band structure, the other parts disappearing. That $G^\sigma(E)$ goes over into the unperturbed Green function $(2\pi N)^{-1}(E - \epsilon_\mathbf{q})^{-1}$ as $I \to 0$ can also be seen directly from (54).

In the limit of zero bandwidth, $\epsilon_\mathbf{q} \to T_0$, the $E^{(1)}_{\mathbf{q}\sigma}, E^{(2)}_{\mathbf{q}\sigma}$ curves become flat and go into $E^{(1)}_{\mathbf{q}\sigma} = T_0$, $E^{(2)}_{\mathbf{q}\sigma} = T_0 + 1$, giving the two levels containing $(1 - n_{-\sigma})$ and $n_{-\sigma}$ states respectively discussed in § 5, the expression (59) going over into that of (40) after $\mathbf{q}$-summation.

Next, a more explicit expression than (59) for $\rho_\sigma(E)$ will be derived. From (55) and (58) one can obtain after a little manipulation the formula

$$\begin{aligned}\rho_\sigma(E) &= |E-T_0-I(1-n_{-\sigma})|\,N^{-1}\sum_{\mathbf{q}}\delta[(E-\epsilon_{\mathbf{q}})(E-T_0-I)+In_{-\sigma}(T_0-\epsilon_{\mathbf{q}})]\\ &= \int_{-\infty}^{\infty}\mathrm{d}t\,|E-T_0-I(1-n_{-\sigma})|\,\delta[(E-t)(E-T_0-I)+In_{-\sigma}(T_0-t)]\,N^{-1}\sum_{\mathbf{q}}\delta[t-\epsilon_{\mathbf{q}}]\\ &= \int_{-\infty}^{\infty}\mathrm{d}t\,P(t)\,\delta\left[\frac{E(E-T_0-I)+In_{-\sigma}T_0}{E-T_0-I(1-n_{-\sigma})}-t\right]\\ &= P\{g(E,n_{-\sigma})\},\end{aligned}\tag{61}$$

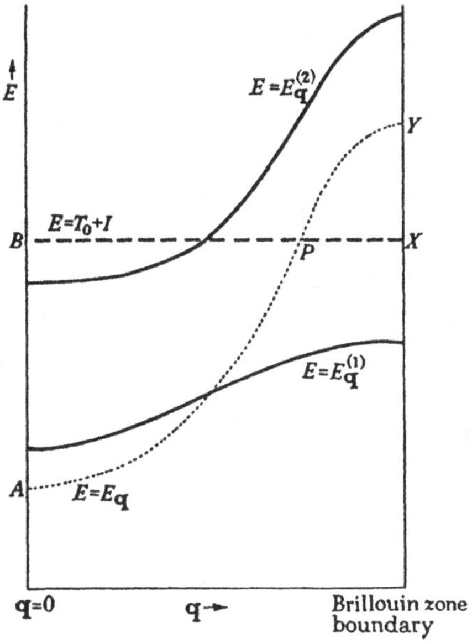

FIGURE 1. A sketch of a typical pseudo-particle band structure $E^{(1)}_{\mathbf{q}\sigma}$, $E^{(2)}_{\mathbf{q}\sigma}$. APY gives the unperturbed band structure while BPX is the line $E = T_0+I$.

where
$$g(E,n_{-\sigma}) = E-In_{-\sigma}-\frac{I^2n_{-\sigma}(1-n_{-\sigma})}{E-T_0-I(1-n_{-\sigma})}\tag{62}$$

and
$$P(E) = N^{-1}\sum_{\mathbf{q}}\delta(E-\epsilon_{\mathbf{q}})\tag{63}$$

is the density of states corresponding to the band structure $\epsilon_{\mathbf{k}}$.

Thus $\rho_\sigma(E)$ is obtained from $P(E)$ by the simple transformation (61), (62). This transformation is illustrated graphically in figure 2 which shows a typical $g(E)$ curve and the projection of $P(E)$ into $\rho_\sigma(E)$. In the limit $I\to 0$ the curve $g(E)$ goes over into the straight line AOB. The splitting of the band into two parts is seen to be due to the infinity of $g(E)$ at $E = T_0+I(1-n_{-\sigma})$.

In order to obtain some feel for the properties of the solution, it is, perhaps, useful to consider the simple example given by the 'square' density of states formula

$$\begin{aligned}P(E) &= 1/\Delta \quad \text{if} \quad T_0-\tfrac{1}{2}\Delta < E < T_0+\tfrac{1}{2}\Delta\\ &= 0 \qquad \text{otherwise},\end{aligned}\tag{64}$$

for which one easily finds

$$\rho_\sigma(E) = 1/\Delta \quad \text{if} \quad E^\sigma_{-1,-1} < E < E^\sigma_{-1,1}$$

or

$$\text{if} \quad E^\sigma_{1,-1} < E < E_{1,1},$$

$$= 0 \quad \text{otherwise,} \tag{65}$$

where $(\alpha, \beta = \pm 1)$

$$E^\sigma_{\alpha\beta} = T_0 + \tfrac{1}{2}I + \tfrac{1}{4}\beta\Delta + \alpha\sqrt{\{(\tfrac{1}{2}I - \tfrac{1}{4}\beta\Delta)^2 + \tfrac{1}{2}\beta\Delta I n_{-\sigma}\}}. \tag{66}$$

One may note that since $(E_{-1,1} - E_{-1,-1}) + (E_{1,1} - E_{1,-1}) = \Delta$, the $\rho_\sigma(E)$ band contains just one state per atom as it must.

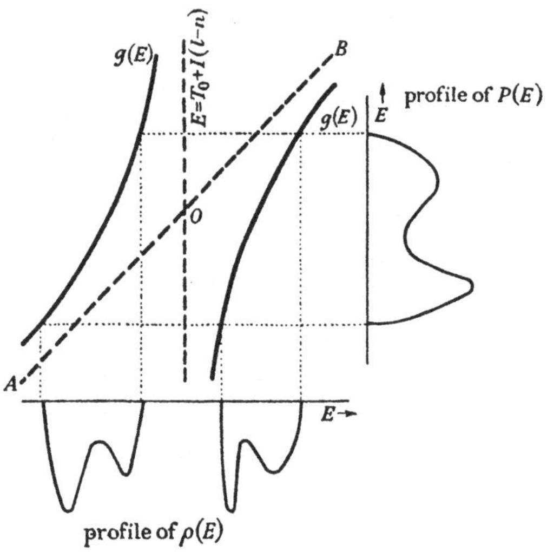

FIGURE 2. A sketch of a typical $g(E, n)$ curve. The projection of the unperturbed density of states function $P(E)$ into the pseudo-particle density of states $\rho_\sigma(E)$ is indicated.

Turning attention now to non-magnetic solutions for which $n_\uparrow = n_\downarrow = \frac{1}{2}n$ (ferromagnetic solutions will be considered in § 9), the Fermi energy μ will be determined by the equation

$$\tfrac{1}{2}n = \int_{-\infty}^{\mu} P\{g(E, \tfrac{1}{2}n)\}\, \mathrm{d}E. \tag{67}$$

Thus μ increases as electrons are added until the lower band is just full. It will then jump discontinuously as further electrons are added and then continue to increase smoothly until the whole band is full. If $E_{\text{max.}}$, $E_{\text{min.}}$ are the highest and lowest energies of the band structure one can see at once from figure 2 that this jump occurs from a value $\mu_<$ to a value $\mu_>$ at a density n_c determined by the equations

$$E_{\text{max.}} = g(\mu_<, \tfrac{1}{2}n_c), \tag{68}$$

$$E_{\text{min.}} = g(\mu_>, \tfrac{1}{2}n_c), \tag{69}$$

$$\tfrac{1}{2}n_c = \int_{-\infty}^{\mu_<} P\{g(E, \tfrac{1}{2}n_c)\}\, \mathrm{d}E,$$

where in (68) the smaller and in (69) the larger root must be taken.

In general $n_c \neq 1$, but in the case of the density of states curve (64), the symmetry between electrons and holes or, to be more precise, the fact that in this case $P(E)$ has the property $P(2T_0 - E) = P(E)$, requires that $n_c = 1$, giving a discontinuity in μ of

$$\mu_> - \mu_< = \sqrt{(I^2 + \tfrac{1}{4}\Delta^2)} - \tfrac{1}{2}\Delta, \tag{70}$$

which goes to 0 as $I \to 0$ and tends to I as $\Delta \to 0$.

8. A two-electron example

In order to obtain a better physical understanding of the solution obtained in the preceding sections, and in particular how the band splits into two parts, it is instructive to consider the problem of two electrons moving and interacting in the manner described by the Hamiltonian (10). Essentially the same problem has been considered by Slater *et al.* (1953) who were mainly interested in the effect of correlations on the condition for ferromagnetism.

Denote by $\psi(i,j)$ the spatial wave function of an eigenstate of the two electron system, $|\psi(i,j)|^2$ measuring the probability of finding one electron on atom i and the other on atom j. Of the $4N^2$ possible states of the system $3N^2$ are spin triplets for which $\psi(i,j) = -\psi(j,i)$, and the other N^2 are singlet states for which $\psi(i,j) = \psi(j,i)$. If $\psi(i,j)$ is an eigenstate with energy E of the Hamiltonian (10), then

$$E\psi(i,j) = \sum_k T_{ik}\psi(k,j) + \sum_k T_{jk}\psi(i,k) + I\delta_{ij}\psi(i,i). \tag{71}$$

Since for the triplet states $\psi(i,i) = 0$, the last term of (71) vanishes for these states, so the triplet states are quite undisturbed by the interaction. This is simply because the Hamiltonian (10) only contains interactions between electrons of opposite spin. Thus attention can be restricted to the singlet states.

In the singlet case we now write

$$\psi(i,j) = N^{-1}\sum_{\mathbf{K}}\sum_{\mathbf{k}} \phi(\mathbf{k},\mathbf{K}) \exp\{i\mathbf{K}.(\mathbf{R}_i + \mathbf{R}_j) + \tfrac{1}{2}i\mathbf{k}.(\mathbf{R}_i - \mathbf{R}_j)\}. \tag{72}$$

Substituting this into (71) and using (7) one obtains

$$E\phi(\mathbf{k},\mathbf{K}) = \{\epsilon_{\mathbf{K}+\frac{1}{2}\mathbf{k}} + \epsilon_{\mathbf{K}-\frac{1}{2}\mathbf{k}}\}\phi(\mathbf{k},\mathbf{K}) + IN^{-1}\sum_{\mathbf{k}'}\phi(\mathbf{k}',\mathbf{K}). \tag{73}$$

Thus solutions with different 'total momentum' $\mathbf{K}$ are not coupled to each other, a consequence of translational symmetry. From (73) one has at once

$$\phi(\mathbf{k},\mathbf{K}) = \frac{IN^{-1}\sum_{\mathbf{k}'}\phi(\mathbf{k}',\mathbf{K})}{E - \epsilon_{\mathbf{K}+\frac{1}{2}\mathbf{k}} - \epsilon_{\mathbf{K}-\frac{1}{2}\mathbf{k}}} \tag{74}$$

whence

$$1 = \frac{I}{N}\sum_{\mathbf{k}}\frac{1}{E - \epsilon_{\mathbf{K}+\frac{1}{2}\mathbf{k}} - \epsilon_{\mathbf{K}-\frac{1}{2}\mathbf{k}}} \tag{75}$$

gives the energy levels for a given $\mathbf{K}$.

The nature of the solutions of equations of the type (75) are well known. The equation has N roots. The right-hand side has infinities at the N energies given by

$$E = \epsilon_{\mathbf{K}+\frac{1}{2}\mathbf{k}} + \epsilon_{\mathbf{K}-\frac{1}{2}\mathbf{k}} \tag{76}$$

for the N values of $\mathbf{k}$, so there are $(N-1)$ roots trapped between these infinities. These $N-1$ roots lie in the unperturbed energy band given by (76). There is one other root. For large enough I this root is quite separate from the band (76), forming a 'bound' state. When I is large compared to the width of the band (76), this root is given by T_0+I as may easily be seen from (75). For small I this 'bound' state does not separate from the band (76).

Thus for large I there are $N(N-1)$ singlet 'scattering' states lying in the unperturbed band and N 'bound' states (one for each of the N values $\mathbf{K}$) with high energy. In the limit $I \to \infty$ the latter states disappear altogether. This is a result of an 'excluded' volume effect of the type familiar from van der Waals's equation. When $I \to \infty$ no two electrons can be on the same site. Thus if one electron is already present (in any one of its N possible states) and another electron is added, then there are only $N-1$ states available to this second electron, whence it follows that there are only $N(N-1)$ possible states available to the two electron system rather than the N^2 possible states for a pair of non-interacting electrons. When I is large but finite the remaining N states reappear with high energy. One may now surmise that when m electrons are already present then only $N-m$ states are available (in the limit $I \to \infty$) to any further electron added to the system, the remaining m states reappearing with high energy when I is finite but large. In this way one can understand how the two bands of figure 1 arise. The lower band is essentially the unperturbed band with some states excluded, these states reappearing in the upper band.

This example reveals a weakness of the approximate solution of § 6. The discussion given above only applies when I is sufficiently large for the 'bound' states to separate, but the solution of § 6 gives a splitting into two bands for all non-zero I. Obviously the approximation is over-estimating the importance of correlation effects for small I, presumably as a consequence of the drastic approximations of equations (45) to (47).

9. The condition for ferromagnetism

In § 3 the condition for ferromagnetism predicted by Hartree–Fock theory was considered. Here the way in which this condition is affected when correlation effects are taken into account (in the approximation of § 6) will be examined.

One expects the condition for ferromagnetism to be more restrictive in a theory which takes into account correlation effects than in Hartree–Fock theory. The reason is simply that ferromagnetism occurs when the (free) energy of the ferromagnetically alined state is less than that of the non-magnetic state. Now, when correlation effects are taken into account it is mainly the correlations between electrons with anti-parallel spin which are being introduced since electrons with parallel spin are already kept apart by the Fermi–Dirac statistics even in the Hartree–Fock approximation. Thus the introduction of correlation effects will lower the energy of non-magnetic states more than that of the ferromagnetic states, and so make the condition for ferromagnetism more stringent. This is indeed found to be the case.

Using the formula (61) for the density of states and the condition (17), n_σ is determined at the absolute zero of temperature by the condition

$$n_\sigma = \int_{-\infty}^{\mu} P\{g(E, n-n_\sigma)\}\, \mathrm{d}E, \tag{77}$$

which is the analogue of (18); μ is determined by the condition (17). One can now take over the discussion of the condition for ferromagnetism in Hartree–Fock theory given in § 3 almost word for word. One finds that the condition that ferromagnetism just be possible is that $n_\uparrow = n_\downarrow = \frac{1}{2}n$ is a double solution of (77). This condition is just (67) together with

$$-\tfrac{1}{2} = \int_{-\infty}^{\mu} \frac{\partial}{\partial n}[P\{g(E, \tfrac{1}{2}n)\}]\, \mathrm{d}E. \tag{78}$$

It is difficult to picture the condition (78) without reference to some specific density of states function $P(E)$. Consider then the density of states function given by (64). In this case Hartree–Fock theory gives according to (21) the condition for ferromagnetism

$$I > \Delta \tag{79}$$

independently of n. To investigate the form taken by the condition (78) in this case one may note that the density of states formula (65) can also be written

$$\rho_\sigma(E) = (1/\Delta)\{\theta(E - E^\sigma_{-1,-1}) - \theta(E - E^\sigma_{-1,1}) + \theta(E - E^\sigma_{1,-1}) - \theta(E - E^\sigma_{1,1})\}, \tag{80}$$

which when substituted into (78) gives

$$-1 = \frac{1}{\Delta}\int_{-\infty}^{\mu} \sum_{\alpha=\pm1} \sum_{\beta=\pm1} \beta \frac{\mathrm{d}E^\sigma_{\alpha\beta}}{\mathrm{d}n}\, \delta(E - E^\sigma_{\alpha\beta})\, \mathrm{d}E. \tag{81}$$

If μ is in the lower band, $E^\sigma_{-1,-1} < \mu < E^\sigma_{-1,1}$ (one need only consider this case because of the symmetry between electrons and holes) then this condition becomes

$$1 < \frac{1}{\Delta}\frac{\partial E^\sigma_{-1,-1}}{\partial n}, \tag{82}$$

or, using the formula (66) with $n_{-\sigma} = \frac{1}{2}n$

$$1 < \frac{\frac{1}{4}I}{\sqrt{\{(\frac{1}{2}I + \frac{1}{2}\Delta)^2 - \frac{1}{4}n\Delta I\}}}. \tag{83}$$

Since μ is in the lower band one must have $n < 1$. But for $n < 1$ the condition (83) cannot be satisfied for any I and Δ. Thus the approximate correlation theory of § 6 predicts that ferromagnetism is not possible for the density of states function of (64) even though Hartree–Fock theory gives the condition (79).

It might now be inquired whether the impossibility of ferromagnetism in a general consequence which can be deduced from the approximate solution of § 6. That the answer to this question is no, can be demonstrated at once by giving an example of a density of states function $P(E)$ for which ferromagnetism is possible.

Consider the density of states function

$$\begin{aligned} P(E) &= 1/\delta \quad \text{if} \quad T_0 - \tfrac{1}{2}\Delta < E < T_0 - \tfrac{1}{2}\Delta + \tfrac{1}{2}\delta \\ &\text{or} \quad\quad \text{if} \quad T_0 + \tfrac{1}{2}\Delta - \tfrac{1}{2}\delta < E < T_0 + \tfrac{1}{2}\Delta \\ &= 0 \quad\quad \text{otherwise,} \end{aligned} \tag{84}$$

which represents two square bands of width $\frac{1}{2}\delta$ symmetrically disposed about T_0. This density of states might be thought of as an approximation to a more general density of states function which has two high peaks at each end of the band and a low density of states in between. By a discussion similar to that given above for the density of states function (64) one may show that for small μ the condition for ferromagnetism is

$$\delta < \frac{\frac{1}{4}\Delta I}{\sqrt{\{(\frac{1}{2}I + \frac{1}{4}\Delta)^2 - \frac{1}{4}n\Delta I\}}}, \tag{85}$$

which can always be satisfied by making δ small enough.

From this example some impression can be gained of what conditions are favourable to ferromagnetism. It is clear that (85) can only be satisfied if δ is somewhat smaller than Δ from which one may infer that it is necessary that the Fermi energy be in a part of the band in which the density of states is rather greater than the mean density of states throughout the band. Further, one may note that since the right-hand side of (85) is a monotonic increasing function of Δ, for fixed δ the condition (85) is more easily satisfied for large Δ. This suggests that the most favourable condition for ferromagnetism is when the Fermi energy lies in a high density of states peak which is well away from the centre of gravity of the band, and that a high density of states peak in the middle of the band would be ineffective in producing ferromagnetism.

Finally, one may note that since for the band structure (85) $\rho(\mu) = 1/\delta$, the condition (85) can formally be written

$$1 < I_{\text{ex.}}\,\rho(\mu) \tag{86}$$

analogous to (21) provided one defines an 'effective intra-atomic exchange energy' $I_{\text{ex.}}$ by

$$I_{\text{ex.}} = I \frac{\frac{1}{4}\Delta}{\sqrt{\{(\frac{1}{2}I + \frac{1}{4}\Delta)^2 - \frac{1}{4}n\Delta I\}}} \tag{87}$$

$I_{\text{ex.}}$ is always less than I, the reduction being due to the weakening of exchange interactions by correlation effects. From (87) one sees that even when I becomes very large $I_{\text{ex.}}$ never becomes much greater than the bandwidth Δ.

The author thanks Dr P. W. Anderson for a very helpful discussion on certain aspects of this work.

References

Bohm, D. & Pines, D. 1953 *Phys. Rev.* **92**, 609, 625.
Condon, E. U. & Shortley, G. H. 1935 *Theory of atomic spectra.* Cambridge University Press.
Edwards, D. M. 1962 *Proc. Roy. Soc.* A, **269**, 338.
Gell-Mann, M. & Brueckner, K. A. 1957 *Phys. Rev.* **106**, 364.
Herring, C. 1952 *Phys. Rev.* **85**, 1003; **87**, 60.
Hubbard, J. 1957 *Proc. Roy. Soc.* A, **240**, 539.
Hubbard, J. 1958 *Proc. Roy. Soc.* A, **243**, 336.
Kubo, R., Izuyama, T. & Kim, D. 1963 *J. Phys. Soc. Jap.* **18**, 1025.
Landau, L. D. 1957 *J. Expt. Theor. Phys.* (*U.S.S.R.*), **32**, 59. Translation: *Soviet Phys. JETP*, **5**, 101.
Pines, D. & Nozieres, P. 1958 *Phys. Rev.* **111**, 442.
Sawada, K., Brueckner, K. A., Fukuda, N. & Brout, R. 1957 *Phys. Rev.* **108**, 507.
Slater, J. C. 1937 *Phys. Rev.* **52**, 198.
Slater, J. C., Statz, H. & Koster, G. F. 1953 *Phys. Rev.* **91**, 1323.
Thompson, E. D. 1960 Thesis (M.I.T.). To be published in *Ann. Phys.*
Watson, R. E. 1960 *Phys. Rev.* **118**, 1036.
Zubarev, D. N. 1960 *Usp. Fiz. Nauk*, **71**, 71. Translation: *Soviet Phys. Usp.* **3**, 320.

THE HUBBARD HAMILTONIAN

M. CYROT
Laboratoire de Magnetisme, CNRS, Grenoble 38042, France

An account is given of the problem of correlated electron motions in transition metals and of Hubbard's analysis of it. The various modes of attack that have been developed since his original work are critically examined. Well-established results are presented, with emphasis on the significance of exact theorems for limiting cases. Recent progress in the theory of the degenerate-band case is described.

1. Introduction

The problem of describing the electronic structure of solids has been attacked along two main lines. On the one hand one may adopt the molecular–orbital approach of Hund and Mulliken, which is a one-electron theory and leads to the band model of Bloch. Interactions between electrons are taken into account only through a self-consistent field, and correlations between the motions of different electrons are neglected. This approach is quite satisfactory for the conduction band of a normal metal. On the other hand there is the Heitler–London model, or localized-electron model, which is a purely atomic description of the solid. This proves to be satisfactory for most insulators, and for the f electrons of rare earth metals. The same cannot be said, however, for the d electrons of transition metals; these undoubtedly form a delocalized d-band. But for many purposes, d electrons in metals exhibit a quasi-atomic behaviour, due to correlations between electron motions neglected in a band model. Thus, one needs to build up a theory of correlations to determine the balance that exists between band-like and atom-like wave-functions.

In practice, the effect of correlations can be extremely pronounced, and can even lead to a complete breakdown of band theory. Let us consider two examples. First let us study the possibility of developing a localized magnetic moment on a transition metal atom. In the band picture, the d electrons hop from atom to atom with a characteristic time h/W (W the band width of the d electrons). Thus it is reasonable to think of the spin being associated with each of the moving d electrons. The spin of an atom, i.e. the total spin of all the electrons on that atom, fluctuates randomly in magnitude and direction. What effect might one expect from the electron interaction? We recall that Hund's first rule for atoms indicates that the intra-atomic interactions will aline the electron spins on an atom. We might expect a tendency to produce the same result in a metal, since if an atom has a spin up it will tend to attract electrons with spins up and repel those with spins down. On this account one would suppose that the total spin on an atom at any one instant tends to be self-perpetuating, so that the spin value can persist for a period long compared to the d-electron hopping time. The electrons on the atoms are always changing about, due to the band motion, but the magnetic moment of the atom persists due to the correlated nature of the electrons' motion. In these circumstances, one can consider the spin as being associated with the atom rather than with the individual electrons. Here we see the possibility of an atomic or a Heisenberg model emerging from the effect of correlations in the band model.

As a second example, let us consider the possibility that there might be insulating behaviour in a not completely empty or filled band. For instance, consider a collection of atoms with one electron each. The band model naturally gives metallic behaviour; the number of electrons on one site can fluctuate, and the frequency of the fluctuations is of the order of the hopping time. Now one general effect of electrostatic interactions is to oppose the build-up of an excess of charge in one place. A consequence of this will be to oppose the jumping of an electron onto an atom which already has one. The effect will be particularly marked if the kinetic energy is small compared to the potential energy. In this limit, which corresponds to taking a large interatomic distance, we must have a set of isolated neutral atoms with insulating behaviour, which is clearly the correct physical description. Thus again from these considerations an atomic model emerges from the effect of correlations on the band theory. We also conclude that the atomic behaviour will be more pronounced if the kinetic energy is small, i.e. if the bandwidth is narrow. Hence correlations will be particularly important in narrow bands; this is the reason for their importance in d bands.

Hubbard has put forward a simple model for treating correlations in narrow bands [1]. However, even his model is not solvable by techniques available today. Accordingly in this paper we will be faced not only with the question: how well does the model describe the actual physics of the correlations? but also: how well do the approximate solutions represent the exact solution? We will begin by describing the model and attempting to answer the first question. Then, after having reviewed such exact results as are known, we will give Hubbard's own solution, bringing out its weak points and describing the attempts that have been made to correct them. A tentative phase diagram for the model will be discussed. In the last part, we examine the effect of degeneracy on correlations.

2. The Hubbard model

The complete hamiltonian for n interacting electrons is so difficult to handle that a model hamiltonian has to be developed. It would be convenient to start with some representation involving Wannier functions, which have a natural role in band theory but are rather localized. However, this is not entirely satisfactory; and contrary to what is done in Hubbard's first paper, we shall base the theory upon the tight binding approximation which is known to be a good starting point for transitional metals. Following Hubbard, and in search of simplicity, we will first completely neglect degeneracy, and in section 6 we will reintroduce the degeneracy and show that special effects arise from it.

Denote by $\phi_\sigma(\boldsymbol{r}-\boldsymbol{R}_i)$ the atomic wave function for an electron on an atom at site $\boldsymbol{R}_i$ and by $c^\dagger_{i\sigma}$ the creation operator for an electron at site $\boldsymbol{R}_i$ with spin σ. In the tight binding approximation the hamiltonian for the electron of the band arising from this atomic shell may be written:

$$H=\sum_{i,j,\sigma} t_{ij}c^\dagger_{i\sigma}c_{j\sigma} + \frac{1}{2}\sum_{\substack{ijkl\\ \sigma\sigma'}}\left\langle i\sigma, j\sigma'\left|\frac{1}{r}\right| k\sigma, l\sigma'\right\rangle c^\dagger_{i\sigma}c^\dagger_{j\sigma'}c_{l\sigma'}c_{k\sigma}, \qquad (1)$$

where

$$t_{ij}=\int \phi^*(\boldsymbol{r}-\boldsymbol{R}_i)\left[-\frac{\hbar^2}{2m}\nabla^2+V\right]\phi(\boldsymbol{r}-\boldsymbol{R}_j), \qquad (2)$$

$$\left\langle i\sigma, j\sigma'\left|\frac{1}{r}\right| k\sigma, l\sigma'\right\rangle = e^2\int \phi^*_\sigma(\boldsymbol{r}-\boldsymbol{R}_i)\phi^*_{\sigma'}(\boldsymbol{r}'-\boldsymbol{R}_j) \times\frac{1}{|\boldsymbol{r}-\boldsymbol{r}'|}\phi_\sigma(\boldsymbol{r}-\boldsymbol{R}_k)\phi_{\sigma'}(\boldsymbol{r}-\boldsymbol{R}_l)\mathrm{d}_3\boldsymbol{r}\mathrm{d}_3\boldsymbol{r}'. \qquad (3)$$

In eq. (2), V represents the nuclear potential acting on the electrons. Thus, the first term of eq. (1) is just the ordinary band hamiltonian. Eq. (3) represents the interaction between electrons. The basic approximation, introduced by Hubbard [1] and Gutzwiller [2] was to neglect in eq. (3) all the interaction terms apart from the one for which $i=j=k=l$. The approximate hamiltonian known as the Hubbard model is

$$H=\sum_{i,j\sigma} t_{ij}c^\dagger_{i\sigma}c_{j\sigma}+\sum_i Un_{i\uparrow}n_{i\downarrow},$$

where we denote by U the only non-zero quantity in eq. (3).

We will not discuss here the important problem of what values are taken by the parameters of the model, namely t_{ij} and U. However, it is obvious that one must not introduce the atomic value for U, which is of the order of 10–20 eV. The effective value is reduced below these figures by many effects, and a good review of the issues involved can be found in Herring's [3] or in Friedel's [4] article. It remains true that this hamiltonian is the simplest that one can construct, even though no exact solution is known. From its simplicity arise the limitations on its applicability to real systems; and thus we have to answer our first question. Clearly a major drawback of the model is its neglect of the long range part of the Coulomb forces. This can be very important; and it is likely that very significant effects are completely missed by the model, in particular effects connected with Mott's [62] picture of the metal–nonmetal transition. We can, however, verify that the essential points about localization of the electrons tending to occur do come to the fore in Hubbard's model, although it takes into account only intra-atomic correlations and completely neglects interatomic ones. A further weakness is that difficulties arise with the sp-band if one attempts to apply the Hubbard hamiltonian to transition metals. Thus, transition metal compounds in which the sp-band does not overlap the d-band are good candidates for application of the theory [5].

3. Exact results

When a model is not exactly solvable, and when consequently approximations must be used, it is particularly valuable to have exact solutions in special cases or corresponding to certain limits. If such are available, the approximations can be tested by comparison with these cases. Unfortunately, very few exact results exist for the non-degenerate Hubbard model. We can list and discuss those known.

(1) In the one-dimensional case, an exact solution exists at zero temperature for any number of electrons. This has been demonstrated by Lieb and Wu [6]. However, not much can be learned from this about the three-dimensional case. There is no metal–nonmetal transition, whatever n and U are (n is the number of electrons per atom). The exact partition function has been obtained by Bari et al. [7] in the infinite-coupling limit ($U \to \infty$).

(2) The small-hopping-integral limit, where t_{ij} can be considered as a perturbation has been considered by Anderson [8] for the case $n = 1$. Direct second-order perturbation theory leads to an effective hamiltonian:

$$H_{\text{eff}} = -\sum_{ij} \frac{t_{ij}^2}{U} \boldsymbol{S}_i \cdot \boldsymbol{S}_j,$$

where $\boldsymbol{S}_i$ and $\boldsymbol{S}_j$ are the spins on sites i and j and are defined by

$$S_i^z = c_{i\uparrow}^\dagger c_{i\uparrow} - c_{i\downarrow}^\dagger c_{i\downarrow},$$
$$S_i^+ = c_{i\uparrow}^\dagger c_{i\downarrow},$$
$$S_i^- = c_{i\downarrow}^\dagger c_{i\uparrow}.$$

Thus the model is equivalent to a Heisenberg model in which the interatomic exchange is replaced by a kinetic term. The model leads to an antiferromagnetic arrangement of moments. Due to the largeness of U compared with t_{ij}, each atom has its own electron, and the antiferromagnetic arrangement permits the wave functions of neighbouring electrons to overlap while remaining orthogonal.

(3) There is an exact theorem for the infinite-U limit with $n = 1-(1/N)$, N being the number of sites. The proof is due to Nagaoka [9]. If U is infinite, there is no superexchange term of the Anderson kind. The hole which exists in this band will gain the maximum energy–zt_{ij} (where z is the number of nearest neighbours)–for a ferromagnetic arrangement of spins. The importance of this theorem is limited by the fact that the infinite-U limit implies

$$N\frac{t_{ij}^2}{U} \ll t_{ij} \quad \text{or} \quad U \gg Nt_{ij},$$

a condition which, in the thermodynamic limit, is never fulfilled. As a consequence it is difficult to draw any conclusions about the possibility of ferromagnetism in the non-degenerate Hubbard model. We will discuss this point further in the following.

(4) High temperature expansions have been carried out by many authors [10, 11]. They give valuable information about the susceptibility and specific heat in this regime.

(5) The low concentration limit, with n going to zero, has been examined by Kanamori [12]. He obtained explicitly the self-energy of the electron Green function, and his results provide a test for the approximate solution described below.

(6) Numerical calculations have been carried out for clusters of atoms [13]. This work has led to only very limited results for the moment, as only the 4-atom case has been worked out.

(7) Let us mention also the existence of sum rules [14] (which give the moments of the spectral weight function to a given order in W/U in terms of certain equal-time correlation functions), and a calculation of the moments of the density of states for the case of one hole in an otherwise half-filled band [35].

4. Hubbard's solution

We shall now review Hubbard's own solution to the Hubbard hamiltonian. We shall not discuss it as first proposed, but only the improved version of it. He divided his argument into two steps.

Let us suppose that by some magic, the $-\sigma$ spin electrons are fixed at particular atoms. If an electron of spin σ is incident on an atom with no electron already present, then that atom must undergo a transition to absorb the σ spin electron, the corresponding resonant energy being t_{ii}. On the other hand, if the atom already has a $-\sigma$ electron present then the resonant energy will be $t_{ii}+U$. The probabilities that any atom does or does not have a $-\sigma$ spin present are $n_{-\sigma}$ and $1-n_{-\sigma}$, respectively. This description amounts to saying that we describe the propagation of an electron of spin σ as though it were moving through an alloy consisting of two species of atoms randomly distributed with probability $n_{-\sigma}$ and $1-n_{-\sigma}$, the binding energies of the s levels on the two atoms being t_{ii} and $t_{ii}+U$. This is the first step of the theory.

In the second we have to take into account the effect of the motion of the $-\sigma$ electrons on the propagation of a σ electron. The motion of the $-\sigma$ spin electrons is such that, so far as σ spin electrons are concerned, any given atom switches back and forth between being resonant at energies t_{ii} and $t_{ii}+U$. The nett effect of this switching depends on whether the rate of alternation is fast or slow compared to U. If the switching rate is slow, the effect is just to broaden the two resonances by an amount h/τ, where τ is the mean life time of an electron on an atom, i.e. h/τ is of the order of the bandwidth W. On the other hand, if the switching rate is fast compared to U, the separate resonances disappear and are replaced by a single resonance at the average energy $t_{ii}+n_{-\sigma}U$, i.e. if U is much smaller than W, one obtains the band behaviour.

The approximation used by Hubbard to treat this "alloy analogy" was subsequently, after its rediscovery by Soven [15], named the "coherent potential approximation" or CPA. It replaces the exact hamiltonian of the alloy by an effective one which refers to an effective medium for the propagation of the electron. This medium is made up of effective atoms with resonant energy $\Sigma(E)$ which depends on the energy of the incident electrons. The propagator of the medium will be immediately given by:

$$F(z)=\frac{1}{N}\sum_k \frac{1}{z-\epsilon_k-\Sigma(z)},$$

with

$$\epsilon_k=\sum_{i,j} t_{ij}\exp i\boldsymbol{k}\cdot(\boldsymbol{R}_i-\boldsymbol{R}_j).$$

The problem is reduced to choosing Σ in the best manner. The physical condition on Σ will be the requirement that if we replace an effective atom by a real one the effective medium produces no further scattering on the average. The scattering of an impurity in a medium is described by the T-matrix, which is

$$\frac{\epsilon-\Sigma}{1-(\epsilon-\Sigma)F},$$

ϵ being the resonant energy level of the real atom. In our case (letting $t_{ii}=0$) $\epsilon=0$ or $\epsilon=U$ with probability $(1-n_{-\sigma})$ and $n_{-\sigma}$. The equation determining Σ is given by the condition that the average T-matrix is zero:

$$-\frac{(1-n_{-\sigma})\Sigma^\sigma(z)}{1+\Sigma^\sigma(z)F^\sigma(z)}+\frac{n_{-\sigma}[U-\Sigma^\sigma(z)]}{1-[U-\Sigma^\sigma(z)]F^\sigma(z)}=0,$$

where we have put a superscript σ to indicate that σ spin electrons are involved.

For our next step we rewrite the self-consistent equation as follows. We introduce the inverse Δ_s^σ of the lifetime of a σ spin electron on a given site in the self-consistent medium:

$$F^\sigma(z)=\frac{1}{z-\Sigma^\sigma(z)-\Delta_s^\sigma},$$

the self-consistent equation can be rewritten in the form

$$\frac{1}{z-\Sigma^\sigma(z)-\Delta_s^\sigma}=\frac{1-n_{-\sigma}}{z-\Delta_s^\sigma}+\frac{n_{-\sigma}}{z-U-\Delta_s^\sigma}.$$

The propagator of the medium is the average of the propagators referring to atom A and atom B embedded in the medium. The atomic level is replaced by an energy band due to the itinerancy of the σ spin electron.

We now want to take into account the fact that a $-\sigma$ spin electron can hop away from the atom where the σ spin electron lands. The finite lifetime of the atomic level will broaden it, and we will have to replace Δ_s^σ by Δ^σ:

$$\Delta^\sigma=\Delta_s^\sigma+\Delta_s^{-\sigma}-\Delta_s^{-\sigma}(U-z).$$

The second term describes the motion of the $-\sigma$ spin electron which leaves the given atom and the last one the arrival of a $-\sigma$ spin electron on the site. This is the so-called "Hubbard III" solution [1]. It was originally applied to a special case where the density-of-states curve has a parabolic form. The main result of the calculation is that the pseudoparticle density of states split into two subbands for a critical ratio of U/W. Thus there is a metal–insulator transition. The gap near the critical value vanishes as

$$\sim\left[\left(\frac{W}{U}\right)_{\text{crit}}-\frac{W}{U}\right]^{\frac{3}{2}}.$$

We will not discuss the effect of long range interactions here.

We now go on to consider the various criticisms which have been made of the Hubbard solution. Hubbard himself pointed out some drawbacks that exist for the half-filled-band case. If in the split band situation one calculates the number of pseudoparticle states in each of the two subbands, one does not in general find that each subband contains exactly one state per atom. This result has important implications because, in the half-filled case, it is tempting to account for the insulating properties of the material by supposing the lower band just full and the upper one empty. Besides this, it has been shown that on the metallic side of the transition there is no well-defined Fermi surface [16], i.e. there is no discontinuity in the occupation number of the k states. This is probably linked to the definition of the Hubbard self-energy, which is independent of k and depends only on the energy.

Probably the most important point not treated by Hubbard is in the domain of magnetism. Hubbard, at least in his first three papers on the matter, restricted the discussion to spatially uniform solutions (i.e. non-magnetic or ferromagnetic states). However, the Anderson limit shows us that in the $n = 1$ case we should expect antiferromagnetism. There has been much controversy about the possibility of antiferromagnetic ordering in Hubbard's coupling schemes. But at least for his first decoupling scheme it has been shown that no antiferromagnetic solution exists in a narrow half-filled band [17]. Recently, Brouers [18] has shown that no antiferromagnetic instability can exist in Hubbard's alloy analogy. This result, however, does not hold for Cyrot's and Lacour-Gayet's [30, 31] modified alloy analogy where the two resonant levels are not 0 and U but two self-consistent ones.

Hubbard presented a single-particle Green function decoupling approximation which was exact for zero interaction energy or zero bandwidth. However, it was subsequently shown [14] that this approximation fails to give the low moments of the spectral weight function correctly to first approximation in the bandwidth. Many authors have generalized the approach and claimed that their procedures were exact to lowest order in W. However, Esterling [19] observes that any solution to the Hubbard model which is correct to lowest order in W would be equivalent to an exact solution to a certain dynamical excluded-volume problem. Since we are some distance from having a complete exact solution of even the static excluded-volume problem, all these claims are probably in error. Esterling's observation strongly decreases the interest of this type of approach, where one tries to find decoupling procedures correct to first order in W.

It has also been shown [20] that Hubbard's simple decoupling scheme fails to reduce to the Hartree–Fock approximation for small interaction energy. Thus its solution is exact only strictly in the limit $U = 0$ or $t_{ij} = 0$.

Let us now review the various treatments which have been based on the Hubbard model. The first, used by Hubbard in his original papers, consists of truncating the hierarchy of Green's functions within the constraint of keeping an exact result in the atomic limit. Many authors have tried to improve the decoupling procedure [21–24]. This is probably the most popular mode of attack; but it leads to approximate solutions which are sometimes difficult to test. Perturbation expansions from the atomic limit [14, 25–27] suffer from the difficulties of getting an exact first-order term, as pointed out above. The third category of solutions comprises those that are equivalent to a Hartree–Fock type solution [28]. In principle they are valid only for $W \gg U$. Many hopes have been put into the functional-integral approach [29–31]. However, difficulties arise in the thermodynamics of the atomic limit in this formulation [32, 33]. The path formulation of Nagoaka [9], used by some authors [34, 35], consists in calculating the quantity of interest (the partition function, etc.) by following the path of a single hole in an otherwise half-filled band. The results, however, rest on further approximations which have been criticized [36]. Finally, the variational approach [28, 37–39] pioneered by Gutzwiller [2] suffers of course from the standard difficulties of this type of treatment. Thus one must conclude at the present time that no one approach is better than the others, and such insight as we have gained into the problem is due to all of them taken together.

5. The attempt to draw a phase diagram

An exact solution being far from our reach, it is important to obtain qualitative results on the behaviour of the Hubbard model. First, what are the likely ground states as a function of W/U and n? Secondly, what happens as the temperature is raised above zero? In dealing with these topics the functional-integral approach has the interest of permitting us to introduce antiferromagnetism and discuss the result in simple physical terms. The conventional Green's function decoupling approaches generally try to treat the coupling only to first order, and are inadequate for antiferromagnetism, which is a second-order effect as we know from superexchange theory.

At zero temperature, and for one electron per atom, approximations in this approach lead to a treatment similar to Slater's [50] which introduces a spin-dependent potential with the double periodicity of the antiferromagnetic lattice. This potential is, on site i [30]

$$V_i^\sigma = -\tfrac{1}{2}U\sigma\langle n_{i\uparrow} - n_{i\downarrow}\rangle.$$

The quantity $\langle n_{i\uparrow} - n_{i\downarrow}\rangle$ can be interpreted as an effective local moment created out of itinerant electrons. It has to be determined self-consistently, and decreases monotonically from the value 1 in the atomic limit to zero in the limit $U = 0$. This reduction of the magnitude of the effective local moment is due to the motional narrowing effect discussed by Hubbard.

The result leads to two critical values for the ratio U/t_{ij}, a_L and $a_I > a_L$. For ratios smaller than a_L the local moment is zero, and the system is metallic and nonmagnetic. For intermediate values the system becomes antiferromagnetic. However, in this approximation there is no gap in the electron spectrum and the system remains metallic. For ratios larger than a_I the system becomes an antiferromagnetic insulator. This behaviour, however, goes against Brinkman and Rice's theory [38] which, on the basis of Gutzwiller's approach [2], argues that no antiferromagnetic metallic state exists. There is no definite answer to this problem at the moment [49, 61]. One can only say that such a phase has been observed recently in V_2O_3 [40].

For the case of non-zero temperature, results are more scarce. We now know that for high enough values of U/W the material can remain insulating above the Néel temperature, contrary to the simple ideas of Slater's treatment. This is due to the fact that the spin-dependent potential makes the system look like a binary alloy; thus a gap in the density of states can persist. On the basis of a Hartree–Fock treatment [28, 41] it was first argued that two phase transitions can exist: a magnetic one, and an insulator–metal transition at higher temperature. However, more exact treatments have since ruled out a true metal–insulator phase transition [29]. Cyrot and Lacour-Gayet [31] have calculated, using the functional approach, the Néel temperature as a function of the ratio U/W (fig. 1). For large values of U they obtain Anderson superexchange, and for small values a Rudermann–Kittel interaction between magnetic moments. More detailed studies of the effect of temperature have recently appeared [42].

Arguing along these lines one would conclude that there can be no first-order transition from a metallic to an insulating paramagnetic phase as observed in V_2O_3. However, one can show through a study of phase stability [43] that a

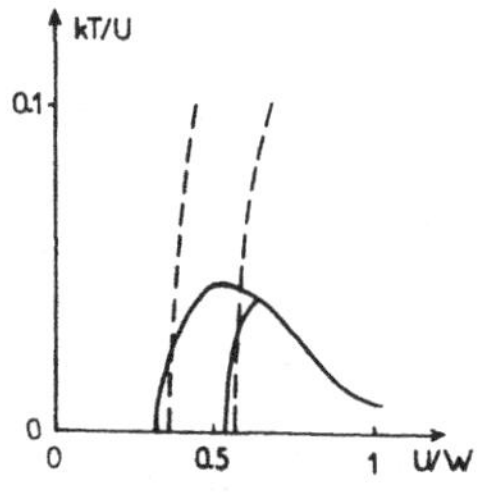

Fig. 1. A theoretical phase diagram for the Hubbard model. The dashed lines correspond to the appearance of a non-zero value for the Hartree–Fock magnetic moment and to the opening up of a gap in the density of states.

first-order transition can be obtained under pressure if the bulk modulus of the material is small enough.

So far we have considered only the case of one electron per atom, $n = 1$. It is interesting to discuss the ground state of the model under the conditions of arbitrary electron concentration, particularly in connexion with the possibility of ferromagnetism in a non-degenerate band. Some authors [44] have conjectured that ferromagnetism can exist only in the presence of orbital degeneracy. Slater et al. [45] called attention to this problem by studying two electrons moving in an otherwise empty band. The result suggested to these authors that the ground state of a metal should be a singlet if the band occupied by the conduction electrons is non-degenerate. The ground state may sometimes be ferromagnetic if the band is degenerate; and in that case, the Hund's rule coupling of electrons on the same atom is responsible for the ferromagnetism. On the other hand, the authors of several approximate theories [46, 34] do predict ferromagnetism at least for some range of electron density, basing their feelings on Nagaoka's exact result for $n = N \pm 1$ and $U \to \infty$. We have here seen at the beginning, however, that his proof needs $U \gg Nt_{ij}$, a condition which is not fulfilled in the thermodynamic limit. In the writer's opinion, no sound proof exists as to the possibility of ferromagnetism in the non-degenerate case. In the following we will show, on the contrary, that degeneracy favours ferromagnetism.

Fukuyama and Ehrenreich [47] have shown that in the coherent potential approximation of Hubbard the static susceptibility at $T = 0$ is non-singular for any electron density. It has also been shown that in the Hubbard alloy analogy, even given an exact solution of the alloy problem, ferromagnetism is impossible [48].

6. Effects of correlation in a degenerate band

Up to now we have discussed only one effect of correlations, namely their tendency to inhibit large fluctuations of charge. In that respect, in fact, there is little difference between the non-degenerate and the degenerate case. This is the main result of Hubbard's paper II where he discusses the degeneracy problem. The physical interpretation which emerges from his analysis is closely related to the theory of "minimum polarity" proposed earlier by Van Vleck. The idea here is that a transition metal could be thought of as a collection of atoms in different configurations, having for example different numbers of d electrons. The band motion can be thought of as due to atoms rapidly changing their configurations. In this model, the states of high ionization for atoms are arbitrarily ruled out as giving an excessive energy to the system. Thus, the configurations of the system are restricted to those which give the minimum polarity.

However, this is only one effect of the correlations–the most discussed one–and is related to the metal–nonmetal transition. There is another effect, which is related to magnetism, as we explained at the beginning, and which is specific to the degenerate case. It results in there being correlations between the atomic configurations of near neighbours. It is believed that this effect was first pointed out by Sommerfield and Bethe, and studied by Van Vleck [44] as a possible basis for the origin of ferromagnetism. This last-mentioned author proposed the following mechanism. Consider two weakly-interacting atoms, each with one electron. The electrons prefer to line up parallel and occupy different orbitals in order to lower their energy by mixing in ionic states with parallel spins. These states lie lowest in energy because of Hund coupling. This circumstance not only produces a ferromagnetic coupling between the spins of the atoms, but also correlation between the atomic configurations of nearest neighbours. In the case of three electrons for a pair of atoms, the ferromagnetic interaction is similar to a double exchange interaction. In that case, no ionic state is considered, and the coupling is independent of U. Without correlation due to fluctuations of charge, there is correlation between atomic configurations.

The correlations between atomic configurations on neighbouring atoms are generally dynamic in nature, and atoms change their configurations with a characteristic time given by the inverse of the bandwidth. However, much as the singlet of the hydrogen molecule can be frozen in a state of antiferromagnetic ordering in a crystal, the atomic configurations can be frozen into a kind of orbital superlattice. For instance, in the doubly degenerate case considered by Van Vleck, it would be energetically favourable for two sublattices to form, each with predominantly one of the orbital states. Laura Roth [51] has studied such a possibility, and Cyrot and Lacroix [52] have further developed this type of theory. The static configuration considered above probably does exist; its dynamic counterpart must be a general phenomenon. This completely new aspect of the degenerate case is in all likelyhood linked not only with ferromagnetism but with the quenching of orbital moment in transition metals.

7. Exact results for the degenerate Hubbard hamiltonian

The degenerate Hubbard hamiltonian is an obvious generalization of eqs. (1)–(3). Let $\phi_{m\sigma}(\boldsymbol{r}-\boldsymbol{R}_i)$ be the atomic wave function of an electron on site $\boldsymbol{R}_i$ in orbital m. We have

$$H = \sum_{\substack{i,j\\ m,m',\sigma}} t_{ij}^{mm'} c_{im\sigma}^{\dagger} c_{jm'\sigma} + \frac{1}{2} \sum_{\substack{ijkl\\ mm'm''m'''\\ \sigma\sigma'}} \left\langle im\sigma, jm'\sigma' \left| \frac{1}{r} \right| km''\sigma, lm'''\sigma' \right\rangle \times c_{im\sigma}^{\dagger} c_{jm'\sigma'}^{\dagger} c_{lm'''\sigma'} c_{km''\sigma},$$

and the straightforward generalization of eqs. (2) and (3). The model hamiltonian is obtained by neglecting all interaction terms except the term $i=j=k=l$. The non-zero quantities corresponding to U are $U_{mm'}$ and $J_{mm'}$.

The following are some exact results, known for the doubly degenerate Hubbard hamiltonian:

(1) In the one-dimensional case, an exact solution [53] for any temperature exists in the limit $U \to \infty$, $J \to \infty$ and $n = N+1$, where n is the number of electrons and N the number of sites. The ground state is ferromagnetic.

(2) In the three-dimensional case, the ground state has been shown to be ferromagnetic [54] for any electron concentration $N+1<n<2N-1$ in the limit U and J infinite. To this result, the limitation that we described as restricting the Nagaoka proof does not apply in the case $n = N+1$, as the ground state there is ferromagnetic, but does apply in the case $n = 2N-1$ as the ground state for $n = 2N$ is antiferromagnetic.

(3) The second-order perturbation from the atomic limit has been worked out for $n = N$ [52], and an effective hamiltonian generalizing Anderson's superexchange has been discussed. This hamiltonian is a function not only of the spin on one site but also the momentum on that site. Approximate solutions of this hamiltonian show that an orbital superlattice is possible and leads to ferromagnetism.

(4) High-temperature expansions have been carried out by Lacroix and Cyrot [55]; they show that degeneracy favours ferromagnetism.

8. The orbital superlattice

Roth [51] was the first to suggest that Van Vleck's mechanism for ferromagnetism leads to the formation of an orbital superlattice, that is to say, that in the doubly degenerate Hubbard model the most favourable configuration is that exhibiting two sublattices each with predominantly one of the orbital states. However, Roth's formalism describes only the very-narrow-band limit when all the spins are aligned in the same direction. There are two alternative methods for studying the possibility of such a phase: (a) one can compare the total energies of each of the various possible states; or (b) one can assume that a state is stable, and test this assumption by computing its response function to a time-independent, spin- and orbit-dependent external potential. A negative value of the response function would indicate thermodynamic instability with respect to the spontaneous distribution corresponding to that induced by the external field. This second method tests only for stability against an infinitesimal perturbation, and is misleading in the case of a first-order transition. The first approach has been applied by Lyon-Caen and Cyrot [52], and the second by Inagaki and Kubo [56, 57]. All these authors conclude that a variety of orbital superlattices are possible (fig. 2).

Lyon-Caen and Cyrot [52] study the properties of such an orbital superlattice. They first show that the occurrence of such a superlattice favours ferromagnetism. In a kind of Hartree–Fock treatment they find that if a superlattice is possible, i.e. if J is not too large compared to U, ferromagnetism occurs before the Stoner criterion is fulfilled. In that case a first-order transition occurs. Too large values of J cease to favour an asymmetry in the occupation of the two orbitals on one atom, because spin up elec-

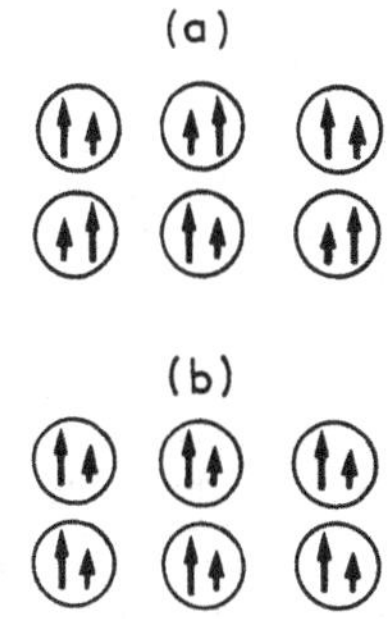

Fig. 2. Orbital superlattices: (a) antiferromagnetic; (b) ferromagnetic.

trons interact only weakly (i.e. $U-J$ is small). New kinds of collective excitation exist in this state. Besides the intraband and interband spin waves, there exist orbital waves which do not reverse any spin and have a dispersion relationship similar to that of antiferromagnetic spin waves.

Kugel and Khomskii [58] have studied the conditions for developing an orbital superlattice in the atomic limit for a perovskite structure. In that case the particular geometrical distribution of charge around the sites leads to a collective distortion of the crystal. This treatment is tantamount to setting up a new kind of approach to the theory of cooperative Jahn–Teller excitations. Its great advantage is that it links all three topics: orbital superlattices, lattice distortions and magnetic long range order. Generalizations of the discussion to triply degenerate bands (t_{2g}) have been given by different authors [59, 60] and other cases have also been examined.

9. Concluding remarks

We have attempted to review the main trends evolving in the study of the Hubbard hamiltonian. However, it should be mentioned that certain generalizations that can be found in the literature have been ignored. The case of an sp- and a d-band together is one such; the introduction of interatomic Coulomb interactions is another. Studies of this kind are generally based on approximate solutions of the simple Hubbard model, and suffer from the same limitations.

It is an unfortunate fact that, since Hubbard's original work, no decisive impulse towards an exact solution has been given in the non-degenerate case. Attempts in this direction using renormalization-group theory have been regrettably few. Nevertheless, interesting results have been obtained in the degenerate case, and in the last few years this has been the most active line of progress. Study of the dynamical correlations between the atomic configurations on neighbouring sites will probably throw a great deal of light on the general problem.

I should like to thank Dr. Lacroix-Lyon-Caen for a critical reading of the manuscript.

References

[1] J. Hubbard, Proc. Roy. Soc. (London), A 276 (1963) 238; 277 (1964) 237; 281 (1964) 401; 285 (1965) 542; 296 (1966) 82 & 100; Proc. Phys. Soc. 84 (1964) 455.
[2] M. Gutzwiller, Phys. Rev. Lett. 10 (1963) 159; Phys. Rev. 137 A (1965) 1726.
[3] C. Herring, in: Magnetism, Vol. IV, G.T. Rado and H. Suhl, eds. (Academic Press, New York, 1966).
[4] J. Friedel, in: Physics of Metals, J.M. Ziman, ed. (Cambridge University Press, 1969) p. 340.
[5] D. Adler, Solid State Phys. 21, p. 1, Academic Press, 1968.
[6] E. Lieb and F. Wu, Phys. Rev. Lett. 20 (1968) 1445.
[7] G. Beni, P. Pincus and T. Holstein, Phys. Rev. B 8 (1973) 312.
[8] P.W. Anderson, Solid State Phys. 14 (1963) 99.
[9] Y. Nagaoka, Phys. Rev. 147 (1966) 392.
[10] G. Beni, P. Pincus and D. Hone, Phys. Rev. B 8 (1973) 3389.
[11] L.N. Bulaevskii and D. Khomskii, Phys. Lett. 41 A, (1972) 257; Sov. Phys. Solid State, 14 (1973) 3015.
[12] J. Kanamori, Progr. Theor. Phys. (Kyoto) 30 (1963) 257.
J. Callaway and D. Edwards, Phys. Rev. 136 A (1964) 1333.
[13] K.H. Heinig, W. Löser and J. Monecke, Phys. Status Solidi b 53 (1972) K 113.
[14] A.B. Harris and R.V. Lange, Phys. Rev. 157 (1967) 295.
[15] P. Soven, Phys. Rev. 156 (1967) 809.
[16] D.M. Edwards and A.C. Hewson, Rev. Mod. Phys. 40 (1967) 810.
[17] D.R. Penn, Phys. Lett. 26 A (1968) 509.
[18] F. Brouers, F. Ducastelle and J. Giner, J. de Phys. (Paris), to be published.
[19] D. Esterling, Phys. Rev. B 2 (1970) 4686.
[20] R.A. Bari and J.A. Kaplan, Phys. Lett. 33 (1970) 400.
[21] R. Tahir-Kheli and H. Jarrett, Phys. Rev. 180 (1968) 544.
[22] L. Roth, Phys. Rev. Lett. 20 (1968) 1431.
[23] R. Kishore and S. Joshi, Phys. Rev. 186 (1969) 484.
[24] T. Arai and M. Parinello, Phys. Rev. Lett. 27 (1971) 1226.
[25] J. Hubbard, Proc. Roy. Soc. (London) A 296 (1966) 100.

[26] J. Kanamori, Progr. Theor. Phys. (Kyoto) 30 (1963) 275.
[27] D. Esterling and R.V. Lange, Rev. Mod. Phys. 40 (1968) 796.
[28] W. Langer, M. Plischke and D. Mattis, Phys. Rev. Lett. 23 (1969) 1448.
[29] J. Kimball and J.R. Schrieffer, Int. Conf. on Magnetism, Chicago (1971).
[30] M. Cyrot, Phys. Rev. Lett. 25 (1970) 871; Phil. Mag. (1971) 1031.
[31] M. Cyrot and P. Lacour-Gayet, J. Phys. C 7 (1974) 400.
[32] R. Bari, Phys. Rev. B 5 (1972) 2736.
[33] R. Hassing and D. Esterling, Phys. Rev. B.
[34] J. Sokoloff, Phys. Rev. B 3 (1971) 3826.
[35] W. Brinkman and T. Rice, Phys. Rev. B 2 (1970) 1324.
[36] D.M. Esterling and H.C. Dublin, Phys. Rev. B 6 (1972) 4276.
[37] T. Kaplan and R. Bari, J. Appl. Phys. 41 (1970) 875.
[38] W. Brinkman and T.M. Rice, Phys. Rev. B 2 (1970) 4302.
[39] K.A. Chao, Phys. Rev. B 4 (1971) 4034.
[40] D. Jerome, J. de Phys. to be published.
[41] M. Cyrot, J. de Phys. 33 (1972) 125.
[42] De Marco, E.N. Economoue and D.C. Licciardo, to be published.
[43] M. Cyrot and P. Lacour-Gayet, Solid State Commun. 11 (1972) 1767.
[44] J.H. Van Vleck, Rev. Mod. Phys. 25 (1959) 2.
[45] J.C. Slater, H. Statz and G.F. Koster, Phys. Rev. 91 (1953) 1323.
[46] D.R. Penn, Phys. Rev. 142 (1966) 350.
[47] H. Fukuyama and H. Ehrenreich, Phys. Rev. B 7 (1973) 3266.
[48] F. Brouers and F. Ducastelle, J. de Phys. 36 (1975) 851.
[49] A.K. Gupta, D.M. Edwards and A.C. Hewson, J. Phys. C 8 (1975) 3207.
[50] J.C. Slater, Phys. Rev. 82 (1951) 538.
[51] L. Roth, Phys. Rev. 149 (1966) 306.
[52] M. Cyrot and C. Lyon-Caen, J. Phys. C 6 (1973) L 274; 36 (1975) 253.
[53] C. Lacroix and M. Cyrot, J. Phys. C, to be published.
[54] C. Lacroix and M. Cyrot, to be published.
[55] C. Lyon-Caen and M. Cyrot, J. Phys. C 8 (1975) 2091.
[56] S. Inagaki and R. Kubo, Int. J. Magn. 4 (1973) 139.
[57] S. Inagaki, J. Phys. Soc. Jap. 39 (1975) 596.
[58] K.I. Kugel and K.I. Khomskii, Sov. Phys. JETP, 37 (1973) 725; Solid State Commun. 13 (1973) 763.
[59] K.I. Kugel and K.I. Khomskii, Sov. Phys. Solid State 17 (1975) 285.
[60] M. Cyrot and C. Lacroix-Lyon-Caen, J. de Phys., to be published.
[61] J.M.D. Coey, these proceedings.
[62] N.F. Mott, Rev. Mod. Phys. 40 (1968) 677.

The Resonating Valence Bond State in La_2CuO_4 and Superconductivity

P. W. ANDERSON

The oxide superconductors, particularly those recently discovered that are based on La_2CuO_4, have a set of peculiarities that suggest a common, unique mechanism: they tend in every case to occur near a metal-insulator transition into an odd-electron insulator with peculiar magnetic properties. This insulating phase is proposed to be the long-sought "resonating-valence-bond" state or "quantum spin liquid" hypothesized in 1973. This insulating magnetic phase is favored by low spin, low dimensionality, and magnetic frustration. The preexisting magnetic singlet pairs of the insulating state become charged superconducting pairs when the insulator is doped sufficiently strongly. The mechanism for superconductivity is hence predominantly electronic and magnetic, although weak phonon interactions may favor the state. Many unusual properties are predicted, especially of the insulating state.

RECENTLY HIGH-TEMPERATURE SUPERCONDUCTIVITY has been observed in a number of doped lanthanum copper oxides near a metal-insulator transition (*1*), a pattern exhibited previously by (Ba,Pb)BiO_3 (*2*). The crystal structure suggests that the Cu^{2+} is in an $S = 1/2$, orbitally nondegenerate state, strongly hybridized with the surrounding oxygen *p*-levels, and this is in agreement with high-temperature magnetic data (*3*) on the stoichiometric, insulating compound La_2CuO_4.

The appropriate model seems to be the basic nearly half-filled Hubbard model (*4*) with moderately large repulsion energy U and antiferromagnetic exchange constant $J = t^2/U$ where t is the site-hopping matrix element. The K_2NiF_4 structure is a well-known case in which the magnetic layers are relatively weakly interacting, and in the temperature range 30 to 70 K we can assume magnetic two-dimensionality. This led me to reexamine the idea of the "resonating valence-bond" (RVB) state (*5*).

Early doubts about the nature of the ground state of the antiferromagnetic Heisenberg Hamiltonian

$$H = J\sum_{innj} \vec{s}_i \cdot \vec{s}_j \qquad (1)$$

of Hulthén (*6*) and Marshall (*7*) (where $\vec{s}_i$ is the spin at site i and $innj$ indicates summation over nearest neighbors i and j) seemed to have been laid to rest by arguments from quantum fluctuations of spin waves in the Néel state (*8*) in >1 dimension, and by experimental observations of antiferromagnetism. In 1973, however, Anderson (*5*) proposed that, at least in the triangular two-dimensional antiferromagnet for $S = 1/2$, and perhaps in other cases, the ground state might be the analog of the precise singlet in the Bethe solution of the linear antiferromagnetic chain (*6*). In both cases, the zeroth order energy of a state consisting purely of nearest neighbor singlet pairs is more nearly realistic than that of the Néel state, and I

Department of Physics, Princeton University, Princeton, NJ 08544.

proposed that higher order corrections, allowing the singlet pairs to move or "resonate" à la Pauling, might make this insulating singlet or RVB state more stable. This state is quite clearly distinct from two other locally stable possibilities, the Néel state and the "spin-Peierls" state of a self-trapped, localized array of singlet pairs. Each of these other states has a broken symmetry relative to the high-temperature paramagnetic state, and would exhibit a phase transition with temperature. Fazekas and Anderson (*9*) improved on my numerical stability estimates, and Hirakawa *et al.* (*10*) have proposed application of the RVB state to specific compounds.

The triangular lattice is not the only possible candidate: the two-dimensional square lattice, for instance, will undoubtedly exhibit an RVB state if either or both of two possibilities occurs: a next nearest neighbor antiferromagnetic interaction strong enough to frustrate the Néel state, or virtual phonon interactions short of being strong enough to allow a spin-Peierls instability. [Hirsch (*11*) has shown numerically that the simple square lattice probably retains a magnetization but that finite U may favor the RVB state.] It is our hypothesis that pure La_2CuO_4 is in an RVB state; this proposal is supported to some extent by the magnetic susceptibility data of Ganguly and Rao (*3*).

It is not easy to calculate with RVB states or to represent them. I want to give here a representation in terms of Gutzwiller-type projections of mobile-electron states, which is probably not particularly useful computationally but is suggestive. This representation was in turn suggested by Rice and Joynt's remarks (*12*) on the Gutzwiller approximation to the Bethe solution.

A single pair of electrons in a mobile valence bond along a lattice vector τ may be written

$$b_\tau^+ \Psi_0 = \frac{1}{\sqrt{N}}\left(\sum_j c_j^{+\uparrow}\, c_{j+\tau\downarrow}\right)\Psi_0 =$$

$$\frac{1}{\sqrt{N}}\left(\sum_k c_{k\uparrow}^+\, c_{-k\downarrow}^+ \exp i\,(k\cdot\tau)\right)\Psi_0 \qquad (2)$$

where b_τ^+ is the electron-pair creation operator, c_j^+ is the single-electron creation operator, and N is the total number of sites. A linear combination of all nearest neighbor bonds may be written

$$b_{nn}^+ = \sum_{\tau=(nn)} b_\tau^+$$

where nn indicates summation over nearest neighbors, or in general, if we want a distribution of bond lengths, the linear combination may be written as

$$b^+ = \sum_k a(k)\, c_{k\uparrow}^+\, c_{-k\downarrow}^+ \qquad (3)$$

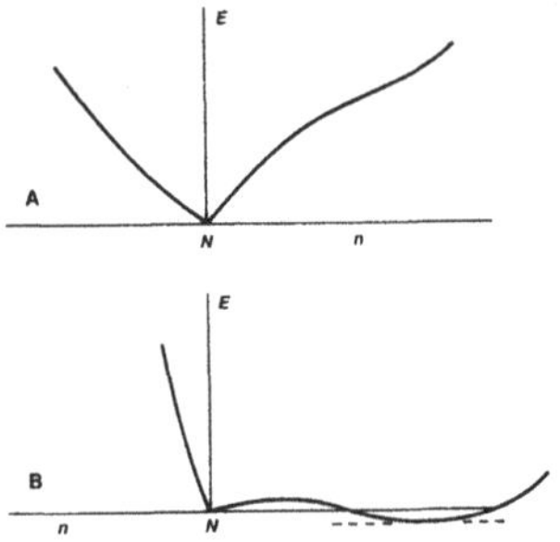

Fig. 1. Total energy E as a function of average occupation number n. (**A**) Insulating case. (**B**, Metallic case. N is the total number of lattice sites.

with the condition on the expansion coefficients

$$\sum_k a(k) = 0 \qquad (4)$$

if we do not allow double occupancy.

Unfortunately, if we now try to make a Bose condensation of N electrons in mobile valence bond states by forming

$$\Psi = (b^+)^{N/2}\, \Psi_0 \qquad (5)$$

this state contains large numbers of empty and doubly occupied sites. To make up a genuine RVB state we may try one of two roughly equivalent projection techniques. The simplest is the straightforward Gutzwiller method: form the projection operator

$$P_d = \prod_i (1 - n_{i\uparrow} n_{i\downarrow}) \qquad (6)$$

where n_i is the occupation number, and

$$\Psi_{RVB} = P_d (b^+)^{N/2} \Psi_0 \qquad (7)$$

is our trial wave function.

We may also use quasifermion operators with the double occupancy projected out (*13*), for example,

$$\underline{c}_{i\uparrow} = c_{i\uparrow}^+ (1 - n_{i\downarrow})$$

and write

$$b^+ = \sum_k a(k) \underline{c}_{k\uparrow}\, \underline{c}_{-k\downarrow} \qquad (8)$$

in Eq. 5.

At this point we resort to the Dyson transformation between product states and Bardeen-Cooper-Schrieffer (BCS) states. As shown by Dyson (*14*), if we note that $(b_k^+)^2 = 0$ (with $b_k^+ = c_{k\uparrow}^+\, c_{-k\downarrow}^+$),

$$1 + a_k b_k^+ = \exp(a_k b_k^+)$$

hence

$$\Psi_{BCS} = \prod_k (\sqrt{1-h_k} + \sqrt{h_k}\, b_k^+)\Psi_0$$

$$\propto \exp\left[\sum_k \sqrt{\frac{h_k}{1-h_k}}\, b_k^+\right]\Psi_0 \qquad (9)$$

where h_k is the variational parameter in the standard BCS treatment.

If we project Ψ_{BCS} on the state with just $N/2$ pairs, we obtain

$$P_{N/2}\Psi_{BCS} = \left[\sum_k \left[\frac{\sqrt{h_k}}{\sqrt{1-h_k}}\right] b_k^+\right]^{N/2} \Psi_0 \qquad (10)$$

or, in other words, Eq. 5 is just a projected BCS function. These transformations are not singular as they are for conventional BCS. Thus our approximate Ψ_{RVB} is also related to the appropriate projection of a BCS function:

$$\Psi_{RVB} = P_{N/2} P_d \prod_k \left[\frac{1}{\sqrt{1+a_k^2}} + \frac{a_k}{\sqrt{1+a_k^2}}\, c_{k\uparrow}^+\, c_{-k\downarrow}^+\right]\Psi_0 \qquad (11)$$

$$\text{or} \quad = P_{N/2} \prod_k \left[\frac{1}{\sqrt{1+a_k^2}} + \frac{a_k}{\sqrt{1+a_k^2}}\, \underline{c}_{k\uparrow}^+\, \underline{c}_{-k\downarrow}^+\right]\Psi_0$$

In the insulating state every site i is filled once so that $|a_k|$ might as well be constant, $a_k = \pm 1$, and the wave function contains a "pseudo-Fermi surface" at which a_k changes sign. There is no reason why this surface should coincide with the Fermi surface of any energy in the system, since the pseudo-Fermi surface is controlled by the condition that it divide k-space into equal halves and by the choice of τ values for the valence bonds. The existence of a pseudo-Fermi surface, I believe, is real and the spin excitations may resemble those of a real Fermi liquid. This would explain the experimental observation of a Fermi-like susceptibility.

On the other hand, considered as a solution to the Hubbard model, there is a gap for any charged excitation. By assumption, U is so large that adding the $(N + 1)$th electron costs an extra $\sim U$ in energy, relative to adding the Nth, and the Fermi energy lies in this gap for the stoichiometric compound. There is a cusp in total energy as a function of occupation number at $n = N$ (Fig. 1A). In this case the $P_{N/2}$ projection operator may not be omitted. One may represent $P_{N/2}$ explicitly by giving a_k a phase $e^{i\theta}$ and writing

$$\Psi_{RVB} = P_d \int d\theta e^{-iN/2\theta} \times \prod_k \left[\frac{1}{\sqrt{1+a_k^2}} + \frac{e^{i\theta} a_k b_k^+}{\sqrt{1+a_k^2}}\right]\Psi_0$$

Thus in this insulating state there is no meaning to the phase of a_k; a wave function with fixed phase contains states of necessarily widely different energy.

Now let us consider the state obtained if we dope the system in order to remove the "half-filled" criterion and make it into a metal. From a "mean field" point of view, as soon as the system is metallized it becomes a superconductor, since the pairing already exists in the RVB state, and an energy $\sim J$ is required to break a valence-bonded pair. As shown in Fig. 1B, as soon as the occupancy leaves N, there is no cusp in the energy, the compressibility becomes finite, and by the standard arguments the state can acquire a fixed θ rather than n.

As a practical matter, the effective mass of quasiparticles will be of order $m^* = m/\delta$, where δ is the fractional doping $n = N(1 - \delta)$. Correspondingly, the coherence length ξ_0 will be of order

$$k_F\xi_0 = \frac{E_F}{\Delta} \ll 1$$

where k_F and E_F are the Fermi wave vector and energy, respectively, Δ is the energy gap, and the kinetic energy is of order $t\delta$. Thus the transition temperature T_c will at first be dominated by phase fluctuations. (In actual physical fact, at first the dopant ions will be screened out by bound quasiparticles, and it will take a finite dopant concentration to metallize the sample.) The maximum T_c, of order or less than $t^2/U = J$, will occur when $t/U \sim \delta$ and kinetic and pair-binding energies match. Pressure will increase t and, within limits, increase T_c as well.

From a theoretical point of view, the most exotic feature of these experimental results (*1*) is that they confirm the existence of a new liquid, only conjectured previously (*3, 5, 10*). This liquid is insulating only by virtue of a "commensurability gap," and therefore resembles the Laughlin state in the fractional quantum Hall effect (*15*). Both of these states may be described as "Mott liquids" since the basic physics is that of the "Mott transition" (*16*), and their key feature is that there is no symmetry breaking vis-à-vis the high-temperature state.

There are several experimental consequences of the above. The key point is the observation of the RVB state in the stoichiometric La_2CuO_4, which should be easy with neutrons, especially if there is a pseudo-Fermi surface. Second, the pseudo-Fermi surface may or may not cross the real Fermi surface, defining lines of zeroes of the gap function. The occurrence of lines of zeroes and of antiferromagnetic correlations in some heavy fermion superconductors suggests a family resemblance to the RVB state, although the parameter values are totally different.

Finally, I would call attention to the numerous unreproducible reports of high-temperature superconductivity in special samples of CuCl. In every case it is reasonable to imagine a surface layer of Cu^{2+} with or without the appropriate degree of oxidation; such reports should sharpen the search for still more RVB superconductors. It is also noteworthy that the first oxide superconductor, Li_2TiO_4 (*17*), closely resembles $NaTiO_2$, the only other likely RVB material.

REFERENCES AND NOTES

1. J. G. Bednorz and K. A. Müller, *Z. Phys. B* **64**, 189 (1986); J. G. Bednorz, M. Takashige, K. A. Müller, *Europhys. Lett.*, in press; S. Uchida, H. Takagi, K. Kitazawa, S. Tanaka, *Jpn. J. Appl. Phys.*, in press; H. Takagi, S. Uchida, K. Kitazawa, S. Tanaka, *ibid.*, in press; R. Cava, R. B. van Dover, B. Batlogg, E. A. Rietman, *Phys. Rev. Lett.* **58**, 408 (1987); C. W. Chu *et al.*, *ibid.*, p. 405; C. W. Chu *et al.*, *Science* **235**, 567 (1987); Z. Zhongxian *et al.*, *Kexue Tongbao (Beijing)*, in press.
2. A. W. Sleight, J. L. Gillson, F. E. Bierstedt, *Solid State Commun.* **17**, 27 (1975).
3. P. Ganguly and C. N. R. Rao, *J. Solid State Chem.* **53**, 193 (1984). See also K. K. Singh, thesis, Indian Institute of Science, Bangalore (1983); C. N. R. Rao, K. J. Rao, J. Gopalakrishnan, *Ann. Rep. Prog. Chem. Sect. C* **233**, 193 (1985).
4. P. W. Anderson, *Phys. Rev.* **115**, 2 (1959).
5. ——, *Mater. Res. Bull.* **8**, 153 (1973).
6. L. Hulthén, *Ark. Mat. Astron. Fys.* **26A**, 1 (1938); H. A. Bethe, *Z. Phys.* **71**, 205 (1931).
7. W. Marshall, *Proc. R. Soc. London Ser. A* **232**, 48 (1955).
8. P. W. Anderson, *Phys. Rev.* **86**, 694 (1952).
9. P. Fazekas and P. W. Anderson, *Philos. Mag.* **30**, 432 (1974).
10. K. Hirakawa, H. Kadowaki, K. Ubikoshi, *J. Phys. Soc. Jpn.* **54**, 3526 (1985); I. Yamada, K. Ubikoshi, K. Hirakawa, *ibid.*, p. 3571.
11. J. E. Hirsch, *Phys. Rev. B* **31**, 4403 (1985); *Phys. Rev. Lett.* **54**, 1317 (1985). Hirsch's simulations in these references are very suggestive confirmation of the RVB model of superconductivity.
12. T. M. Rice and R. Joynt, paper presented at the International Conference on Valence Fluctuations, Bangalore, 1987.
13. I am indebted for this suggestion to B. S. Shastry and T. V. Ramakrishnan.
14. F. J. Dyson, personal communication.
15. R. B. Laughlin, *Phys. Rev. Lett.* **50**, 1395 (1983).
16. N. F. Mott, *Proc. Phys. Soc. London Sect. A* **62**, 416 (1949).
17. D. C. Johnson, H. Prakash, W. H. Zachariasen, R. Viswanathan, *Mater. Res. Bull.* **8**, 777 (1973).
18. I would like to acknowledge the hospitability of the Council for Scientific and Industrial Research and the Tata Institute at Bombay, in hosting my visit to Bangalore, where the key parts of this work were done. Discussions with many people there were helpful, especially T. V. Ramakrishnan, B. S. Shastry, C. M. Varma, J. Hirsch, and T. M. Rice. Also I would like to acknowledge discussions with I. Affleck and T. H. Geballe, and that my faith in the RVB state was sustained for many years by P. Fazekas. The manuscript was prepared while P.W.A. was a Fairchild Scholar at the California Institute of Technology. This work was partially supported by NSF grant DMR 851-8163.

23 January 1987; accepted 3 February 1987

Reprinted from Mod. Phys. Lett. B4 (1990) 759–766

SO_4 SYMMETRY IN A HUBBARD MODEL

CHEN NING YANG
Institute for Theoretical Physics, State University of New York, Stony Brook, NY 11794-3840, USA

and

S. C. ZHANG
IBM Research Division, Almaden Research Center, San Jose, CA 95120-6099, USA

Received 28 May 1990

For a simple Hubbard model, using a particle-particle pairing operator η and a particle-hole pairing operator ζ, it is shown that one can write down two commuting sets of angular momenta operators $\mathbf{J}$ and $\mathbf{J}'$, both of which commute with the Hamiltonian. These considerations allow the introduction of quantum numbers j and j', and lead to the fact that the system has $SO_4 = (SU_2 \times SU_2)/Z_2$ symmetry. j is related to the existence of superconductivity for a state and j' to its magnetic properties.

In a recent paper[1] it was found that a pairing operator η is useful for considering the Hamiltonian in a simple Hubbard model on an $L \times L \times L$ lattice, where L = even. We shall extend such considerations in the present paper. All notations are the same as in Ref. 1. We introduce here a Hamiltonian H' and a momentum operator $\mathbf{P}'$ which are trivially different from the H and $\mathbf{P}$ of Ref. 1, in order to bring out more *symmetries* of the system:

$$H' = T' + V' , \tag{1}$$

$$T' = -2\varepsilon \sum_{\mathbf{k}} (\cos k_x + \cos k_y + \cos k_z)(a_{\mathbf{k}}^+ a_{\mathbf{k}} + b_{\mathbf{k}}^+ b_{\mathbf{k}}) , \tag{2}$$

$$V' = 2W \sum_{\mathbf{r}} \left(a_{\mathbf{r}}^+ a_{\mathbf{r}} - \frac{1}{2} \right) \left(b_{\mathbf{r}}^+ b_{\mathbf{r}} - \frac{1}{2} \right) . \tag{3}$$

$$\mathbf{P}' = \Sigma \left(\mathbf{k} - \frac{1}{2} \boldsymbol{\pi} \right) (a_{\mathbf{k}}^+ a_{\mathbf{k}} + b_{\mathbf{k}}^+ b_{\mathbf{k}}) \, (\mathrm{mod.} 2\pi) . \tag{4}$$

(1) *The operators J_x, J_y, and J_z* — It is easy to verify that $\eta^+ \eta - \eta \eta^+ = \Sigma(a^+ a + b^+ b) - M$, where $M = L^3$. Calculating the commutator of this commutator with η we obtain

PACS Nos: 74.20.-z, 05.30.Fk.

Theorem 1. Defining

$$\eta^+ = J_x + iJ_y\,,\quad \eta = J_x - iJ_y\,,\quad J_z = \frac{1}{2}\Sigma\,(a^+a + b^+b) - \frac{1}{2}M\,, \tag{5}$$

one finds that J_x, J_y, J_z commute with each other like the components of an angular momentum. Hence the eigenvalue of $\mathbf{J}^2$ is $j(j+1)$ where $2j$ = integer $\geqq 0$. Furthermore (as can be easily checked),

$$[T',\mathbf{J}]_- = [V',\mathbf{J}]_- = [H',\mathbf{J}]_- = [\mathbf{P}'\mathbf{J}]_- = 0\,. \tag{6}$$

(2) *The operators J'_x, J'_y and J'_z* — We now define a particle-hole pairing operator,

$$\zeta = \Sigma\, a_{\mathbf{k}}b^+_{\mathbf{k}} = \Sigma\, a_{\mathbf{r}}b^+_{\mathbf{r}}\,. \tag{7}$$

Then

$$\zeta\zeta^+ - \zeta^+\zeta = -\Sigma\, a^+a + \Sigma\, b^+b\,.$$

Theorem 2. Defining

$$\zeta^+ = J'_x + iJ'_y\,,\quad \zeta = J'_x - iJ'_y\,,\quad J'_z = \frac{1}{2}\Sigma\, a^+a - \frac{1}{2}\Sigma\, b^+b\,, \tag{8}$$

one finds that J'_x, J'_y, J'_z commute with each other like the components of an angular momentum. Hence the eigenvalue of $\mathbf{J}'^2$ is $j'(j'+1)$ where $2j'$ = integer $\geqq 0$. Furthermore all 3 components of $\mathbf{J}$ commute with all 3 components of $\mathbf{J}'$, and

$$[T',\mathbf{J}']_- = [V',\mathbf{J}']_- = [H',\mathbf{J}']_- = [\mathbf{P}'\mathbf{J}']_- = 0\,. \tag{9}$$

ζ is the usual spin lowering operator and $\mathbf{J}'$ is the usual "spin" operator.

(3) *Explicit eigenfunctions of H'* — We can find many eigenstates of H' with Theorems 1 and 2 as follows. We diagonize $\mathbf{J}^2$, $\mathbf{J}'^2$, J_z, J'_z, H' and $\mathbf{P}'$ simultaneously. These states can be sorted out into multiplets $\{j, j'\}$, each comprising of $(2j+1)\,(2j'+1)$ states, as illustrated in Fig. 1, where N_a and N_b are eigenvalues of Σa^+a and Σb^+b,

$$j_z = \frac{1}{2}(N_a + N_b - M)\,,\quad j_z = \frac{1}{2}(N_a - N_b)\,. \tag{10}$$

As explained in Fig. 1, $j+j'$ = integer, i.e., not all representations of $SU_2 \times SU_2$ are present. This means that the true symmetry of the problem is $(SU_2 \times SU_2)/Z_2 = SO_4$.

Consider now the states in one spot on the bottom row of Fig. 1. For these states, $N_a = 0$. The operators H' and $\mathbf{P}'$ for such states are easily diagonizable since for such states, there are no a-particle — b-particle interac-

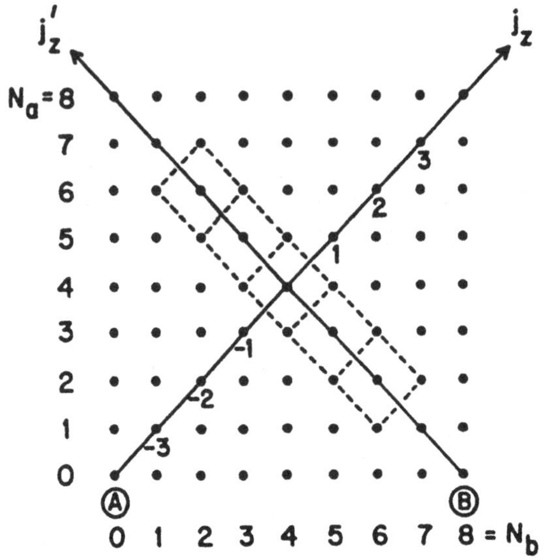

Fig. 1. (N_a, N_b) diagram for $M = 8$. The relationship between (j_z, j'_z) with (N_a, N_b) is given by Eq. (10). Each multiplet $\{j, j'\}$ is represented by a rectangular set of states centered at $j_z = j'_z = 0$ in this diagram. The number of states in the multiplet is $(2j + 1)(2j' + 1)$. Illustrated is the multiplet $\left\{\frac{1}{2}, \frac{5}{2}\right\}$. All states of a multiplet share the same eigenvalue of H' and $\mathbf{P}'$. The lowest corner in the multiplet is where $j_z = -j$, $j'_z = -j'$. One can generate all states of a multiplet by starting from its lowest corner and repeatedly operate on it with $\eta^+ = J_x + iJ_y$ (which increases j_z) and with $\zeta^+ = J'_x + iJ'_y$ (which increases j'_z). Obviously $j + j' =$ integer. Notice that for fixed j and j', there are in general a large number of multiplets $\{j, j'\}$, except for $\{M/2, 0\}$ and $\{0, M/2\}$, each of which occurs only once. For the former, the lowest corner is the point A where $N_a = N_b = 0$ which is a single state. For the latter, the lowest corner is B where $N_a = 0$, $N_b = M$ which is also a single state.

tions, so that the problem reduces to that of N_b noninteracting fermions. One can thus trivially write down the eigenstates of H' and $\mathbf{P}'$ in momentum space. There are $\binom{M}{N_b}$ such states. Operating with η^+ and ζ^+ on these states generates $\binom{M}{N_b}$ multiplets $\{j, j'\}$. Now obviously

$$j = \frac{1}{2}(M - N_b)\ , \quad j' = \frac{1}{2}N_b\ .$$

Thus we can easily write down explicitly the eigenfunctions for H' and $\mathbf{P}'$ for $\binom{M}{N_b}$ multiplets $\left\{\frac{1}{2}(M - N_b), \frac{1}{2}N_b\right\}$. The total number of such states is $\Sigma\binom{M}{N_b}(M - N_b + 1)(N_b + 1)$, where the summation extends from $N_b = 0$ to M. The summation is equal to $2^{M-2}(M^2 + 3M + 4)$. This is an enormous number of eigenstates, but still very small compared to the total number of eigenstates which is 4^M. We remark here that the eigenstates ψ_N of Ref. 1 are special cases of the states discussed in this section.

The eigenstates of H' constructed above obviously do not depend on W and are simultaneous eigenstates of T' and V'. We believe they are the only W-independent eigenstates of H', but we do not know how to prove this statement except in special cases.

(4) *ODLRO* — We shall show

Theorem 3. For any state ψ for which $j^2 - j_z^2 = O(M^2)$, there is ODLRO.

The 2-particle reduced density matrix ρ_2 has matrix element

$$\langle b_s a_s | \rho_2 | b_r a_r \rangle = \psi^+ a_r^+ b_r^+ a_s b_s \psi \ .$$

Thus

$$\Sigma\, e^{i\pi\cdot(r-s)} \langle b_s a_s | \rho_2 | b_r a_r \rangle = \psi^+ \eta^+ \eta \psi = \psi^+ (J_x + iJ_y)(J_x - iJ_y)\psi$$
$$= j^2 - j_z^2 + j + j_z \ .$$

Using

$$\langle b_{r'} a_r | \phi \rangle = M^{-1/2} e^{i\pi\cdot r} \delta(\mathbf{r} - \mathbf{r}')$$

as a trial wave function for ρ_2, we find the expectation value of ρ_2 to be

$$\langle \rho_2 \rangle = \frac{1}{M}(j^2 - j_z^2) + O(1) = O(M) \geqq 0.$$

Thus the largest eigenvalue of ρ_2 is $O(M)$ and the state has ODLRO.[2]

In Ref. 1 we had showed that the states ψ_N have ODLRO. That fact is a special case of the above theorem, because for ψ_N, $j = M/2$, and $j_z = -M/2 + N$.

In the above discussions, the pairs are particle-particle pairs. If the particle is charged e, then the state exhibits[2] flux quantization in units of $ch/2e$. If $j'^2 - j_z'^2 = O(M^2)$, the system exhibits particle-hole ODLRO. There is no superconductivity for such a system.[2,3] Thus *j is related to superconductivity and j′ to magnetic properties.*

(5) *Unitary Operators U_b and X* — We define these two operators as follows:

$$U_b a_r U_b^{-1} = a_r \ , \quad U_b b_r U_b^{-1} = e^{i\pi\cdot r} b_r^+ \ , \quad U_b^2 = 1 \ , \tag{11}$$

and

$$X a_r X^{-1} = e^{i\pi\cdot r} a_r \ , \quad X b_r X^{-1} = e^{i\pi\cdot r} b_r \ , \quad X^2 = 1 \ . \tag{12}$$

Operator X is well known and operator U_b has been discussed in the literature.[4] We observe that

$$U_b b_k U_b^{-1} = b_{\pi-k}^+ \ , \tag{13}$$

and

$$\zeta = U_b \eta U_b^{-1} . \tag{14}$$

Theorem 4. Writing $H'(W)$ for H', we have

$$U_b H'(W) U_b^{-1} = H'(-W) , \tag{15}$$

$$U_b (\Sigma b^+ b) U_b^{-1} = M - \Sigma b^+ b , \quad U_b (\Sigma a^+ a) U_b^{-1} = \Sigma a^+ a . \tag{16}$$

Theorem 5.

$$X H'(W) X^{-1} = -H'(-W) , \tag{17}$$

$$X(\Sigma a^+ a) X^{-1} = \Sigma a^+ a , \quad X(\Sigma b^+ b) X^{-1} = \Sigma b^+ b . \tag{18}$$

It follows that

$$(XU_b)(H'(W))(XU_b)^{-1} = -H'(W) ,$$

$$(XU_b)(\Sigma a^+ a)(XU_b)^{-1} = \Sigma a^+ a ,$$

$$(XU_b)(\Sigma b^+ b)(XU_b)^{-1} = M - \Sigma b^+ b . \tag{19}$$

Denoting by Spm (W, N_a, N_b) the spectrum of $H'(W)$ for given N_a and N_b, we have, by Theorem 4,

Theorem 6.

$$\begin{aligned} \text{Spm}\,(W, N_a, N_b) &= \text{Spm}\,(-W, N_a, M - N_b) \\ &= \text{Spm}\,(-W, M - N_a, N_b) \\ &= \text{Spm}\,(W, M - N_a, M - N_b) . \end{aligned} \tag{20}$$

By Theorem 5, we have

Theorem 7.

$$\text{Spm}\,(W, N_a, N_b) = -\text{Spm}(-W, N_a, N_b) . \tag{21}$$

Combining these two results we obtain

$$\begin{aligned} \text{Spm}\,(W, N_a, N_b) &= -\text{Spm}\,(W, N_a, M - N_b) \\ &= -\text{Spm}\,(W, M - N_a, N_b) \\ &= \text{Spm}\,(W, M - N_a, M - N_b) . \end{aligned} \tag{22}$$

(6) *Limit* $M \to \infty$ — We shall now put $\varepsilon = 1$ in (2). Diagonalizing $\mathbf{J}^2$, $\mathbf{J}'^2$, J_z, J'_z, H', $\mathbf{P}'$, we have also diagonalized N_a and N_b because of (10). Let the lowest eigenvalue of H' at a fixed N_a, N_b be denoted by $E_0(W, N_a, N_b)$. Now keeping fixed the values of

$$N_a/M = \rho_a \ , \quad N_b/M = \rho_b$$

we approach the limit $M \to \infty$. It can be proved, by a method used in Ref. 5, that $M^{-1}E_0$ approaches a limit which we shall denote by $f(W, \rho_a, \rho_b)$. f is the lowest eigenvalue of H' per site at fixed densities ρ_a and ρ_b.

The function f has many symmetries. Because of Theorems 1 and 2,

$$f(W, \rho_a, \rho_b) = f(W, \rho_b, \rho_a) = f(W, 1 - \rho_a, 1 - \rho_b) = f(W, 1 - \rho_b, 1 - \rho_a) \ . \tag{23}$$

Because of (20),

$$f(W, \rho_a, \rho_b) = f(-W, \rho_a, 1 - \rho_b) = f(-W, 1 - \rho_a, \rho_b) \ . \tag{24}$$

These symmetries are illustrated in Fig. 2.

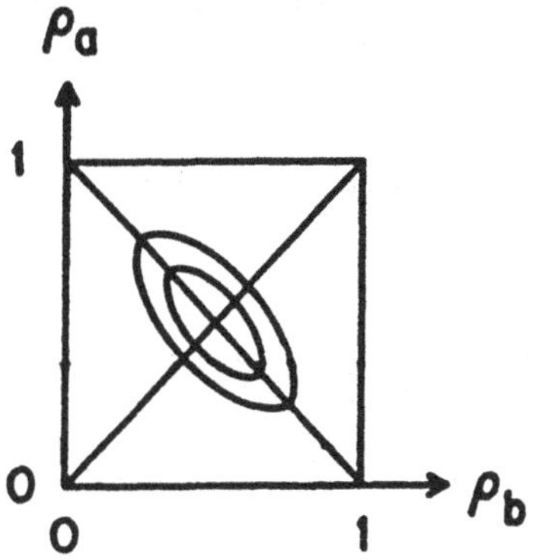

Fig. 2. Equi-f contours in ρ_a, ρ_b plane (schematic). Because of (23), these contours are reflection symmetrical with respect to the $\rho_a = \rho_b$ axis and the $\rho_a + \rho_b = 1$ axis. Because of Theorem 8, these contours are convex. One can obtain the $(-W)$ contours from the (W) contours by a rotation through 90° around the center of the square.

Theorem 8. $f(W, \rho_a, \rho_b)$ as a function of ρ_a and ρ_b is continuous and concaves upwards.

Theorem 9. $f(W, \rho_a, \rho_b)$ as a function of W concaves downwards.
These two theorems can be proved using the methods of Ref. 5.

Theorem 8 and Eq. (23) show that the minimum of $f(W, \rho_a, \rho_b)$ for fixed W is $f(W, 1/2, 1/2)$. This minimum value may be shared by f at other values of (ρ_a, ρ_b) than (1/2, 1/2). Let the region of (ρ_a, ρ_b) where this is true be denoted by R, and call the states that have this minimum value of f *lowest states.* (23) shows that R is reflection symmetrical with respect to the axis: $\rho_a = \rho_b$, and with respect to the axis: $\rho_a + \rho_b = 1$. Using Theorem 8 we can show

Theorem 10. The region R in (ρ_a, ρ_b) where $f(W, \rho_a, \rho_b) = f(W, 1/2, 1/2)$ is convex. Possible schematic shapes of R are illustrated in Fig. 3.

Each of the *lowest state* belongs to a multiplet $\{j, j'\}$. Within that multiplet the leading state (i.e. where $j_z = j, j'_z = j$,) is also a *lowest state*. Hence it must be in the $j_z \geqq 0, j'_z \geqq 0$ quadrant of R. Thus

Theorem 11. All the *lowest states* on the boundary of R have $j = |j_z|, j' = |j'_z|$. Finally we remark that for the points $\rho_a = 0$ (or $\rho_b = 0$,) the system is devoid of a (or b) particles. Hence the value of $f(W, 0, \rho_b)$ and $f(W, \rho_a, 0)$ can be easily evaluated. (23) then allows one to write down $f(W, 1, \rho_b)$ and $f(W, \rho_a, 1)$. Thus the value of f on the boundary of the square in Fig. 2 is known.

We now define $g(W, \rho_a, \rho_b)$ to be highest eigenvalue of H' per site. Equation (22) then shows that

$$g(W, \rho_a, \rho_b) = -f(W, \rho_a, 1-\rho_b) = -f(W, 1-\rho_a, \rho_b) \,. \tag{25}$$

More generally we define the free energy per site by

$$F(\beta, W, \rho_a, \rho_b) = \lim(-M\beta)^{-1} \ln(\text{p.f.}) \tag{26}$$

where

$$(\text{p.f.}) = \text{trace of block of } \exp(-\beta H') \text{ belonging to given } \rho_a, \rho_b \,, \tag{27}$$

and the limit is for $M \to \infty$. Then

$$F(\infty, W, \rho_a, \rho_b) = f(W, \rho_a, \rho_b) \,,$$

$$F(-\infty, W, \rho_a, \rho_b) = g(W, \rho_a, \rho_b) \,. \tag{28}$$

The function F has many symmetries. Theorems 1 and 2 show that

$$\begin{aligned} F(\beta, W, \rho_a, \rho_b) &= F(\beta, W, \rho_b, \rho_a) \\ &= F(\beta, W, 1-\rho_a, 1-\rho_b) \\ &= F(\beta, W, 1-\rho_b, 1-\rho_a) \,. \end{aligned} \tag{29}$$

Equation (20) shows that

$$F(\beta, W, \rho_a, \rho_b) = F(\beta, -W, \rho_a, 1-\rho_b) = F(\beta, -W, 1-\rho_a, \rho_b) \,. \tag{30}$$

Equation (21) shows that

$$F(\beta, W, \rho_a, \rho_b) = -F(-\beta, -W, \rho_a, \rho_b) \,. \tag{31}$$

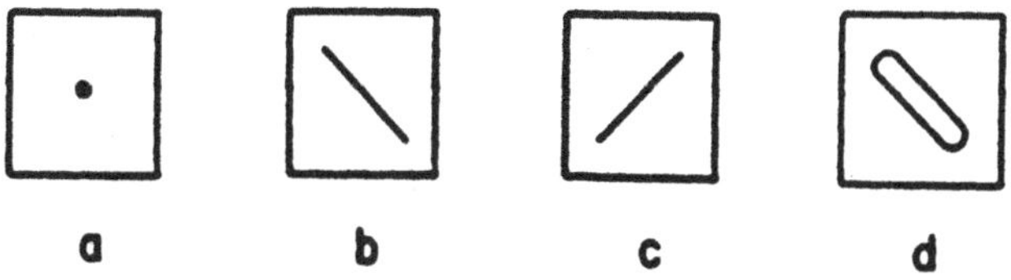

Fig. 3. Possible shapes for R. R is convex and is reflection symmetrical with respect to the $\rho_a = \rho_b$, and the $\rho_a + \rho_b = 1$ axes. For case c there is particle-particle ODLRO at low temperatures in the open line segment. For case d there is particle-particle ODLRO at low temperatures *inside* of the region R. These cases exhibit superconductivity.

These two last equations together show that

$$F\left(0, W, \rho_a, \frac{1}{2}\right) = F\left(0, W, \frac{1}{2}, \rho_b\right) = 0 \; . \tag{32}$$

Acknowledgments

One of us (CNY) is supported in part by the National Science Foundation under grant number PHY 8908495.

References

1. C. N. Yang, *Phys. Rev. Lett.* **63** (1989) 2144.
2. C. N. Yang, *Rev. Mod. Phys.* **34** (1962) 694.
3. W. Kohn and D. Sherrington, *Rev. Mod. Phys.* **42** (1970) 1.
4. H. Shiba, *Prog. Theor. Phys.* **48** (1972) 2171.
5. C. N. Yang and C. P. Yang, *Phys. Rev.* **147** (1966) 303.

Chapter 2

RIGOROUS RESULTS AND EXACT SOLUTIONS FOR LARGE SYSTEMS

Apart from the physical parameters U and t, the ingredients which determine the possible different behaviors of the systems described by the model-hamiltonian (1.1) are the dimension D of the lattice Λ on which the model is defined, the temperature T, and the filling of the band, δ. So far the exact results that have been obtained fix at least one, if not all, of the above quantities.

For $D = 1$, the model has been solved exactly in 1968 by Lieb and Wu[2.1], using the *Bethe Ansatz* approach. In fact in one dimension, as shown twenty years later by Shastry[2.2], it is possible to construct a one-parameter family of transfer matrices commuting among themselves and with the Hubbard hamiltonian, and these identify a new 2–D integrable classical model — corresponding to two coupled six-vertex models — which is related to a lattice-covering statistical model.

The Lieb and Wu solution provides the exact ground state energy of the model once a set of four integral equations are solved. This could be done explicitly at half-filling, giving an insulating ground state at any $U > 0$. Subsequent work of Ovchinnikov[2.3] succeeded in determining, at $T = 0$ and for a half-filled band, the spin-wave excitation spectrum, in particular evidentiating that a finite energy is required for creating an electron-hole pair for $U > 0$ (optical gap). At the same time, Takahashi[r2.1] calculated the zero-temperature susceptibility at $\delta = \frac{1}{2}$, showing that it is a monotonic function of U.

For $\delta \neq \frac{1}{2}$ the system of integral equations is not analytically solved. Nevertheless, by approximating it with a suitable set of linear algebraic equations, it was possible[2.4] to evaluate both the ground-state energy and the magnetic susceptibility for any δ, the latter turning out to be a decreasing function of δ. The same ground-state energy was successively obtained analytically in a rigorous way in [3.9], for the large U limit. As for the excitation spectrum, so far it was explored only numerically[r2.2], and gives no optical gap for a non half-filled band.

Even in $D = 1$, rigorous results concerning finite-temperature properties for the infinite system are still very few. One can list mainly the low-temperature behavior of the specific heat, investigated again by Takahashi[2.5], and the strong-coupling

($U \to \infty$) partition function for arbitrary electron density[3.10].

The $D = 1$ case is essentially the only case in which an exact solution for the Hubbard model has been worked out, except naturally the two limiting conditions $t = 0$ and $U = 0$, for which the model turns out to be exactly solvable in any dimension. Also, van Dongen and Vollhardt[2.6] obtained the exact solution and the thermodynamical properties of the model for the (unphysical) situation in which the hopping is unconstrained, with infinite-range, finding once more an insulating ground state only for a half-filled band. However, such ground state is pathological, in that the double-occupancy expectation value has always the minimum allowed value, independently of the interaction strength.

Finally, a few general rigorous results are known to hold for any dimension D. The first, in order of time, states rigorous upper and lower bounds for the ground-state energy[2.7] of the model. The second, due to Lieb[2.8], consists in two theorems about the spin angular momentum of the ground state for, positive and negative U respectively. In particular, it is shown that at half-filling certain versions of the repulsive model give rise to ferromagnetism, thus generating a first example of itinerant magnetism with finite potential (as opposed to Nagaoka's result[3.6], which holds only in the infinite U limit). Subsequently Kubo and Kishi[2.9] gave rigorous bounds for the spin susceptibility in the finite temperature case.

[r2.1] M. Takahashi, *Prog. Theor. Phys.* **42**, 1098 (1969); **43**, 1619 (1970).
[r2.2] C.F. Coll III, *Phys. Rev.* **B9**, 2150 (1970).

VOLUME 20, NUMBER 25 PHYSICAL REVIEW LETTERS 17 JUNE 1968

ABSENCE OF MOTT TRANSITION IN AN EXACT SOLUTION OF THE SHORT-RANGE, ONE-BAND MODEL IN ONE DIMENSION

Elliott H. Lieb* and F. Y. Wu
Department of Physics, Northeastern University, Boston, Massachusetts
(Received 22 April 1968)

The short-range, one-band model for electron correlations in a narrow energy band is solved exactly in the one-dimensional case. The ground-state energy, wave function, and the chemical potentials are obtained, and it is found that the ground state exhibits no conductor-insulator transition as the correlation strength is increased.

The correlation effect of electrons in a partially filled energy band has been a subject of interest for many years.[1-4] A realistic model which takes this correlation into consideration, and which is hopefully amenable to mathematical treatment, is the short-range, one-band model considered by a number of authors.[2-5] In this model, one pictures the electrons in a narrow energy band hopping between the Wannier states of neighboring lattice sites, with a repulsive interaction energy between two electrons of opposite spins occupying the same lattice site. The central problems of interest have been (a) the possible existence of a "Mott transition" between conducting and insulating states as the strength of the interaction is increased, and (b) the magnetic nature (ferromagnetic or antiferromagnetic) of the ground state. While previous treatments of this model have always been approximate, we have succeeded in solving the model exactly in the one-dimensional case. Our exact result shows that the Mott transition does occur in the ground state of the one-dimensional model. Furthermore, a general theorem of Lieb and Mattis[6] on one-dimensional systems tells us that the ground state is necessarily antiferromagnetic.

It may be argued that the absence of a Mott transition in one dimension is irrelevant for the study of real three-dimensional systems because of the folkloristic dictum that there are never any phase transitions in one dimension with short-range interactions. In actual fact, the dictum is only true for nonzero temperature; the ground state is another matter. Generally speaking, when a Hamiltonian is considered to be a function of some parameter, U (which in our case is the electron-electron repulsion), singularities with respect to U usually do appear in the ground-state wave function, energy, polarizability, etc., even in one dimension. A good example of this is the one-dimensional Heisenberg chain (to which the present model is very close) which, when considered as a function of the anisotropy parameter, does have two singularities in the ground state and, presumably, no singularities for nonzero temperatures.[7,8]

Consider a crystal (one-, two-, or three-dimensional) of N_a lattice sites with a total of $N \leqslant 2N_a$ electrons. We suppose that the electrons can hop between the Wannier states of neighboring lattice sites, and that each site is capable of accommodating two electrons of opposite spins, with an interaction energy $U>0$. The Hamiltonian to consider is then[2-5]

$$H=T\sum_{\langle ij\rangle}\sum_{\sigma}c_{i\sigma}^{\dagger}c_{j\sigma}+U\sum_{i}c_{i\uparrow}^{\dagger}c_{i\uparrow}c_{i\downarrow}^{\dagger}c_{i\downarrow}, \quad (1)$$

where $c_{i\sigma}^{\dagger}, c_{i\sigma}$ are, respectively, the creation and annihilation operators for an electron of spin σ in the Wannier state at the ith lattice site, and the sum

$$\sum_{\langle ij\rangle}$$

is restricted to nearest-neighbor sites.

First of all, it can be shown that the energy spectrum of H is invariant under the replacement of T by $-T$.[9] Therefore, for simplicity we shall take, in appropriate units, $T=-1$. Since the numbers M of down-spin electrons and M' of up-spin electrons are good quantum numbers ($M+M'=N$), we may designate the ground-state energy of H by $E(M,M';U)$. It is then easy to derive the following relations [by considering holes instead of particles in (1)]:

$$\begin{aligned}E(M,M';U)&=-(N_a-M-M')U\\&\quad+E(N_a-M,N_a-M';U)\\&=MU+E(M,N_a-M';-U)\\&=M'U+E(N_a-M,M';-U).\end{aligned} \quad (2)$$

Without loss of generality, therefore, we may take

$$S_z\equiv\tfrac{1}{2}(N-2M)\geqslant 0 \text{ and } N\leqslant N_a$$

(less than half-filled band).

It can similarly be shown that the maximum energy $G(M,M';U)$ is related to the ground-state energy by

$$G(M,M';U)=M'U-E(N_a-M,M';U).$$

Therefore, a knowledge of the ground-state energies also tells us about the maximum energies.

For a one-dimensional system, the lattice sites can be numbered consecutively from 1 to N_a. Let $f(x_1,x_2,\cdots,x_M,x_{M+1},\cdots,x_N)$ represent the amplitude in ψ for which the down spins are at the sites $x_1,\cdots,x_M$, and the up spins at $x_{M+1},\cdots,x_N$. Then the eigenvalue equation $H\psi=E\psi$ leads to

$$-\sum_{i=1}^{N}\sum_{s=\pm1} f(x_1,\cdots,x_i+s,\cdots,x_N)$$
$$+U\sum_{i<j}\delta(x_i-x_j)f(x_1\cdots x_N)$$
$$=Ef(x_1\cdots x_N), \tag{3}$$

where it is understood that we require a solution of the form

$$f(x_1,x_2,\cdots,x_M|x_{M+1},x_{M+2},\cdots,x_N) \tag{4}$$

which is antisymmetric in the first M and the last $N-M$ variables.

In each region defined by $1\leqslant x_{Q1}\leqslant x_{Q2}\leqslant\cdots x_{QN}\leqslant N$, we make the following Ansatz for f:

$$f(x_1,\cdots,x_M|x_{M+1},\cdots,x_N)$$
$$=\sum_P[Q,P]\exp(i\sum_{j=1}^{N}k_{Pj}x_{Qj}), \tag{5}$$

where $P=(P1,P2,\cdots,PN)$ and $Q=(Q1,Q2,\cdots,QN)$ are two permutations of the numbers $(1,2,\cdots,N)$, $\{k_1,k_2,\cdots k_N\}$ is a set of N unequal real numbers, and $[Q,P]$ is a set of $N!\times N!$ coefficients to be determined.

The coefficients $[Q,P]$ are not all independent. The condition of single valuedness (or continuity) of f and the requirement that (5) be a solution of (3) lead to the following:

$$E=-2\sum_{j=1}^{N}\cos k_j \tag{6}$$

and, for all Q and P, the coefficients $[Q,P]$ must be chosen to satisfy the relations

$$[Q,P]=Y_{nm}{}^{ab}[Q,P']. \tag{7}$$

In (7), $Y_{nm}{}^{ab}$ is an operator defined by

$$Y_{nm}{}^{ab}=\frac{-\frac{1}{2}iU}{\sin k_n-\sin k_m+\frac{1}{2}iU}$$
$$+\frac{\sin k_n-\sin k_m}{\sin k_n-\sin k_m+\frac{1}{2}iU}P^{ab}, \tag{8}$$

where, for $j=i+1$,

$$Qi=a=Q'j,\quad Qj=b=Q'i,$$
$$Qk=Q'k \text{ for all } k\neq i,j;$$
$$Pi=m=P'j,\quad Pj=n=P'i,$$
$$Pk=P'k \text{ for all } k\neq i,j;$$

and P^{ab} is an operator which exchanges $Qi=a$ and $Qj=b$.

It is fortunate that the Ansatz (5) and the algebraic consistency conditions (7) and (8) have, in essence, appeared before in the study of the one-dimensional delta-function gas for particles in a continuum. The first solution of that problem was for bosons (symmetric f) by Lieb and Liniger[10] but this case is not relevant here, besides which the consistency conditions there are trivial to solve. The two-component fermion case was solved by McGuire[11] for $M=1$, but again (7) is trivial because of translational invariance. The next development was the solution of the case $M=2$ by Flicker and Lieb[12] by an inelegant algebraic method which could not be easily generalized. However, the case $M=2$ is the simplest one which displays the full difficulty of the problem. Shortly thereafter, Gaudin[13] published the solution of the general-M problem. The method of his brilliant solution did not appear for some time and is now available as his thesis.[14] In the meantime, Yang[15] also discovered the method of solution (essentially the same as Gaudin's) and published it with considerable detail. Here, we have followed Yang's notation with slight modification.

The important point is that our Eqs. (7) and (8) are the same as for the continuum gas except for the replacement of k by $\sin k$ in the latter. This has no effect on the beautiful algebraic analysis which finally leads to the following conditions which determine the set $\{k_{1,2},\cdots,k_N\}$:

$$N_a k_j = 2\pi I_j + \sum_{\beta=1}^{M} \theta(2\sin k_j - 2\Lambda_\beta),\quad j=1,2,\cdots,N, \tag{9}$$

where the Λ's are a set of real numbers related to the k's through

$$-\sum_{j=1}^{N} \theta(2\Lambda_\alpha - 2\sin k_j) = 2\pi J_\alpha - \sum_{\beta=1}^{M} \theta(\Lambda_\alpha - \Lambda_\beta),\quad \alpha=1,2,\cdots,M, \tag{10}$$

$$\theta(p) \equiv -2\tan^{-1}(2p/U),\quad -\pi \leqslant \theta < \pi, \tag{11}$$

and I_j = integers (or half-odd integers) for M = even (or odd), J_α = integers (or half-odd integers) for M' = odd (or even). An immediate consequence is

$$\sum_{j=1}^{N} k_j = \frac{1}{N_a}\left(\sum_j I_j + \sum_\alpha J_\alpha\right). \tag{12}$$

For the ground state, J_α and I_j are consecutive integers (or half-odd integers) centered around the origin and satisfying $\sum_j k_j = 0$.

In the limit of $N\to\infty$, $N_a\to\infty$, $M\to\infty$ with the ratios N/N_a, M/N_a kept finite, the real numbers k and Λ are distributed continuously between $-Q$ and $Q\leqslant\pi$ and $-B$ and $B\leqslant\infty$, with density functions $\rho(k)$ and $\sigma(\Lambda)$, respectively. Equations (9) and (10) then lead to the coupled integral equations for the distribution function $\rho(k)$ and $\sigma(\Lambda)$:

$$2\pi\rho(k) = 1 + \cos k \int_{-B}^{B} \frac{8U\sigma(\Lambda)d\Lambda}{U^2 + 16(\sin k - \Lambda)^2}, \tag{13}$$

$$\int_{-Q}^{Q} \frac{8U\rho(k)dk}{U^2 + 16(\Lambda - \sin k)^2} = 2\pi\sigma(\Lambda) + \int_{-B}^{B} \frac{4U\sigma(\Lambda')d\Lambda'}{U^2 + 4(\Lambda - \Lambda')^2}, \tag{14}$$

where Q and B are determined by the conditions

$$\int_{-Q}^{Q} \rho(k)dk = N/N_a, \tag{15}$$

$$\int_{-B}^{B} \sigma(\Lambda)d\Lambda = M/N_a. \tag{16}$$

The ground-state energy (6) now becomes

$$E = -2N_a \int_{-Q}^{Q} \rho(k)\cos k\, dk. \tag{17}$$

We have established the following:

(a) Equations (13)-(16) have a unique solution which is positive for all allowed B and Q.

(b) M/N is a monotonically increasing function of B reaching a maximum of $\frac{1}{2}$ at $B=\infty$. This is the antiferromagnetic case, $S_z=0$, and corresponds to the absolute ground state.

(c) N/N_a is a monotonically increasing function of Q, reaching a maximum of 1 (half-filled band) at $Q=\pi$.

For $B=\infty$ and $Q=\pi$, (13)-(16) can be solved in closed form by Fourier transforms with the result

$$\sigma(\Lambda) = (2\pi)^{-1}\int_0^\infty \operatorname{sech}(\tfrac{1}{4}\omega U) \times \cos(\omega\Lambda) J_0(\omega) d\omega, \tag{18}$$

$$\rho(k) = (2\pi)^{-1} + \pi^{-1}\cos k \times \int_0^\infty \frac{\cos(\omega \sin k) J_0(\omega) d\omega}{1 + \exp(\frac{1}{2}\omega U)}, \tag{19}$$

$$E \equiv E(\tfrac{1}{2}N_a, \tfrac{1}{2}N_a; U) = -4N_a \int_0^\infty \frac{J_0(\omega)J_1(\omega)d\omega}{\omega[1+\exp(\frac{1}{2}\omega U)]}, \tag{20}$$

where J_0 and J_1 are Bessel functions.

To investigate whether the ground state is conducting or insulating, we compute the chemical potentials μ_+ and μ_- as defined in a forthcoming paper by Mattis[16]:

$$\mu_+ \equiv E(M+1,M;U)-E(M,M;U),$$
$$\mu_- \equiv E(M,M;U)-E(M-1,M;U). \quad (21)$$

If μ_+ and μ_- are equal, the system has the property of a conductor. If, on the other hand, we find $\mu_+>\mu_-$, then the system shares the property of an insulator. We can compute μ_- directly from (9) and (10) by replacing $M\to M-1$ and $N\to N-1$, while letting all the k's, Λ's, and their distribution functions change slightly. The procedure is quite similar to the calculation of the excitation spectrum of the continuum gas.[10] If $N<\frac{1}{2}N_a$, we can compute μ_+ in the same way and thereby find that $\mu_+=\mu_-$ for all U. If, however, N is exactly $\frac{1}{2}N_a$, then we must compute μ_+ by using the first line of (2) which tells us that

$$\mu_+ = U-\mu_- \text{ (half-filled band).} \quad (22)$$

The calculation of μ_- can be done in closed form for a half-filled band with the result

$$\mu_- -2 = -4\int_0^\infty \frac{J_1(\omega)d\omega}{\omega[1+\exp(\frac{1}{2}\omega U)]}$$
$$= -4\sum_{n=1}^{\infty}(-1)^n[(1+\tfrac{1}{4}n^2U^2)^{\frac{1}{2}}-\tfrac{1}{2}nU]. \quad (23)$$

It can be established from (22) and (23) that, indeed, $\mu_+>\mu_-$ for $U>0$, and

$$\lim_{U\to 0}\mu_\pm = 0.$$

Therefore, we conclude that the ground state for a half-filled band is insulating for any nonzero U, and conducting for $U=0$. That is, there is no Mott transition for nonzero U. This absence of a Mott transition is also reflected by the fact that the ground-state energy and the ground-state wave function are analytic in U on the real axis (except at the origin).

We have also investigated the excitation spectrum $E(p)$ for a given total momentum $\sum_j k_j=p$ and a given value of S_z. Just as in the case of a continuum gas for which the spectrum can be regarded as consisting of several elementary excitations,[10,15] we find three types of excitations: (I) a "hole" state in the Λ distribution, (II) a "hole" state in the k distribution, and (III) a "particle" state in the k distribution. While the $S_z=0$ spin-wave state may have any of these three types of spectra, the $S_z=1$ spin-wave state is always associated with the type-I spectrum. The type-I excitation has the lowest energy and is characterized by a double periodicity similar to that of an antiferromagnetic chain.[7] In the limit $U\to 0$, it goes over to $E(p)=|\sin p|$, while the type-II and -III spectra have the identical limiting form $E(p)=|2\sin(\frac{1}{2}p)|$. Detailed discussions of these matters will be given elsewhere.

We are grateful to Dr. D. C. Mattis for helpful advice and suggestions and E. L. would like to thank Dr. J. Zittartz for interesting him in the problem.

---

*Work partially supported by National Science Foundation Grant No. GP-6851.

[1]J. H. Van Vleck, Rev. Mod. Phys. 25, 220 (1953).
[2]M. C. Gutzwiller, Phys. Rev. Letters 10, 159 (1963), and Phys. Rev. 134, A923 (1964), and 137, A1726 (1965).
[3]J. Hubbard, Proc. Roy. Soc. (London), Ser. A 276, 238 (1963), and 277, 237 (1964).
[4]G. Kemeny, Ann. Phys. (N.Y.) 32, 69, 404 (1965).
[5]See also the review article by C. Herring, in Magnetism, edited by G. T. Rado and H. Suhl (Academic Press, Inc., New York, 1966), Vol. IV, Chap. 10.
[6]E. Lieb and D. C. Mattis, Phys. Rev. 125, 164 (1962). The proof of the theorem given in this reference can be adopted to our Hamiltonian (1).
[7]J. des Cloizeaux and M. Gaudin, J. Math. Phys. 7, 1384 (1966).
[8]C. N. Yang and C. P. Yang, Phys. Rev. 150, 321, 327 (1966).
[9]The proof assumes A and B sublattices and uses the unitary transformation

$$\exp[i\pi\sum_\sigma\sum_{i\in A}c_{i\sigma}^\dagger c_{i\sigma}]$$

which changes $c_{i\sigma}$ to $-c_{i\sigma}$ and $c_{i\sigma}^\dagger$ to $-c_{i\sigma}^\dagger$ for $i\in A$. This transformation does not change the number operator.
[10]E. Lieb and W. Liniger, Phys. Rev. 130, 1605 (1963); E. Lieb, Phys. Rev. 130, 1616 (1963).
[11]J. B. McGuire, J. Math. Phys. 6, 432 (1965).
[12]M. Flicker, thesis, Yeshiva University, 1966 (unpublished); M. Flicker and E. Lieb, Phys. Rev. 161, 179 (1967).
[13]M. Gaudin, Phys. Letters 24A, 55 (1967).
[14]M. Gaudin, thesis, University of Paris, 1967 (unpublished).
[15]C. N. Yang, Phys. Rev. Letters 19, 1312 (1967).
[16]D. Mattis, to be published.

VOLUME 56, NUMBER 23 PHYSICAL REVIEW LETTERS 9 JUNE 1986

Exact Integrability of the One-Dimensional Hubbard Model

B. Sriram Shastry
Theory Group, Tata Institute of Fundamental Research, Bombay 400 005, India
(Received 24 March 1986)

The 1D Hubbard model is shown to be an exactly integrable system. A "covering" model of 2D statistical mechanics which I proposed recently was shown to provide a one-parameter family of transfer matrices, commuting with the Hamiltonian of the Hubbard model. I show in this work that any two transfer matrices of a family commute mutually. At the root of the commutation relation is the ubiquitous Yang-Baxter factorization condition. The form of the R operator is displayed explicitly.

PACS numbers: 05.50.+q, 64.60.Cn, 75.10.Lp

The 1D Hubbard model is of considerable interest in solid-state physics. It is exactly solvable by the Bethe *Ansatz*.[1] In this work I establish its exact integrability as well. Exact integrability, as is well known, is a very powerful result and is encountered in several classic models in statistical mechanics, such as the 2D Ising, the *XYZ*, and the eight-vertex models. A commonly accepted feature of exactly integrable models is the existence of an infinite number of conserved "currents," and of their mutual commutation. This is usually established, for a quantum Hamiltonian, by the identification of a "covering" lattice-statistical model with the property that a one-parameter family of transfer matrices commutes with the Hamiltonian. The mutual commutation of two transfer matrices (of the same family) establishes the exact integrability.

In a previous work,[2] I identified a new model in 2D classical statistical mechanics, and showed that it is a "covering" model for the 1D Hubbard model (in the above sense). In this work, I prove that two transfer matrices of the same family commute mutually. This is shown by a demonstration that the ubiquitous star-triangle[3,4] (Yang-Baxter) relation holds in this case as well.

The model considered here consists of two six-vertex models, obeying the free Fermi condition and coupled in a particular way by a diagonal vertex. The transfer matrix is written as $T=\mathrm{tr}_g \mathrm{Y}$, with

$$\mathrm{Y}=L_{N,g}L_{N-1,g}\cdots L_{1,g},\quad L_{n,m}=I_m l_{n,m} I_m,\ \text{with } l_{n,m}=S_{n,m}T_{n,m},$$
$$S_{n,m}=\tfrac{1}{2}(a+b)+\tfrac{1}{2}(a-b)\sigma_n^z\sigma_m^z+c(\sigma_n^+\sigma_m^-+\mathrm{H.c.}),\tag{1}$$

and $I_m=\exp(\frac{1}{2}h\sigma_m^z\tau_m^z)$. $T_{n,m}$ has the same form as $S_{n,m}$ with σ's replacing τ's.

The model thus consists of two species of Pauli matrices σ and τ residing at sites $n=1,\ldots,N$, and periodic boundary conditions are assumed. The spin at site g is an auxiliary variable (the ghost spin) corresponding to a horizontal arrow, and Y is the monodromy matrix. The model is characterized by three distinct parameters a/c, b/c, and h, with $a^2+b^2=c^2$ (we set $c=1$ in the following). In Ref. 2 I showed that T commutes with the Hubbard Hamiltonian H provided that

$$(2/ab)\sinh(2h)=U,\tag{2}$$

and further that H is a logarithmic derivative of T. (Here U is the Coulomb constant in H). Let us note that $L_{n,m}$ is asymmetric in n and m.

Next let us consider two transfer matrices (T and T') with parameters a_1,b_1,h_1 and a_2,b_2,h_2, both obeying the free Fermi condition, and examine the commutator. Clearly

$$\mathrm{YY}'=(L_{N,g_1}L'_{N,g_2})\cdots(L_{1,g_1}L'_{1,g_2}),$$
$$\mathrm{Y'Y}=(L'_{N,g_1}L_{N,g_2'})\cdots(L'_{1,g_1}L_{1,g_2'}).$$

Taking the trace over g_1 and g_2 of the two equations and subtracting, we obtain the commutator $[T,T']$. Baxter[3] noted that the commutator vanishes provided that the two expressions are similarity transforms of one another, and found a *local* relation that is sufficient. This may be written in the form

$$L_{3,2}L'_{3,1}R_{12}=R_{12}L'_{3,2}L_{3,1}\tag{3}$$

(with $g_1\to 2$, $g_2\to 1$, and $n\to 3$). The operator R_{12} does not depend on the site 3 (or n), and provided that it is invertible, the commutator vanishes. The remainder of this paper is concerned with the demonstration that Eq. (3) holds in the present model, and with the determination of R_{12}.

As a prelude to the calculation let us first consider the infinitesimal case $\theta_2=\epsilon$ ($a_2=\cos\theta_2$, $b_2=\sin\theta_2$, $c_2=1$), to first order in ϵ, and h_2 determined from Eq. (2). This gives $L'_{3,1}=P_{31}[1+\epsilon H_{3,1}]$ where P_{31} is the permutation operator and

$$H_{3,1}=(\sigma_3^+\sigma_1^-+\mathrm{H.c.})$$
$$+(\sigma\to\tau)+\tfrac{1}{8}U(\sigma_3^z\tau_3^z+\sigma_1^z\tau_1^z).$$

Thus to order ϵ, $T' = T(0)[1+\epsilon H + O(\epsilon^2)]$, where H is the Hubbard Hamiltonian $\sum H_{n,n+1}$. We know already (from Ref. 2) that T must commute with T' to this order, and examine if this commutation relation can be used to extract R_{12} [to $O(\epsilon)$]. Writing $R_{12} = P_{12}V_{12}$ and $V_{12} = U_{12} + \epsilon W_{12} + O(\epsilon^2)$ we find from Eq. (3) two equations to order ϵ: $L_{21}U_{12} = U_{13}L_{31} \Rightarrow U_{12} = L_{21}^{-1}$, and

$$L_{21}H_{32}L_{21}^{-1} - L_{31}^{-1}H_{32}L_{31} = W_{13}L_{31} - L_{21}W_{12}. \quad (4)$$

The left-hand side of Eq. (4) is also encountered in Ref. 2, where it is noted to be $M_{2,1}^{\dagger} - M_{3,1}$, with M a non-Hermitean operator [Eq. (8) of Ref. 2]. This remarkable separation of variables requires only the free Fermi condition and may be used to write the general solution for W,

$$L_{21}W_{12} = -M_{2,1}^{\dagger} + f_1, \quad W_{13}L_{31} = -M_{3,1} + f_1,$$

where f_1 is an arbitrary operator depending on site 1 only. Writing $M_{3,1} = L_{3,1}^{-1}Q_{3,1}$, we have

$$W_{12} = -L_{21}^{-1}Q_{2,1}^{\dagger}L_{2,1}^{-1} + L_{2,1}^{-1}f_1,$$
$$W_{13} = -L_{31}^{-1}Q_{3,1}L_{3,1}^{-1} + f_1L_{3,1}^{-1}. \quad (5)$$

Demanding the equality of the two expressions $(2 \rightarrow 3)$ we find a constraint

$$L_{3,1}^{-1}(B_{3,1})L_{3,1}^{-1} = \tfrac{1}{2}[L_{3,1}^{-1}, f_1] \quad (6)$$

with $B = (Q^{\dagger} - Q)/2$. However, I showed in Ref. 2 that provided that Eq. (2) holds, $B_{3,1} = \alpha[L_{3,1}I_1^{-4}]$, where $\alpha = (c^2+2b^2)/4ab$. Therefore, $f_1 = -2\alpha I_1^{-4}$, and R_{12} exists and can be found to $O(\epsilon)$.

We now turn to the general problem, Eq. (3), and seek to determine R. The symmetries of the model lead us to require the commutation of R_{12} with the operators (symmetry generators) (1) $\sigma_1^x\sigma_2^x\tau_1^x\tau_2^x$ (conjugation of all arrows), (2) $\sigma_1^z + \sigma_2^z$ (conservation of particles of σ species), and (3) $\tau_1^z + \tau_2^z$ (τ species). Further, R should be invariant with respect to the interchange of σ's and τ's.

The most general form of R subject to the above may be written as

$$\begin{aligned} R_{12} = {} & g_0 + g_1\sigma_1^z\tau_1^z + g_2(\sigma_1^z\sigma_2^z + \tau_1^z\tau_2^z) + g_3\sigma_2^z\tau_2^z + g_4(\sigma_1^+\sigma_2^- + \tau_1^+\tau_2^- + \text{H.c.}) + g_5(\sigma_1^z\tau_2^z + \sigma_2^z\tau_1^z) \\ & + g_6\sigma_1^z\sigma_2^z\tau_1^z\tau_2^z + g_7[(\sigma_1^+\sigma_2^- - \sigma_2^+\sigma_1^-)\tau_1^z + (\tau_1^+\tau_2^- - \tau_2^+\tau_1^-)\sigma_1^z] + g_8(\sigma_1^+\sigma_2^- + \text{H.c.})(\tau_1^+\tau_2^- + \text{H.c.}) \\ & + g_9[(\sigma_1^+\sigma_2^- - \sigma_2^+\sigma_1^-)\tau_2^z + (\tau_1^+\tau_2^- - \tau_2^+\tau_1^-)\sigma_2^z] + g_{10}[(\sigma_1^+\sigma_2^- + \text{H.c.})\tau_1^z\tau_2^z + (\tau_1^+\tau_2^- + \text{H.c.})\sigma_1^z\sigma_2^z] \\ & + g_{11}(\sigma_1^+\sigma_2^- - \sigma_2^+\sigma_1^-)(\tau_1^+\tau_2^- - \tau_2^+\tau_1^-). \end{aligned} \quad (7)$$

We have twelve parameters $g_0, \ldots, g_{11}$ at our disposal and wish to satisfy the (linear) operator equation (3).

The algebra involved is tedious. I used the symbolic manipulation package REDUCE2 in order to perform the calculations on a computer. The problem and the method used to solve it seem to be general enough to justify some discussion. The basic idea used was to convert Eq. (3) into a partial differential equation, by the use of a representation of spin operators as partial derivatives on the space of polynomials.[5,6] [Such a representation is natural[6] if we construct unnormalized spin coherent-states $|z\rangle = \exp(zs^-)|\uparrow\rangle$ and represent abstract spinors by "wave functions" $\psi(z) = \langle z|\psi\rangle$.] Specifically for $s=\frac{1}{2}$, at a site i, we may write $\sigma_i^+ \rightarrow \partial/\partial x_i$, $\sigma_i^- \rightarrow x_i - x_i^2\,\partial/\partial x_i$, $\sigma_i^Z \rightarrow 1 - 2x_i\,\partial/\partial x_i$, and the wave functions $|\uparrow\rangle \rightarrow 1$, $|\downarrow\rangle \rightarrow x_i$. Associating x_1, x_2, x_3 with the three σ's and y_1, y_2, y_3 with the three τ's, we see that manifold of 64 states is spanned by the wave functions $\pi(x_i)^{n_i}\pi(y_i)^{m_i}$, with $n_i, m_i = 0, 1$. We use the symmetries mentioned above, and the transformations generated by $\sigma_1^x\sigma_2^x$ or $\tau_1^x\tau_2^x$, which reverse all arrows on one sublattice, thereby negating the "fields" h_1 and h_2, and negate all g_n's with n odd. This enables us to restrict the considerations to the nine (irreducible) wave functions x_1, x_2, x_3, x_1y_1, x_1y_2, x_1y_3, x_2y_2, x_2y_3, and x_3y_3. The Yang-Baxter operator [left-hand side of Eq. (3) minus the right-hand side] is applied to these states and the coefficients of all the states in the resultant are required to vanish. Also the conjugates ($h_i \rightarrow -h_i$, $g_{2n+1} \rightarrow -g_{2n+1}$) are required to vanish. A total of 144 linear homogeneous equations result. Eleven of them suffice to determine the g's: This is an overdetermined set of linear equations. By explicit calculation, I checked that all the expressions vanish identically with a choice of parameters.

Let us define $a_3 \equiv a_1a_2 + b_1b_2$, $a_4 \equiv a_1a_2 - b_1b_2$, $b_3 \equiv b_2a_1 - b_1a_2$, $b_4 \equiv b_2a_1 + b_1a_2$, $s_+ \equiv \sinh(h_2+h_1)$, $s_- \equiv \sinh(h_2-h_1)$, $c_+ \equiv \cosh(h_2+h_1)$, $c_- \equiv \cosh(h_2-h_1)$. Further, let $K_1 \equiv g_0 - g_6$ and $K_2 \equiv g_4 - g_{10}$. In terms of these, we find the relations $g_3 = g_1$, $g_9 = g_7$,

$$\begin{aligned} & g_7 = \tfrac{1}{2}s_-b_4K_1; \quad g_8 = c_+b_3K_2; \quad g_{11} = s_+b_4K_2; \\ & g_4 + g_{10} = c_-b_3K_1; \quad g_1 + g_5 = \tfrac{1}{2}s_-a_4K_1; \quad g_1 - g_5 = \tfrac{1}{2}s_+b_3K_2/(b_2^2 - b_1^2); \\ & g_0 + g_6 + 2g_2 = c_-a_3K_1; \quad g_0 + g_6 - 2g_2 = c_+b_4K_2/(b_2^2 - b_1^2). \end{aligned} \quad (8)$$

These ten equations enable us to express all the g's in terms of K_1 and K_2. The final equation reads

$$\frac{K_2}{K_1}=\frac{\sinh(h_2-h_1)}{\sinh(h_2+h_1)}\frac{(b_2^2-b_1^2)}{(a_2b_2-a_1b_1)}=\frac{\cosh(h_2-h_1)}{\cosh(h_2+h_1)}\frac{(b_2^2-b_1^2)}{(a_2b_2+a_1b_1)}. \tag{9}$$

Consistency requires a single constraint (obtained by equating the two expressions),

$$a_1b_1/a_2b_2=\sinh(2h_1)/\sinh(2h_2). \tag{10}$$

This condition, together with the free Fermi condition $(a_1^2+b_1^2=1=a_2^2+b_2^2)$ is then sufficient to guarantee the commutation of the two transfer matrices. From Eq. (2) we see that Eq. (10) is not a new constraint; it is automatically fulfilled if we require that the two transfer matrices commute with H [i.e., Eq. (2)]. Thus, the one-parameter family found by requiring commutation with H is fundamental. This situation is identical to the one encountered in the eight-vertex model.[3,7]

The structure of the R operator is rather unwieldy and is best summarized in the following equations, describing its operation on the set of states $\psi_0=1$, $\psi_1=x_1$, $\psi_2=x_2$, $\psi_3=x_1y_1$, $\psi_4=x_1y_2$, $\psi_5=x_2y_2$, $\psi_6=x_2y_1$:

$$R\psi_0=K_1(c_-a_3+s_-a_4)\psi_0,\quad R\psi_1=K_1\psi_1+K_1(c_-b_3+s_-b_4)\psi_2,\quad R\psi_2=K_1\psi_2+K_1(c_-b_3-s_-b_4)\psi_1,$$

$$\begin{aligned}
R\psi_3&=\frac{K_2}{b_2^2-b_1^2}(c_+b_4+s_+b_3)\psi_3+K_2(\psi_4+\psi_6)+K_2(c_+b_3+s_+b_4)\psi_5,\\
R\psi_4&=K_2(\psi_3+\psi_5)+\frac{K_2}{b_2^2-b_1^2}(c_+b_4-s_+b_3)\psi_4+K_2(c_+b_3-s_+b_4)\psi_6,\\
R\psi_5&=K_2(c_+b_3+s_+b_4)\psi_3+K_2(\psi_4+\psi_6)+\frac{K_2}{b_2^2-b_1^2}(c_+b_4+s_+b_3)\psi_5.
\end{aligned} \tag{11}$$

The other equations may be inferred by use of appropriate symmetries. Let us note the symmetry of R; $R_{21}=R_{12}^\dagger$.

Finally, I remark on a curious side result. If we rotate the lattice through $\pi/2$ (and exchanging $a \leftrightarrow b$), the column-to-column transfer matrix may be shown to commute with another Hamiltonian. Linearizing Eq. (3) about $\theta_2=\theta_1$ we note that the first-order term in the variation has the form of Sutherland[7] for the commutation of the transfer matrix (column-to-column) with a Hamiltonian. The latter is the sum of two-body terms, each of the form of R_{12} [Eq. (7)] with the first-order variations of the g's substituted. The Hamiltonian is not Hermitean ($\delta g_7\neq 0$), but appears to be intimately related to the Hubbard model.

The work reported here should be useful in understanding further the properties of the 1D Hubbard model. Generalization of the statistical model to allow for the eight-vertex configurations obeying the free-terms condition seems interesting.

[1]E. H. Lieb and F. Y. Wu, Phys. Rev. Lett. **20**, 1445 (1968).

[2]B. S. Shastry, Phys. Rev. Lett. **56**, 1529 (1986).

[3]R. J. Baxter, Phys. Rev. Lett. **26**, 832 (1971), and Ann. Phys. **70**, 193 (1972).

[4]R. J. Baxter, *Exactly Solved Models in Statistical Mechanics* (Academic, London, 1982), Chaps. 9 and 10.

[5]H. Weyl, *The Theory of Groups and Quantum Mechanics* (Methuen, London, 1931), Chap. 3.

[6]B. S. Shastry, G. S. Agarwal, and I. Rama Rao, Pramana **11**, 85 (1978); V. Singh has kindly pointed out that the representation derived here is a special case of the one quoted in Ref. 5. The latter is most simply derived by use of Schwinger's coupled-boson representation for spins, followed by a partial differential representation for bosons.

[7]B. Sutherland, J. Math. Phys. **11**, 3183 (1970).

SOVIET PHYSICS JETP VOLUME 30, NUMBER 6 JUNE, 1970

EXCITATION SPECTRUM IN THE ONE-DIMENSIONAL HUBBARD MODEL

A. A. OVCHINNIKOV

L. Ya. Karpov Physical-Chemical Institute

Submitted June 12, 1969

Zh. Eksp. Teor. Fiz. 57, 2137-2143 (December, 1969)

The spectrum of the lowest (quasi-homopolar) excitations in the one-dimensional Hubbard model are investigated within the framework of the exact method developed in articles[8-10]. The excitations are classified according to spin and momentum. The singlet states are states of the bound type. It is shown that both singlet and triplet excitations start from zero, i.e., they do not have a gap. The magnitude of the gap is determined for the spectrum of quasi-ionic states to which an optical transition is possible. Its dependence on the parameter characterizing the electron interaction is investigated.

1. INTRODUCTION

In order to describe the metal-dielectric transition associated with an increase of the repulsion between electrons, Hubbard[1] proposed a model of a Fermi lattice gas having an interaction of the electrons only at one center. In the case of a one-dimensional cyclic chain, the corresponding Hamiltonian has the following form:

$$H = \sum_{m,n}^{N} \sum_{\sigma} T_{m,n} a_{m\sigma}^{+} a_{n\sigma} + \frac{\gamma}{2} \sum_{n}^{N} \sum_{\sigma} a_{n-\sigma}^{+} a_{n-\sigma} a_{n\sigma}^{+} a_{n\sigma}, \tag{1}$$

where $a_{n\sigma}^{+}$ and $a_{n\sigma}$ denote the creation and annihilation operators for an electron with spin σ in atom n; all $T_{m,n} = 0$ except $T_{n\pm 1,n} = -\beta$ $(\beta > 0)$.

The Hamiltonian (1) was used in article[2] in order to explain the appearance of a gap in the optical spectrum of long polymers with conjugated bonds. In this connection it was shown, within the framework of the generalized Hartree-Fock method, that an excited state to which an optical transition is possible is separated from the ground state by a gap for arbitrary values of the parameter γ. For a suitable choice of γ it was possible to obtain agreement with the experimentally observed dependence of the magnitude of the first transition on the length of the chain. In addition to the excitations of the indicated type, the Hamiltonian (1) has below a gap a set of singlet and triplet quasi-homopolar excitations.[3,4] Here, as shown in the work by Kohn[5] and Bulaevskiĭ[3], an optical transition to these states is forbidden or very weak. Meanwhile these states play a major role in the determination of the physical and chemical properties of long systems with conjugated bonds. For example, the fact that the spectrum of the triplet excitations starts from zero leads, for infinitely long chains, to an appreciable paramagnetism of these molecules.[6]

The goal of the present article is a determination of the spectrum of the lowest quasi-homopolar excitations of the Hamiltonian (1) and their classification. We shall use the exact expression for the wave function of the Hamiltonian which was obtained in articles[7-9], where Bethe's idea[10] was extended.

Let us consider an eigenfunction of the Hamiltonian (1) with the number of electrons equal to the number of sites, i.e., N, and with the z-component of the total spin equal to zero (we shall assume N to be even). We shall seek it in the form

$$\Psi_Q(n_1, n_2, \ldots, n_N) = \sum_{P} [Q, P] \exp\left\{ i \sum_{j=1}^{N} k_{P_j} n_{Q_j} \right\},$$
$$1 \leqslant n_{Q_1} \leqslant n_{Q_2} \leqslant \ldots \leqslant n_{Q_N} \leqslant N. \tag{2}$$

Here $k_1, k_2, \ldots, k_N$ denotes the set of quasimomenta for which the equation will be written down; $(Q_1, Q_2, \ldots, Q_N)$ and $(P_1, P_2, \ldots, P_N)$ denote permutations among the coordinates and momenta respectively. The summation in (2) is carried out over all permutations of the momenta k_i; the $[Q, P]$ are coefficients which simultaneously depend on Q and P and which are represented by a square matrix of order $N! \times N!$, which must be determined. The Schrödinger equation gives the following relation between these coefficients:

$$[Q, P] = Y_{nm}^{ab} [Q, P'], \tag{3}$$

where the operator Y_{nm}^{ab} has the form[10]

$$Y_{nm}^{ab} = -\frac{i\gamma/2 + (\sin k_n - \sin k_m) P^{ab}}{\sin k_n - \sin k_m + i\gamma/2};$$
$$Q_i = a = Q_j', \quad Q_j = b = Q_i',$$
$$P_i = m = P_j', \quad P_j = n = P_i', \tag{4}$$

$Q_k = Q_k'$, $P_k = P_k'$ for $k \neq i, j$ and the operator P^{ab} interchanges the sites Q_i and Q_j. In this connection the characteristic energy of the system is expressed in terms of the quasimomenta k_i in the following way:

$$E = -2\beta \sum_{j=1}^{N} \cos k_j. \tag{5}$$

By successively applying the operator Y_{mn}^{ab}, one can express any arbitrary coefficient $[Q, P]$ in terms of (a vector of dimension N!) the coefficient $[Q, I]$, where I denotes the identity permutation among the momenta $k_1, k_2, \ldots, k_N$.

Utilization of the conditions for the cyclic nature and symmetry of the wave function leads to a system of equations for the coefficients $[Q, I]$. Omitting the subsequent calculations which are rather completely given in the article by Yang,[9] let us write down the transcendental equations for the quasimomenta k_i arising upon the solution of this system

$$Nk_j = 2\pi I_j + \sum_{\beta=1}^{N/2} \varphi(j\beta), \tag{6a}$$

$$\sum_{j=1}^{N} \varphi(j\alpha) = 2\pi J_\alpha + \sum_{\beta=1}^{N/2} \psi(\beta\alpha) + \pi, \tag{6b}$$

$$e^{i\varphi(j\beta)} = \frac{\sin k_j - \Lambda_\beta + ic/2}{\sin k_j - \Lambda_\beta - ic/2}, \tag{7a}$$

$$e^{i\psi(\beta\alpha)} = \frac{\Lambda_\beta - \Lambda_\alpha + ic}{\Lambda_\beta - \Lambda_\alpha - ic}, \quad c = \frac{\gamma}{2\beta}. \tag{7b}$$

Here Λ_α $(\alpha = 1, 2, \ldots, N/2)$ denotes a set of numbers, all of which are different, and which in general may be complex. The phases $\psi(\alpha\beta)$ and $\varphi(j\beta)$ are determined so that

$$-\pi < \operatorname{Re}\psi(\alpha\beta),\ \operatorname{Re}\varphi(j\beta) < \pi,$$

I_j $(j = 1, 2, \ldots, N)$ and J_α $(\alpha = 1, 2, \ldots, N/2)$ are integers; they label the eigenstates of the system. For example, the total momentum Q of the system is expressed in terms of them in the following manner:

$$Q = \sum_{j=1}^{N} k_j = \frac{2\pi}{N}\Big(\sum_{j=1}^{N} I_j + \sum_{\alpha=1}^{N/2} J_\alpha\Big). \tag{8}$$

2. SPECTRUM OF THE TRIPLET EXCITATIONS

Let us consider the solution of the system of Eqs. (6) and (7) in the limit $\gamma \to \infty$. As is well-known, in this limit all eigenstates of the Hamiltonian (1) are divided into groups of almost degenerate states: homopolar, ionic, doubly ionic, etc. The first group consists of 2^N states with almost zero energy. The splitting of the energy levels among this group is described by the Heisenberg spin Hamiltonian. The second group consists of $2^N N$ states with energy $\sim\gamma$. A lowest excited state, to which an optical transition is possible, is found among this group. The third group contains $N(N-1)2^{N-1}$ states with energy $\sim 2\gamma$ and so forth. We will primarily be interested in the first group of states. Since the excited states of the spin Hamiltonian are well-known, then this makes it possible to classify the quasihomopolar states of the Hamiltonian (1) according to spin and momentum.

As $\gamma \to \infty$, Eqs. (6a), (6b) and (7a), (7b) go over into the following system of equations:

$$Nk_j = 2\pi I_j + \sum_{\beta=1}^{N/2} p_\beta, \quad \xi_\alpha = \operatorname{ctg}\frac{p_\alpha}{2} = -\frac{2\Lambda_\alpha}{c},$$

$$Np_\beta = 2\pi J_\beta + \sum_{\alpha(\neq\beta)} \psi(\alpha\beta), \quad 0 < p_\beta < 2\pi,$$

$$\operatorname{ctg}\frac{\psi(\beta\alpha)}{2} = \frac{1}{2}(\xi_\alpha - \xi_\beta). \tag{9}$$

This system agrees with the system of equations for the case of the spin Hamiltonian.[4] For the ground state of the system it is necessary to choose J_α and I_j in the following way:

$$J_\alpha = 1, 3, 5, \ldots, N-1, \tag{10}$$

$$I_j = -N/2, -N/2+1, \ldots, N/2-1. \tag{11}$$

For the quasi-homopolar levels all k_j are real, and for convenience one can reduce them to the interval $(-\pi, \pi)$.

In order to determine the excited triplet states, following[11] let us choose J_α in the form

$$J_\alpha = 0, 2, 4, \ldots, 2n-2, 2n+1, \ldots, N-1, \tag{12}$$

where n is a certain number which determines the total quasimomentum of the system. The solution of Eqs. (6) and (7) is obtained by changing to a continuous distribution of the numbers k_j and Λ_α. In this connection one can use the formal equation $\rho(k) = dj/dk_j$ for the density of the numbers k_j in the interval $(-\pi, \pi)$ and $\sigma(\Lambda) = d\alpha/d\Lambda_\alpha$ for the density of the numbers Λ_α over the entire axis $(-\infty, \infty)$. Carrying out the required differentiation in Eqs. (6) and (7) under the conditions (11) and (12), we obtain the following system of equations for the triplet states:

$$\rho(k) = \frac{1}{2\pi} + \frac{\cos k}{2\pi}\int_{-\infty}^{\infty} \frac{4c\sigma(\Lambda)\,d\Lambda}{c^2 + 4(\Lambda - \sin k)^2}, \tag{13}$$

$$\int_{-\pi}^{\pi} \frac{4c\rho(k)\,dk}{c^2 + 4(\Lambda - \sin k)^2} = 2\pi\sigma(\Lambda) + \int_{-\infty}^{\infty} \frac{2c\sigma(\Lambda')\,d\Lambda'}{c^2 + (\Lambda - \Lambda')^2} + \frac{2\pi}{N}\delta(\Lambda - \Lambda_n),$$

$$E = -2N\beta\int_{-\pi}^{\pi} \rho(k)\cos k\,dk. \tag{14}$$

Here Λ_n is equal to its own unperturbed value, i.e., it is obtained from the solution of Eqs. (6) and (7) by utilization of the numbers J_α and I_j, just as for the ground state (10) and (11). Taking the Fourier transform of the function $\sigma(\Lambda)$, one can easily obtain an expression for $\rho(k)$ and $\sigma(\Lambda)$. Omitting this calculation, we cite the answer for the energy of the triplet states

$$E_t(q) = E_0 + 2\beta\int_0^{\infty} \frac{J_1(\omega)\cos\omega\Lambda_n\,d\omega}{\omega\,\operatorname{ch}(\omega c/2)}. \tag{15}$$

Here E_0, the energy of the ground state which was first determined by Lieb and Wu,[10] is given by

$$E_0 = -4N\beta\int_0^{\infty} \frac{J_1(\omega)J_0(\omega)\,d\omega}{\omega(1 + e^{\omega c})},$$

$J_0(\omega)$ and $J_1(\omega)$ are Bessel functions. The quantity Λ_n is expressed in terms of the quasimomentum of the system $q = 2\pi n/N$ in the following way:

$$q = \frac{\pi}{2} + \int_0^{\infty} \frac{J_0(\omega)}{\omega}\,\frac{\sin\omega\Lambda_n}{\operatorname{ch}(\omega c/2)}\,d\omega. \tag{16}$$

The system (15) and (16) parametrically determines the $E_t(q)$ dependence. The function $E_t(q)$ possesses a double periodicity and reaches a maximum at $q = \pi/2$. If $\gamma \to \infty$

$$\varepsilon_t(q) = E_t(q) - E_0 \cong (4\pi\beta^2/2\gamma)\,|\sin q|,$$

which agrees with the expression for the triplet excitations[11] in the Heisenberg model with an exchange integral equal to $4\beta^2/\gamma$.

3. SPECTRUM OF THE SINGLET EXCITATIONS

As was shown in[4] the lowest singlet states of an antiferromagnetic Heisenberg chain necessarily belong to the bound state type, i.e., they correspond to complex momenta in the spin system. Our calculation of the spectrum of the singlet quasi-homopolar excitations of the Hamiltonian (1) will be entirely based on an analogy with a similar calculation for the spin Hamiltonian.

Let us choose sets of numbers I_j and J_α in the

following way. Let us leave the set I_j unchanged, as given by Eq. (11), but

$$J_\alpha = 1, 3, \ldots, 2\beta_1 - 1, 2\beta_1 - 1, \ldots, 2\beta_2 - 3, 2\beta_2 + 1, \ldots, N - 1. \quad (17)$$

According to[4] two complex-conjugate numbers $\Lambda_a = \lambda + i\kappa$ and $\Lambda_b = \lambda - i\kappa$ will correspond to two identical numbers J_{β_1}. We note that the total quasi-momentum of such a system will be determined in the following way:

$$q = 2\pi(J_{\beta_1} - J_{\beta_2}) / N. \quad (18)$$

One can choose all remaining Λ_α to be real. From the imaginary part of Eq. (6c) for $\alpha = a$ we have $\kappa = c/2$. Changing to a continuous distribution of the numbers Λ_α and k_j and introducing the corresponding densities according to the formulas of the preceding Section, we obtain the following system of equations:

$$\rho(k) = \frac{1}{2\pi} + \frac{\cos k}{2\pi} \int_{-\infty}^{\infty} \frac{4c\sigma(\Lambda)\, d\Lambda}{c^2 + 4(\Lambda - \sin k)^2} - \frac{1}{2N\pi} T(k),$$

$$T(k) = \left[2\pi\delta(\sin k - \Lambda) - \frac{2c}{c^2 + (\sin k - \lambda)^2} \right] \cos k$$

$$+ \cos k \sum_{m=1,2} \left[\frac{4c}{c^2 + 4(\bar{\Lambda}_{\beta_m} - \sin k)^2} - 2\pi\delta(\sin k - \bar{\Lambda}_{\beta_m}) \right]; \quad (19)$$

$$\int_{-\pi}^{\pi} \frac{4c\rho(k)\, dk}{c^2 + 4(\sin k - \Lambda)^2} = 2\pi\sigma(\Lambda) + \int_{-\infty}^{\infty} \frac{2c\sigma(\Lambda')\, d\Lambda'}{c^2 + (\Lambda - \Lambda')^2} + \frac{D(\Lambda)}{N},$$

$$D(\Lambda) = -4\pi\delta(\Lambda - \lambda) + \frac{4c}{4(\Lambda - \lambda)^2 + c^2} + \frac{12c}{4(\lambda - \Lambda)^2 + 9c^2}$$

$$+ \sum_{m=1,2} \left[2\pi\delta(\Lambda - \bar{\Lambda}_{\beta_m}) - \frac{2c}{c^2 + (\Lambda - \bar{\Lambda}_{\beta_m})^2} \right]. \quad (20)$$

In connection with the derivation of these equations we added to the system of real numbers Λ_α two additional numbers $\bar{\Lambda}_{\beta_1}$ and $\bar{\Lambda}_{\beta_2}$ which satisfy the same equations as the number Λ_α for $J_{\beta_1} = 2\beta_1 - 1$, $J_{\beta_2} = 2\beta_2 - 1$. The function $\sigma(\Lambda)$ is represented out of the density of real numbers Λ_α together with the two additional numbers $\bar{\Lambda}_{\beta_1}$ and $\bar{\Lambda}_{\beta_2}$.

The solution of the system of equations is obtained by transition to the Fourier transform for the function $\sigma(\Lambda)$. Omitting the calculations, we write down an expression for the energy of the singlet quasi-homopolar excitations

$$E_s(q) = E_0 + 8\beta \int_0^{\infty} \frac{d\omega J_1(\omega)}{\omega\, \mathrm{ch}(\omega c/2)} (\cos\omega\bar{\Lambda}_{\beta_1} + \cos\omega\bar{\Lambda}_{\beta_2} - \cos\omega\lambda). \quad (21)$$

In this connection, just as in[4], the following restriction is imposed on $\bar{\Lambda}_{\beta_1}$, $\bar{\Lambda}_{\beta_2}$, and λ:

$$|\bar{\Lambda}_{\beta_1}| > 1, \ |\bar{\Lambda}_{\beta_2}| > 1, \ |\lambda| > 1. \quad (22)$$

The condition for solvability of the system of equations for the number Λ_α (here it is required that $\Lambda_\alpha \neq \Lambda_\beta$ for $\alpha \neq \beta$) at once gives the equation

$$\lambda = \bar{\Lambda}_{\beta_1}. \quad (23)$$

The real part of Eq. (6b) for $\alpha = a$ together with Eq. (23) leads to the relation

$$2\beta_1 = \beta_2. \quad (24)$$

Finally Eqs. (18) and (24) make it possible to relate the total momentum q of the system to Λ_{β_1}:

$$|q| = \pi - 2 \int_0^{\infty} \frac{d\omega J_0(\omega) \sin\omega\bar{\Lambda}_{\beta_1}}{\omega\, \mathrm{ch}(\omega c/2)}. \quad (25)$$

Equation (25) together with the equation which follows from (21) and (23),

$$\varepsilon_s(q) = 8\beta \int_0 \frac{d\omega J_1(\omega) \cos\omega\bar{\Lambda}_{\beta_1}}{\omega\, \mathrm{ch}(\omega c/2)} \quad (26)$$

give the parametric dependence of the energy of the singlet excitations on the quasimomenta. Here one should keep in mind the limiting condition $|\bar{\Lambda}_{\beta_1}| > 1$. It leads to the result that the singlet excitation spectrum has a termination point at

$$q_0 = \pi - 2 \int_0^{\infty} d\omega \frac{J_0(\omega) \sin\omega}{\omega\, \mathrm{ch}(\omega c/2)}. \quad (27)$$

For $\gamma \to \infty$ the value $|q_0| = \pi/2$. If $\gamma = 0$ then $q_0 = 0$, which indicates the absence of bound states in this limit. For small q the spectra of singlet and triplet excitations have identical slopes:

$$\varepsilon_{t,s}(q) = |q| \frac{2\beta I_1(\pi/c)}{I_0(\pi/c)}, \quad (28)$$

where I_1 and I_0 are Bessel functions of imaginary argument. For large values of q the singlet levels always lie above the triplet levels. For sufficiently large but not infinite values of N, the energy of the first triplet level tends to zero in the following way:

$$\varepsilon_t(N) = \frac{4\pi\beta}{N} \frac{I_1(2\pi\beta/\gamma)}{I_0(2\pi\beta/\gamma)} \quad (29)$$

Let us make several remarks about the energy of the singlet quasi-ionic states. A strong optical transition takes place precisely to these states. The quasi-ionic states possess a nonvanishing current. The energy of the lowest current state and, consequently, the gap in the optical spectrum in the one-dimensional Hubbard model were calculated in the article by Lieb and Wu.[10] For its determination they obtained an energy $E_+ = E_0 + \mu_+$ for the ground state of the system containing N + 1 electrons and an energy $E_- = E_0 + \mu_-$ for the ground state of the system containing N − 1 electrons. The gap in the spectrum of the quasi-ionic states is then determined in the following way:

$$\Delta E = E_+ - E_- = \mu_+ - \mu_-. \quad (30)$$

In order to determine the spectrum of the quasi-ionic states it is necessary to determine the energy of a system containing N + 1 or N − 1 electrons and having a total momentum q. This computation is quite

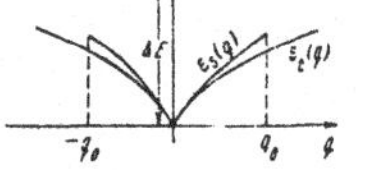

Different types of excitations of the system. $\epsilon_S(q)$ is the spectrum of singlet homopolar excitations for small q, as given by Eqs. (25) and (26); q_0 given by Eq. (27) is the point of termination of the spectrum; $\epsilon_t(q)$ is the spectrum for the homopolar triplet excitations which are described by Eqs. (15) and (16); $\epsilon_i(q)$ is the spectrum for the ionic excitations, and ΔE given by Eq. (33) is the gap in the spectrum of the ionic states.

analogous to the one given in the text. Without giving it in detail, in the figure we show the general form of the spectra for the lowest excited states. Different types of excitations of the system. $\epsilon_s(q)$ is the spectrum of singlet homopolar excitations for small q, as given by Eqs. (25) and (26); q_0 given by Eq. (27) is the point of termination of the spectrum; $\epsilon_t(q)$ is the spectrum for the homopolar triplet excitations which are described by Eqs. (15) and (16); $\epsilon_i(q)$ is the spectrum for the ionic excitations, and ΔE given by Eq. (33) is the gap in the spectrum of the ionic states.

Lieb and Wu[10] arrived at the following expression for the gap ΔE:

$$\Delta E = \gamma - 4\beta + 8\beta \sum_{n=1}^{\infty} (-1)^n [(1 + c^2 n^2)^{1/2} - cn]. \quad (31)$$

It is possible to give a more convenient expression for ΔE. For this purpose let us represent the series in (31) in terms of an integral along a contour C_0 which encompasses the real axis from c to ∞:

$$\sum_{n=1}^{\infty} (-1)^n [(1 + c^2 n^2)^{1/2} - nc] = \frac{1}{2ic} \int_{C_0} \frac{dz}{\operatorname{sh}(\pi z/c)} (\sqrt{z^2 + 1} - z). \quad (32)$$

Deforming the contour C_0 until it coincides with the imaginary axis, we can represent ΔE in the form

$$\Delta E = \frac{16\beta^2}{\gamma} \int_1^{\infty} \frac{\sqrt{y^2 - 1}\, dy}{\operatorname{sh}(\pi y/c)}. \quad (33)$$

For $\gamma \to \infty$ the gap is given by $\Delta E \approx \gamma - 4\beta + (8\beta^2/\gamma) \ln 2 + \ldots$. If the strength of the electron interaction is decreased, i.e., if $\gamma \to 0$, then

$$\Delta E \approx 8\pi^{-1} \sqrt{\gamma\beta}\, e^{-2\pi\beta/\gamma}. \quad (34)$$

We note that to within the pre-exponential factor this expression agrees with the expression given in article[2] for the gap as $\gamma \to 0$.

In conclusion the author thanks Ya. B. Zel'dovich, I. M. Khalatnikov, I. M. Lifshitz, and E. Lieb (USA) for interesting discussions of this work.

[1] J. Hubbard, Proc. Roy. Soc. (London), Ser. **A276**, 238 (1963), and **277**, 237 (1964).

[2] I. A. Misurkin and A. A. Ovchinnikov, ZhETF Pis. Red. **4**, 248 (1966) [JETP Lett. **4**, 167 (1966)].

[3] L. N. Bulaevskiĭ, Zh. Eksp. Teor. Fiz. **51**, 230 (1966) Sov. Phys.-JETP **24**, 154 (1967).

[4] A. A. Ovchinnikov, Zh. Eksp. Teor. Fiz. **56**, 1354 (1969) [Sov. Phys.-JETP **29**, 727 (1969)].

[5] W. Kohn, Phys. Rev. **133**, A171 (1964).

[6] L. A. Blyumenfel'd, A. A. Berlin, A. A. Slinkin, and A. É. Kalmanson, Zhurn. strukt. khim. **1**, 1031 (1960).

[7] M. Gaudin, Phys. Letters **24A**, 55 (1967).

[8] C. N. Yang, Phys. Rev. Letters **19**, 1312 (1967).

[9] E. H. Lieb and F. Y. Wu, Phys. Rev. Letters **20**, 1445 (1968).

[10] H. Bethe, Z. Physik **71**, 205 (1931).

[11] J. des Cloizeaux and J. J. Pearson, Phys. Rev. **128**, 2131 (1962).

Translated by H. H. Nickle
246

PHYSICAL REVIEW B VOLUME 6, NUMBER 3 1 AUGUST 1972

Magnetic Susceptibility at Zero Temperature for the One-Dimensional Hubbard Model*

H. Shiba†
Department of Physics, University of California, Los Angeles, California 90024
(Received 22 February)

The magnetic susceptibility at absolute-zero temperature for the one-dimensional Hubbard model is studied exactly as a function of the concentration of electrons by using Lieb and Wu's theory for this system. Our analysis essentially follows Griffiths's method for the magnetic susceptibility of the one-dimensional Heisenberg antiferromagnet, and is a generalization of Takahashi's calculation to the systems with an arbitrary concentration of electrons. The ground-state energy and the magnitude of local moments at each site are also studied. Combined with the results on the susceptibility, they should suggest how the effect of the Coulomb interaction on the properties of the system at low temperatures changes with the concentration of electrons.

I. INTRODUCTION

One-dimensional systems are fascinating for various reasons. They are usually easier to handle mathematically than higher-dimensional systems. One can often give exact statements without resorting to approximations.[1] Moreover, in some cases and for some properties, they are remarkably different from higher-dimensional systems. The one-dimensional Hubbard model, a model of interacting itinerant electrons, is not an exception.

The one-dimensional Hubbard Hamiltonian has the form

$$\mathcal{H} = \mathcal{H}_0 + \mathcal{H}_1 ;$$
$$\mathcal{H}_0 = -\sum_{ij\sigma} t_{ij}\, c_{i\sigma}^\dagger c_{j\sigma} , \tag{1.1}$$
$$\mathcal{H}_1 = U \sum_i c_{i\uparrow}^\dagger c_{i\uparrow} c_{i\downarrow}^\dagger c_{i\downarrow} ,$$

where t_{ij} is assumed to be t for $|i-j|=1$ (nearest-neighbor hopping) and zero otherwise. For this model Lieb and Wu[2] first gave an exact analysis on the ground state by essentially the same approach as that for the one-dimensional Heisenberg spin system[3,4] and for the one-dimensional fermion gas with δ-function interactions.[5]

In our Hamiltonian (1.1) there are three fundamental parameters for the thermodynamic properties of the system, that is, the strength of the Coulomb interaction relative to the transfer integral U/t, the concentration of electrons N/N_a (N and N_a are the total number of electrons and lattice points, respectively), and the temperature of the system k_BT. Let us review previous work on this system and then make the purpose of the paper clear, using these parameters.

A. Case (i): Half-Filled Band ($N/N_a = 1$)

1. *Absolute-Zero Temperature* ($k_BT = 0$)

The ground-state energy was obtained in an analytic form by Lieb and Wu.[2] According to them, the ground state is antiferromagnetic and insulating. Following this work, the spin-wave spectrum and the magnetic susceptibility at zero temperature were calculated by Ovchinnikov[6] and Takahashi,[7] respectively. By these calculations the properties of the one-dimensional half-filled Hubbard model were clarified almost completely as far as the absolute-zero temperature is concerned.

2. *Finite Temperature* ($k_BT \neq 0$)

Unfortunately, no exact solution is available for finite-temperature properties of the infinite system. But the thermodynamic properties of finite chains were calculated exactly by Shiba and Pincus.[8] Based on this calculation, we can guess a gradual "transition" from the paramagnetic and metallic state at high temperatures to the antiferromagnetic and insulating state at low temperatures.

B. Case (ii): System with $N/N_a \neq 1$

According to the Lieb-Mattis theorem[9] the ground state of our system is a singlet irrespective of the concentration of electrons. Even if $N/N_a \neq 1$, Lieb and Wu's theory should be useful and, in fact, it predicts a metallic ground state. But the dependence of the ground-state energy and other quantities at $T=0$ on the concentration N/N_a has not yet been examined.

This paper is devoted exactly to the case (ii). Since the Hubbard model is often discussed in connection with the origin of the itinerant-electron magnetism, the magnetic properties of this Hamiltonian are especially interesting. Therefore, a special emphasis in this paper is placed on the study of the magnetic susceptibility. In other words, the dependence of the magnetic susceptibility of the system on U/t and N/N_a is our main concern. Many elaborate but approximate theories[10–12] have been proposed on the role of the correlation effect in metallic magnetism, based on the three-dimensional version of the Hubbard Hamiltonian. We believe that *exact* calculations of the properties of the one-dimensional Hubbard model must be interesting. For the half-filled case, Takahashi calculated the magnetic susceptibility at zero temperature, as mentioned before. The present paper is an extension of his work to arbitrary concentrations of electrons.

In Sec. II we give a short summary of Lieb and Wu's work and then calculate the ground-state energy as well as the magnitude of magnetic moments at each site as a function of N/N_a and U/t. This section may be regarded as the introductory part of Sec. II. In Sec. III we study the magnetic susceptibility of our system without any restrictions to the concentration of electrons, using Lieb and Wu's formulation and following Griffiths's analysis[13] on the magnetic susceptibility at zero temperature of the one-dimensional Heisenberg antiferromagnet. A brief discussion is given in Sec. IV.

II. GROUND-STATE ENERGY AS A FUNCTION OF THE CONCENTRATION OF ELECTRONS

Lieb and Wu[2] gave an excellent analysis of the ground state of the system (1.1) and a good starting point to the exact discussion on the behaviors at zero temperature. According to their conclusion, the lowest state of the one-dimensional Hubbard model with a fixed magnetization is described by the coupled equations for two "distribution functions," $\rho(k)$ and $\sigma(\Lambda)$,

$$2\pi\rho(k)=1+\cos k\int_{-B}^{B}\frac{8u\,\sigma(\Lambda)\,d\Lambda}{u^2+16(\sin k-\Lambda)^2}\,, \tag{2.1}$$

$$\int_{-Q}^{Q}\frac{8u\,\rho(k)\,dk}{u^2+16(\Lambda-\sin k)^2}=2\pi\sigma(\Lambda)+\int_{-B}^{B}\frac{4u\,\sigma(\Lambda')\,d\Lambda'}{u^2+4(\Lambda-\Lambda')^2}\,, \tag{2.2}$$

where $u=U/t$, and the parameters B and Q are determined by the conditions

$$\int_{-Q}^{Q}dk\,\rho(k)=N/N_a\,, \tag{2.3}$$

$$\int_{-B}^{B}d\Lambda\,\sigma(\Lambda)=M/N_a\,, \tag{2.4}$$

with the total number of down-spin electrons M. Once these coupled equations are solved, the lowest energy is obtained by the formula

$$E=-2tN_a\int_{-Q}^{Q}dk\cos k\,\rho(k)\quad. \tag{2.5}$$

Using this formulation we will calculate the magnetic susceptibility at zero temperature in Sec. III. In this section we discuss the ground-state energy and the magnitude of local moments at each site (defined later) at arbitrary values of N/N_a.

First of all, from the Lieb-Mattis theorem we know that the ground state of our system (1.1) is a singlet, that is, $M/N=\frac{1}{2}$, which corresponds to $B=\infty$.[2] Now we can reduce the coupled integral equations (2.1) and (2.2) into a single one. Introducing the Fourier transform of $\sigma(\Lambda)$ and substituting it into Eqs. (2.1) and (2.2), we get

$$2\pi\rho(k)=1+\cos k\int_{-Q}^{Q}dk'\,\rho(k')\int_{-\infty}^{\infty}d\omega\,\frac{e^{i\omega(\sin k-\sin k')}}{e^{|\omega|u/2}+1}\,. \tag{2.6}$$

It is convenient for later purposes to define the function

$$\begin{aligned}R(x)&\equiv\frac{1}{4\pi}\int_{-\infty}^{\infty}dy\,\frac{1}{e^{|y|}+1}\,e^{ixy/2}\\&=\frac{1}{4\pi}\int_{-\infty}^{\infty}dt\,\frac{\operatorname{sech}\frac{1}{2}\pi t}{1+(x+t)^2}\\&=\frac{1}{\pi}\sum_{n=1}^{\infty}(-1)^{n+1}\frac{2n}{x^2+(2n)^2}\end{aligned} \tag{2.7}$$

as in the study of the one-dimensional antiferromagnetic Heisenberg model.[13] Thus, Eq. (2.6) can be written in the form

$$2\pi\rho(k)=1+\cos k\int_{-Q}^{Q}dk'\,\rho(k')\times\frac{8\pi}{u}R\left(\frac{4(\sin k-\sin k')}{u}\right)\quad. \tag{2.8}$$

Although this equation is difficult to solve in a compact form except for the half-filled case ($Q=\pi$), it is easy to obtain the solution by the iteration method or numerically, or to examine some limiting cases. In fact, approximating Eq. (2.8) by a set of 41 coupled linear algebraic equations, we calculated N/N_a in Eq. (2.3) and the ground-state energy per site E/N_a. The results are shown in Figs. 1 and 2. From Fig. 1 the relation between the concentration of electrons N/N_a and Q in the ground state is found. As easily noted, $U/t=0$ is a singular point. From Fig. 2, which shows the ground-state energy as a function of N/N_a, we can point out some features of the effect

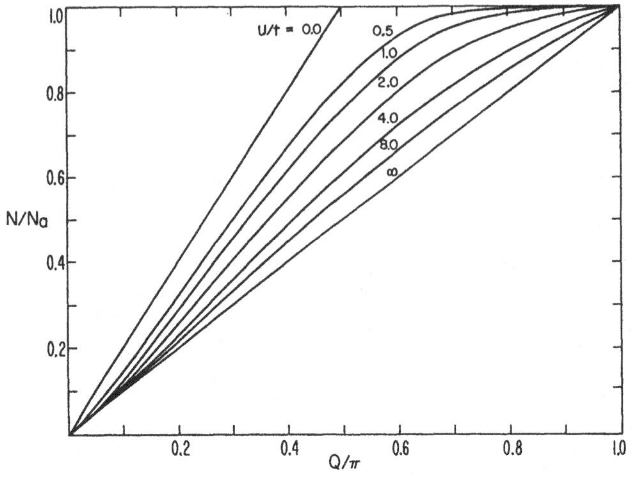

FIG. 1. Relation between N/N_a and Q at some typical values of U/t.

of correlation.

(i) At a low density of electrons ($N/N_a \lesssim 0.4$) the effect of correlation on the ground-state energy is not so large, because electrons occupy the states at the bottom of the band, where there is a high density of states, and therefore they can avoid each other without much cost of their kinetic energies. The effect of correlation is the most evident in the half-filled case.

(ii) When U/t increases, the system can gain energy only by migration processes of electrons through vacant sites, and thus the position of the minimum of the ground state as a function of N/N_a shifts from $N/N_a = 1.0$ to 0.5. Although our system is a simple one-dimensional one, with a single orbital, this result should be, in principle, suggestive of the effect of correlation on the cohesive energy of transition metals.[14]

(iii) The energy of the lowest state with the maximum total spin coincides with the $U/t = \infty$ curve in

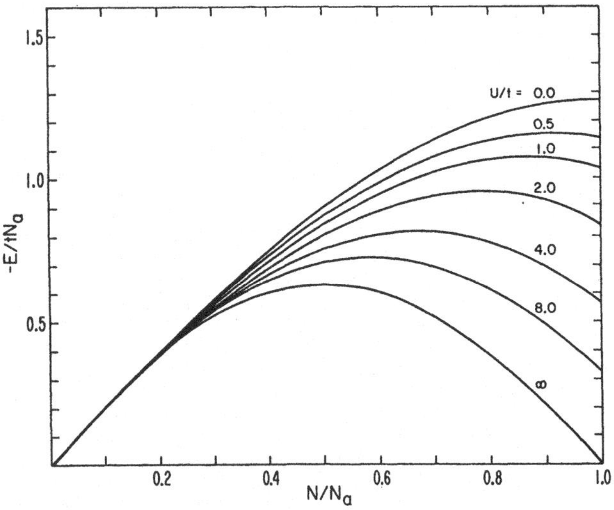

FIG. 2. Concentration dependence of the ground-state energy at typical values of U/t. At $U/t = 0$, the ground-state energy is given by $-(4/\pi)\sin\{\frac{1}{2}\pi N/N_a\}$ while $E/tN_a = -(2/\pi)\sin\{\pi N/N_a\}$ at $U/t = \infty$.

Fig. 2. Physically, this is natural.

Let us study the strong U limit ($u \gg 1$), starting from Eq. (2.8) and expanding in power of $1/u$. We can easily obtain

$$2\pi\rho(k) = 1 + \cos k\,\frac{4}{u}\ln 2\,\frac{Q}{\pi} + O\left(\frac{1}{u^2}\right) \quad . \qquad (2.9)$$

Substituting this equation into (2.3) and (2.5) and determining Q up to the order of $1/u$, we have

$$\frac{E}{tN_a} \cong -\left[\frac{2}{\pi}\sin\left(\pi\frac{N}{N_a}\right) + \frac{4\ln 2}{u}\left(\frac{N}{N_a}\right)^2 \times\left(1 - \frac{\sin(2\pi N/N_a)}{2\pi N/N_a}\right)\right] \quad . \qquad (2.10)$$

In the half-filled case ($N/N_a = 1$), Eq. (2.10) reproduces the ground-state energy of the one-dimensional Heisenberg antiferromagnet with the exchange coupling $J = 2t^2/U$.[3,4]

Another quantity, which is useful to understand the ground state and is easy to derive from Eqs. (2.5) and (2.6), is the magnitude of local moments at each site introduced in Ref. 8:

$$L_0 = (1/N_a)\sum_j \langle(\vec{S}_j)^2\rangle \,, \qquad (2.11)$$

where $\vec{S}_j$ is the spin operator at the jth site: $\vec{S}_j = \sum_{\sigma\sigma'}\langle\sigma|\vec{S}|\sigma'\rangle c^\dagger_{j\sigma} c_{j\sigma'}$ and the average $\langle\cdots\rangle$ is taken in the ground state. This quantity is related to the ground-state energy by

$$L_0 = \frac{3}{4}\frac{N}{N_a} - \frac{3}{2}\frac{1}{N_a}\frac{\partial E(U)}{\partial U} \,, \qquad (2.12)$$

as easily proved. Note that in Eq. (2.12) Q also depends on U. In fact, from the condition of the fixed number of electrons we obtain

$$\frac{\partial Q}{\partial U} = -\frac{1}{2\rho(Q)}\int_{-Q}^{Q} dk\,\frac{\partial\rho(k)}{\partial U} \quad . \qquad (2.13)$$

By using this expression, the derivative $\partial E(U)/\partial U$ is written in the form

$$\frac{1}{N_a}\frac{\partial E(U)}{\partial U} = 2\int_{-Q}^{Q} dk\,(\cos Q - \cos k)\,\frac{\partial\rho(k)}{\partial u} \,, \qquad (2.14)$$

where $\partial\rho(k)/\partial u$ is obtained as the solution of the integral equation

$$2\pi\frac{\partial\rho(k)}{\partial u} = \cos k\int_{-Q}^{Q} dk'\,\rho(k')\left(\frac{4}{u}\right)^2 \Phi\left(\frac{4(\sin k - \sin k')}{u}\right) + \cos k\int_{-Q}^{Q} dk'\,\frac{\partial\rho(k')}{\partial u} \times\left[\frac{8\pi}{u}R\left(\frac{4(\sin k - \sin k')}{u}\right) - \frac{4\pi}{u}R\left(\frac{4(\sin k - \sin Q)}{u}\right) - \frac{4\pi}{u}R\left(\frac{4(\sin k + \sin Q)}{u}\right)\right] \,, \qquad (2.15)$$

with

$$\Phi(x) \equiv \sum_{n=1}^{\infty}(-1)^{n+1} n\,\frac{x^2 - (2n)^2}{[x^2 + (2n)^2]^2} \quad . \qquad (2.16)$$

Thus, it is easy to get the quantity L_0.

Figure 3 shows the dependence of the magnitude of local moments at each site on the concentration of electrons at various values of U/t. In the noninteracting system ($U/t = 0$), L_0 is given by

$$L_0 = \tfrac{3}{4}(N/N_a)(1 - \tfrac{1}{2}N/N_a) \,, \qquad (2.17)$$

while in the strong U limit we obtain

$$L_0 \simeq \frac{3}{4}\frac{N}{N_a} - \frac{6\ln 2}{u^2}\left(\frac{N}{N_a}\right)^2\left(1 - \frac{\sin(2\pi N/N_a)}{2\pi N/N_a}\right) \quad . \qquad (2.18)$$

Again the effect of correlation on L_0 is small in the low density of electrons, and it becomes evident when N/N_a approaches unity.

III. MAGNETIC SUSCEPTIBILITY AT ZERO TEMPERATURE

The magnetic susceptibility gives important information on this system. For the half-filled case, Takahashi[7] calculated the susceptibility and showed how it changes from the Pauli paramagnetic behavior to that of localized spins as U/t increases. It is interesting to study the magnetic susceptibility for $N/N_a < 1$. The Griffiths method,[13] which Takahashi followed in his analysis, is still found useful even when $N/N_a < 1$.

In the study of the susceptibility we can assume that the magnetization induced by an external field is small, and therefore B remains quite large in Eqs. (2.1)–(2.4). The increase of energy due to the magnetization of the system, which is directly connected with the magnetic susceptibility, is determined by the asymptotic behavior of $\sigma(\Lambda)$ in the region $\Lambda \gg 1$.

Let us rewrite Eqs. (2.1)–(2.4) into a convenient form for this purpose. Integrating both sides of Eq. (2.2) over Λ from $-\infty$ to $+\infty$, and using the relations (2.3) and (2.4), we get

$$S \equiv \tfrac{1}{2}N/N_a - M/N_a = \int_B^{\infty} d\Lambda\,\sigma(\Lambda) \quad . \qquad (3.1)$$

The integral equation for $\sigma(\Lambda)$ is obtained by substituting Eq. (2.1) into (2.2) in the form

$$2\pi\,\sigma(\Lambda) = (1/2\pi)\,g_Q^{(0)}(\Lambda) - \int_{-B}^{B} d\Lambda'\,S_Q(\Lambda,\Lambda')\,\sigma(\Lambda') \,, \qquad (3.2)$$

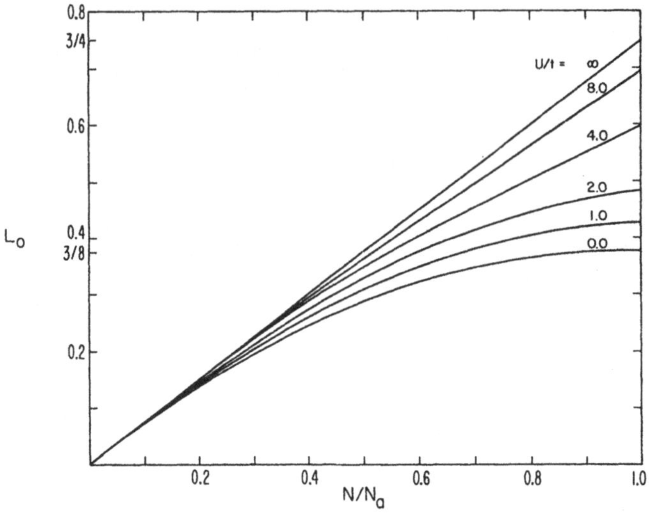

FIG. 3. Magnitude of local moments at each site L_0 versus the concentration of electrons N/N_a. When $U/t=\infty$, L_0 is equal to $\frac{3}{4}N/N_a$. On the other hand, L_0 is given by $\frac{3}{4}(N/N_a)(1-\frac{1}{2}N/N_a)$, when $U/t=0$.

where

$$g_Q^{(n)}(\Lambda) \equiv \int_{-Q}^{Q} dk \cos^n k \frac{8u}{u^2+16(\sin k-\Lambda)^2} \quad (3.3)$$

and the kernel is defined by

$$S_Q(\Lambda,\Lambda') = \frac{4u}{u^2+4(\Lambda-\Lambda')^2} - \int_{-Q}^{Q} \frac{dk}{2\pi} \cos k \frac{8u}{u^2+16(\Lambda-\sin k)^2} \times \frac{8u}{u^2+16(\Lambda'-\sin k)^2} . \quad (3.4)$$

It is useful to introduce a new function $\sigma^{(n)}(\Lambda)$ as the solution of the integral equation

$$\sigma^{(n)}(\Lambda) = (1/2\pi) g_Q^{(n)}(\Lambda) - \int_{-\infty}^{\infty} d\Lambda' S_Q(\Lambda,\Lambda')\sigma^{(n)}(\Lambda') . \quad (3.5)$$

Then $\sigma^{(n)}(\Lambda)$ can be expressed in terms of the resolvent kernel $\mathcal{S}_Q(\Lambda,\Lambda')$ defined by

$$\begin{aligned}\mathcal{S}_Q(\Lambda,\Lambda') &= S_Q(\Lambda,\Lambda') - \int_{-\infty}^{\infty} d\Lambda'' S_Q(\Lambda,\Lambda'')\mathcal{S}_Q(\Lambda'',\Lambda') \\ &= S_Q(\Lambda,\Lambda') - \int_{-\infty}^{\infty} d\Lambda'' \mathcal{S}_Q(\Lambda,\Lambda'') S_Q(\Lambda'',\Lambda') . \end{aligned} \quad (3.6)$$

In fact, it is easy to show

$$\sigma^{(n)}(\Lambda) = \frac{1}{2\pi} g_Q^{(n)}(\Lambda) - \int_{-\infty}^{\infty} \frac{d\Lambda'}{2\pi} \mathcal{S}_Q(\Lambda,\Lambda') \frac{1}{2\pi} g_Q^{(n)}(\Lambda') . \quad (3.7)$$

By using these newly introduced functions and integrating the product of $\mathcal{S}_Q(\Lambda_1,\Lambda)$ and Eq. (3.2) over Λ from $-\infty$ to $+\infty$, we obtain

$$\sigma(\Lambda) = \sigma^{(0)}(\Lambda) + \int_{|\Lambda'|>B} \frac{d\Lambda'}{2\pi} \mathcal{S}_Q(\Lambda,\Lambda')\sigma(\Lambda') . \quad (3.8)$$

Similarly, the integration of the product of $\sigma(\Lambda)$ and Eq. (3.5) over Λ from $-B$ to B yields

$$\frac{E}{tN_a} = \frac{E_0(Q)}{tN_a} + 2\int_{|\Lambda|>B} \frac{d\Lambda}{2\pi} \sigma(\Lambda)(2\pi)^2\sigma^{(2)}(\Lambda) , \quad (3.9a)$$

$$\frac{N}{N_a} = \frac{N_0(Q)}{N_a} - \int_{|\Lambda|>B} \frac{d\Lambda}{2\pi} \sigma(\Lambda)(2\pi)^2\sigma^{(1)}(\Lambda) , \quad (3.9b)$$

where $E_0(Q)$ and $N_0(Q)$ are the ground-state energy and the total number of electrons at a fixed value of Q, respectively, i.e.,

$$\frac{E_0(Q)}{tN_a} = -\frac{2}{\pi}\sin Q - 2\int_{-\infty}^{\infty} \frac{d\Lambda}{2\pi} \sigma^{(0)}(\Lambda) g_Q^{(2)}(\Lambda) , \quad (3.10a)$$

$$\frac{N_0(Q)}{N_a} = \frac{Q}{\pi} + \int_{-\infty}^{\infty} \frac{d\Lambda}{2\pi} \sigma^{(0)}(\Lambda) g_Q^{(1)}(\Lambda) . \quad (3.10b)$$

The quantities $E_0(Q)$ and $N_0(Q)$ were actually calculated in Sec. II. In order to evaluate the second term in Eqs. (3.9a) and (3.9b) we have to know $\sigma^{(n)}(\Lambda)$ and $\sigma(\Lambda)$ for $\Lambda \gg 1$. Starting from Eqs. (3.6) and (3.7), solving the latter by iteration and rearranging terms, we find that

$$\sigma^{(n)}(\Lambda)=\frac{1}{u}\int_{-Q}^{Q}\frac{dk}{2\pi}\cos^n k\,\mathrm{sech}\frac{2\pi(\Lambda-\sin k)}{u}+\int_{-\sin Q}^{\sin Q}\frac{dt}{2\pi}\int_{-\sin Q}^{\sin Q}\frac{dt'}{2\pi}\,\frac{2\pi}{u}\,\mathrm{sech}\frac{2\pi(\Lambda-t)}{u}\times L_Q(t,t')\int_{-Q}^{Q}\frac{dk}{2\pi}\cos^n k\,\frac{4}{u}R\left(\frac{4(\sin k-t')}{u}\right),\tag{3.11}$$

where $L_Q(t,t')$ is the solution of the equation

$$L_Q(t,t')=\delta(t-t')+\int_{-\sin Q}^{\sin Q}dt''\,\frac{4}{u}R\left(\frac{4(t-t'')}{u}\right)L_Q(t'',t')\;.\tag{3.12}$$

Now it is clear that for $\Lambda\gg1$, $\sigma^{(n)}(\Lambda)$ decays exponentially. In fact, the asymptotic form of $\sigma^{(n)}(\Lambda)$ in $|\Lambda|\gg1$ is given by

$$\sigma^{(n)}(\Lambda)\simeq(2/u)\,e^{-2\pi|\Lambda|/u}I_Q^{(n)}(u)\,,\tag{3.13}$$

where

$$I_Q^{(n)}(u)=\int_{-Q}^{Q}\frac{dk}{2\pi}\cos^n k\,e^{2\pi(\sin k)/u}+\int_{-\sin Q}^{\sin Q}dt\,e^{2\pi t/u}\int_{-\sin Q}^{\sin Q}dt'\,L_Q(t,t')\times\int_{-Q}^{Q}\frac{dk}{2\pi}\cos^n k\,\frac{4}{u}R\left(\frac{4(\sin k-t')}{u}\right)\;.\tag{3.14}$$

The next task is to simplify Eq. (3.8) for $|\Lambda|>B\gg1$.

Defining the function $P(\Lambda)$ by

$$\sigma(\Lambda+B)\equiv\frac{2}{u}\,I_Q^{(0)}(u)P(\Lambda)\,e^{-2\pi B/u}\tag{3.15}$$

and using the asymptotic expression for $\sigma^{(0)}(\Lambda)$ [Eq. (3.13)], we have

$$P(\Lambda)=e^{-2\pi\Lambda/u}+\int_{-\infty}^{\infty}\frac{d\Lambda'}{2\pi}P(\Lambda')\times[\mathcal{S}_Q(B+\Lambda,B+\Lambda')+\mathcal{S}_Q(B+\Lambda,-B-\Lambda')]\tag{3.16}$$

for $\Lambda>0$. But this equation is further simplified in the case where $B\gg1$, Λ and $\Lambda'>0$, since we can apply the following approximations:

$$\begin{aligned}\mathcal{S}_Q(B+\Lambda,B+\Lambda')+\mathcal{S}_Q(B+\Lambda,-B-\Lambda')&\simeq\mathcal{S}_Q(B+\Lambda,B+\Lambda')\\&\simeq(8\pi/u)R[4(\Lambda-\Lambda')/u]\;.\end{aligned}\tag{3.17}$$

Here we ignored exponentially small terms. Substituting this into Eq. (3.16), we obtain

$$P(\Lambda)=e^{-2\pi\Lambda/u}+\int_0^{\infty}d\Lambda'\,\frac{4}{u}R\left(\frac{4(\Lambda-\Lambda')}{u}\right)P(\Lambda')\tag{3.18}$$

or

$$P\left(\frac{u}{4}x\right)=e^{-\pi x/2}+\int_0^{\infty}dx'\,R(x-x')\,P\left(\frac{u}{4}x'\right).\tag{3.18'}$$

Now Eqs. (3.9a), (3.9b), and (3.1) can be written in the form

$$\frac{E}{N_a t}=\frac{E_0(Q)}{N_a t}+\frac{8b_0}{u}\,e^{-4\pi B/u}I_Q^{(2)}(u)\,I_Q^{(0)}(u)\,,\tag{3.19a}$$

$$\frac{N}{N_a}=\frac{N_0(Q)}{N_a t}-\frac{4b_0}{u}\,e^{-4\pi B/u}I_Q^{(1)}(u)\,I_Q^{(0)}(u)\,,\tag{3.19b}$$

$$S=a_0e^{-2\pi B/u}I_Q^{(0)}(u)\,,\tag{3.19c}$$

where a_0 and b_0 are given by

$$a_0=\tfrac{1}{2}\int_0^{\infty}dx\,p(\tfrac{1}{4}ux)\tag{3.20a}$$

and

$$b_0=\pi\int_0^{\infty}dx\,e^{-\pi x/2}\,p(\tfrac{1}{4}ux)\,,\tag{3.20b}$$

respectively. The important point here is that Eq. (3.18′) is exactly the same as that in Griffiths's work,[13] his equation (43), and that a_0 and b_0 are the same quantities with the same notation. So there is actually no need to solve (3.18) or (3.18′). Substituting Eq. (3.19c) into (3.19a) and (3.19b), we find that

$$\frac{E}{tN_a}=\frac{E_0(Q)}{tN_a}+S^2\frac{(2\pi)^2}{u}\frac{I_Q^{(2)}(u)}{I_Q^{(0)}(u)}\,,\tag{3.21a}$$

$$\frac{N}{N_a}=\frac{N_0(Q)}{N_a}-S^2\frac{(2\pi)^2}{2u}\frac{I_Q^{(1)}(u)}{I_Q^{(0)}(u)}\;.\tag{3.21b}$$

Here the use was made of the relation $b_0/a_0^2=\frac{1}{2}\pi^2$ conjectured by Griffiths[13] and proved rigorously by Yang and Yang.[15] In Eqs. (3.21a) and (3.21b) there is a deviation of Q due to the magnetization S from the value in the singlet ground state. The change of Q is determined by the condition that the total number of electrons should be constant. Substituting the deviation of Q obtained in this way into Eq. (3.21a), we find that

$$\frac{E}{tN_a}=\frac{E_0(Q)}{tN_a}+(2\pi S)^2\left[\frac{1}{u}\frac{I_Q^{(2)}(u)}{I_Q^{(0)}(u)}+\frac{1}{2u}\frac{I_Q^{(1)}(u)}{I_Q^{(0)}(u)}\left(\frac{\partial E_0(Q)}{\partial Q}\Big/\frac{\partial N_0(Q)}{\partial Q}\right)\right]\;.\tag{3.22}$$

Here Q is related to N/N_a through the relation we found in Sec. II, i.e.,

$$N/N_a=N_0(Q)/N_a\;.$$

The second term in Eq. (3.22) represents the increase of the energy due to the magnetization. Adding the Zeeman term and minimizing the total energy with respect to S, we find our final expression for the susceptibility

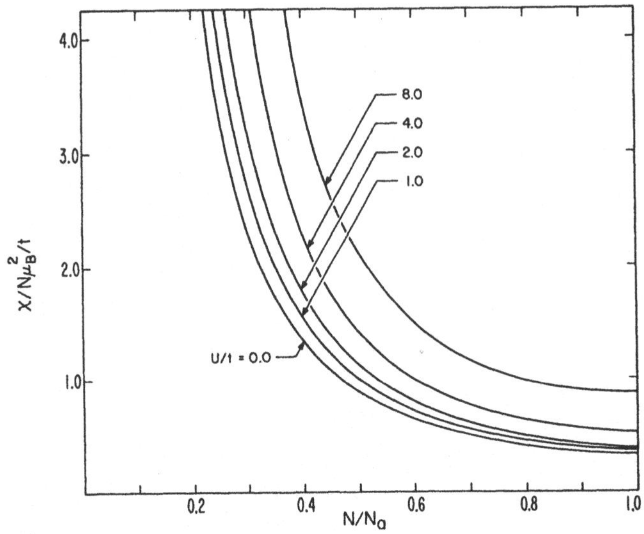

FIG. 4. Magnetic susceptibility per electron $\chi/(N\mu_B^2/t)$ vs N/N_a. All curves diverge as a function of N/N_a only at $N/N_a = 0$.

$$\frac{\chi}{N_a \mu_B^2/t} = \left[2\pi^2 \left(\frac{1}{u} \frac{I_Q^{(2)}(u)}{I_Q^{(0)}(u)} + \frac{1}{2u} \frac{I_Q^{(1)}(u)}{I_Q^{(0)}(u)} \frac{\partial E_0(Q)}{\partial Q} \Big/ \frac{\partial N_0(Q)}{\partial Q} \right) \right]^{-1} . \quad (3.23)$$

By using the formula (3.23) we can calculate the magnetic susceptibility at arbitrary values of U/t and N/N_a. Details of the numerical calculation are described in the Appendix. Figure 4 shows our results for the susceptibility per electron versus N/N_a at typical values of U/t, while Fig. 5 shows the dependence of the susceptibility on U/t with fixed values of N/N_a. From these figures it is evident that with the decrease of the concentration of electrons and/or with the increase of U/t the system is more easily magnetized. This tendency is consistent with the results for the energy (Fig. 2) and can be explained by the fact that when N/N_a decreases and/or U/t increases, one can get a small amount of magnetization without much cost of energy. In Fig. 4, all the curves of the susceptibility diverge at $N/N_a = 0$ simply because of the divergence of the density of states at band edges.

Let us study some limiting cases.

a. Half-filled case $(Q = \pi)$. It is easy to find

$$I_Q^{(0)} = I_0(2\pi/u) , \qquad I_Q^{(1)} = 0 ,$$

and

$$I_Q^{(2)} = (u/2\pi) I_1(2\pi/u) ,$$

where $I_\nu(x)$ is the Bessel function of imaginary argument of the order ν. Therefore, (3.23) gives

$$\frac{\chi}{N_a \mu_B^2/t} = \frac{1}{\pi} \frac{I_0(2\pi/u)}{I_1(2\pi/u)} , \qquad (3.24)$$

which is Takahashi's result[7] for the half-filled case.

b. Strong U limit $(U/t \gg 1)$. In this limit the approximation

$$I_Q^{(n)} \simeq \int_{-Q}^{Q} \frac{dk}{2\pi} \cos^n k$$

is appropriate. Furthermore, we have

$$\frac{1}{N_a} \frac{\partial E_0}{\partial Q} \simeq -\frac{2}{\pi} \cos Q$$

and

$$\frac{1}{N_a} \frac{\partial N_0}{\partial Q} \simeq \frac{1}{\pi} .$$

Therefore, the susceptibility in the strong U limit is proportional to u in the way

$$\frac{\chi}{N_a \mu_B^2/t} \simeq \frac{u}{\pi^2} \left(1 - \frac{\sin 2Q}{2Q} \right)^{-1} \simeq \frac{u}{\pi^2} \left(1 - \frac{\sin(2\pi N/N_a)}{2\pi N/N_a} \right)^{-1} . \qquad (3.25)$$

Note exactly the same factor $1 - [\sin(2\pi N/N_a)]/(2\pi N/N_a)$ appeared in Eq. (2.10). When $N/N_a = 1$, the right-hand side of Eq. (3.25) represents the Griffiths formula for the susceptibility of the one-dimensional antiferromagnet with the nearest-neighbor coupling $J \simeq 2t^2/U$, while in the low-density limit Eq. (3.25) gives

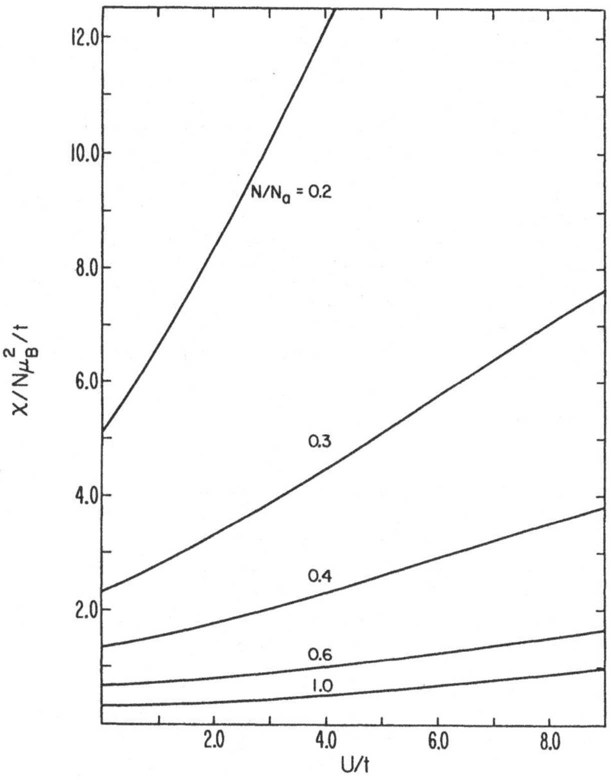

FIG. 5. Susceptibility per electron $\chi/(N\mu_B^2/t)$ vs U/t at some fixed values of N/N_a.

$$\frac{\chi}{N_a \mu_B^2/t} \simeq \frac{3}{2\pi^4} \frac{u}{(N/N_a)^2} \ . \tag{3.26}$$

c. Weak U limit ($U/t \ll 1$). As evident in Fig. 1, the relation between Q and N/N_a is singular in this limit. Carefully examining each factor, we obtain, from Eq. (3.23),

$$\frac{\chi}{N_a \mu_B^2/t} \simeq \frac{1}{\pi} \frac{1}{\sin Q}$$

$$\simeq \frac{1}{\pi} \left[\sin\left(\frac{\pi}{2} \frac{N}{N_a}\right) \right]^{-1} , \tag{3.27}$$

which is nothing but the Pauli paramagnetism of the one-dimensional noninteracting electron system.

IV. DISCUSSION

In this paper we have studied the properties of the one-dimensional Hubbard model at absolute-zero temperature as a function of the concentration of electrons N/N_a and the strength of correlation relative to the hopping integral U/t, calculating the magnetic susceptibility as well as the ground-state energy and the magnitude of local moments at each site. Our calculations are based on Lieb and Wu's theory, and therefore they are exact.

Since our model is a one-dimensional system, some of our conclusions should be inherent in one dimensionality, while others are applicable irrespective of dimensionality. The three-dimensional version of our Hubbard model is believed to have a ferromagnetic or antiferromagnetic ground state with a long-range ordering under certain conditions of the concentration of electrons and the strength of correlation. In fact, Penn[16] showed such possibilities, employing the random phase approximation. On the other hand, in our one-dimensional system, the ground state is always a singlet and is "smooth" as a function of N/N_a (mathematically speaking, $N/N_a = 1$ is a singular point). This is due to the fact that our model is a one-dimensional one with a single orbital. But as far as general aspects of the effect of correlation are concerned, our results should be suggestive beyond one dimensionality.

Our study in this paper has been restricted to absolute-zero temperature. Combining the results with the conclusions on the finite-temperature properties of the half-filled case,[8] we can give some conjectures on finite-temperature properties of the one-dimensional Hubbard model with $N/N_a < 1$.

(i) The high-temperature peak of the specific heat per atom,[8] which was found in the half-filled case for $U/t \gtrsim 4$, must be observed, as far as N/N_a is not too small. But the height of the peak will decrease with the decrease of N/N_a, because the probability of finding doubly occupied states becomes small. As for the low-temperature peak the coefficient of the linear increase with temperature must become large with the decrease of N/N_a, and the height of the peak will decrease.

(ii) When N/N_a decreases, the susceptibility per electron increases at low temperatures, and it must decrease rapidly with temperature.

ACKNOWLEDGMENTS

The author is indebted to Professor P. A. Pincus for valuable discussions and his critical reading of the manuscript. It is a pleasure to thank Professor T. Holstein and Professor R. Orbach for their hospitality.

APPENDIX

Here an outline of our numerical calculation of the susceptibility is described.

The quantities $I_Q^{(n)}(u)$, $\partial E_0(Q)/\partial Q$, and $\partial N_0(Q)/\partial Q$ appearing in Eq. (3.23) can be expressed in terms of the solutions of integral equations. In fact, we can write Eq. (3.14) in the form

$$I_Q^{(n)}(u)=\int_{-Q}^{Q}\frac{dk}{2\pi}\cos^n k\,\psi(k)\,, \tag{A1}$$

where $\psi(k)$ is the solution of the equation

$$\psi(k)=e^{2\pi(\sin k)/u}+\int_{-Q}^{Q}dk'\cos k'\times\frac{4}{u}R\left(\frac{4(\sin k-\sin k')}{u}\right)\psi(k')\,, \tag{A2}$$

as one can easily check by the iteration method. As for $\partial E_0(Q)/\partial Q$ and $\partial N_0(Q)/\partial Q$, the same technique as in the calculation of L_0 is useful. From Eqs. (2.3) and (2.5) we obtain

$$\frac{\partial E_0(Q)}{\partial Q}=-4\cos Q\,\rho(Q)-2\int_{-Q}^{Q}dk\cos k\,\frac{\partial\rho(k)}{\partial Q} \tag{A3}$$

and

$$\frac{\partial N_0(Q)}{\partial Q}=2\rho(Q)+\int_{-Q}^{Q}dk\,\frac{\partial\rho(k)}{\partial Q}\,. \tag{A4}$$

Here use was made of the relation $\rho(k)=\rho(-k)$. Performing the differentiation of both sides of Eq. (2.9) by Q, we find the equation

$$2\pi\frac{\partial\rho(k)}{\partial Q}=\cos k\,\rho(Q)\left[\frac{8\pi}{u}R\left(\frac{4(\sin k-\sin Q)}{u}\right)+\frac{8\pi}{u}R\left(\frac{4(\sin k+\sin Q)}{u}\right)\right]+\cos k\int_{-Q}^{Q}dk'\frac{\partial\rho(k')}{\partial Q}\frac{8\pi}{u}R\left(\frac{4(\sin k-\sin k')}{u}\right)\,. \tag{A5}$$

In our numerical calculations the integral equations (A2) and (A5) were replaced again by 41 coupled linear algebraic equations and then $I_Q^{(n)}$, $\partial E_0(Q)/\partial Q$, and $\partial N_0(Q)/\partial Q$ were evaluated.

---

*Work supported in part by the National Science Foundation, GP-21290, and the Office of Naval Research, under Contract No. N00014-69-A-0200-4032.

†Present address: Department of Physics, Faculty of Science, Osaka University, Toyonaka, Japan.

[1]See, for instance, L. H. Lieb and D. C. Mattis, *Mathematical Physics in One Dimension* (Academic, New York, 1966).

[2]L. H. Lieb and F. Y. Wu, Phys. Rev. Letters 20, 1445 (1968).

[3]H. Bethe, Z. Physik 71, 205 (1931).

[4]L. Hulthen, Ark. Mat. Astron. Fys. 26A, No. 11 (1938).

[5]C. N. Yang, Phys. Rev. Letters 19, 1312 (1967).

[6]A. A. Ovchinnikov, Zh. Eksperim. i Teor. Fiz. 57, 2144 (1970) [Sov. Phys. JETP 30, 1160 (1970)].

[7]M. Takahashi, Progr. Theoret. Phys. (Kyoto) 42, 1098 (1969); 43, 1619 (1970).

[8]H. Shiba and P. A. Pincus, Phys. Rev. B 5, 1966 (1972); H. Shiba, Progr. Theoret. Phys. (Kyoto) (to be published).

[9]L. H. Lieb and D. C. Mattis, Phys. Rev. 125, 164 (1962).

[10]J. Hubbard, Proc. Roy. Soc. (London) A276, 238 (1963); A281, 401 (1964).

[11]M. C. Gutzwiller, Phys. Rev. Letters 10, 159 (1963).

[12]J. Kanamori, Progr. Theoret. Phys. (Kyoto) 30, 275 (1963).

[13]R. B. Griffiths, Phys. Rev. 133, A768 (1964).

[14]J. Friedel, in *Theory of Magnetism in Transition Metals*, edited by W. Marshall (Academic, New York, 1967), p. 283.

[15]C. N. Yang and C. P. Yang, Phys. Rev. 151, 258 (1966).

[16]D. R. Penn, Phys. Rev. 142, 350 (1966).

Progress of Theoretical Physics, Vol. 52, No. 1, July 1974

Low-Temperature Specific-Heat of One-Dimensional Hubbard Model

Minoru TAKAHASHI

Department of Physics, College of General Education
Osaka University, Osaka

(Received February 18, 1974)

Low-temperature specific heat per site (C) of one-dimensional Hubbard model is investigated by the method of non-linear integral equations. For the half-filled case we show $\lim_{H\to 0}\lim_{T\to 0} C/T=\pi I_0(\pi/2U)/(6I_1(\pi/2U))$, where T is temperature, H is magnetic field, U is the coupling constant, and I_0 and I_1 are modified Bessel functions. Although this equation yeilds $\lim_{T,H\to 0} C/T=\pi/6$ in the limit $U\to 0+$, the true value of $\lim_{T,H\to 0} C/T$ at $U=0$ is $\pi/3$. This means that $\lim_{T,H\to 0} C/T$ is a discontinuous function of U at $U=0$. This discontinuity disappears when the band is not half filled.

§ 1. Introduction

Low-temperature behavior of Hubbard model is interesting physically, and difficult to treat rigorously. The one-dimensional case of this model has been investigated by many physicists. Its thermodynamic potential density is difined by

$$\omega(U,T,A,H)=-T\lim_{N_a\to\infty}\{\ln(\mathrm{Tr}\exp(-T^{-1}(\mathcal{H}-A\sum_{i=1}^{N_a}(n_{i\uparrow}+n_{i\downarrow})))/N_a\},\tag{1·1a}$$

where $\mathcal{H}$ is the Hamiltonian:

$$\mathcal{H}=-\sum_{i=1}^{N_a}\sum_{\sigma}(c_{i\sigma}^\dagger c_{i+1\sigma}+c_{i+1\sigma}^\dagger c_{i\sigma})+4U\sum_{i=1}^{N_a}n_{i\uparrow}n_{i\downarrow}-\mu_0 H\sum_{i=1}^{N_a}(n_{i\uparrow}-n_{i\downarrow}),$$

$$c_{N_a+1\sigma}\equiv c_{1\sigma}\,,\qquad n_{i\sigma}\equiv c_{i\sigma}^\dagger c_{i\sigma}\,.\tag{1·1b}$$

Here we have following symmetry relations through appropriate unitary transformations:

$$\begin{aligned}\omega(U,T,A,H)&=\omega(U,T,A,-H)=4U-2A+\omega(U,T,4U-A,H)\\&=\mu_0H-A+\omega(-U,T,\mu_0H-2U,\mu_0^{-1}(A-2U)).\end{aligned}\tag{1·2}$$

The first identity is obtained by changing up-spin and down-spin, the second by changing the creation and annihilation operators and the third by changing the creation and annihilation operators in the up-spin band. If we know the value of ω in the region $U\geq 0$, $H\geq 0$ and $A\leq 2U$, we easily obtain the value of ω outside of this region through the relations (1·2). Then we restrict ourselves to calculate ω in this region. Other thermodynamic quantities such as energy and entropy per site (e, s), specific heat per site $(C_{H,A})$ and densities of up-spin and down-spin electrons $(n_\uparrow, n_\downarrow)$ are obtained by the differentiations of ω:

$$n_\uparrow + n_\downarrow = \frac{\partial \omega}{\partial A}, \qquad n_\uparrow - n_\downarrow = \frac{1}{\mu_0}\frac{\partial \omega}{\partial H}, \qquad e = -T^2 \frac{\partial}{\partial T}\left(\frac{\omega}{T}\right) + \frac{\partial \omega}{\partial A} A,$$

$$S = -\frac{\partial \omega}{\partial T}, \qquad C_{H,A} = -T\frac{\partial^2 \omega}{\partial T^2}, \qquad \chi = -\frac{\partial^2 \omega}{\partial H^2}. \tag{1·3}$$

In a previous paper[1] the author derived a set of non-linear integral equations for the calculation of thermodynamic potential density ω. We used Bethe ansatz, which was first applied to this model by Lieb and Wu,[2] and some assumptions on the distributions of quasi-momenta k and parameters Λ on the complex plane. Recently Shiba and Pincus[3] calculated the energy levels of this model in the case of finite atomic numbers (such as six or five) and thermodynamic quantities. Their method is not useful to investigate the low-temperature properties of the model in the thermodynamic limit. For example, magnetic susceptibility of the finite system becomes zero or infinity in the limit of zero temperature. But this is not valid in the thermodynamic limit because magnetic susceptibility has finite values at $T=0$ in the half-filled state.[4),5)]

In the following sections we investigate the low-temperature behavior of this system, using the set of integral equations given in Ref. 1), and come to the conclusion that in the half-filled case low-temperature specific heat is proportional to temperature and coefficient is given analytically:

$$\lim_{H\to 0}\lim_{T\to 0} C/T = \pi I_0(\pi/2U)/(6I_1(\pi/2U)).$$

It should be noted that this is inversely proportional to the magnon velocity[6] at $T=0$:

$$v = 2I_1(\pi/2U)/I_0(\pi/2U),$$

and proportional to the magnetic susceptibility[5] at $T=0$:

$$\chi = \mu_0^2 I_0(\pi/2U)/(\pi I_1(\pi/2U)).$$

§ 2. Integral equations

The eigenvalue problem of one-dimensional Hubbard model described by the Hamiltonian (1·1b) can be treated by the method of Bethe's hypothesis. According to Lieb and Wu, we must solve a set of equations for N quasi-momenta k and M parameters Λ where N is the number of fermions and M is the number of down-spin fermions,

$$e^{ik_jN_a} = -\prod_{\alpha=1}^{M}\left(\frac{k_j - \Lambda_\alpha - 2iU}{k_j - \Lambda_\alpha + 2iU}\right), \qquad j = 1, 2, \cdots, N,$$

$$\prod_{j=1}^{N}\left(\frac{\Lambda_\alpha - k_j + iU}{\Lambda_\alpha - k_j - iU}\right) = -\prod_{\beta=1}^{M}\left(\frac{\Lambda_\alpha - \Lambda_\beta + 2iU}{\Lambda_\alpha - \Lambda_\beta - 2iU}\right), \qquad \alpha = 1, 2, \cdots, M.$$

In the previous paper[1] the author assumed that the k's and Λ's form bound states on the complex plane, and derived a set of non-linear integral equations for the distribution of the bound states at given temperature T, magnetic field H and chemical potential A:

$$\ln \zeta(k) = \kappa_0(k)/T + \int_{-\infty}^{\infty} s(\Lambda - \sin k)\ln((1+\eta_1'(\Lambda))/(1+\eta_1(\Lambda)))d\Lambda\,, \quad (2\cdot1a)$$

$$\ln \eta_1(\Lambda) = s^* \ln(1+\eta_2(\Lambda)) - \int_{-\pi}^{\pi} dk \cos k\, s(\Lambda - \sin k)\ln(1+\zeta^{-1}(k)), \quad (2\cdot1b)$$

$$\ln \eta_1'(\Lambda) = s^* \ln(1+\eta_2'(\Lambda)) - \int_{-\pi}^{\pi} dk \cos k\, s(\Lambda - \sin k)\ln(1+\zeta(k)), \quad (2\cdot1c)$$

$$\ln \eta_n(\Lambda) = s^* \ln(1+\eta_{n-1}(\Lambda))(1+\eta_{n+1}(\Lambda)), \qquad n=2,3,\cdots, \quad (2\cdot1d)$$

$$\ln \eta_n'(\Lambda) = s^* \ln(1+\eta'_{n-1}(\Lambda))(1+\eta'_{n+1}(\Lambda)), \qquad n=2,3,\cdots, \quad (2\cdot1e)$$

$$\lim_{n\to\infty} \frac{\ln \eta_n}{n} = \frac{2\mu_0 H}{T}, \quad (2\cdot1f)$$

$$\lim_{n\to\infty} \frac{\ln \eta_n'}{n} = \frac{4U-2A}{T}, \quad (2\cdot1g)$$

where $s(\Lambda) \equiv \mathrm{sech}(\pi x/2U)/4U$, $f^*g \equiv \int_{-\infty}^{\infty} f(\Lambda-\Lambda')g(\Lambda')d\Lambda'$,

$$\kappa_0(k) = -2\cos k - 4\int_{-\infty}^{\infty} s(\Lambda - \sin k)\,(\mathrm{Re}\sqrt{1-(\Lambda-Ui)^2})d\Lambda\,. \quad (2\cdot1h)$$

Function $\zeta(k)$ is the ratio of hole density and particle density of unbound quasi-momenta. Function $\eta_n(\Lambda)$ is that of n-th order bound state of Λ. Function $\eta_n'(\Lambda)$ is that of bound state of the $n\Lambda$'s and $2nk$'s. Thermodynamic potential per site is given by

$$\omega(T,A,H) = -T\int_{-\pi}^{\pi} \ln(1+\zeta^{-1}(k))\frac{dk}{2\pi} - T\sum_{n=1}^{\infty}\int_{-\infty}^{\infty} \ln(1+\eta_n^{-1}(\Lambda))$$

$$\times \mathrm{Re}\,\frac{1}{\sqrt{1-(\Lambda-nUi)^2}}\cdot\frac{d\Lambda}{\pi} \quad (2\cdot2a)$$

$$= E_0 - A - T\left\{\int_{-\pi}^{\pi} \rho_0(k)\ln(1+\zeta(k))dk + \int_{-\infty}^{\infty} \sigma_0(\Lambda)\ln(1+\eta_1(\Lambda))d\Lambda\right\}, \quad (2\cdot2b)$$

where E_0, $\rho_0(k)$, $\sigma_0(\Lambda)$ are the ground state energy per site, distribution function of the k's and that of the Λ's at $T=0$, $A=2U$, $\mu_0 H=0$, respectively:

$$\sigma_0(\Lambda) \equiv \int_{-\pi}^{\pi} s(\Lambda - \sin k)\frac{dk}{2\pi}, \quad (2\cdot2c)$$

$$\rho_0(k) \equiv \frac{1}{2\pi} + \cos k\int_{-\infty}^{\infty} d\Lambda\, a_1(\Lambda - \sin k)\sigma_0(\Lambda), \quad (2\cdot2d)$$

$$a_n(\Lambda) \equiv \frac{nU}{\pi(\Lambda^2 + (nU)^2)}, \quad (2\cdot2e)$$

$$E_0 \equiv -2\int_{-\pi}^{\pi}\cos k\,\rho_0(k)dk\,. \tag{2·2f}$$

One should note that Eqs. (2·1) and (2·2) are valid only at $U \geq 0$, $A \leq 2U$ and $\mu_0 H \geq 0$. The other cases can be treated through Eqs. (1·2).

From Eqs. (2·1c), (2·1e) and (2·1g) we have

$$\ln(1+\eta_n') \geq 2n(2U-A)/T\,, \qquad n=1,2,3,\cdots.$$

At $2U-A \gg T$, we can replace $\ln\eta_n'$ by $\ln(1+\eta_n')$ in Eqs. (2·1c), (2·1e) and (2·1g) and obtain

$$\ln(1+\eta_n') = 2n(2U-A)/T + \int_{-\pi}^{\pi} a_n(\Lambda-\sin k)\ln(1+\zeta(k))\cos k\,dk$$
$$+O(\exp(-(2U-A)/T)), \qquad n=1,2,\cdots. \tag{2·3a}$$

Substituting case $n=1$ of this equation into (2·1a), we have

$$\kappa(k) = \kappa_0(k) + 2U - A + T\int_{-\pi}^{\pi} R(\sin k - \sin k')\ln(1+\exp(\kappa(k')/T))\cos k'dk'$$
$$-T\int_{-\infty}^{\infty} s(\Lambda-\sin k)\ln(1+\exp(\varepsilon_1(\Lambda)/T))d\Lambda + O(T\exp(-(2U-A)/T)), \tag{2·4a}$$

where $R \equiv s * a_1$, $\kappa = T\ln\zeta$, $\varepsilon_1 = T\ln\eta_1$.

At $2U-A = O(T)$, function $\kappa(k)$ is always negative. Then the last term of (2·1c) is of the order of $T^{1/2}\exp(\kappa^{(0)}(\pi)/T)$ at low temperatures. Then we have

$$1+\eta_j' = (\mathrm{sh}\{(j+1)(2U-A)/T\}/\mathrm{sh}\{(2U-A)/T\})^2 + O(T^{1/2}\exp(\kappa^{(0)}(\pi)/T)), \tag{2·3b}$$

where $\kappa^{(0)}$ is κ at zero temperature (hereafter we put (0) for the functions at zero temperature). Substituting this into (2·1a), we have

$$\kappa(k) = \kappa_0(k) + T\ln(2\,\mathrm{ch}\{(2U-A)/T\}) - T\int_{-\infty}^{\infty} s(\Lambda-\sin k)$$
$$\times \ln\left(1+\exp\frac{\varepsilon_1(\Lambda)}{T}\right)d\Lambda + O(T^{3/2}\exp(\kappa^{(0)}(\pi)/T)). \tag{2·4b}$$

At $\mu_0 H \gg T$, we have

$$\ln(1+\eta_n) = a_{n-1}^* \ln(1+\eta_1) + 2(n-1)\mu_0 H/T\,, \qquad n=2,3,\cdots. \tag{2·3c}$$

Substituting this into (2·1b), we have

$$\varepsilon_1(\Lambda) = TR^* \ln(1+\exp(\varepsilon_1(\Lambda)/T)) + \mu_0 H - T\int_{-\pi}^{\pi} dk\cos k\, s(\Lambda-\sin k)$$
$$\times \ln(1+\exp(-\kappa(k)/T)) + O(T\exp(2\mu_0 H/T)). \tag{2·4c}$$

Equations (2·4a) and (2·4c) are transformed as follows:

$$\kappa(k) = -2\cos k - A - \mu_0 H - T\int_{-\infty}^{\infty} a_1(\sin k - \Lambda)\ln(1+\exp(-\varepsilon_1(\Lambda)/T))d\Lambda\,, \tag{2·5a}$$

$$\varepsilon_1(\Lambda) = T\int_{-\infty}^{\infty} a_2(\Lambda-\Lambda')\ln(1+\exp(-\varepsilon_1(\Lambda')/T))d\Lambda'$$
$$-T\int_{-\pi}^{\pi} a_1(\Lambda - \sin k)\ln(1+\exp(-\kappa(k)/T))\cos k\, dk + 2\mu_0 H\,. \tag{2·5b}$$

From Eq. (2·2a) we have

$$\omega(T, A, H) = -T\int_{-\pi}^{\pi} \ln(1+\exp(-\kappa(k)/T))\frac{dk}{2\pi}\,. \tag{2·5c}$$

Here we have neglected the terms which are of the order of $e^{-2\mu_0 H/T}$ or $e^{-(4U-2A)/T}$ Equations (2·4) or (2·5) are useful to obtain thermodynamic potential at $2U - A \gg T$ and $2\mu_0 H \gg T$.

As shown in Fig. 1, (A, H) plane is devided into several regions by the low-temperature properties. The number of fermions per site n has the following properties at zero temperature:

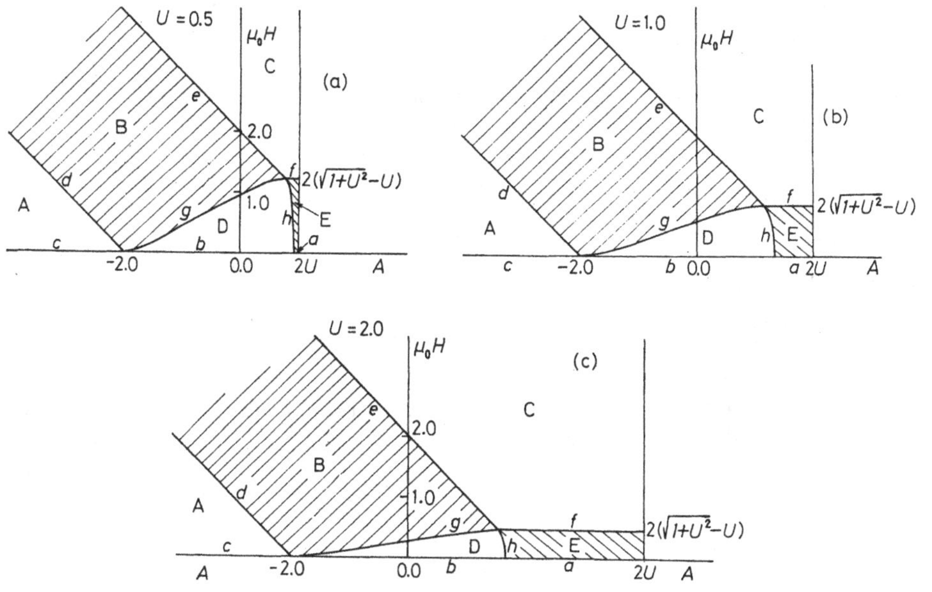

Fig. 1. Characteristic regions of low-temperature specific heat for various values of U. On lines d, e, f, g and h, low-temperature specific heat is proportional to $T^{1/2}$. In regions B, D and E, it is proportional to T. In regions A and C, it is proportional to $T^{-3/2}\exp(-\alpha/T)$.

a) $U=0.5$
b) $U=1.0$
c) $U=2.0$

$$n=1 \quad \text{at regions } C \text{ and } E,$$
$$0<n<1 \quad \text{at regions } B \text{ and } D,$$
$$n=0 \quad \text{at region } A.$$

On lines a, b and c, magnetization is zero.

§ 3. Case $\mu_0 H \gg T$

a) $A \leq -2-\mu_0 H$ *(Region A)*

In this region, density of fermions is zero at zero temperature. From Eq. (2·5a), we have

$$\kappa(k) = -2\cos k - A - \mu_0 H - T\exp(-2\mu_0 H/T).$$

Substituting this into (2·5c), we obtain

$$\omega(T,A,H) = -\pi^{-1}T^{3/2}\int_0^\infty \ln\left(1+\exp\left(\frac{2+A+\mu_0 H}{T}\right)e^{-x^2}\right)dx.$$

b) $\varepsilon_1^{(0)}(0)\geq 0$, $A>-2-\mu_0 H$ *(Region B)*

Here the number of fermions per site, n, satisfies $1>n>0$. At zero temperature all fermions have up-spin. From Eq. (2·5c) we have

$$\omega(T,A,H)-\omega(0,A,H) = -T\int_{-\pi}^{\pi}\ln(1+\exp(-|\kappa(k)|/T))\frac{dk}{2\pi} - \int_{-Q}^{Q}\delta\kappa(k)\frac{dk}{2\pi},$$

where Q and $-Q(Q>0)$ are zeroes of $\kappa^{(0)}(k)$. From Eqs. (2·5a) and (2·5b) we obtain

$$\delta\kappa = -T\int_{-\infty}^{\infty} a_1(\sin k-\Lambda)\ln(1+\exp(-\varepsilon_1(\Lambda)/T))d\Lambda,$$

$$\varepsilon_1(\Lambda) = -2\int_{-Q}^{Q} a_1(\Lambda-\sin k)\cos^2 k\,dk + 2\mu_0 H + O(T^2) + O(T^{3/2}\exp(-\varepsilon_1^{(0)}/T)).$$

Then we have

$$\omega(T,A,H) = \omega(0,A,H) - \frac{T^2}{2\pi}\frac{1}{2\sin Q}\cdot\frac{\pi^2}{3} - T^{3/2}2g(0)\sqrt{\frac{2}{\varepsilon_1^{(0)\prime\prime}(0)}}\int_0^\infty \ln(1+\exp(-\varepsilon_1^{(0)\prime}(0)/T)\cdot e^{-x^2})dx,$$

where

$$g(\Lambda) = \int_{-Q}^{Q} a_1(\Lambda-\sin k)\frac{dk}{2\pi}.$$

c) $\mu_0 H \geq 2(\sqrt{1+U^2}-U)$, $A \geq 2-\mu_0 H$ *(Region C)*

At zero temperature, density of fermion is one, and all fermions have up-spin. Substituting Eq. (2·5a) into Eq. (2·5c), we have

$$\omega(T,A,H) = -A-\mu_0 H - T\int_{-\pi}^{\pi} \ln(1+\exp(\kappa(k)/T))\frac{dk}{2\pi}$$

$$-T\int_{-\infty}^{\infty} 2\left(\mathrm{Re}\,\frac{1}{\sqrt{1-(\Lambda-Ui)^2}}\right)\ln(1+\exp(-\varepsilon_1(\Lambda)/T))d\Lambda. \qquad (3\cdot1)$$

From Eqs. (2·5a) and (2·5b), we obtain

$$\kappa(k) = -2\cos k - A - \mu_0 H - O(T^{3/2}\lambda),$$

$$\varepsilon_1(\Lambda) = -4\,\mathrm{Re}(\sqrt{1-(\Lambda-Ui)^2}-U)+2\mu_0 H + O(T^{3/2}\lambda)+O(T^{3/2}\mu),$$

$$\lambda \equiv \exp(-(4(\sqrt{1+U^2}-U)-2\mu_0 H)/T), \qquad \mu = \exp((2-A-\mu_0 H)/T).$$

Substituting these into Eq. (3·1), we have

$$\omega(T,A,H) = -A-\mu_0 H - \pi^{-1}T^{3/2}\int_0^{\infty}\ln(1+\mu e^{-x^2})dx$$

$$-4T^{3/2}(\sqrt{1+U^2}-U)(1+U^2)^{-1/4}\int_0^{\infty}\ln(1+\lambda e^{-x^2})dx. \qquad (3\cdot2)$$

On the boundary of this region we have

$$\omega = \begin{cases} -A-\mu_0 H - \pi^{-1}T^{3/2}\zeta\left(\frac{3}{2}\right)\left(1-\frac{1}{\sqrt{2}}\right)\frac{\sqrt{\pi}}{2} & \text{at } \mu_0 H = 2(\sqrt{1+U^2}-U), \\ -A-\mu_0 H - 4T^{3/2}(\sqrt{1+U^2}-U)(1+U^2)^{-1/4}\zeta\left(\frac{3}{2}\right)\left(1-\frac{1}{\sqrt{2}}\right)\frac{\sqrt{\pi}}{2} & \\ & \text{at } A = 2-\mu_0 H. \end{cases} \qquad (3\cdot3)$$

d) $\varepsilon_1^{(0)}(0)<0$, $\kappa^{(0)}(\pi)>0$ *(Region D)*

From Eq. (2·5c), we have

$$\omega(T,A,H)-\omega(0,A,H) = -\frac{\pi T^2}{6\kappa'(Q)} + \int_{-Q}^{Q}\frac{dk}{2\pi}\delta\kappa(k). \qquad (3\cdot4)$$

Function $\delta\kappa(k)$ is determined by

$$\delta\kappa(k) = \int_{-B}^{B} a_1(\sin k-\Lambda)\delta\varepsilon_1(\Lambda)d\Lambda - \frac{\pi^2 T^2}{6\varepsilon_1'(B)}\{a_1(\sin k-B)+a_1(\sin k+B)\},$$

$$(3\cdot5\text{a})$$

$$\delta\varepsilon(\Lambda) + \int_{-B}^{B} a_2(\Lambda-\Lambda')\delta\varepsilon(\Lambda')d\Lambda' = \int_{-Q}^{Q} dk\cos k\, a_1(\sin k-\Lambda)\delta\kappa(k)$$

$$-\frac{\pi^2 T^2\cos Q}{6\kappa'(Q)}\{a_1(\sin Q-\Lambda)+a_1(\sin Q+\Lambda)\}$$

$$+\frac{\pi^2T^2}{6\varepsilon_1'(B)}\{a_2(B-\Lambda)+a_2(B+\Lambda)\}, \tag{3·5b}$$

where Q and B are zeroes of $\kappa^{(0)}(k)$ and $\varepsilon_1^{(0)}(\Lambda)$, respectively. From these equations we obtain

$$\omega(T,A,H)-\omega(0,A,H)=-\frac{\pi^2T^2}{3}\left[\frac{\sigma_1^{(0)}(B)}{\varepsilon_1^{(0)\prime}(B)}+\frac{\rho^{(0)}(Q)}{\kappa^{(0)\prime}(Q)}\right]+O(T^3), \tag{3·6a}$$

where $\rho^{(0)}(k)$ and $\sigma_1^{(0)}(\Lambda)$ are the distribution functions of k and Λ at zero temperature and determined by

$$\rho_1^{(0)}(k)=\frac{1}{2\pi}+\cos k\int_{-B}^{B}a_1(\Lambda-\sin k)\sigma_1^{(0)}(\Lambda)d\Lambda, \tag{3·6b}$$

$$\sigma_1^{(0)}(\Lambda)+\int_{-B}^{B}a_2(\Lambda-\Lambda')\sigma_1^{(0)}(\Lambda')d\Lambda'=\int_{-Q}^{Q}a_1(\Lambda-\sin k)\rho^{(0)}(k)dk. \tag{3·6c}$$

The equations for $\sigma_1^{(0)}$ and $\varepsilon_1^{(0)\prime}$ are written as

$$\sigma_1^{(0)}(\Lambda)-\int_{|\Lambda|>B}R(\Lambda-\Lambda')\sigma_1^{(0)}(\Lambda')d\Lambda'=\int_{-Q}^{Q}s(\Lambda-\sin k)\rho^{(0)}(k)dk, \tag{3·6d}$$

$$\varepsilon_1^{(0)\prime}(\Lambda)-\int_{|\Lambda|>B}R(\Lambda-\Lambda')\varepsilon_1^{(0)\prime}(\Lambda')d\Lambda'=\int_{-Q}^{Q}s(\Lambda-\sin k)\kappa^{(0)\prime}(k)dk. \tag{3·6e}$$

The right-hand sides of these equations are

$$\exp\left(-\frac{\pi|\Lambda|}{2U}\right)(2U)^{-1}\int_{-Q}^{Q}dk\exp\left(-\frac{\pi\sin k}{2U}\right)\rho^{(0)}(k)$$

and

$$\mathrm{sign}(\Lambda)\exp\left(-\frac{\pi|\Lambda|}{2U}\right)(2U)^{-1}\int_{-Q}^{Q}dk\exp\left(-\frac{\pi\sin k}{2U}\right)\kappa^{(0)\prime}(k)$$

at $|\Lambda|\gg 1$, U. Then we have

$$\frac{\sigma_1^{(0)}(B)}{\varepsilon_1^{(0)\prime}(B)}=\frac{\int_{-Q}^{Q}dk\exp(\pi\sin k/2U)\rho^{(0)}(k)}{\int_{-Q}^{Q}dk\exp(\pi\sin k/2U)\kappa^{(0)\prime}(k)}+O(B^{-2})$$

and

$$\omega(T,A,H)-\omega(0,A,H)=-\frac{\pi^2T^2}{3}\left\{\frac{\rho^{(0)}(Q)}{\kappa^{(0)\prime}(Q)}+\frac{\int_{-Q}^{Q}dk\exp(\pi\sin k/2U)\rho^{(0)}(k)}{\int_{-Q}^{Q}dk\exp(\pi\sin k/2U)\kappa^{(0)\prime}(k)}\right.$$

$$\left.+O(\{\ln(\mu_0H)\}^{-2})\right\}, \tag{3·7}$$

when μ_0H is very small. From this equation we obtain

$$\lim_{H\to 0}\lim_{T\to 0}C_{A,H}/T=\frac{2\pi^2}{3}\left\{\frac{\rho^{(0)}(Q)}{\kappa^{(0)\prime}(Q)}+\frac{\int_{-Q}^{Q}dk\exp(\pi\sin k/2U)\rho^{(0)}(k)}{\int_{-Q}^{Q}dk\exp(\pi\sin k/2U)\kappa^{(0)\prime}(k)}\right\}. \tag{3·8}$$

e) $2(\sqrt{1+U^2}-U)>\mu_0 H \gg T,\ \kappa^{(0)}(\pi)\le 0$ (*Region E*)

From Eq. (2·2b) we have

$$\omega(T,A,H)-\omega(0,A,H)=-2T^{3/2}\rho_0(\pi)\sqrt{\frac{2}{-\kappa(\pi)}}\int_0^\infty \ln\left(1+\exp\left(\frac{\kappa(\pi)}{T}\right)\cdot e^{-x^2}\right)dx$$
$$-T\int_{-\infty}^{\infty}\sigma_0(\Lambda)\ln\left(1+\exp\left(-\frac{|\varepsilon_1(\Lambda)|}{T}\right)\right)d\Lambda-T\int_{|\Lambda|>B}\sigma_0(\Lambda)\delta\varepsilon_1(\Lambda)d\Lambda+O(T^4), \tag{3·9}$$

where $\delta\varepsilon\equiv\varepsilon-\varepsilon^{(0)}$. From Eq. (2·4b) we have

$$\delta\varepsilon(\Lambda)-\int_{|\Lambda'|>B}R(\Lambda-\Lambda')\delta\varepsilon_1(\Lambda')d\Lambda'$$
$$=-\pi^2T^2(R(\Lambda-B)$$
$$+R(\Lambda+B))/(6\varepsilon_1'(B))$$
$$+O((B-B')^2),$$

where B and B' are zeroes of ε_1 and $\varepsilon_1^{(0)}$, respectively. Summing the second and the third terms of r.h.s. of (3·9), we have

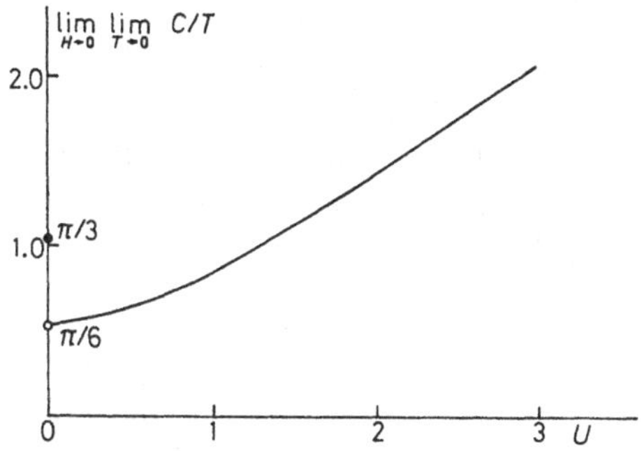

Fig. 2. Coefficient of T-linear low-temperature specific heat in the half-filled case ($A=2U$), and $\mu_0 H=0$.

$$-\frac{\pi^2T^2}{3}\frac{\sigma_1^{(0)}(B)}{\varepsilon^{(0)\prime}(B)}+O(T^3), \tag{3·10}$$

where $\sigma_1^{(0)}$ and $\varepsilon_1^{(0)\prime}$ are determined by

$$\sigma_1^{(0)}(\Lambda)-\int_{|\Lambda'|>B}R(\Lambda-\Lambda')\sigma_1^{(0)}(\Lambda')d\Lambda'=\sigma_0(\Lambda), \tag{3·11a}$$

$$\varepsilon_1^{(0)\prime}(\Lambda)-\int_{|\Lambda'|>B}R(\Lambda-\Lambda')\varepsilon_1^{(0)\prime}(\Lambda')d\Lambda'=2\int_{-\pi}^{\pi}s(\Lambda-\sin k)\sin k\,dk. \tag{3·11b}$$

At $|\Lambda|\gg\max.(1,1/U)$, r.h.s. of (3·11a) and (3·11b) are $(2U)^{-1}I_0(\pi/2U)e^{-\pi|\Lambda|/2U}$ and $\mathrm{sign}(\Lambda)\cdot 2\pi U^{-1}I_1(\pi/2U)e^{-\pi|\Lambda|/2U}$, respectively. Then we have

$$\frac{\sigma_1^{(0)}(B)}{\varepsilon_1^{(0)\prime}(B)}=\frac{I_0(\pi/2U)}{4\pi I_1(\pi/2U)}+O(B^2)$$

and

$$\omega(T,A,H)-\omega(O,A,H)=-2T^{3/2}\rho_0(\pi)\sqrt{\frac{2}{-\kappa^{(0)\prime}(\pi)}}$$
$$\times\int_0^\infty \ln(1+\exp(\kappa^{(0)}(\pi)/T)e^{-x^2})-\frac{\pi T^2}{12}\cdot\frac{I_0(\pi/2U)}{I_1(\pi/2U)}$$
$$+O((\ln\mu_0 H)^{-2})+O(T^3). \tag{3·12}$$

The coefficient of T-linear specific heat at $A=2U$ is

$$\lim_{H\to 0}\lim_{T\to 0} C_{A,H}/T=\frac{\pi}{6}\cdot\frac{I_0(\pi/2U)}{I_1(\pi/2U)}.$$

This value is shown in Fig. 2 as a function of U.

§ 4. Case $\mu_0 H=O(T)$

a) $\kappa^{(0)}(\pi)\leq 0$ (*near line a*)

From Eqs. (2·1) we have

$$\ln\eta_1(\Lambda)=s*\ln(1+\eta_2(\Lambda))-\frac{2}{T}\int_{-\pi}^{\pi}dk\cos^2 k\, s(\Lambda-\sin k)+O\left(T^{1/2}\exp\frac{\kappa^{(0)}(\pi)}{T}\right),$$

$$\ln\eta_n(\Lambda)=s*\ln(1+\eta_{n-1}(\Lambda))(1+\eta_{n+1}(\Lambda)),\qquad n=2,3,\cdots,$$

$$\lim_{n\to\infty}\frac{\ln\eta_n}{n}=\frac{2\mu_0 H}{T}.$$

Thus Eq. (2·2b) can be written as follows:

$$\omega(T,A,H)=E_0-A-2T^{3/2}\rho_0(\pi)\sqrt{\frac{2}{-\kappa^{(0)''}(\pi)}}\int_0^{\infty}\ln\left(1+\exp\frac{\kappa^{(0)}(\pi)}{T}\cdot e^{-x^2}\right)dx$$
$$-\frac{T^2}{2}\cdot\frac{I_0(\pi/2U)}{I_1(\pi/2U)}C\left(\frac{2\mu_0 H}{T}\right)+O(T^4),\tag{4·1}$$

where $C(y)$ is determined by

$$C(y)=\int_{-\infty}^{\infty}e^{-\pi x/2}\ln(1+\eta_1(x))dx,$$

$$\ln\eta_1(x)=-e^{-\pi x/2}+\int_{-\infty}^{\infty}\frac{1}{4}\operatorname{sech}\frac{\pi(x-x')}{2}\ln(1+\eta_2(x'))dx',$$

$$\ln\eta_n(x)=\int_{-\infty}^{\infty}\frac{1}{4}\operatorname{sech}\frac{\pi(x-x')}{2}\ln(1+\eta_{n-1}(x'))(1+\eta_{n+1}(x'))dx',$$
$$n=2,3,\cdots,$$

$$\lim_{n\to\infty}\frac{\ln\eta_n}{n}=y.\tag{4·2}$$

b) $\kappa^{(0)}(\pi)>0$ (*near line b*)

From Eq. (2·2b) we have

$$\omega(T,A,H)-\omega(0,A,H)=\int_{-Q}^{Q}\frac{dk}{2\pi}\delta\kappa(k)-\frac{\pi}{12}\cdot\frac{T^2}{\kappa^{(0)'}(Q)}+O(T^3).\tag{4·3}$$

The equation for $\delta\kappa\equiv\kappa-\kappa^{(0)}$ is

$$\delta\kappa(k)-\int_{-Q}^{Q}R(\sin k-\sin k')\delta\kappa(k')\cos k'dk'$$

$$= -T\int_{-\infty}^{\infty} s(\Lambda - \sin k)\ln(1+\eta_1(\Lambda))d\Lambda$$

$$+\frac{\pi^2 T^2}{6\kappa^{(0)}(Q)}(R(\sin k - \sin Q) + R(\sin k + \sin Q)) + O(T^4). \qquad (4\cdot4)$$

After some calculations we obtain

$$\omega(T,A,H) - \omega(0,A,0) = -\frac{\pi^2 T^2 \rho^{(0)}(Q)}{3\kappa^{(0)\prime}(Q)} - 2\pi T^2 C\left(\frac{2\mu_0 H}{T}\right)$$

$$\times\left(\int_{-Q}^{Q}\exp\left(\frac{\sin k}{U}\right)\rho^{(0)}(k)dk\Big/\int_{-Q}^{Q}\exp\left(\frac{\sin k}{U}\right)\kappa^{(0)\prime}(k)dk\right) + O(T^3), \quad (4\cdot5)$$

where $\rho^{(0)}$, $\kappa^{(0)}$ and $C(y)$ are defined in Eqs. (3·6) and (4·2). Functions similar to $C(y)$ defined in (4·2) appeared in the investigation of the low-temperature specific heat of Heisenberg-Ising ring at $|\Delta|\leq 1$.[7] From the result of numerical calculation in Ref. 7) we conjecture

$$C(0) = \pi/6 \quad \text{and} \quad C''(0) = 1/2\pi\,. \qquad (4\cdot6)$$

If these equations are true, we obtain

$$\lim_{T\to 0}\lim_{H\to 0} C/T = \lim_{H\to 0}\lim_{T\to 0} C/T\,,$$

$$\lim_{T\to 0}\lim_{H\to 0}\chi = \lim_{H\to 0}\lim_{T\to 0}\chi\,.$$

§ 5. Discussions and summary

From the theory of non-interacting fermions, thermodynamic potential per site at $U=0$ is

$$\omega(T,A,H) = \frac{1}{2\pi}\left\{\int_{-\pi}^{\pi}\ln(1+\exp(-2\cos k - \mu_0 H - A/T))dk\right.$$

$$\left. + \int_{-\pi}^{\pi}\ln(1+\exp(-2\cos k + \mu_0 H - A/T))dk\right\}. \qquad (5\cdot1)$$

From this equation we obtain

$$\lim_{H\to 0}\lim_{T\to 0} C/T = \pi/3$$

at $A=2U=0$ and $\mu_0 H=0$. This value differs from $\lim_{U\to 0}\lim_{H\to 0}\lim_{T\to 0} C/T = \pi/6$. One can interpret this discontinuity of the coefficient of T-linear specific heat at $U=0$ as follows. In the half-filled case at $U>0$ one-particle excitation spectrum has a energy gap $-\kappa^{(0)}(\pi) = 2U - 2 + 4\int_0^\infty d\omega\, J_1(\omega)/\omega(1+e^{2U\omega})$. Then this excitation does not contribute to the coefficient of T-linear specific heat. But at $U=0$, gap is zero and this excitation does contribute to the coefficient. In the case $n<1$ one finds no such discontinuity, because both magnon excitation and one-particle excitation contribute to the coefficient of T-linear specific heat.

References

1) M. Takahashi, Prog. Theor. Phys. **47** (1972), 69.
2) E. Lieb and F. Y. Wu, Phys. Rev. Letters **20** (1968), 1445.
3) H. Shiba and P. A. Pincus, Phys. Rev. **B5** (1972), 1966.
4) M. Takahashi, Prog. Theor. Phys. **42** (1969), 1098; **43** (1970), 860, 1619.
5) H. Shiba, Phys. Rev. **B6** (1972), 930.
6) A. A. Ovchinikov, Zhur. Eksp. i Theoret. Fiz. **57** (1969), 2137.
7) M. Takahashi, Prog. Theor. Phys. **50** (1973), 1519, and to be published.

PHYSICAL REVIEW B VOLUME 40, NUMBER 10 1 OCTOBER 1989

Exact solution and thermodynamics of the Hubbard model with infinite-range hopping

Peter van Dongen and Dieter Vollhardt
Institut für Theoretische Physik C, Technische Hochschule Aachen, D-5100 Aachen, Federal Republic of Germany
(Received 2 March 1989)

The Hubbard model with unconstrained hopping of the particles on a lattice is solved exactly. It is shown that in this case the kinetic energy commutes with the interaction part, i.e., the model is essentially trivial. The thermodynamics is worked out explicitly. One finds that the results of the quasichemical approximation for the occupation probability of lattice sites are exact for this model. The ground state is insulating at half-filling and $U>0$ and is conducting otherwise.

The Hubbard model[1] is the simplest model for interacting, itinerant electrons on a lattice. Nevertheless, except for dimension $d=1$,[2] it is still much too difficult to be solved exactly. In this situation it is natural to investigate the model in simple, yet perhaps unrealistic, limits.[3] The most common is that of an infinite-range (rather than zero-range) interaction, which leads to a trivial mean-field theory. Another limit is that of high dimensions, $d\to\infty$,[4] which has only recently been introduced for the Hubbard model,[5] and which may be expected to lead to new insight into the properties of the model in finite dimensions.

In this paper we discuss, and solve exactly, a version of the Hubbard model where the *hopping* of particles, rather than their interaction, has infinite range and occurs with equal probability. This is rather artificial in view of the physics underlying the derivation of the Hubbard model where hopping is assumed to be of short range (tight binding). Nevertheless, since the exact solution of this model has not been available so far,[6] we will discuss it here.

The Hubbard model has the form[1]

$$H=H_0+H_I-\mu\sum_{\mathbf{k},\sigma}n_{\mathbf{k}\sigma}\,, \tag{1a}$$

$$H_0=\sum_{ij\sigma}t_{ij}c_{i\sigma}^{\dagger}c_{j\sigma}=\sum_{\mathbf{k},\sigma}\varepsilon_{\mathbf{k}}n_{\mathbf{k}\sigma}\,, \tag{1b}$$

$$H_I=U\sum_{i}\hat{n}_{i\uparrow}\hat{n}_{i\downarrow}=U\sum_{\mathbf{k}}\rho_{\mathbf{k}\uparrow}\rho_{-\mathbf{k}\downarrow}\,, \tag{1c}$$

where the kinetic energy H_0 and the on-site interaction H_I have been expressed both in position and momentum representation. Here $c_{i\sigma}^{\dagger}$ ($a_{\mathbf{k}\sigma}^{\dagger}$) creates a particle with spin σ at site i (with momentum $\mathbf{k}$), respectively, and

$$\rho_{\mathbf{k}\sigma}=L^{-1/2}\sum_{\mathbf{q}}a_{\mathbf{q}\sigma}^{\dagger}a_{\mathbf{q}+\mathbf{k},\sigma}$$

is the Fourier component of the local density. We assume $n_\uparrow=n_\downarrow=n/2$ with μ as the common chemical potential. While H_0 is diagonal in $\mathbf{k}$ space, H_I is diagonal in position space since it is given by the number operator of doubly occupied sites $D=\sum_i D_i$, with $D_i=n_{i\uparrow}n_{i\downarrow}$. We now assume the hopping to have infinite range, with $t_{ij}\equiv -t$ for all i,j. In this case the actual lattice structure, as well as the dimensionality of the system, becomes unimportant. From (1b) it follows that

$$\varepsilon_{\mathbf{k}}=-tL\delta_{\mathbf{k},0}\,, \tag{2}$$

where L is the number of lattice sites. Clearly, the operator

$$A_\sigma\equiv L^{-1/2}\sum_i c_{i\sigma}=a_{\mathbf{k}=0,\sigma}$$

still obeys Fermi statistics with $A_\sigma^{\dagger}A_\sigma=n_{\mathbf{k}=0,\sigma}$. Note, that — in contrast to naive expectation — the hopping constant must *not* be scaled with L for the average kinetic energy

$$\bar{\varepsilon}_0=\sum_{k<k_F,\sigma}\varepsilon_{\mathbf{k}}=-2tL$$

to remain extensive. Equation (2) implies that the two particles in the state $\mathbf{k}=0$ carry the entire kinetic energy of the system, while states with $\mathbf{k}\neq 0$ do not contribute and are degenerate. The question is then whether $\mathbf{k}=0$ contributes significantly to the interaction terms at all. Below we will show that this is not the case, i.e., that the expectation value of $[H_0,H_I]$ vanishes in the thermodynamic limit, such that H_0,H_I may be diagonalized separately.

To calculate the expectation value of H, (1a), with (2), we introduce the partition function via a time-ordered exponential[7]

$$Z(t,U)=\mathrm{tr}\left[T\exp\left[-\int_0^\beta d\tau H(\tau)\right]\right] \tag{3a}$$

$$=Z_0\left\langle\exp\left[-\int_0^\beta d\tau H_I(\tau)\right]\right\rangle_0 . \tag{3b}$$

Here $H(\tau)$ is given by (1a) with $n_{\mathbf{k}\sigma}=n_{\mathbf{k}\sigma}(\tau)$ and $H_I=H_I(\tau)$, and $\langle\ \rangle_0$ is the thermal average in terms of $Z_0=Z(t,0)$. In principle Z can be calculated from (3b) in a diagrammatic expansion, using Wick's theorem. The grand canonical potential $\Omega=-\beta^{-1}\ln Z$ follows as $\Omega=\Omega_0-\beta^{-1}W_t(\beta,U)$, where $\Omega_0=-\beta^{-1}\ln Z_0$ and

$$W_t(\beta,U)\equiv\left\langle\exp\left[-\int_0^\beta d\tau H_I(\tau)\right]\right\rangle_0^c \tag{4}$$

is obtained from $\langle\ \rangle_0$ by retaining only connected diagrams. The subscript t indicates the explicit dependence of W_t on the hopping constant. The propagators in these diagrams are calculated with respect to Z_0 and have the form

$$\langle Ta_{\mathbf{k}\sigma}(\tau)a^\dagger_{\mathbf{k}'\sigma'}(\tau')\rangle_0=\delta_{\mathbf{k}\mathbf{k}'}\delta_{\sigma\sigma'}G_{\mathbf{k}\sigma}(\tau-\tau') , \tag{5}$$

where

$$G_{\mathbf{k}\sigma}(\tau)=\begin{cases} e^{-(\varepsilon_{\mathbf{k}}-\mu)\tau}(1-f_{\mathbf{k}\sigma}) & (\tau>0) , \\ -e^{-(\varepsilon_{\mathbf{k}}-\mu)\tau}f_{\mathbf{k}\sigma} & (\tau\le 0) \end{cases} \tag{6}$$

with $f_{\mathbf{k}\sigma}=\{\exp[\beta(\varepsilon_{\mathbf{k}}-\mu)]+1\}^{-1}$ as the Fermi function. Equation (6) implies that $G_{\mathbf{k}\sigma}$ is independent of $\mathbf{k}$ for all $\mathbf{k}\neq 0$, while for $\mathbf{k}=0$,

$$G_{\mathbf{k}=0,\sigma}(\tau)=0, \quad \tau<0 \text{ or } 0<\tau<\beta \tag{7}$$

and $G_{0\sigma}(0)=-G_{0\sigma}(\beta)=-1$. If one now calculates W_t, (4), in perturbation theory, one has

$$W_t=\sum_{m=1}^{\infty}\frac{(-1)^m}{m!}\int_0^\beta d\tau_1\cdots\int_0^\beta d\tau_m\langle H_I(\tau_1)\cdots H_I(\tau_m)\rangle_0^c . \tag{8}$$

Clearly, W_t is completely determined by contributions from $\mathbf{k}\neq 0$ since the propagator for $\mathbf{k}=0$ vanishes according to (7). One may therefore just as well put $t=0$ in (4) but nevertheless include the momentum $\mathbf{k}=0$ in the $\mathbf{k}$ sums, since the error introduced thereby is only of relative order $1/L$:

$$W_t=W_{t=0}+O(1/L) .$$

In the thermodynamic limit W_t is therefore seen to be independent of t. Hence the partition function $Z(t,U)$ in (3a) may be expressed in terms of the partition function for $t=0$ as

$$Z(t,U)=Z(0,U)\exp(2\beta tL) \tag{9a}$$

with

$$Z(0,U)=\sum_{\{n_{i\sigma}\}}\exp\left[\beta\mu\sum_{i,\sigma}n_{i\sigma}-\beta U\sum_i D_i\right] . \tag{9b}$$

The Hamiltonian in (9) is site diagonal, such that

$$Z(0,U)=(1+2z+z^2e^{-\beta U})^L , \tag{10}$$

where $z=e^{\beta\mu}$ is the fugacity. In this way one finally finds

$$\Omega=L[-2t-\beta^{-1}\ln(1+2z+z^2e^{-\beta U})] , \tag{11}$$

which is exact in the thermodynamic limit. The contributions from H_0 and H_I are seen to decouple completely, since $\mathbf{k}=0$ (the only relevant state for the kinetic energy) is irrelevant for the interaction energy in the thermodynamic limit. Hence, both parts of the Hamiltonian may be diagonalized individually. In this sense the model is essentially trivial.

All thermodynamic quantities of the model are determined by (11). In particular, the chemical potential μ is obtained from the density

$$n=N/L=-L^{-1}(\partial\Omega/\partial\mu)_\beta$$

as

$$\mu=U+k_BT\ln(\{[(1-n)^2+n(2-n)e^{-\beta U}]^{1/2}-(1-n)\}/(2-n)) , \tag{12}$$

which for $n=1$ reduces to $\mu=U/2$. For $n\le 1$ the density of doubly occupied sites, $\bar d=D/L=\langle H_I\rangle/U$, is found as[8]

$$\bar d(n,\beta U)=\frac{[(n-1)^2+n(2-n)e^{-\beta U}]^{1/2}-1+n(1-e^{-\beta U})}{2(1-e^{-\beta U})} . \tag{13}$$

In particular, for $n=1$

$$\bar d=\frac{1}{2}\frac{1}{1+e^{\beta U/2}} . \tag{14}$$

The parameter $\bar d$ only depends on temperature and interaction via the single parameter βU. Equation (13) may also be written as

$$\frac{\bar d(1-n+\bar d)}{(n/2-\bar d)^2}=e^{-\beta U} . \tag{15}$$

This expression has the form known from the law of mass action, with the Boltzmann factor $e^{-\beta U}$ regulating the equilibrium between the total concentration of doubly occupied sites ($\bar d$) and empty sites ($1-n+\bar d$) relative to that of the singly occupied sites $[(n/2-\bar d)^2]$. Equation (15) is typical for a result obtained within the "quasichemical approximation" in the theory of mixtures.[9] It is interesting to note that (15) is also identical to the result of the Gutzwiller approximation[10] for the conventional Hubbard model, if $e^{-\beta U}$ is replaced by g^2, where g is a variational parameter entering the trial wave function used in the approach. Indeed, this approximation is precisely of the quasichemical type and is known to yield the correct result for expectation values of the ground-state energy, etc., in terms of the trial wave function in $d=\infty$.[5] Equation (14) had already been obtained by Seiler *et al.*[11] within a phenomenological extension of the Gutzwiller approximation[10] to finite temperatures. In the model discussed here (15) is *exact*; hence, the quasichemical approximation, with equilibrium constant $e^{-\beta U}$, is found to be exact in this case. This is not surprising since hopping is completely unconstrained in the model and every site can be reached without restriction.

At $T=0$ (13) implies that $\bar d(U)=0$ for $n\le 1$ and $\bar d(U)=n-1$ for $n>1$, *independent* of the strength of U, i.e., the unrestricted hopping allows the system to assume a ground state with the least possible number of doubly occupied sites. Hence, $\bar d(U)$ has a kink at $n=1$. On the other hand, for $T>0$ one obtains

$$\bar d(U)\simeq\left[\frac{n}{2}\right]^2-\left[\frac{n(2-n)}{4}\right]^2\beta U, \quad \beta U\to 0 \tag{16a}$$

and

$$\bar{d}(U)=\begin{cases}\dfrac{n^2}{4(1-n)}e^{-\beta U}, & n<1,\ \beta U\to\infty\ ,\\ \frac{1}{2}e^{-\beta U/2}, & n=1,\ \beta U\to\infty\ .\end{cases} \tag{16b}$$

Hence for $T>0$, $\bar{d}(U)$ has the form one should expect it to have in the conventional Hubbard model, namely $\bar{d}(0)=(n/2)^2$ and an exponential decrease for $U\to\infty$. Note that the half-filled case ($n=1$) leads to a different dependence on U than that for $n<1$.

The pressure, $P=-\Omega/L$, of the model is given by

$$P=\begin{cases}2t+nk_BT, & n\to 0,\ T\text{ fixed}\\ \ln\left[1-\dfrac{n}{2}\right]^{-2}k_BT, & n\to 2,\ T\text{ fixed}\end{cases} \tag{17}$$

and by

$$P=2t+\tfrac{1}{2}U+\ln 2(k_BT),\quad n=1,\quad T\to 0\ . \tag{18}$$

Hence, P remains finite even for $n\to 0$. This unrealistic dependence is caused by the two particles at $\mathbf{k}=0$, which produce the complete extensive kinetic energy.

The entropy is given by

$$S=k_BL[\beta U\bar{d}-\beta\mu n+\ln(1+2z+z^2e^{-\beta U})]\ . \tag{19}$$

It approaches a *finite* value both for $T\to 0$ (due to the large degeneracy of the ground state) and for $T\to\infty$ (due to the finite number of possible states on a lattice). From the internal energy

$$E=\Omega+TS+\mu N=L(-2t+U\bar{d})$$

the specific heat follows as

$$C_v=-k_BLU\beta^2\left[\frac{\partial\bar{d}}{\partial\beta}\right]_n\ . \tag{20}$$

At low temperatures ($\beta U\to\infty$) it approaches zero exponentially fast

$$C_v=\begin{cases}\dfrac{n^2}{4(1-n)}k_BL(\beta U)^2e^{-\beta U} & (n<1)\ ,\\ \frac{1}{4}k_BL(\beta U)^2e^{-\beta U/2} & (n=1)\ ,\end{cases} \tag{21}$$

while at high temperatures ($\beta U\to 0$) it vanishes as T^{-2} according to

$$C_v=[\tfrac{1}{4}n(2-n)]^2k_BL(\beta U)^2\ . \tag{22}$$

Finally, we remark on the conduction properties of the model. Introducing independent chemical potentials μ_σ, i.e., particle numbers N_σ, for the spins, the criterion of Mattis[12] and Lieb and Wu[2] states that the ground state is insulating if the difference of the energies $E(N_\uparrow,N_\downarrow)$

$$\Delta\mu\equiv\lim_{T\to 0}\left[E\left[\frac{N}{2}+1,\frac{N}{2}\right]+E\left[\frac{N}{2}-1,\frac{N}{2}\right]-2E\left[\frac{N}{2},\frac{N}{2}\right]\right] \tag{23}$$

is positive, and is conducting for $\Delta\mu=0$. In the model investigated here $\Delta\mu$ is entirely determined by $\bar{d}(N_\uparrow,N_\downarrow)$

$$\Delta\mu=UL\lim_{T\to 0}\left[\bar{d}\left[\frac{N}{2},\frac{N}{2}-1\right]+\bar{d}\left[\frac{N}{2}-2,\frac{N}{2}-1\right]-2\bar{d}\left[\frac{N}{2}-1,\frac{N}{2}-1\right]\right]\ . \tag{24}$$

Since $\bar{d}$ has a kink at half-filling [cf. the discussion above Eq. (16a)] it follows that in this case $\Delta\mu>0$ (insulating ground state), while $\Delta\mu=0$ (conducting ground state) for all other fillings. It is interesting to note that the exact ground state of the ordinary Hubbard model in $d=1$ is insulating too at half-filling.[2] However, in the latter case the origin of the insulating behavior is different, being a consequence of the antiferromagnetic correlations by the perfect nesting property of the lattice for $U>0$.

In summary, we have presented the exact solution and thermodynamics of the Hubbard model in the case where hopping of particles may occur to any lattice site with equal transition rate. In this situation the two terms of the Hamiltonian commute, i.e., do not compete with each other, and can thus be diagonalized independently. This is true also for general interactions and kinetic energies, as long as $\varepsilon_{\mathbf{k}}\neq 0$ only for a nonextensive number of $\mathbf{k}$ states. Although the model is, therefore, essentially a trivial one, its solution allows for some interesting conclusions: (i) the result for the interaction energy, i.e., the number of doubly occupied sites $\bar{d}L$, has a form known from the law of mass action with equilibrium constant $e^{-\beta U}$, (ii) in the ground state $\bar{d}(U)=0$ for $n\leq 1$ and arbitrary U, (iii) at any $T>0$ with T fixed $\bar{d}(U)$ has a form that one expects from the conventional Hubbard model, (iv) the ground state is insulating in the half-filled case for all $U>0$ and is conducting for other fillings.

ACKNOWLEDGMENTS

One of us (D.V.) thanks L. Stodolsky, C. Castellani, F. D. M. Haldane, and A. E. Ruckenstein for useful discussions. We are also grateful to F. Gebhard and W. Metzner for many helpful discussions. This work was supported in part by the Sonderforschungsbereich (SFB) 341 of the Deutsche Forschungsgemeinschaft.

[1]M. C. Gutzwiller, Phys. Rev. Lett. **10**, 159 (1963); J. Hubbard, Proc. R. Soc. London, Ser. A **276**, 238 (1963); J. Kanamori, Prog. Theor. Phys. **30**, 275 (1963).

[2]E.H. Lieb and F. Y. Wu, Phys. Rev. Lett. **20**, 1445 (1968).

[3]For an early review of the Hubbard model, see M. Cyrot, Physica **91 B**, 141 (1977).

[4]In this limit and for $U\to\infty$, where the Hubbard model is equivalent to the Heisenberg model, the Néel state is known

to be the exact ground state [T. Kennedy, E. H. Lieb, and B. S. Shastry, Phys. Rev. Lett. **61**, 2582 (1988)].

[5]W. Metzner and D. Vollhardt, Phys. Rev. Lett. **62**, 324 (1989).

[6]The model was considered previously by one of us (D.V.), who also discussed it with participants of the Aspen workshop on "Heavy Fermions and Valence Fluctuations" in 1985; at that occasion F. D. M. Haldane mentioned that he had been looking into the same model independently. At the time no conclusion was reached and no published work resulted. The model itself seems to have been discussed first by J. D. Patterson [Phys. Rev. B **6**, 1041 (1972)] who investigated it numerically for very small systems. We are grateful to K. A. Penson for this information. However, such an approach cannot lead to significant insight since, in view of the importance of the $\mathbf{k}=0$ state, a study of small systems is clearly inadequate.

[7]See, for example, J. W. Negele and H. Orland, *Quantum Many-Particle Systems* (Addison-Wesley, Menlo Park, 1988).

[8]The case $1<n\le 2$ is obtained by particle-hole symmetry as $\bar{d}(2-n)=\bar{d}(n)+1-n$.

[9]E. A. Guggenheim, *Mixtures* (Oxford University, New York, 1952), p. 38.

[10]M. C. Gutzwiller, Phys. Rev. A **137**, 1726 (1965); for a simple exposition see D. Vollhardt, Rev. Mod. Phys. **56**, 99 (1984).

[11]K. Seiler, C. Gros, T. M. Rice, K. Ueda, and D. Vollhardt, J. Low Temp. Phys. **64**, 195 (1986).

[12]D. Mattis (unpublished) (cited in Ref. 2).

Volume 36A, number 2 PHYSICS LETTERS 16 August 1971

GROUND STATE ENERGY OF HUBBARD MODEL

W. D. LANGER *
Niels Bohr Institute, Copenhagen, Denmark
and
D. C. MATTIS ‡
Belfer Graduate School of Science, Yeshiva University, New York, New York 10033, USA

Received 26 June 1971

We find rigorous upper and lower bounds to the ground state energy of the Hubbard-Gutzwiller model.

Several years ago Hubbard [1], in a series of papers, proposed a model for interacting electrons in solids in which the kinetic energy was expressed in terms of nearest neighbor interactions and the Coulomb repulsion in terms of a delta function interaction between electrons of opposite spin on the same site. It is only in one dimension that an exact solution of this model can be found [2,3] and the exact ground state energy has been calculated by Lieb and Wu [4]. (Their result provides an opportunity, lacking in three dimensions, to compare approximate solutions with an exact one). Reference to some of the more recent studies of the ferromagnetic and paramagnetic states in this model can be found listed in Young [5] and Roth [6]. The antiferromagnetic state has also been studied [7,8-10] but has not yielded to solutions much beyond Hartree-Fock. Such approximate solutions, being essentially variational in character, provide upper bounds on the true ground state energy (or for that matter the free energy at finite temperatures). It is one of the objects of this paper to derive a lower bound as well, to the ground state energy of the Hubbard Hamiltonian in an arbitrary number of dimensions. Write the Hamiltonian as

$$H = -\sum_{ij\sigma} T_{ij}\, c^{+}_{i\sigma} c_{j\sigma} + U\sum_{i} n_{i\uparrow} n_{i\downarrow} - \mu \sum_{i\sigma} n_{i\sigma} + \tfrac{1}{2} UN \tag{1}$$

where the sum on i, j is over nearest neighbors and $c^{+}_{i\sigma}$, $c_{i\sigma}$ are the creation and annihilation operators for an electron of spin σ at site i. T_{ij} is the kinetic energy in the band and U is the Coulomb repulsion between the particles on the same site. The chemical potential, μ, is introduced to conserve the number of particles.

Any variational solution to the Hamiltonian will, of course, give an upper bound to the ground state energy. In this paper we will be calculating our results for the important special case of a half filled band where the ground state is an ordered antiferromagnet [7,8-10].

From ref. [7] we derive the following results

$$1 = U(2N)^{-1} \sum_{k} (\epsilon_k^2 + \tfrac{1}{4} U^2 x^2)^{-1/2} \tag{2}$$

$$E_{\mathrm{A}}(U,x) = -N^{-1} \sum_{\text{all } k} (\epsilon_k^2 + \tfrac{1}{4} U^2 x^2)^{1/2} + \tfrac{1}{4} U(1+x^2) \tag{3}$$

where x is the optimal amount of antiferromagnetic ordering, and ϵ_k is the Fourier transform of T_{ij}, i.e. for simple cubic lattices $\epsilon_k \sim [\cos k_x + \cos k_y + \cos k_z]$. Here $E_{\mathrm{A}}(U,x)$ is the variational ground state energy per site. Eqs. (2) and (3) are for a half filled band ($\mu = \frac{1}{2}U$) and the generalization to arbitrary electron density is straightforward [7]. We now derive a lower bound for the Hubbard model, separating the Hamiltonian into two parts $H = H_1 + H_2$. Now if we knew the true eigenfunction of H, call it ψ_0, then we would also have E_0 the true ground state energy. Taking ψ_0 as a trial function for H_1 and H_2, clearly

* NSF - NATO Postdoctoral Fellow.
‡ Research supported by a grant of the United States Air Force, AFOSR 69-1642B.

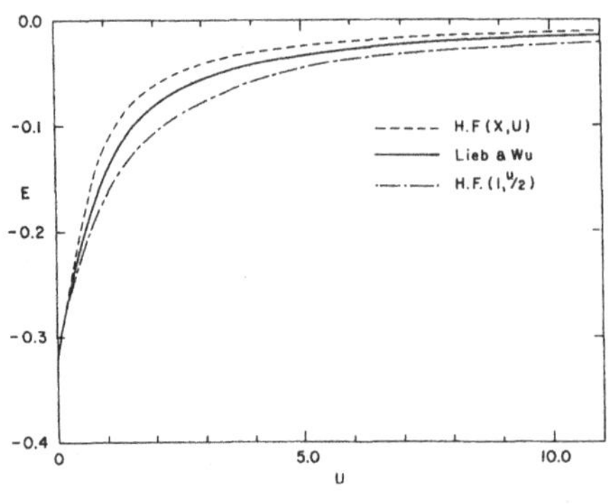

Fig. 1. The true ground state energy for the Hubbard model in one dimension, as given by Lieb and Wu, plotted as function of U in units such that the bandwidth is one. Also shown are our upper and lower bounds.

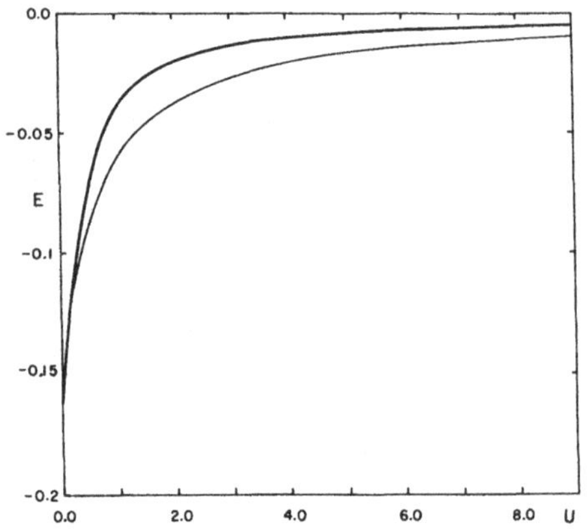

Fig. 2. Upper and lower bounds on the ground state energy of the Hubbard model in a three dimensional simple cubic lattice (bandwidth is the unit of energy).

$$E_0 = (\langle \psi_0 | H_1 | \psi_0 \rangle + \langle \psi_0 | H_2 | \psi_0 \rangle) \geqslant E_1 + E_2 \qquad (4)$$

where E_1 and E_2 are the true ground state energies for H_1 and H_2; it follows that $E_A(U, x) \geqslant$ $\geqslant E_0 \geqslant E_1 + E_2$. If we choose to split the Hamiltonian into spin-up and spin-down parts, with spins-down immobilized at one of the antiferromagnetic sublattices, we can solve for E_1 and E_2 exactly. We set

$$H_1 = -\sum_{ij} T_{ij}\, c_{i\uparrow}^{+} c_{j\uparrow} + \tfrac{1}{2} U \sum_i n_{i\uparrow} n_{i\downarrow} - \mu \sum_i n_{i\uparrow} + \tfrac{1}{2} UN \qquad (5)$$

where $n_{i\downarrow} = \frac{1}{2}(1 + \exp(iQ \cdot R_i))$. H_2 is the same with spins reversed, i.e. $E_1 = E_2$. When this is diagonalized by elementary techniques [11], the lowest eigenvalue turns out to be: $\frac{1}{2} E_A(\frac{1}{2}U, 1)$. Thus we can obtain our upper and lower bounds using only the Hartree-Fock solutions: $E_A(U, x) \geqslant E_0 \geqslant E_A(\frac{1}{2}U, 1)$.

In fig. 1 we have plotted the upper bound (here labelled H.F.(X, U)) and lower bound (labelled H. F. $(1, U/2)$ for 1D as a function of U, in units of the bandwidth; together with the exact ground state energy from Lieb and Wu's result [4]. In 3D the bounds shown in fig. 2 were calculated using the relation $\sum \to \int \rho(\epsilon) d\epsilon$, where, for the density of states, ρ, we used expressions derived by Jelitto [12] for the simple cubic lattice.

One of us, W.L., would like to thank Prof. A. Bohr for his hospitality at the Niels Bohr Institute, and the National Science Foundation for financial support.

References

[1] J. Hubbard, Proc. Roy. Soc. (London), Ser. A276 (1963) 238; 277 (1963) 237 and 281 (1964) 401.
[2] M. Gaudin, Phys. Lett. 24A (1967) 55.
[3] C. N. Yang, Phys. Rev. Lett. 19 (1967) 1312.
[4] E. Lieb and F. Wu, Phys. Rev. Letters 20 (1968) 1445.
[5] W. Young, Phys. Rev. B2 (1970) 167; J. Phys. and Chem. Solids 31 (1970) 865.
[6] L. Roth, Phys. Rev. 184 (1969) 451.
[7] W. Langer, M. Plischke and D. Mattis, Phys. Rev. Letters 23 (1969) 1448, Errata 24 (1970) 635.
[8] J. des Cloizeaux, J. Phys. Radium 20 (1959) 606, 751.
[9] D. R. Penn, Phys. Rev. 142 (1966) 350; M. Cyrot, Phys. Rev. Letts. 25 (1970) 871.
[10] B. Johansson and K. F. Berggren, Phys. Rev. 181 (1969) 855.
[11] D. C. Mattis and W. D. Langer, Phys. Rev. Lett. 25 (1970) 376.
[12] R. Jelitto, J. Phys. Chem. Solids 30 (1969) 609.

* * * * *

VOLUME 62, NUMBER 10 PHYSICAL REVIEW LETTERS 6 MARCH 1989

Two Theorems on the Hubbard Model

Elliott H. Lieb

Departments of Physics and Mathematics, Princeton University, P. O. Box 708, Princeton, New Jersey 08544

(Received 12 December 1988)

In the attractive Hubbard model (and some extended versions of it), the ground state is proved to have spin angular momentum $S=0$ for every (even) electron filling. In the repulsive case, and with a bipartite lattice and a half-filled band, the ground state has $S=\frac{1}{2}||B|-|A||$, where $|B|$ ($|A|$) is the number of sites in the B (A) sublattice. In both cases the ground state is unique. The second theorem confirms an old, unproved conjecture in the $|B|=|A|$ case and yields, with $|B|\neq|A|$, the first provable example of itinerant-electron ferromagnetism. The theorems hold in all dimensions without even the necessity of a periodic lattice structure.

PACS numbers: 75.10.Lp, 71.20.Ad, 74.65.+n

The importance of the Hubbard model of itinerant electrons is increasingly being appreciated. Because of the model's subtlety, rigorous results and exact solutions are clearly useful bench marks, but these are rare. Two theorems about the ground states are stated and proved here. Parts of them are resolutions of old conjectures while other parts are new. In particular, the assertion that certain versions of the model show ferromagnetic behavior for the half-filled band is, I believe, new and yields the first provable example of itinerant-electron ferromagnetism with finite forces and without *ad hoc* assumptions.

After some preliminary definitions, the two theorems are stated. Each is followed by some remarks about their significance. Finally, the proofs are given. The proofs utilize a new kind of reflection positivity which does not involve the usual spatial reflections but rather reflections in spin space. In fact, spatial symmetry plays no role whatsoever and therefore the theorems apply in the widest generality to any collection of sites; all dimensions and topologies are included. The Hubbard model is not known to satisfy any kind of spatial reflection positivity or infrared bounds and this unfortunate fact has prevented the application of the usual proof techniques[1-3] for establishing the existence of long-range order in periodic lattices.

The Hubbard model on a finite lattice Λ is defined by the Hamiltonian

$$H=\sum_{\sigma}\sum_{x,y\in\Lambda} t_{xy}c^\dagger_{x\sigma}c_{y\sigma}+\sum_{x\in\Lambda}U_x n_{x\uparrow}n_{x\downarrow}, \qquad (1)$$

with the following notation. The operators $c_{x\uparrow}$ and $c_{x\downarrow}$ and their adjoints $c^\dagger_{x\sigma}$ satisfy the usual fermion anticommutation relations $\{c^\dagger_{x\sigma},c_{y\tau}\}=\delta_{xy}\delta_{\sigma\tau}$ and $\{c_{x\sigma},c_{y\tau}\}=0$. The hopping matrix elements t_{xy} are required to be real and satisfy $t_{xy}=t_{yx}$, but no other *a priori* assumption is made about them (e.g., the condition $t_{xx}=0$ is not assumed, which means that an x-dependent single-particle potential $\sum_{x\sigma}t_{xx}n_{x\sigma}$ is allowed). The reality of the t_{xy}'s is consistent with their interpretation as overlap matrix elements of real operators in real, localized orbitals. The number operators are $n_{x\sigma}=c^\dagger_{x\sigma}c_{x\sigma}$, while U_x is the on-site energy which, for theorem 1, is allowed to depend on the site x. The word "lattice" is certainly a misnomer because no particular topology (i.e., periodicity or dimensionality) is assumed; the generality assumed here is that Λ is merely a collection of sites. The number of these sites is denoted by $|\Lambda|$. There is said to be a *bond* between sites x and y if $t_{xy}\neq 0$, and Λ is said to be *connected* if there is a connected path of bonds between every pair of sites. Obviously it is no loss in generality to assume that Λ is connected, and this will always be done here. Finally, Λ is said to be *bipartite* if the sites of Λ can be divided into two disjoint sets A and B such that $t_{xy}=0$ whenever $x\in A$ and $y\in A$ or $x\in B$ and $y\in B$. The bipartite condition is really an assertion about the t_{xy}'s and not about Λ, but the terminology is conventional; it implies, in particular, that $t_{xx}=0$ for each $x\in\Lambda$. The symbols $|A|$ and $|B|$ denote the number of sites in A and B in this case, whence $|\Lambda|=|A|+|B|$. The number of electrons is denoted by N; necessarily $N\leq 2|\Lambda|$.

The aim here is to study the ground state (or states if there is more than one) of H for a given N. Of central importance is the total spin S which is a conserved quantity. The spin operators are the quadratic operators

$$S^z=\tfrac{1}{2}\sum_{x\in\Lambda}(n_{x\uparrow}-n_{x\downarrow}),\quad S^+=(S^-)^\dagger=\sum_{x\in\Lambda}c^\dagger_{x\uparrow}c_{x\downarrow}, \qquad (2)$$

and $(S_{\rm op})^2=(S^z)^2+\frac{1}{2}S^+S^-+\frac{1}{2}S^-S^+$, with eigenvalues $S(S+1)$.

Theorem 1 (attractive case).—Assume $U_x\leq 0$ for every x (but U_x is not necessarily constant) and that N is even. No extra assumption about Λ or the t_{xy}'s is made. Then (a) among the ground states of H there is one with spin $S=0$; (b) if $U_x<0$ for every x, the ground state is unique (and hence has $S=0$).

Remarks.—(1) The theorem is "obvious" if all the U_x's are very large, for then the ground state consists of paired electrons on $N/2$ sites of Λ. In the other extreme that each $U_x=0$, theorem 1 is also obvious because one just fills the lowest $N/2$ levels of the $|\Lambda|\times|\Lambda|$ Hermitian hopping matrix $T=\{t_{xy}\}$.

(2) Theorem 1 can be considerably generalized to what is called an *extended Hubbard model*, and more. We can add any *real* operator M to H provided that M satisfies the following two conditions. (A real operator is polynomial in the $c_{x\sigma}$'s and $c_{x\sigma}^{\dagger}$'s with real coefficients.) (i) M commutes with the spin operators S^z and $S^{\pm}$ and conserves both spin-up and spin-down particle numbers; (ii) M can be written as $M^{\uparrow}+M^{\downarrow}-M^{\uparrow\downarrow}$. Here $M^{\uparrow}$ ($M^{\downarrow}$) is real, Hermitian, involves only spin-up (spin-down) operators, and $M^{\uparrow}$ is identical to $M^{\downarrow}$ when the spins are flipped (i.e., $c_{x\uparrow}$ and $c_{x\downarrow}$ are interchanged). The up-down interaction $M^{\uparrow\downarrow}$ can be written as a sum of terms of the form (in which μ merely denotes a summation index) $M^{\uparrow\downarrow}=\sum_{\mu}V_{\mu}^{\uparrow}(V_{\mu}^{\downarrow})^{\dagger}$ in which each V_{μ} operator is real (but not necessarily Hermitian) and involves only operators for one kind of electron. Again, $V_{\mu}^{\uparrow}$ must be the spin reflection of $V_{\mu}^{\downarrow}$ for each μ. The necessary changes in the proof are straightforward (see Ref. 4). It is also easy to extend the proof to some multiband Hubbard models; the details are left to the reader.

Theorem 2 (repulsive case).—Assume $U_x=U=$ positive constant, independent of x. Assume $|\Lambda|$ is even, Λ is bipartite (so that the t_{xy}'s couple only A and B) and $|B|\geq|A|$. No other assumption about Λ or the t_{xy}'s is made. Let $N=|\Lambda|$ (half-filled band). Then the ground state of H is unique [apart from the trivial $(2S+1)$-fold degeneracy] and has spin $S=\frac{1}{2}\times(|B|-|A|)$.

Remarks.—(3) Theorem 2 is considerably more subtle than theorem 1. The assumptions are more stringent. The theorem has long been assumed to be true in the $|B|=|A|$ case ("the half-filled band has spin zero"), but its proof has been elusive.

(4) The fact that $2S=|B|-|A|$ should be no surprise. In the limit $U=0$ we fill the levels of the matrix $T=\{t_{xy}\}$ and one might hastily conclude that S must be zero in this limit. If so, theorem 2 would be contradicted by a continuity argument with respect to U. However, the rank of T is at most $2|A|$ and so T has at least $|\Lambda|-2|A|=|B|-|A|$ zero eigenvalues. The remaining eigenvalues of T come in plus-minus pairs, so that T has at most $|A|$ negative eigenvalues. To achieve $2S=|B|-|A|$, we fill the negative levels twice with opposite spins and place the remaining electrons in the zero levels with a common spin, say, spin up. Thus the ground state is degenerate when $U=0$, but $S=\frac{1}{2}(|B|-|A|)$ is among them. Therefore, there is no contradiction with the continuity argument mentioned above. If, on the other hand, U is very large we know[5] from second-order perturbation theory that H is effectively an isotropic spin-$\frac{1}{2}$ Heisenberg antiferromagnet with Hamiltonian

$$h=(2/U)\sum_{x,y}t_{xy}^2(\mathbf{S}_x\cdot\mathbf{S}_y-\tfrac{1}{4}).$$

For such models it is also known[6] that the ground state is unique and has $2S=|B|-|A|$.

(5) It is easy to construct many regular, periodic lattices in every dimension greater than one with $|B|\neq|A|$. A classic (high-T_c superconductor) two-dimensional example is to start with a square lattice and then intercalate one site in the middle of each bond. The original vertices of the squares are A sites (copper) and the intercalated sites are B sites (oxygen). The half-filled band has three electrons per unit cell and theorem 2 says that then the total net magnetization is $S=\frac{1}{2}(2-1)=\frac{1}{2}$ times the number of unit cells. This was already observed for this lattice in the large-U limit by Mattis.[7]

Whether or not this example is physically realizable is less important than the fact that theorem 2 applied to a periodic lattice with $|B|>|A|$ yields, for the first time, a natural, provable example of an *itinerant-electron model of ferromagnetism.* I use the word ferromagnetism here only in the sense that the spin is extensive, i.e., it is proportional to the number of particles (or cells). Spatial ordering is not implied. A more accurate appelation might be unsaturated ferromagnetism. Still more accurately, ferrimagnetism might be the right word—but technically that word implies a spatial ordering that I am not prepared to prove. In one dimension $|B|=|A|$ by definition and therefore $S=0$; this conclusion coincides with the known result[8] that S is always zero in one dimension with nearest-neighbor nonpositive hopping $t_{xy}\leq 0$ and for *any many-body potential.* There is also the example of Thouless and Nagaoka[9-11] with $N=|\Lambda|-1$, $U=\infty$, and $2S=|\Lambda|-1$, but infinite potentials are crucial for this example.

(6) Theorem 2 also has some extensions similar to some of those described in remark (2) above; hole-particle symmetry is required.

Proof of theorem 1.—S^2 and S^z are conserved and I work in the $S^z=0$ subspace since all competitors have a representative there. That is to say, each eigenstate with a given S value can (by the well known properties of angular momentum) be rotated in spin space to a state with $S_z=0$ without changing its energy. Then there are $n=\frac{1}{2}N$ electrons of each type, spin up and spin down. Let $\{\psi^{\alpha}\}$ be any orthonormal basis for *one species* of n spinless fermions; there are

$$m=\binom{|\Lambda|}{n}$$

of these and I require that they be *real* (i.e., each ψ^{α} is a *real*, homogeneous polynomial of order n in the $c_x^{\dagger}$'s acting on the vacuum). A ground state ψ can then be written as $\psi=\sum_{\alpha,\beta}W_{\alpha\beta}\psi_{\uparrow}^{\alpha}\otimes\psi_{\downarrow}^{\beta}$ with $W_{\alpha\beta}$ as coefficients. This $W_{\alpha\beta}$ is here viewed as a $m\times m$ matrix. Because all operators and basis vectors are real and because the Hamiltonian is symmetric between the up and the down spins, it is obvious that if $W_{\alpha\beta}$ corresponds to a ground state then so does $W_{\beta\alpha}^{*}=(W^{\dagger})_{\alpha\beta}$, and hence (by linearity) so does $W+W^{\dagger}$ and $i(W-W^{\dagger})$. Thus, for convenience, we may henceforth assume $W=W^\dagger$ (but $W_{\alpha\beta}$ is not assumed to be real). The norm of ψ squared is $\langle\psi|\psi\rangle = \sum_{\alpha,\beta}|W_{\alpha\beta}|^2 = \mathrm{Tr}W^2$ and I assume that this is unity. The hopping energy part of $\langle\psi|H|\psi\rangle$ is easily found to be $2\,\mathrm{Tr}KW^2$ where

$$K_{\alpha\beta}=\langle\psi^\beta|\sum_{x,y}t_{xy}c_x^\dagger c_y|\psi^\alpha\rangle .$$

Clearly K is real and symmetric since each t_{xy} is real. The on-site energy is given by $-\sum_x U_x \mathrm{Tr}(WL_xWL_x)$ with $(L_x)_{\alpha\beta}=\langle\psi^\beta|n_x|\psi^\alpha\rangle$, which is also real and symmetric. The total energy is then

$$E(W)=\langle\psi|H|\psi\rangle=2\,\mathrm{Tr}KW^2+\sum_x U_x \mathrm{Tr}(WL_xWL_x)\,, \quad (3)$$

and the equation for W, corresponding to the eigenvalue equation $H\psi=e\psi$, is

$$KW+WK+\sum_x U_xL_xWL_x=eW. \quad (4)$$

Now consider the positive semidefinite matrix $|W|$ defined by $|W|^2=W^2$. Obviously, $\mathrm{Tr}W^2=\mathrm{Tr}|W|^2$. Moreover, in an orthonormal basis (not to be confused with the ψ^α basis for the electrons) in which the Hermitian $m\times m$ matrix W is diagonal, with diagonal elements w_i, the Hermitian matrix $|W|$ is also diagonal with elements $|w_i|$. In this diagonal basis I compute

$$\mathrm{Tr}WL_xWL_x=\sum_{i,j}w_iw_j|(L_x)_{ij}|^2\le\sum_{i,j}|w_i||w_j||(L_x)_{ij}|^2$$
$$=\mathrm{Tr}|W|L_x|W|L_x .$$

Since $U_x\le 0$, I conclude that $E(W)\ge E(|W|)$ and therefore that *among the ground states there is one satisfying* $W=|W|\ge 0$. This is the "spin-space reflection positivity" mentioned in the second paragraph. (This part of the proof is an adaptation of that given in Ref. 4.) I choose this positive (possibly semidefinite) matrix W and now make the choice that the ψ^α's are the natural x-space basis for the electrons, i.e., α denotes n points in Λ and $\psi^\alpha=\prod_{x\in\alpha}c_x^\dagger|0\rangle$. (Some arbitrary convention for the sign of the ψ^α's can be made here.) Since $W\ge 0$, it follows that $W_{\alpha\alpha}>0$ for at least one α, for otherwise W vanishes identically. However, the vector $\phi^\alpha=\psi_\uparrow^\alpha\otimes\psi_\downarrow^\alpha$ satisfies $(S_{\mathrm{op}})^2\phi^\alpha=0$ and therefore this ground state ψ has a nonzero projection onto the eigenspace of $(S_{\mathrm{op}})^2$ in which $S=0$. This would be impossible if conclusion (a) were false and thus conclusion (a) must be true.

To prove conclusion (b) I shall prove that necessarily a Hermitian W satisfies $W=|W|$ or $W=-|W|$ for every ground state when $U_x<0$ for every x. This will prove conclusion (b) for the following reason. If there were two normalized, Hermitian ground state W's, say W^1 and W^2 with $W^1\ne\pm W^2$, then for every real constant d, the Hermitian matrix $W^1+dW^2\equiv W_d$ is not zero and defines (after normalization) a ground state, by virtue of the linearity of the eigenvalue Eq. (4). It is easy to verify that there must be a d for which W_d is neither positive nor negative semidefinite, and this contradicts the assertion that the ground state W_d satisfies $W_d=\pm|W_d|$.

With W given then, consider the Hermitian, positive semidefinite matrix $R\equiv|W|-W$ which is also a multiple of a ground state and satisfies (4). If Q denotes the kernel of R, i.e., $Q=\{$vectors V such that $RV=0\}$, then the assertion $W=\pm|W|$ is implied by the following statement which I shall prove: Q is either just the zero vector or else every vector is in Q. Let V be in Q and take the expectation of (4) in this state V, i.e.,

$$\langle V|KR+RK+\sum_x U_xL_xRL_x|V\rangle=e\langle V|R|V\rangle .$$

Since $RV=0$ and, for all x, $U_x<0$ and $\langle V|L_xRL_x|V\rangle\ge 0$ by the positive semidefiniteness of R, I conclude that $\langle V|L_xRL_x|V\rangle=0$ for all x. Since R is positive semidefinite, I conclude that $RL_xV=0$. Thus each L_x maps Q into Q. Now let the matrices in (4) act on V (without taking expectation values). Since $RL_xV=0$ and $RV=0$, I conclude that $RKV=0$. Thus K also maps Q into Q. As before, let α denote a collection of n points in Λ and define $L^\alpha=\prod_{x\in\alpha}L_x$, which is the projector onto the basis vector μ^α in C^m [with components $(\mu^\alpha)_\gamma=\delta_{\alpha\gamma}$] and which has matrix elements $(L^\alpha)_{\gamma\delta}=\delta_{\alpha\gamma}\delta_{\gamma\delta}$. Note that the L_x's commute with each other and so the ordering of the L_x's is unimportant in the definition of L^α. Each L^α maps Q into Q because each L_x does. Since Λ is connected by T, it is easy to see that the α's are connected by K, i.e., for all α and β there are indices $\gamma_1,\gamma_2,\ldots,\gamma_p$ for some integer p such that the ordinary (not matrix) product $G_{\beta\alpha}\equiv K_{\beta\gamma_1}K_{\gamma_1\gamma_2}\cdots K_{\gamma_p\alpha}$ is not zero. If Q is not just the zero vector and $V\ne 0$ is in Q then $L^\alpha V\ne 0$ for some fixed choice of α (because $\sum_\alpha L^\alpha$ is the identity). Then $L^\alpha V=z\mu^\alpha$ for some nonzero constant z and with μ^α being the aforementioned basis vector. But then the vector $F\equiv L^\beta KL^{\gamma_1}K\cdots L^{\gamma_p}KL^\alpha V$ is in Q and, in fact, $F=zG_{\beta\alpha}\mu^\beta$. In short, Q contains a complete set of vectors (i.e., every μ^β) because $G_{\beta\alpha}$ is nonzero for every β by virtue of the connectivity. Thus every vector is in Q since Q is a linear space, which implies that $W=\pm|W|$ and which, in turn, implies uniqueness of the ground state. Q.E.D.

Proof of theorem 2.—First make the conventional unitary hole-particle transformation for the spin-up electrons followed by a sign change on the B sublattice, i.e., $c_{x\uparrow}\to\epsilon(x)c_{x\uparrow}^\dagger$ and $c_{x\uparrow}^\dagger\to\epsilon(x)c_{x\uparrow}$ with $\epsilon(x)=1$ for $x\in A$, $\epsilon(x)=-1$ for $x\in B$. The spin-down electrons are unaltered, i.e., $c_{x\downarrow}\to c_{x\downarrow}$. Then $n_{x\uparrow}\to 1-n_{x\uparrow}$ and the transformed H is $\tilde H+UN_\downarrow$ with

$$\tilde H=-\sum_\sigma\sum_{x,y\in\Lambda}t_{xy}c_{x\sigma}^\dagger c_{y\sigma}-U\sum_{x\in\Lambda}n_{x\uparrow}n_{x\downarrow}\,, \quad (5)$$

and with $N_\sigma=\sum_x n_{x\sigma}$. The original number operators N_σ transform as $N_\downarrow\to\tilde N_\downarrow=N_\downarrow$ and $N_\uparrow\to\tilde N_\uparrow=|\Lambda|-N_\uparrow$. The original condition $N=|\Lambda|$ becomes $N_\uparrow=N_\downarrow$. The spin operators (2) become the *pseudo-spin operators*

$$\tilde{S}^z = \tfrac{1}{2}(|\Lambda| - N_\uparrow - N_\downarrow), \quad \tilde{S}^+ = \sum_{x \in \Lambda} \epsilon(x) c_{x\uparrow} c_{x\downarrow}, \tag{6}$$

and $S^2 \to (\tilde{S}^z)^2 + \tfrac{1}{2}\tilde{S}^+\tilde{S}^- + \tfrac{1}{2}\tilde{S}^+\tilde{S}^+ \equiv (\tilde{S})^2$. The $\tilde{S}$ operators commute with $\tilde{H}$, but so do the spin operators S given in (2). The S operators are the transforms of the pseudo-spin operators in the original variables and are of no special physical interest. The $\tilde{S}$ operators are the ones of interest as far as $\tilde{H}$ is concerned.

As before, I can work in the $\tilde{N}_\uparrow = \tilde{N}_\downarrow = |\Lambda|/2$ subspace, which implies that $N_\uparrow = N_\downarrow = |\Lambda|/2$. The uniqueness part of theorem 2 is then a consequence of conclusion (b) of theorem 1, which also states that the unique ground state $\tilde{\psi}$ of $\tilde{H}$ has $S = 0$. This last fact is of secondary importance. The real problem is to prove that $2\tilde{S} = |B| - |A|$. The shortest proof is to return to H and the $S^z = 0$ subspace (in the *original variables).* For each $U > 0$ the ground state $\psi(U)$ is nondegenerate as has been shown and I want to prove that $2S = |B| - |A|$. The nondegeneracy of the ground state for *all* $U > 0$ implies that the S of this unique ground state must be independent of U, for otherwise continuity in U would imply a degeneracy for some value of $U > 0$. However, when U is very large, H, as stated before, is equivalent (to leading order in U) to h, the Heisenberg antiferromagnetic Hamiltonian defined in remark (4). As stated there, h *also* has a unique ground state[6] (for $S^z = 0$) and this state has $2S = |B| - |A|$. The uniqueness property of h is crucial for it implies a gap (however small it may be) in the spectrum of h. Thus, for large enough U the S value of the ground state of h is identical to that of H. Q.E.D.

The partial support of the U.S. National Science Foundation Grant NO. PHY85-15288 A02 is gratefully acknowledged.

---

[1]F. J. Dyson, E. H. Lieb, and B. Simon, J. Stat. Phys. **18**, 335 (1978).

[2]J. Fröhlich, R. Israel, E. H. Lieb, and B. Simon, Commun. Math. Phys. **62**, 1 (1978).

[3]J. Frohlich, R. Israel, E. H. Lieb, and B. Simon, J. Stat. Phys. **22**, 297 (1980). The proof of theorem 5.5 on p. 334 is incorrect.

[4]T. Kennedy, E. H. Lieb, and S. Shastry, J. Stat. Phys. (to be published).

[5]P. W. Anderson, Phys. Rev. **115**, 2 (1959).

[6]E. H. Lieb and D. C. Mattis, J. Math. Phys. **3**, 749 (1962). This paper uses a Perron-Frobenius argument to prove that in a certain basis the ground state ψ for the connected case is a positive vector and it is unique. By comparing this ψ with the ψ for a simple soluble model (which is also positive) in the $2S^z = |B| - |A|$ subspace, we concluded $2S \leq |B| - |A|$. However, if the comparison is made in the $S^z = 0$ subspace, the same argument shows $2S = |B| - |A|$. This simple remark was overlooked in the 1962 paper, but it is crucial for the present work.

[7]D. C. Mattis, Phys. Rev. B **38**, 7061 (1988).

[8]E. H. Lieb and D. C. Mattis, Phys. Rev. **125**, 164 (1962).

[9]Y. Nagaoka, Phys. Rev. **147**, 392 (1966).

[10]D. J. Thouless, Proc. Phys. Soc. London **86**, 893 (1965).

[11]E. H. Lieb, in *Phase Transitions, Proceedings of the Fourteenth Solvay Conference* (Wiley-Interscience, New York, 1971), pp. 45–63.

PHYSICAL REVIEW B VOLUME 41, NUMBER 7 1 MARCH 1990

Rigorous bounds on the susceptibilities of the Hubbard model

Kenn Kubo and Tatsuya Kishi
Institute of Physics, University of Tsukuba, Ibaraki 305, Japan
(Received 7 November 1989)

Rigorous bounds on the susceptibilities of the single-band Hubbard model which hold in all dimensions are presented. In the attractive model the spin susceptibility is bounded above by $(4|U|)^{-1}$ where $U(<0)$ is the on-site interaction potential. In the half-filled repulsive model on a bipartite lattice the charge and the on-site pairing susceptibilities are bounded above by U^{-1}. The present result implies that the susceptibilities never diverge in the above-mentioned circumstances and also the absence of corresponding long-range order.

The single-band Hubbard model is an important model in the solid-state theory as the model is thought to involve the essential features of interacting electrons in solids. It is widely believed that the model simulates many interesting phenomena such as ferromagnetism, antiferromagnetism, and metal-insulator transition in appropriate circumstances.[1] Quite recently the possibility of superconductivity in this model has attracted intense interest.[2] In spite of its apparent simplicity and long and intense effort for its understanding, much is left to be clarified except for the one-dimensional case where the exact solution was obtained by Lieb and Wu.[3] According to the exact solution the half-filled repulsive model has a gap in the one-particle excitation spectrum[3] and the charge susceptibility vanishes at zero temperature.[4] In the attractive model electrons form singlet bound states which lead to a gap in the excitation spectrum[3,5] and the vanishing spin susceptibility at $T=0$.[5] In two or three dimensions, on the other hand, the ground state of the half-filled repulsive model is widely believed to have the antiferromagnetic long-range order (AFLRO).[6] The existence of an energy gap is expected in this case as well as in the attractive model. No rigorous result, however, on the existence of AFLRO or energy gap or on the susceptibilities at zero temperature has been known so far.

In this paper we report rigorous upper bounds for some susceptibilities which hold in all dimensions. The first result assures that the spin susceptibility is bounded by $(4|U|)^{-1}$ in the attractive model where $U(<0)$ is the on-site interaction potential. The second result states that the on-site pairing and the charge susceptibilities are bounded by U^{-1} in the half-filled repulsive $(U>0)$ model on a bipartite lattice. As is well known, the Fermi-liquid theory predicts finite spin and charge susceptibilities at $T=0$ where T denotes temperature.[7] On the other hand an exponential decrease of susceptibilities with temperature is expected from the existence of an energy gap. Unfortunately, obtained bounds, which are independent of T, are too loose to say anything on the interesting question whether the model simulates a Fermi liquid or not. Obtained bounds, however, lead to the conclusion that no phase transition leading to corresponding long-range order occurs in above-mentioned circumstances.

Quite recently Lieb proved theorems on the spin state of the ground state of the Hubbard model which hold in all dimensions.[8] According to his result the attractive model with an even number of electrons has a singlet ground state as well as the half-filled repulsive model on a bipartite lattice with the same number of sites in each sublattice. Our result and method are closely related with Lieb, although his argument is confined to the ground state and ours is concerned with finite-temperature properties.

We consider the following Hamiltonian on a finite lattice Λ with $|\Lambda|$ sites

$$H=\sum_{\sigma}\sum_{\alpha\beta\in\Lambda} t_{\alpha\beta}C^{\dagger}_{\alpha\sigma}C_{\beta\sigma}+U\sum_{\alpha\in\Lambda} n_{\alpha\uparrow}n_{\alpha\downarrow}-\sum_{\sigma}\sum_{\alpha\in\Lambda}\mu_{\alpha\sigma}n_{\alpha\sigma}, \tag{1}$$

where $C_{\alpha\sigma}(C^{\dagger}_{\alpha\sigma})$ is the annihilation (creation) operator of an electron with spin σ at the lattice site α and $n_{\alpha\sigma}=C^{\dagger}_{\alpha\sigma}C_{\alpha\sigma}$. The hopping matrix elements $t_{\alpha\beta}$ are assumed to be real and satisfy $t_{\alpha\beta}=t_{\beta\alpha}$. We consider the thermal average over the grand canonical ensemble in the following and the chemical potentials $\mu_{\alpha\sigma}$ are included in H. The spin operators are represented by fermion operators as $S^z_\alpha=(n_{\alpha\uparrow}-n_{\alpha\downarrow})/2$, $S^+_\alpha=C^{\dagger}_{\alpha\uparrow}C_{\alpha\downarrow}$, and $S^-_\alpha=C^{\dagger}_{\alpha\downarrow}\times C_{\alpha\uparrow}$. The density and the on-site pairing operators are given by $n_\alpha=n_{\alpha\uparrow}+n_{\alpha\downarrow}$ and $p_\alpha=C^{\dagger}_{\alpha\uparrow}C^{\dagger}_{\alpha\downarrow}$, respectively. The spin susceptibility with the wave vector $\mathbf{q}$ is given by

$$\chi_{\mathbf{q}}=\beta(S^z_{\mathbf{q}},S^z_{-\mathbf{q}}), \tag{2}$$

with $S^z_{\mathbf{q}}=|\Lambda|^{-1/2}\sum_\alpha S^z_\alpha e^{-i\mathbf{q}\cdot\alpha}$ and (A,B) denotes the Duhamel two-point function

$$(A,B)=\int_0^1 dx\langle e^{\beta xH}Ae^{-\beta Hx}B\rangle .$$

The thermal average is a grand canonical one, i.e., $\langle A\rangle=\Xi^{-1}\mathrm{Tr}(e^{-\beta H}A)$, $\Xi=\mathrm{Tr}(e^{-\beta H})$, where the trace operation is done over the Fock space of electrons on the finite lattice Λ. First we consider the attractive model.

Theorem 1. Assume U is negative and $\mu_{\alpha\sigma}=\mu_\alpha$ (no external magnetic field). No extra assumptions are necessary. Then we can bound the spin susceptibility as

$$\chi_{\mathbf{q}}\leqq(4|U|)^{-1}. \tag{3}$$

Remark 1. Inequality (3) implies the absence of divergence of the spin susceptibility. Using the Falk-Bruch inequality[9] and (3) we obtain an upper bound to the correlation function $\langle S_q^z S_{-q}^z \rangle$ as

$$\langle S_q^z S_{-q}^z \rangle \leq \tfrac{1}{4} C_q^{1/2} |U|^{-1/2} \coth(\beta C_q^{1/2} |U|^{1/2}) , \quad (4)$$

where C_q is any function of **q** which has only to satisfy $C_q \geq \langle (S_q^z, [H, S_{-q}^z]) \rangle$.[10] If we assume the translational and inversion symmetry of the Hamiltonian, we can adopt, for example, $C_q = (2|\Lambda|)^{-1} \sum_k |2E_k - E_{k-q} - E_{k+q}|$, where $E_k = \sum_\beta t_{\alpha\beta} e^{ik\cdot(\alpha-\beta)}$. As a result $\langle S_q^z S_{-q}^z \rangle$ remains bounded when $|\Lambda| \to \infty$ and therefore there is no magnetic long-range order at any temperature. For the uniform spin correlation function we have $\lim_{q\to 0} \langle S_q^z S_{-q}^z \rangle \leq (4\beta|U|)^{-1}$.

Next we consider the repulsive model on a bipartite lattice.

Theorem 2. Assume U is positive and $\mu_{\alpha\sigma} = U/2$. The lattice Λ is assumed to be bipartite, i.e., $t_{\alpha\beta} \neq 0$ only for pairs of α and β belonging to different sublattices (A and B). Then the charge and the on-site pairing susceptibilities satisfy the following inequalities:

$$\beta(\delta n_q, \delta n_{-q}) \leq U^{-1} \quad (5)$$

and

$$\beta(p_q, p_{-q}) \leq U^{-1} , \quad (6)$$

where $\delta n_q = n_q - \langle n_q \rangle$ and $n_q(p_q)$ is the Fourier transform of $n_\alpha(p_\alpha)$ with wave vector **q**.

Remarks 2. Under the above assumptions the unitary transformation $C_{\alpha\sigma}(C_{\alpha\sigma}^\dagger) \to \eta_\alpha C_{\alpha\sigma}^\dagger(\eta_\alpha C_{\alpha\sigma})$, where $\eta_\alpha = 1$ for $\alpha \in A$ and -1 for $\alpha \in B$, does not change the Hamiltonian.[3] As $n_{\alpha\sigma}$ is transformed to $1 - n_{\alpha\sigma}$, $\langle n_{\alpha\sigma} \rangle = \frac{1}{2}$ is deduced, i.e., the system is half-filled.

Remarks 3. According to (5) and (6) the charge and the on-site pairing susceptibility do not diverge at a finite temperature. Also by using the Falk-Bruch inequality we can prove the absence of corresponding long-range order at any temperature. We conclude, therefore, no phase transition leading to charge-density wave or on-site Cooper pairing occurs in the half-filled repulsive model on a bipartite lattice.

Proof of Theorem 1. Proof follows straightforwardly that of the Gaussian domination in the quantum spin systems.[10] Define for a set of real numbers $\{h_\alpha\}$,

$$\Xi(\{h_\alpha\}) = \mathrm{Tr}\left[\exp\left\{K_\uparrow + K_\downarrow - \tfrac{1}{2}\beta|U| \sum_{\alpha\in\Lambda} (n_{\alpha\uparrow} - n_{\alpha\downarrow} - h_\alpha)^2\right\}\right] , \quad (7)$$

where

$$K_\sigma = -\beta \sum_{\alpha\beta\in\Lambda} t_{\alpha\beta} C_{\alpha\sigma}^\dagger C_{\beta\sigma} + \beta \sum_{\alpha\in\Lambda} [\mu_{\alpha\sigma} - (U/2)] n_{\alpha\sigma} .$$

We leave $\mu_{\alpha\sigma}$ dependent on σ. Using the Trotter formula we rewrite as $\Xi(\{h_\alpha\}) = \lim_{n\to\infty} a_n$, where

$$a_n = \mathrm{Tr}\left\{\left[\exp(K_\uparrow/n)\exp(K_\downarrow/n) \prod_{\alpha\in\Lambda} \exp\left(-\frac{\beta U}{2n} \sum_{\alpha\in\Lambda} (n_{\alpha\uparrow} - n_{\alpha\downarrow} - h_\alpha)^2\right)\right]^n\right\} .$$

The operator identity $\exp(-A^2) = (4\pi)^{-1/2} \int \exp(ikA)\exp(-k^2/4) dk$ leads to

$$a_n = (4\pi)^{-n|\Lambda|/2} \int d^{n|\Lambda|}k \exp\left(-\sum_{i=1}^{n} \sum_\alpha (k_{\alpha,i}^2/4 + i\epsilon_n k_{\alpha,i} h_\alpha)\right) \beta_{n\uparrow} \beta_{n\downarrow}^* , \quad (8)$$

where $\beta_{n\sigma}$ is the trace of a product of $n(|\Lambda|+1)$ operators as

$$\beta_{n\sigma} = \mathrm{Tr}_\sigma\left[\prod_{i=1}^{n} \left\{\exp(K_\sigma/n) \prod_\alpha \exp(i\epsilon_n k_{\alpha,i} n_{\alpha\sigma})\right\}\right] ,$$

$\epsilon_n = (\beta|U|/2n)^{1/2}$ and Tr_σ denotes the trace operation over the Fock space of σ-spin electrons on the lattice Λ. Here we have made use of the fact that K_σ and $n_{\alpha\sigma}$ operate as the identity operator in the Fock space of $-\sigma$-spin electrons and are represented by real matrices since $t_{\alpha\beta}$ is real. Applying Schwarz inequality to (8) and the operator identity in the reversed way and noting the Fock spaces for up and down spins are identical, we have $|a_n|^2 \leq \delta_{n\uparrow}\delta_{n\downarrow}$, where

$$\delta_{n\sigma} = \mathrm{Tr}\left\{\left[\exp(K_\sigma/n)\exp(K'_{-\sigma}/n) \prod_{\alpha\in\Lambda} \exp\left(-\frac{\beta U}{2n} \sum_{\alpha\in\Lambda} (n_{\alpha\uparrow} - n_{\alpha\downarrow})^2\right)\right]^n\right\}$$

and K'_σ is given by replacing $\mu_{\alpha\sigma}$ with $\mu_{\alpha-\sigma}$ in K_σ as

$$K'_\sigma = -\beta \sum_{\alpha\beta\in\Lambda} t_{\alpha\beta} C_{\alpha\sigma}^\dagger C_{\beta\sigma} + \beta \sum_{\alpha\in\Lambda} [\mu_{\alpha-\sigma} - (U/2)] n_{\alpha\sigma} .$$

Taking the limit $n \to \infty$ we obtain

$$\Xi(\{h_\alpha\})^2 \leq \Xi_\uparrow \Xi_\downarrow , \quad (9)$$

where

$$\Xi_\sigma = \mathrm{Tr}\left[\exp\left\{K_\sigma + K'_{-\sigma} - \tfrac{1}{2}\beta|U| \sum_{\alpha\in\Lambda} (n_{\alpha\uparrow} - n_{\alpha\downarrow})^2\right\}\right] .$$

For $\mu_{\alpha\uparrow} = \mu_{\alpha\downarrow} = \mu$, we have $\Xi(\{h_\alpha\}) \leq \Xi$ for any real $\{h_\alpha\}$. Expanding $\Xi(\{h_\alpha\})$ up to second order in $\{h_\alpha\}$ and extending the result to complex $\{h_\alpha\}$ using the relation (A,B)

$=(B,A)$, we have

$$\left[\sum_\alpha h_\alpha S^z_\alpha, \sum_\alpha h^*_\alpha S^z_\alpha\right] \leqq (4\beta|U|)^{-1} \sum_\alpha |h_\alpha|^2 . \qquad (10)$$

If we choose $h_\alpha = |\Lambda|^{-1/2} e^{-i\mathbf{q}\cdot\alpha}$ we obtain inequality (3), QED.

Proof of Theorem 2. We make use of a unitary transformation which transforms the repulsive model to an attractive one, i.e., $C_{\alpha\downarrow}(C^\dagger_{\alpha\downarrow}) \to \eta_\alpha C^\dagger_{\alpha\downarrow}(\eta_\alpha C_{\alpha\downarrow})$, while the spin-up operators are not altered.[3,8] Then the Hamiltonian (1) is transformed to $\tilde{H}$ where

$$\tilde{H} = \sum_\sigma \sum_{\alpha\beta\in\Lambda} t_{\alpha\beta} C^\dagger_{\alpha\sigma} C_{\beta\sigma} - U \sum_{\alpha\in\Lambda} n_{\alpha\uparrow} n_{\alpha\downarrow} - \sum_{\alpha\in\Lambda} [(\mu_{\alpha\uparrow} - U) n_{\alpha\uparrow} - \mu_{\alpha\downarrow} n_{\alpha\downarrow}] - \sum_{\alpha\in\Lambda} \mu_{\alpha\downarrow} .$$

If $U > 0$ and $\mu_{\alpha\uparrow} = \mu_{\alpha\downarrow} = U/2$, $\tilde{H}$ is the attractive model with the chemical potential $-U/2$, for which Theorem 1 holds. As n_α is transformed to $1 + 2S^z_\alpha$ the charge susceptibility $\beta(\delta n_{\mathbf{q}}, \delta n_{-\mathbf{q}})$ is transformed to $4\beta(S^z_{\mathbf{q}}, S^z_{-\mathbf{q}})$ of $\tilde{H}$ and we obtain inequality (5). On the other hand, the pairing operator p_α is transformed to $\eta_\alpha S^+_\alpha$. The rotational invariance of $\tilde{H}$ in the spin space and Theorem 1 immediately leads to inequality (6), QED.

Remarks 4. It is obvious that the proofs for Theorems 1 and 2 can be generalized to the cases where the on-site interaction potential U varies with site on the lattice. In those cases inequality (10) is modified and we cannot necessarily express the result in a compact form with a single wave vector $\mathbf{q}$.

Remarks 5. Putting $h_\alpha \equiv 0$ and $\mu_{\alpha\sigma} = \mu + \epsilon_\sigma B$ [$\epsilon_\sigma = 1(-1)$ for $\sigma = \uparrow(\downarrow)$] in inequality (9) we can immediately prove $\langle(\sum_\alpha S^z_\alpha)^2\rangle \leqq \langle(\sum_\alpha \delta n_\alpha)^2\rangle/4$ for $U < 0$, i.e., the spin fluctuation is suppressed compared to the charge fluctuation in the attractive model. Also the one-site relation $(S^z_\alpha, S^z_\alpha) \leqq (\delta n_\alpha, \delta n_\alpha)/4$ holds for $U < 0$. Inequalities in the opposite direction hold in the half-filled repulsive model on a bipartite lattice.

We thank S. Takada for helpful advice and encouragement. We are indebted to T. A. Kaplan and M. Takahashi for useful discussions.

[1]See, for example, papers in *Electron Correlation and Magnetism in Narrow-Band Systems,* edited by T. Moriya (Springer-Verlag, Berlin, 1981), and references therein.

[2]For example, C. Gros, R. Joynt, and T. M. Rice, Z. Phys. B **68**, 425 (1987); H. Q. Lin, J. E. Hirsch, and D. J. Scalapino, Phys. Rev. B **37**, 7359 (1988); H. Shimahara and S. Takada, J. Phys. Soc. Jpn. **57**, 1044 (1987); M. Imada, *ibid.* **56**, 3793 (1987); **57**, 3128 (1988); S. R. White, D. J. Scalapino, R. L. Sugar, N. E. Bickeres, and R. T. Scalettar, Phys. Rev. B **39**, 839 (1989).

[3]E. H. Lieb and F. Y. Wu, Phys. Rev. Lett. **20**, 1445 (1968).

[4]T. Usuki, N. Kawakami, and A. Okiji, Phys. Lett. A **135**, 476 (1989); N. Kawakami, T. Usuki, and A. Okiji, *ibid.* **137**, 287 (1989).

[5]M. Takahashi, Prog. Theor. Phys. **42**, 1098 (1969); **43**, 1619 (1970); **44**, 348 (1970); **45**, 756 (1971); **52**, 103 (1974).

[6]K. Kubo and M. Uchinami, Prog. Theor. Phys. **54**, 1289 (1975); J. E. Hirsch, Phys. Rev. B **31**, 4403 (1985).

[7]L. D. Landau, Zh. Eksp. Teor. Fiz. **30**, 1058 (1957) [Sov. Phys. JETP **3**, 920 (1957)].

[8]E. H. Lieb, Phys. Rev. Lett. **62**, 1201 (1989).

[9]H. Falk and L. W. Bruch, Phys. Rev. **180**, 442 (1969).

[10]F. J. Dyson, E. H. Lieb, and B. Simon, J. Stat. Phys. **18**, 335 (1978).

Chapter 3

APPROXIMATE RESULTS IN THE STRONG-COUPLING LIMIT

The failure of the attempts in finding exact statistical mechanics solutions to the model in dimensions greater than one has stimulated the growth of several approximate methods. The latter, being intended to describe correctly the behavior of the model in some specific physical situation, of course have in the exact solution in one dimension a crucial test.

The ratio $U/|t|$ of the two parameters entering the Hubbard hamiltonian (1.1) is the natural quantity to consider in the different physical regions. The so-called strong coupling limit is characterized by $U/|t| \gg 1$, and is the most appropriate in dealing with strongly correlated electron systems.

Within this framework, standard many-body techniques can be applied[1.2], but two analytical approaches turned out to be particularly successful. The Gutzwiller variational approach[3.1], on the one hand gives an interesting variational trial wave function for the ground state, and by the use of the so-called Gutzwiller approximation (which, incidentally, has been proved to be exact only in the limit of infinite D[4.2]) extrapolates weak-coupling results to the strong-coupling region. On the other hand, the auxiliary field (or slave boson) approach[r3.1] consists in enlarging the Fock space at each site by adding a set of virtual bosons. After suitable constraints have been imposed — the virtual boson acts as projection operator onto the singly occupied and empty electronic states, as the doubly occupied states are forbidden in the $U = \infty$ limit. A reformulation of the latter approach was proposed in [3.2], designed to describe also the finite U regime of the Hubbard model, and was shown to reproduce some results originally derived within the Gutzwiller approximation scheme, as well as other types of mean field solutions[r3.2].

In the strong coupling limit, and with the above analytical methods, the two questions mentioned in the previous chapters, namely the metal-insulator transition and the magnetic behavior of the model, have received partial answer.

As for the Mott transition, the question was approached by showing first that at half-filling[r1.1],[3.3], in the limit of large U and to second order in t/U, H reduces to an antiferromagnetic Heisenberg hamiltonian

$$\mathcal{H} = -J \sum_{\mathbf{i},\mathbf{j}} \mathbf{S}_\mathbf{i} \cdot \mathbf{S}_\mathbf{j} \quad , \tag{3.1}$$

$J = t^2/U$ being the exchange coupling, and $\mathbf{S}_\mathbf{i}$ the spin at site $\mathbf{i}$. The ground state is therefore insulating. Brinkmann and Rice[3.4], by use of the Gutzwiller variational approach, were then able to determine, within the paramagnetic phase, the value of the ratio t/U at which the metal-insulator transition occurs, depending on the dimension D. The same value was successively obtained via the slave-boson technique[3.2]. It is worth recalling that the exact one-dimensional solution exhibits Mott transition at half-filling already at $U = 0$. For a less than half-filled band, the Gutzwiller wavefunction is always metallic, being of no use in studying the possible metal-insulator transition.

At half-filling and for large U, hamiltonian (3.1) should of course describe also the magnetic properties of the Hubbard model. Unfortunately, the ground state for the Heisenberg model is known exactly only for $D = 1$[r3.3], and in general the ground state of a quantum antiferromagnet can be highly non-trivial. For a less than half-filled band, it is possible to show that, following the derivation which leads to (3.1), to the localized spin degrees of freedom described by $\mathcal{H}$ is added a hopping term, responsible for the itinerant degrees of freedom of the Hubbard model[3.5] (t–J model). The hole motion now disturbs the spin background, and could destroy the long-range antiferromagnetic order. This possibility has first been studied for $U = \infty$ by Nagaoka[3.6]. He rigorously proved that the ground state of an Hubbard-like system of electrons with filling $\delta = \frac{1}{2}(1 - \frac{1}{N})$, N being the number of sites in Λ, is ferromagnetically ordered, as the superexchange term in (3.1) is vanishing with respect to the hopping term. The next question is then whether the Nagaoka's ferromagnetic state is stable or not with respect to the number of extra holes created, and if such state still exists at finite U. In fact, Nagaoka's theorem holds only for a single hole, and for $U \gg N|t|$, the latter condition being very delicate to fulfill in the thermodynamic limit. Besides, Kanamori[4.4] proved that for a large enough concentration of holes the ground state is instead paramagnetic.

Recently, a number of studies[r3.4],[3.7],[3.8],[5.6] have been focused on the above problem, i.e. hole motion in a quantum antiferromagnet, mainly aiming to construct a variational Gutzwiller-like wavefunction — describing typically one spin down in a sea of spin up's — for the ground state, which gives a lower energy than the ferromagnetic state. The question is of relevant physical meaning, in that one would like to understand if there may exist some coupling between holes in an otherwise half-filled band and magnetic degrees of freedom, which could be responsible for the high-T_c superconducting properties of some materials.

While it is clear in one dimension that the ground state describes a gas of spinless fermions, hence is never ferromagnetic for any filling[r1.2], it can be seen that in $D = 2, 3$, for $U = \infty$, the ferromagnetic state becomes unstable only for a large enough concentration of holes[3.7],[r3.4]. This conclusion is somewhat in disagreement with results obtained by exact diagonalization on very small clusters[r3.5], which would suggest that the Nagaoka's state is singular, being unstable for even two holes. However it has been pointed out[3.8] that, especially close to half-filling, the Gutzwiller wavefunction is a poor representation of the physics of a single spin flip.

Both the Mott transition, and the discussion on the stability of Nagaoka's state, at least in one dimension seem to be well understood. Nevertheless the knowledge of the physics underlying the model is far from complete even for $D = 1$. Indeed, the strong coupling limit allows also in this case to obtain some further results, as the ground state energy[3.9] or the partition function[3.10] for any filling; however many relevant quantities requiring the knowledge of the wavefunction, remain to be studied even in this limit. For example, for the discussion of the momentum distribution see [r3.6].

[r3.1] S.E. Barnes, *J. Phys.* **F6**, 1375 (1976) and **7**, 2637 (1977); N. Read and D. Newns, *J. Phys.* **C16**, 3273 (1983); P. Coleman *Phys. Rev.* **B21**, 3035 (1984).

[r3.2] P. Wolfe, *Int. J. Mod. Phys.* **B3**, 1833 (1989).

[r3.3] D.C. Mattis, *The Theory of Magnetism I* (Springer, Berlin, 1988).

[r3.4] A.E. Ruckenstein and S. Schmitt-Rink, *Int J. Mod. Phys.* **B3**, 1810 (1989).

[r3.5] M. Takahashi, *J. Phys. Soc. Jpn.* **51**, 3475 (1982); Y. Fang, A.E. Ruckenstein, E. Dagotto, and S. Schmitt-Rink, *Phys. Rev.* **B40**, 7406 (1989).

[r3.6] M. Ogata and S. Shiba, *Phys. Rev.* **B41**, 2326 (1990).

Reprinted from THE PHYSICAL REVIEW, Vol. 137, No. 6A, A1726–A1735, 15 March 1965
Printed in U. S. A.

Correlation of Electrons in a Narrow s Band

MARTIN C. GUTZWILLER
IBM Watson Laboratory, Columbia University, New York, New York
(Received 22 October 1964)

The ground-state wave function for the electrons in a narrow s band is investigated for arbitrary density of electrons and arbitrary strength of interaction. An approximation is proposed which limits all the calculations to counting certain types of configurations and attaching the proper weights. The expectation values of the one-particle and two-particle density matrix are computed for the ferromagnetic and for the nonferromagnetic case. The ground-state energy is obtained under the assumption that only the intra-atomic Coulomb interaction is of importance. Ferromagnetism is found to occur if the density of states is large at the band edges rather than in the center, and if the intra-atomic Coulomb repulsion is sufficiently strong. The relation of this approximation to certain exact results for one-dimensional models is discussed.

INTRODUCTION

In order to understand transition metals as well as certain insulating crystals with unfilled d shells a ground-state wave function has to be found for the situation which was described particularly by Van Vleck.[1] Its main feature is the localization of the $3d$ wave functions around the nuclei. The Hamiltonian becomes, therefore, different from the one commonly used for metals. The single-particle terms correspond more closely to a tight-binding picture and the two-particle terms describe a very short-range interaction rather than the long-range Coulomb interaction which is studied in the ordinary theory of electron correlation in metals. The discussion of such a Hamiltonian has been taken up recently by a number of authors,[2] each using a different approach and obtaining qualitatively different results. These results are mainly concerned with the occupation probabilities for the electrons in reciprocal space, and with a comparison of the ground-state energies for a ferromagnetic and for a paramagnetic ground state.

The present work attempts to break away from the limitation of low particle density which was inherent in the diagrammatic analysis of GI and GII,[2] although the wave function which was proposed therein is believed to be a good approximation at all densities. The success of the present attempt depends on a relatively simple proposition to which this paper is dedicated and which seems to hold at least in the case of a narrow s band with a strong intra-atomic Coulomb repulsion (cf. GI and HI). This proposition states the following: The main features of the behavior of spin-up electrons, such as their occupation probability in reciprocal space, can be understood qualitatively if it is assumed that the spin-down electrons occupy a band of zero width; and vice versa.

Unfortunately, it is not clear to what extent this proposition is true, except that the domain of validity is larger than one might expect at first. In the present paper, the main argument in favor of this claim comes from an exact result about the occupation probabilities in reciprocal space for the wave function of GI. This result states that for very strong interaction the occupation probability is constant inside the Fermi surface and depends only on the total number of electrons of opposite spin. A very simple approximate treatment of this result is given, in addition to the exact derivation, in the Appendix, and is generalized to the case of intra-atomic Coulomb repulsion of arbitrary strength. The computation of any specific expectation value depends then directly on the number of contributing configurations and their relative weights.

Section 1 states the problem and defines the terms. Section 2 discusses the manner in which various configurations in the lattice contribute to the various density functions, in order to show the reasonableness of our main proposition. Section 3 gives precise rules for calculating the various expectation values for arbitrary numbers of electrons and arbitrary strength of interaction. These rules are illustrated on some examples. Section 4 applies the method to the computation of the energy and goes on to find a criterion for the occurrence of ferromagnetism. The final section is devoted to a discussion of the present results in the light of some exact results in one dimension.

1. GENERAL WAVE FUNCTION FOR THE NARROW s BAND

The case of one narrow s band is described exactly as in GI. There are L lattice sites numbered by a small latin index f, g, h, or j. To each site there belongs, for a given spin, only one orbital $\varphi(x-g)$ of the Wannier type, i.e.,

$$\int \varphi^*(x-f)\varphi(x-h)dx=\delta_{fh}. \qquad (1)$$

Bloch waves $\psi_k(x)$, with wave vectors called k or l, are constructed by forming

$$\psi_k(x)=L^{-1/2}\sum_g \exp(ikg)\varphi(x-g). \qquad (2)$$

Each of these wave functions is to be multiplied by a

[1] J. H. Van Vleck, Rev. Mod. Phys. **25**, 220 (1953).
[2] Martin C. Gutzwiller, Phys. Rev. Letters **10**, 159 (1963); and Phys. Rev. **134**, A923 (1964), to be referred to as GI and GII; J. Hubbard, Proc. Roy. Soc. (London) **A276**, 238 (1963); **277**, 238 (1964), to be referred to as HI and HII; J. Kanamori, Progr. Theoret. Phys. (Kyoto) **30**, 275 (1963).

spin-wave function, indicated by an arrow, ↑ or ↓, as index. Corresponding to $\varphi(x-g)$, there is a Fermion creation operator $a_g^\dagger$, in the usual manner; and corresponding to $\psi_k(x)$, there is a Fermion creation operator $a_k^\dagger$, with the relation (2) now becoming

$$a_k^\dagger = L^{-1/2}\sum_g \exp(ikg)a_g^\dagger. \tag{3}$$

Let $G=(g_1,\cdots g_m)$ denote a set of lattice sites to be occupied by ↑ particles, and $\Gamma=(\gamma_1,\cdots,\gamma_\mu)$ a set of lattice sites to be occupied by ↓ particles; a configuration $\Phi_{G\Gamma}$ belongs to these two sets, namely,

$$\Phi_{G\Gamma}=\prod_G a_{g\uparrow}^\dagger \prod_\Gamma a_{\gamma\downarrow}^\dagger \Phi_0, \tag{4}$$

where Φ_0 is the vacuum state. The correlated state Ψ is expressed in terms of the amplitudes $A_{G\Gamma}$ by the expansion

$$\Psi=\sum_{G\Gamma} A_{G\Gamma}\Phi_{G\Gamma}, \tag{5}$$

where the summation goes over all different sets G and Γ.

The amplitudes $A_{G\Gamma}$ can be chosen arbitrarily, as real or even complex numbers. In order to compute the energy expectation values for Ψ, only the first- and second-order density functions, ρ_1 and ρ_2, are needed. But, quite generally, it is of interest to examine the nth-order density function ρ_n in terms of its arguments, which we now choose to be the lattice sites and the spin directions, e.g.,

$$\rho_n(h_1\uparrow,\cdots,h_n\uparrow;\, f_1\uparrow,\cdots,f_n\uparrow) = (n!)^{-1}(\Psi|a_{h_1\uparrow}^\dagger\cdots a_{h_n\uparrow}^\dagger a_{f_1\uparrow}\cdots a_{f_n\uparrow}|\Psi). \tag{6}$$

The calculation of the diagonal elements of ρ_n, where the set $(h_1,\cdots,h_n)$ coincides with the set $(f_1,\cdots,f_n)$, involves only the absolute value of the amplitudes $A_{G\Gamma}$. However, in computing the off-diagonal elements of ρ_n, where the set $(h_1,\cdots,h_n)$ does not coincide with the set $(f_1,\cdots,f_n)$, it is necessary to know something about the "relative phases between adjacent configurations." This expression refers to the complex number $\exp(i\alpha)$ of absolute value 1 by which the amplitudes of two configurations differ, if their sets G and Γ coincide, except that one configuration contains $(f_1\uparrow,\cdots,f_n\uparrow)$, where the other contains $(h_1\uparrow,\cdots,h_n\uparrow)$.

The requirement of antisymmetry imposes very strong restrictions on the choice of the relative phases. These restrictions are not easy to enforce. Therefore, it is reasonable to take one particular set of relative phases which has proven its value in a related problem and stick to that set all through the development. The domain of variation for the variational ansatz (5) is thereby drastically reduced, but it is felt that it would lead into too many complications to proceed differently.

The set of phases to be adopted arises from some uncorrelated wave function Φ which is the antisymmetrized product of Bloch functions. In the case of a simple s band, let this uncorrelated function Φ be given by

$$\Phi=\prod_{(k)} a_{k\uparrow}^\dagger \prod_{(\kappa)} a_{\kappa\downarrow}^\dagger\Phi_0. \tag{7}$$

The set (k) is supposed to be the volume contained inside some Fermi surface S, and similarly the set (κ) is the inside of some Fermi surface Σ in reciprocal space. The coefficients in an expansion of the type (5) are given by the product of two determinants[3]:

$$\left(L^{-1/2}e^{ikg}\middle|\begin{matrix}k_1\cdots k_m\\ g_1\cdots g_m\end{matrix}\right)\left(L^{-1/2}e^{i\kappa\gamma}\middle|\begin{matrix}\kappa_1\cdots\kappa_\mu\\ \gamma_1\cdots\gamma_\mu\end{matrix}\right).$$

If the sets (k) and (κ) are chosen in such a way as to include always k and $-k$ simultaneously, and similarly with κ and $-\kappa$, each of the two determinants is easily shown to be a real number times some fixed power of i. This power depends on the order in which k and $-k$ appear in the set $(k_1\cdots k_m)$ and on some other conventions which are irrelevant to the further discussion.

The amplitudes $A_{G\Gamma}$ of the correlated wave function Ψ are now written as

$$A_{G\Gamma}=B_{G\Gamma}\eta^\nu\left(L^{-1/2}e^{ikg}\middle|\begin{matrix}k_1\cdots k_m\\ g_1\cdots g_m\end{matrix}\right)\times\left(L^{-1/2}e^{i\kappa\gamma}\middle|\begin{matrix}\kappa_1\cdots\kappa_\mu\\ \gamma_1\cdots\gamma_\mu\end{matrix}\right), \tag{8}$$

where ν is the number of identical lattice sites among the sets $(g_1,\cdots g_m)$ and $(\gamma_1,\cdots,\gamma_\mu)$. $B_{G\Gamma}$ is assumed to be real and positive. The parameter η gives a weight to each configuration depending on the amount of crowding. Obviously, the uncorrelated wave function Φ is obtained by setting $B=1$ and $\eta=1$. The correlated wave function Ψ of GI is obtained by setting $B=1$ but $\eta<1$. By letting B differ from one, the diagonal and the off-diagonal elements of ρ_n can be given certain simple properties, which may not follow from setting $B=1$.

2. CASE FOR INFINITELY HEAVY SPIN-DOWN ELECTRONS

The probability for finding the electrons in a configuration G and Γ is given by

$$|A_{G\Gamma}|^2=B_{G\Gamma}^2\eta^{2\nu}\left(w(g'-g'')\middle|\begin{matrix}g_1\cdots g_m\\ g_1\cdots g_m\end{matrix}\right)\times\left(\omega(\gamma'-\gamma'')\middle|\begin{matrix}\gamma_1\cdots\gamma_\mu\\ \gamma_1\cdots\gamma_\mu\end{matrix}\right), \tag{9}$$

[3] Determinants are written, where possible, using the following abbreviation:

$$\left(f(x,y)\middle|\begin{matrix}x_1\cdots x_n\\ y_1\cdots y_n\end{matrix}\right)=\begin{vmatrix}f(x_1,y_1)\,f(x_1,y_2)\cdots f(x_1,y_n)\\ f(x_2,y_1)\,f(x_2,y_2)\cdots f(x_2,y_n)\\ \vdots\qquad\vdots\\ f(x_n,y_1)\,f(x_n,y_2)\cdots f(x_n,y_n)\end{vmatrix}.$$

where $w(g'-g'')$ and $\omega(\gamma'-\gamma'')$ are the propagation functions

$$w(h-f)=(1/L)\sum_{(k)}\exp(ikh-ikf),$$
$$\omega(h-f)=L^{-1}\sum_{(\kappa)}\exp(i\kappa h-i\kappa f). \tag{10}$$

Apart from contributing a sign to $A_{G\Gamma}$, the determinants in (8) contribute to the absolute value of the amplitude of the determinants appearing in (9). The latter describe the effect of the so-called exchange hole. Their value is small if the sites $(g_1,\cdots,g_m)$ or $(\gamma_1,\cdots,\gamma_\mu)$ are close to one another.

The correlation tends to magnify this effect, if it is not compensated by a judicious choice of the coefficients $B_{G\Gamma}$. For instance consider the case $B_{G\Gamma}=1$ and $\eta=0$. If Eq. (9) is summed over all sets Γ, it follows that

$$\sum_\Gamma |A_{G\Gamma}|^2=\left(w(g'-g'')\middle|\begin{matrix}g_1\cdots g_m\\ g_1\cdots g_m\end{matrix}\right)\times\left(\delta_{\gamma'\gamma''}-\omega(\gamma'-\gamma'')\middle|\begin{matrix}g_1\cdots g_m\\ g_1\cdots g_m\end{matrix}\right). \tag{11}$$

The second determinant gives simply the probability of not finding any of the ↓ particles in $g_1, \cdots, g_m$. This probability shows exactly the same effect of the exchange hole, in that it is small for a set of sites $(g_1,\cdots,g_m)$ close to one another. The probability of finding ↑ particles at $(g_1,\cdots,g_m)$ independently of the location of the ↓ particles, provided there is no crowding, shows therefore an enhanced effect of the exchange hole.

As long as the Hamiltonian does not contain any Coulomb interaction terms between atomic orbitals of different lattice sites, as happens to be the case for (35), there seems to be no reason *a priori* for reducing this enhanced effect of the exchange hole. Also, the enhanced exchange hole has a simple explanation: Since a ↓ particle cannot share a lattice site with a ↑ particle, there is a tendency to have unlike neighbors (↑ and ↓) rather than like ones (↑ and ↑). Two spin-up electrons tend to stay away from each other. This situation would change, of course, if we introduce some terms into the Hamiltonian which tend to favor parallel alignment of spins on neighboring atoms.

It would seem, however, that this enhanced exchange hole effect has a bad influence on the kinetic energy and crystal potential terms which are determined by $\rho_1(h\uparrow,f\uparrow)$. To make this more evident, compare the expression for $\eta=1$ and $B_{G\Gamma}=1$, viz.,

$$\rho_1(h\uparrow,f\uparrow)=\sum_{(g_2\cdots g_m)}\left(w(g'-g'')\middle|\begin{matrix}hg_2\cdots g_m\\ fg_2\cdots g_m\end{matrix}\right)=w(h-f), \tag{12}$$

with the expression for $\eta=0$, *viz.*,

$$\rho_1(h\uparrow,f\uparrow)=\sum_{(g_2\cdots g_m)}\left(w(g'-g'')\middle|\begin{matrix}hg_2\cdots g_m\\ fg_2\cdots g_m\end{matrix}\right)\times\sum_{\Gamma\neq(fGh)}B_{fG\Gamma}B_{hG\Gamma}\left(\omega(\gamma'-\gamma'')\middle|\begin{matrix}\gamma_1\cdots\gamma_\mu\\ \gamma_1\cdots\gamma_\mu\end{matrix}\right), \tag{13}$$

where the summation over Γ includes only such sets of lattices sites $\gamma_1, \cdots, \gamma_\mu$ which do not coincide with any of the sites $f, g_2, \cdots, g_m, h$. First, observe that the sum over Γ has $\binom{L-m-1}{\mu}$ terms when $h\neq f$, and $\binom{L-m}{\mu}$ when $h=f$. The value of $\rho_1(f\uparrow,f\uparrow)$ is fixed by the normalization to $m/L=w(f-f)$. The value of $\rho_1(h\uparrow,f\uparrow)$ will usually be smaller than $w(h-f)$, if $h\neq f$, because the sum over Γ has fewer terms, namely, by a factor $(L-m-\mu)/(L-m)$, as a consequence of $\eta=0$. The correlation affects the transfer of electrons between lattice sites more than their chance of being at any particular lattice site. Second, note that the sum over Γ in (13) has the least harmful effect on ρ_1 if it does not depend on G, because then the sum over G approximates most closely the formula (12). The coefficients $B_{G\Gamma}$ should, therefore, be chosen so as to compensate at least partially the exchange hole effect which is inherent in the last determinant of (13). In this manner, the particular features of the spin-down electrons are completely smeared out in the sum over Γ, and the specific choice of the coefficients $B_{G\Gamma}$ does not have to be described in detail in order to obtain a reasonable result for the summation over Γ. (For additional discussion of the best values for $B_{G\Gamma}$, cf. Sec. 5.)

It is then conceivable that $\rho_1(h\uparrow,f\uparrow)$ becomes largely independent of the particular choice of the wave vectors κ appearing in ω, i.e., independent of the region in reciprocal space occupied by spin-down electrons, and that a typical value of $\rho_1(h\uparrow,f\uparrow)$ is obtained by putting

$$\sum_{\Gamma\neq G}|B_{G\Gamma}|^2\left(\omega(\gamma'-\gamma'')\middle|\begin{matrix}\gamma_1\cdots\gamma_\mu\\ \gamma_1\cdots\gamma_\mu\end{matrix}\right)=1, \tag{14}$$

$$\sum_{\Gamma\neq(fGh)}B_{fG\Gamma}B_{hG\Gamma}\left(\omega(\gamma'-\gamma'')\middle|\begin{matrix}\gamma_1\cdots\gamma_\mu\\ \gamma_1\cdots\gamma_\mu\end{matrix}\right)=\frac{L-m-\mu}{L-m}. \tag{15}$$

The Eqs. (14) and (15) may be considered, either as giving the average over some sample of the quantity on the left, or as imposing conditions to be satisfied by the coefficients $B_{G\Gamma}$. Clearly, in order to put any of these two possibilities on a more secure basis, one has either to specify the sample to be averaged over or to show that Eqs. (14) and (15) can be solved. Since the author has not been able to satisfy completely any such requirement, the reasonableness of (14) and (15) is confirmed in two quite different ways.

Let us consider the consequences of (14) and (15). If (14) and (15) are inserted into (13), one finds with the help of (12)

$$\rho_1(h\uparrow,f\uparrow)=w(h-f) \qquad \text{for } h=f,$$
$$=[(L-m-\mu)/(L-m)]w(h-f) \text{ for } h\neq f. \quad (16)$$

By taking the Fourier transform of (16), we find the occupation probability in reciprocal space:

$$n_{k\uparrow}=\sum_g e^{-ikg}\rho_1(g\uparrow,0\uparrow)=1-(\mu/L) \qquad \text{for } k\in(k),$$
$$=\mu m/L(L-m) \text{ for } k\notin(k). \quad (17)$$

This piecewise constant function with a discontinuity at the Fermi surface of the uncorrelated electron gas corresponding to the state Φ of (7) is plotted in Fig. 1.

Equation (17) has been written under the assumption that $m+\mu<L$. If one has $m+\mu>L$, however, the whole theory can be written in terms of holes rather than particles, and the formula (17) gives then an expression for the density of holes in reciprocal space where m and μ are the total numbers of spin-up and spin-down holes in the lattice.

Now, formula (17) is also obtained if we put $B_{G\Gamma}=\text{const}$ in (13) and then average the last determinant in (13) over all possible sets (κ). This means that we do not try to compensate for the enhanced exchange hole by letting $B_{G\Gamma}$ vary with the configurations G and Γ, but we average over all possible distributions (κ) or spin-down electrons in reciprocal space. Indeed, one finds without difficulty that

$$\sum_{(\kappa)}\left(\omega(\gamma'-\gamma'')\middle|\begin{matrix}\gamma_1\cdots\gamma_\mu\\ \gamma_1\cdots\gamma_\mu\end{matrix}\right)=\left(L^{-1}\sum_\kappa e^{i\kappa(\gamma'-\gamma'')}\middle|\begin{matrix}\gamma_1\cdots\gamma_\mu\\ \gamma_1\cdots\gamma_\mu\end{matrix}\right)=1, \quad (18)$$

where we have to divide by $L!/(L-\mu)!\mu!$, the number of different sets (κ), in order to obtain the average. All terms in the summation over Γ in (13) are then the same; there are $(L-m)!/\mu!(L-m-\mu)!$ of them for $h=f$, and $(L-m-1)!/\mu!(L-m-\mu-1)!$ of them for $h\neq f$. The summation over Γ in (13) gives therefore $(\text{const})^2(L-m)!(L-\mu)!/L!(L-m-\mu)!$ for $h=f$, $(\text{const})^2(L-m-1)!(L-\mu)!/L!(L-m-\mu-1)!$ for $h\neq f$. After fixing the constant by normalizing $\rho_1(h\uparrow,h\uparrow)$ to m/L and after summing over all lattice sites $g_2,\cdots,g_m$, we find again (16). Thus, formula (17) is the correct average over all possible sets of wave vectors κ for the spin-down electrons.

The other way of confirming the reasonableness of (16) is based on a result which was mentioned in GI without proof. This result is derived in the Appendix and states the following. If all $B_{G\Gamma}$ are equal, the first part of (17), referring to k inside the Fermi surface, is true whatever the set (κ) for the spin-down electrons may be; but there is seemingly no such simple confirmation for the second part of (17), referring to k outside the Fermi surface. Obviously, the second part of (17) holds only for the average over all sets (κ), whereas the first part of (17) holds for each set (κ) separately.

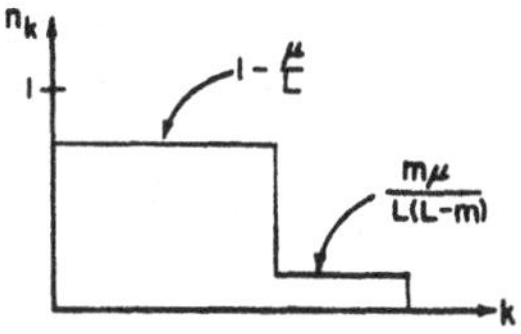

FIG. 1. Occupation probabilities in reciprocal space along a particular direction.

The author considers the result derived in the Appendix as very strong evidence for his claim that the spin-up electrons behave in a way which is largely independent of the way the spin-down electrons behave. In particular, the spin-down electrons might just as well be infinitely heavy as far as the spin-up electrons are concerned. The rest of this paper is then concerned with the physical consequences if this principle is stretched beyond the confirmation which has been presented.

3. METHOD FOR ESTIMATING VARIOUS DENSITY FUNCTIONS IN THE PRESENCE OF CORRELATION

Formulas (14) and (15) express the idea that the coefficients can be chosen such as to give simple values to the density functions ρ_n, the choice being limited by the total number of configurations which contribute to a given ρ_n. It is believed that the properties of ρ_n obtained in this manner are typical for the correlated wave functions of interest in the narrow-band problem. The presence of the spin-down electron merely restricts the freedom of movement for the spin-up electrons without, however, destroying the phase relations among the latter.

In order to give the weighting factor η^ν in (8) a well-defined meaning, we assume that configurations with different values of ν have the same average weight apart from the factor η^ν. Then it is postulated that

$$\rho_n(h_1\uparrow,\cdots,h_n\uparrow,f_1\uparrow,\cdots,f_n\uparrow)=\text{const}\left(w(h-f)\middle|\begin{matrix}h_1\cdots h_n\\ f_1\cdots f_n\end{matrix}\right),$$
$$(19)$$
$$\rho_n(h_1\downarrow,\cdots,h_n\downarrow,f_1\downarrow,\cdots,f_n\downarrow)=\text{const}\left(\omega(h-f)\middle|\begin{matrix}h_1\cdots h_n\\ f_1\cdots f_n\end{matrix}\right),$$

where the constants depend on m, μ, η, and the number of different lattice sites appearing in (h) and (f). Explicit values for these constants are obtained by counting how many configurations Γ or G contribute toward computing the left-hand side of (19), and attaching the proper weight η^ν to each contributing configuration.

Example 1. $n=m$, $h_1=f_1$, $\cdots$, $h_m=f_m$. If there are ν doubly occupied sites, there is a total of $m!(L-m)!/\nu!(m-\nu)!(\mu-\nu)!(L-m-\mu+\nu)!$ configurations contributing to ρ_m. The constant appearing in (19) is given apart from a normalization factor by

$$\sum_{\nu=0} \eta^{2\nu} m!(L-m)!/\nu!(m-\nu)!(\mu-\nu)!(L-m-\mu+\nu)!$$
$$=(L-m)!F(-m,\,-\mu;\,L-m-\mu+1;\,\eta^2)/\mu!(L-m-\mu)!. \quad (20)$$

If we had taken $n<m$, the constant appearing in (19) would still have been the same, since according to (6) we always have

$$\rho_n(h_1\uparrow,\cdots,h_n\uparrow;\,f_1\uparrow,\cdots,f_n\uparrow)$$
$$=m!/n!(m-n)!\sum_{g_{n+1},\cdots,g_m}\rho_m(h_1\uparrow,\cdots,h_n\uparrow,g_{n+1}\uparrow,\cdots;\,f_1\uparrow,\cdots,f_n\uparrow,g_{n+1}\uparrow,\cdots). \quad (21)$$

In particular, we obtain the normalization from

$$\rho_0=(\Psi|\Psi)=\sum_{g_1\cdots g_m}\rho_m(g_1\uparrow,\cdots,g_m\uparrow;\,g_1\uparrow,\cdots,g_m\uparrow)$$
$$=\text{const}\sum_{g(1\cdots g_m)}\left(w(g'-g'')\middle|\begin{matrix}g_1\cdots g_m\\ g_1\cdots g_m\end{matrix}\right)=\text{const}, \quad (22)$$

where the constant is just given by (20). All quantities to be computed henceforth have, therefore, to be divided by (20).

Example 2. $n=m$, $h_1\neq f_1$, $h_2=f_2$, $\cdots$, $h_m=f_m$. There are a number of possibilities to be distinguished which lead to different contributions to the constant in (19). The sites h_1 and f_1 can both be doubly occupied; one but not the other can be doubly occupied; and, finally, none of them may be doubly occupied. One obtains therefore, a sum of expressions similar to (20):

$$(L-m)!/\mu!(L-m-\mu)!$$
$$\times\{(L-m-\mu)F(1-m,\,-\mu;\,L-m-\mu;\,\eta^2)/(L-m)$$
$$+2\mu\eta F(1-m,\,1-\mu;\,L-m-\mu+1;\,\eta^2)/(L-m)$$
$$+\mu(\mu-1)\eta^2F(1-m,\,2-\mu;\,L-m-\mu+2;\,\eta^2)/(L-m-\mu+1)\}. \quad (23)$$

With the help of (12) and (21) we now get for ρ_1 the expression

$$\begin{aligned}\rho_1(h\uparrow,f\uparrow)&=w(h-f) \quad \text{for } h=f,\\ &=qw(h-f) \quad \text{for } h\neq f,\end{aligned} \quad (24)$$

where q is defined as the quotient (23)/(20), as a generalization of (16). Formula (24) again leads to a piecewise constant occupation probability in reciprocal space, as in Fig. 1, with $q+(1-q)m/L$ inside and $(1-q)m/L$ outside the Fermi surface. Since the hypergeometric function becomes 1 when its argument is zero, the result (16) is recovered for $\eta=0$. Also with the help of the Gauss formula[4]

$$F(a,b;c;1)=\Gamma(c)\Gamma(c-a-b)/\Gamma(c-a)\Gamma(c-b), \quad (25)$$

it is immediately checked that (23) = (20) for $\eta=1$. The evaluation of q for the values of η between 0 and 1 can be performed with the help of the formulas in the Appendix B. These formulas are asymptotically correct for very large L, m, and μ, which is our case of interest.

It goes without saying that a formula similar to (24) is obtained for $\rho_1(h\downarrow,f\downarrow)$. The only necessary modification is to exchange m and μ in (20) and (23).

In order to be able to compute the expectation value of the energy we have to know $\rho_2(g\uparrow,g\downarrow;\,g\uparrow g\downarrow)$. More generally, we postulate formulas analogous to (19), namely,

$$\rho_{n+1}(h_1\uparrow,\cdots,h_n\uparrow,g\downarrow;\,f_1\uparrow,\cdots,f_n\uparrow,g\downarrow)=\text{const}\left(w(h-f)\middle|\begin{matrix}h_1\cdots h_n\\ f_1\cdots f_n\end{matrix}\right);$$
$$\rho_{n+1}(g\uparrow,h_1\downarrow,\cdots,h_n\downarrow;\,g\uparrow,f_1\downarrow,\cdots,f_n\downarrow)=\text{const}\left(\omega(h-f)\middle|\begin{matrix}h_1\cdots h_n\\ f_1\cdots f_n\end{matrix}\right), \quad (26)$$

with the constants again determined by counting the weighted configurations which contribute to the left-hand side. The validity of (26) is examined in Sec. 5. where we argue that Eq. (26) is better realized in three dimensions than in one dimension.

Example 3. $n=m$, $h_1=f_1\neq g$, $\cdots$, $h_m=f_m\neq g$. The constant multiplying (26) is found to be

$$(L-m-1)!F(-m,\,1-\mu;\,L-m-\mu+1;\,\eta^2)/(\mu-1)!(L-m-\mu)!, \quad (27)$$

which has to be divided by (20) for normalization.

Example 4. $n=m$, $h_1=f_1=g$, $h_2=f_2\neq g$, $\cdots$, $h_m=f_m\neq g$ gives for the constant in (26) the value

$$(L-m)!\eta^2F(1-m,\,1-\mu;\,L-m-\mu+2;\,\eta^2)/(\mu-1)!(L-m-\mu+1)!, \quad (28)$$

to be divided again by (20) for normalizations. The consistency of this generalization is shown by checking that

$$\sum_g \rho_{m+1}(h_1\uparrow,\cdots,h_m\uparrow,g\downarrow;\,f_1\uparrow,\cdots,f_m\uparrow,g\downarrow)=\mu\rho_m(h_1\uparrow,\cdots,h_m\uparrow;\,f_1\uparrow,\cdots,f_m\uparrow). \quad (29)$$

As an application of the last formulas, we obtain

$$\rho_2(g\uparrow,g\downarrow;\,g\uparrow,g\downarrow)=\frac{m\mu}{L(L-m-\mu+1)}\times\eta^2\frac{F(1-m,\,1-\mu;\,L-m-\mu+2;\,\eta^2)}{F(-m,\,-\mu;\,L-m-\mu+1;\,\eta^2)}. \quad (30)$$

It is satisfying that this formula is symmetric in m and μ, although its derivation is not symmetric in spin-up and spin-down electrons.

[4] E. T. Whittaker and G. N. Watson, *A Course of Modern Analysis* (Cambridge University Press, New York, 1927), p. 281.

Another formula of interest is obtained for $\rho_2(g\uparrow,\gamma\downarrow;g\uparrow,\gamma\downarrow)$, where $g\neq\gamma$. Its derivation is somewhat more tricky. We find that

$$(20)\rho_2(g\uparrow,\gamma\downarrow;g\uparrow,\gamma\downarrow)$$

$$=(27)\sum_{(g_2\cdots g_m)\neq\gamma}\left(w(g'-g'')\left|\begin{matrix}gg_2\cdots g_m\\ gg_2\cdots g_m\end{matrix}\right.\right)$$

$$+(28)\sum_{(g_3\cdots g_m)}\left(w(g'-g'')\left|\begin{matrix}g\gamma g_3\cdots g_m\\ g\gamma g_3\cdots g_m\end{matrix}\right.\right)$$

$$=(m\mu/L^2)(20)+w(g-\gamma)w(\gamma-g)[(27)-(28)]. \quad (31)$$

The difference $(27)-(28)$ vanishes only for $\eta=1$. The second term represents an increase over the purely statistical value $m\mu/L^2$. Since the formula (31) is not symmetric in spin-up and spin-down electrons, it might be safer to claim its validity only in the case $w(g-\gamma)=\omega(g-\gamma)$, i.e., $m=\mu$.

Formula (31) shows explicitly that it is quite erroneous to assume statistical independence among spin-up and spin-down electrons at adjacent sites. Formula (31) might suggest the possible usefulness of a mixed density function, such as $\rho_1(g\uparrow;\gamma\downarrow)$ corresponding to an expectation value according to the definition (6), which one might assume to vanish at first. Such a mixed density function would describe a ground state in which configurations of different Z components for the total spin participate and have well-defined relative phases. Spin-flip excitations would be present in such a ground state, indicating a tendency toward antiferromagnetism, which is, of course, just described by formula (31).

The same phenomenon appears if we calculate the second-order density $\rho_2(h\uparrow,\gamma\downarrow;f\uparrow,\gamma\downarrow)$ with $\gamma\neq f\neq h\neq\gamma$ or with $\gamma=f\neq h$. The calculations are very tedious and the results very lengthy, unless the simplifications of Appendix B are used. The result can be written for $\gamma\neq f\neq h\neq\gamma$ as

$$\rho_2(h\uparrow,\gamma\downarrow,f\uparrow,\gamma\downarrow)=(\mu/L)\rho_1(h\uparrow,f\uparrow)+(m\mu-\nu L)$$
$$\times[w(h-\gamma)w(\gamma-f)-L^{-1}w(h-f)]/m(L-m), \quad (32)$$

and for $\gamma=f\neq h$ as

$$\rho_2(h\uparrow,\gamma\downarrow;f\uparrow,\gamma\downarrow)=(\mu/L)\rho_1(h\uparrow,f\uparrow)$$
$$+\left\{\frac{-\mu}{L}+\frac{[1-(\mu/L)]\nu}{\eta(m-\nu)}\right\}\frac{qw(h-f)}{[1+\nu/\eta(m-\nu)]}, \quad (33)$$

where the relation between ν and η as derived in Appendix B has to be inserted. The second terms in (32) and (33) would give a hint as to the values of the mixed density $\rho_1(h\uparrow,f\downarrow)$ for $h\neq f$ and $h=f$, if we were to express them as $\rho_1(h\uparrow;\gamma\downarrow)\rho_1(\gamma\downarrow;f\uparrow)$.

It is now part of the main proposition of this paper to rule out, at least in a first approximation, the use of such a mixed first-order density function. The additional terms which appear in (31), (32), and (33) are interpreted instead in the simplest imaginable way as the direct consequence of the spin-down electrons which act like inert obstructions, distributed at random, to the movement of the spin-up electrons. In this view, it would seem rather artificial to write a relation like

$$\rho_2(h\uparrow,\gamma\downarrow;f\uparrow\gamma\downarrow)=\rho_1(h\uparrow,f\uparrow)\rho_1(\gamma\downarrow,\gamma\downarrow)$$
$$\pm\rho_1(\gamma\downarrow,f\uparrow)\rho_1(h\uparrow,\gamma\uparrow). \quad (34)$$

One might have thought of (34) as a possible representation of (30) through (33), but it turns out to be unfeasible.

4. CALCULATION OF THE ENERGY EXPECTATION VALUE

With the help of (24) and (30) we can compute the expectation value of the energy for the Hamiltonian

$$H=\sum_k(a_{k\uparrow}{}^\dagger a_{k\uparrow}+a_{k\downarrow}{}^\dagger a_{k\downarrow})\epsilon_k+C\sum_g a_{g\uparrow}{}^\dagger a_{g\downarrow}{}^\dagger a_{g\downarrow}a_{g\uparrow}. \quad (35)$$

This Hamiltonian was proposed in GI, and its physical significance has been discussed in HI and GII. It represents, in a certain sense, the opposite of the Hamiltonian which is usually investigated in the study of free electrons with Coulomb repulsion, and it is believed to be a good model for the situation in a d band.

If we eliminate the weighting factor η with the help of (B4), so as to express everything in terms of the number ν of doubly occupied sites, we obtain in the case of $m=\mu$ the formula (this case corresponds to the nonferromagnetic state and is indicated by the index N)

$$\langle H\rangle_N=2mq\bar{\epsilon}+\nu C, \quad (36)$$

where $\bar{\epsilon}=m^{-1}\sum_{(k)}\epsilon_k$ is the average energy of the electrons without correlation. If we normalize $\sum_k\epsilon_k=0$, we have $\bar{\epsilon}<0$. The factor $q<1$, which was defined in (24), gives the discontinuity of the occupation probability in reciprocal space at the Fermi surface. The number of doubly occupied lattice sites is then obtained by minimizing $\langle H\rangle_N$ with respect to ν.

The condition for ν becomes (with $\bar{\nu}=\nu/L$ and $\bar{m}=m/L$)

$$dq/d\bar{\nu}=-C/2\bar{m}\bar{\epsilon}. \quad (37)$$

If this relation is used to eliminate C from (36), the energy expectation value $\langle H\rangle_N$ becomes

$$\langle H\rangle_N=2m\bar{\epsilon}(q-\bar{\nu}dq/d\bar{\nu}). \quad (38)$$

The expectation value of the energy has been increased from its value $2m\bar{\epsilon}<0$ without interaction by the factor $(q-\bar{\nu}dh/d\bar{\nu})<1$. With the help of the curves for $q(\bar{\nu})$ as obtained in Appendix B, we can now plot $q-\bar{\nu}dq/d\bar{\nu}$ as a function of $\bar{\nu}$. The slope at $\bar{\nu}=0$ is still infinite, indicating a very strong dependence of $\langle H\rangle$ on the number of doubly occupied sites, but the slope at $\bar{\nu}=\bar{m}^2$ is finite. The initial value for $\bar{\nu}=0$ is the same as for q, viz., $(1-2\bar{m})/(1-\bar{m})$, as shown in Fig. 2. For a given $\bar{\nu}$ and

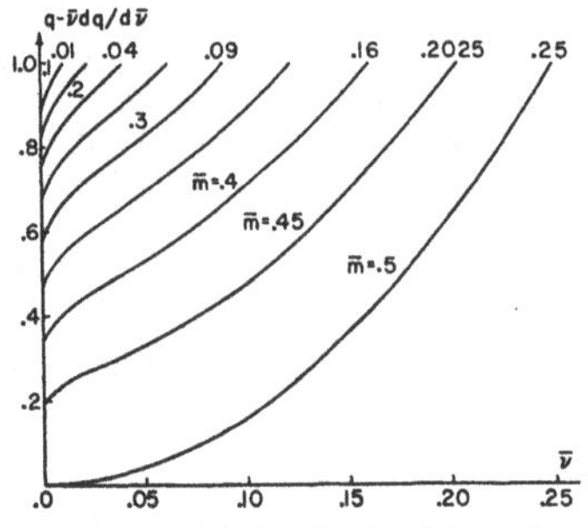

FIG. 2. Plot of $q-\bar{\nu}dq/d\bar{\nu}$, the reduction of the total energy from its value without interaction, versus $\bar{\nu}$, the density of doubly occupied sites, for various values of $\bar{m}=\bar{\mu}$, the density of spin-up and spin-down electrons. (The slopes at $\bar{\nu}=0$ are infinite.)

$\bar{m}$, the value of the interaction constant C may be obtained by using (37) and Fig. 3.

The question of the possibility of a ferromagnetic ground state can now be answered. A ferromagnetic state has the expectation value of H given by

$$\langle H\rangle_F = 2m\epsilon_F\,, \tag{39}$$

where ϵ_F is the average energy of the $2m$ electrons if they all have the same spin. The condition for a ferromagnetic ground state is therefore simply that ϵ_F be smaller than the average energy (38) of the electrons in the nonferromagnetic state, i.e.,

$$\epsilon_F < \bar{\epsilon}(q-\bar{\nu}dq/d\bar{\nu})\,, \tag{40}$$

where both ϵ_F and $\bar{\epsilon}$ are negative in our normalization.

The domain of existence for a ferromagnetic ground state can be obtained from (40), if we plot the ratio $\epsilon_F/\bar{\epsilon}$ as a function of $\bar{m}$ for a given band structure, viz., a given density-of-states curve, and if we draw into the same plot the function $(1-2\bar{m})/(1-\bar{m})$, i.e., the minimum value of $(q-\bar{\nu}dq/d\bar{\nu})$. Wherever $\epsilon_F/\bar{\epsilon}$ is larger than $(q-\bar{\nu}dq/d\bar{\nu})$, a ferromagnetic ground state exists for a sufficiently large value of C, e.g., for a constant density of states, we find in terms of the total bandwidth Δ that $\epsilon_F=-\Delta(1-2\bar{m})/2$ and $\bar{\epsilon}=\Delta(1-\bar{m})/2$. The condition for a ferromagnetic ground state is just not satisfied, because $|\epsilon_F|$ is too small relative to $|\bar{\epsilon}|$ for all $\bar{m}$.

It is easy to see the following: If the density of states is large at the band edges, the ratio $\epsilon_F/\bar{\epsilon}$ tends to be larger than $(1-2\bar{m})/(1-\bar{m})$, and ferromagnetism would appear to be possible if the intra-atomic Coulomb repulsion is strong enough. If the density of states curve is large at the center, its ratio $\epsilon_F/\bar{\epsilon}$ is smaller than $(1-2\bar{m})/(1-\bar{m})$, and a ferromagnetic ground state is excluded. The former case arises in a one-dimensional crystal, whereas the latter is typical of a three-dimensional crystal, although some structures in three dimensions, such as the fcc lattice, may present both aspects (cf. GI).

5. DISCUSSION

The remarks at the end of the preceding section show that the present approximate theory may lead to a ferromagnetic ground state in one dimension. Such a result is in disagreement with a theorem of Lieb and Mattis,[5] according to which the ground state in a one-dimensional system always has vanishing total spin momentum. Although the arguments of Lieb and Mattis are not completely suited to the Hamiltonian (35), their reasoning can be adapted to this simple model Hamiltonian (35). It is, therefore, worthwhile to examine at which point our procedure fails, at least, in one dimension, and what may be done to correct this situation. Also, some of the relevant statements may be true in more than one dimension, and it is of interest to point them out.

A sufficient condition for the coefficients $B_{G\Gamma}$ in (8) to generate a wave function of vanishing total spin momentum in any dimension is the following: The sets $(k_1,\cdots,k_m)$ and $(\kappa_1,\cdots,\kappa_\mu)$ are identical, and the values of $B_{G\Gamma}$ depend only on the set of occupied lattice sites regardless of how this set has been divided up into the subsets G and Γ. These conditions are a consequence of the state (7) having zero total spin if the sets $(k_1\cdots k_m)$ and $(\kappa_1\cdots\kappa_\mu)$ coincide, and of the fact that the total spin momentum operator only shifts the individual spins around the occupied lattice sites but does not shift the electrons themselves.

Even after the coefficients $B_{G\Gamma}$ have been restricted in this manner in order to obtain a state of vanishing total spin momentum, there are enough parameters available to make the requirements (19) and (26) seem reasonable, provided most of the lattice sites $(h_1,\cdots,h_n)$ and $(f_1,\cdots,f_n)$ coincide, as in the various examples of Sec. 3. In particular, the occupation probabilities in reciprocal space as given by (24) appear to be compatible with the requirement of vanshing total spin momentum.

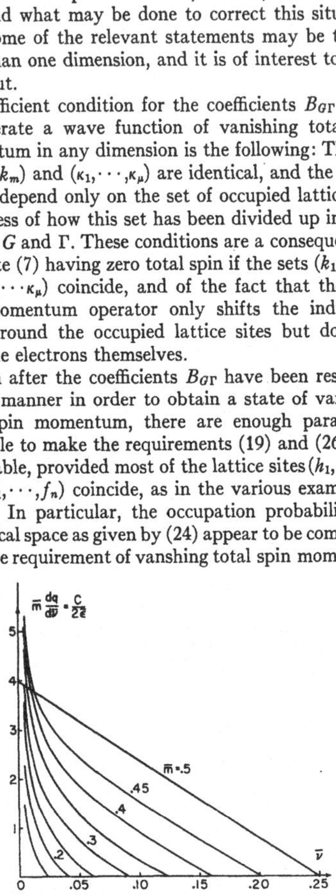

FIG. 3. Plot of $\bar{m}dq/d\bar{\nu}=C/2\bar{\epsilon}$, the ratio of the interaction strength C to twice the average energy per electron $\bar{\epsilon}$ in case of no interaction, versus $\bar{\nu}$, the density of doubly occupied sites. Given the ratio $C/2\bar{\epsilon}$, one can find $\bar{\nu}$ from this figure for a fixed value of $\bar{m}$. Figure 2 can then be used to find the reduction of the average energy per electron $\bar{\epsilon}$ due to the correlation.

[5] Elliott Lieb and Daniel Mattis, Phys. Rev. 125, 164 (1962).

In one dimension, the coefficients $A_{G\Gamma}$ have an additional important property. If the common set $(k_1,\cdots,k_m)=(\kappa_1,\cdots,\kappa_\mu)$ consists of all wave vectors inside some segment in reciprocal space which is centered at the origin, each determinant in (8) is positive if $0\leq g_1<g_2<\cdots<g_m<L$, $0\leq\gamma$, $<\gamma_2<\cdots<\gamma_\mu<L$, and $k_1<k_2<\cdots<k_m$. Therefore, the amplitude $A_{G\Gamma}$ is positive. We can prove this statement exactly as the statements made earlier in this section by simply remarking that the amplitudes $A_{G\Gamma}$ arise from the known ground state in the case of no interaction. According to Lieb and Mattis, the correlated state Ψ seems to be a good candidate for the ground state with interaction.

It is quite clear what causes the trouble in our approach for the one-dimensional case. The difficulty comes from the long tail of the occupation probability in reciprocal space outside the Fermi surface. That tail goes clear out to the zone boundary according to (24), which is particularly serious in one dimension where the density of states in the simple cases is large at the band edges rather than in the center. If we could modify (24) so as to make the occupation probability vanish at the zone boundary, a lot could be gained in one dimension. This could be achieved by a more careful choice of the coefficients B which appear in (13).

The sum over all configurations Γ of spin-down electrons in (13) depends on the set of points $(f,g_2,\cdots,g_m,h)$, i.e., the configuration of spin up electrons with the initial site h and the final site f. For the discussion, we can simplify this dependence by regarding the second line in (13) as a product, the first factor being a function of the set $(g_2,\cdots,g_m)$ of spin-up electrons which remain at their sites, and the second factor being a function of the initial site h and the final site f of the spin-up electron being transferred. It turns out that the first dependence does not influence the constancy of the occupation probability in reciprocal space outside the Fermi surface, if there is no dependence on f and h. We are, therefore, led to consider the sum over Γ in (13), primarily as it depends on f and h, and to disregard its dependence on $(g_2,\cdots,g_m)$. This dependence on f and h comes over and above the main dependence of the second line in (13) on whether or not $f=h$. Such an additional dependence, e.g., on whether or not f and h are nearest neighbors seems particularly indicated in one dimension, where any pair of sites is not surrounded by many third sites close by. This dependence on the exact relative positions of f and h should be much stronger in one dimension than in three.

The result in one dimension can only be improved if one is willing to go through more involved calculations which consist in evaluating sums over Γ, like the second line in (13), at least approximately for certain more specific assumptions about the coefficients B. It may be, however, that such improvements are not badly needed in three dimensions, where the averaging in any small neighborhood is likely to give better results than in one dimension.

ACKNOWLEDGMENT

The author would like to thank Professor David Paul for interesting discussions and helpful suggestions.

APPENDIX A

In the case $B=1$, $\eta=0$, we can write the correlated wave function as

$$\Psi=\prod_g(1-n_{g\uparrow}n_{g\downarrow})\Phi, \tag{A1}$$

where Φ is the independent electron wave function (7). In order to obtain the first-order density function, we have to evaluate $(\Psi|a_{h\uparrow}a_{f\uparrow}^\dagger|\Psi)$, which becomes, for $f\neq h$, simply equal to

$$(\Phi|a_{h\uparrow}\prod_g(1-n_{g\uparrow}n_{g\downarrow})a_{f\uparrow}^\dagger|\Phi)$$

$$=(1/L)\sum_{kl>k_F}e^{-ikf+ilh}(\Phi_{l\uparrow}|\prod_g(1-n_{g\uparrow}n_{g\downarrow})|\Phi_{k\uparrow}). \tag{A2}$$

The index $l\uparrow$ or $k\uparrow$ on Φ indicates the addition of an electron of wave vector l or k and spin-up to the state Φ. The product over all lattice sites g is now expanded, and we examine a particular term $(\Phi_{l\uparrow}|n_{g_1\uparrow}n_{g_1\downarrow}\cdots n_{g_\nu\uparrow}n_{g_\nu\downarrow}|\Phi_{k\uparrow})$. If we expand the wave functions $\Phi_{l\uparrow}$ and $\Phi_{k\uparrow}$ into configurations, this expectation value can be expressed as a sum over products of determinants, one term for each arrangement of the remaining $m+1-\nu$ spin-up electrons at the sites $g_{\nu+1}\cdots g_{m+1}$ and the $\mu-\nu$ spin-down electrons at the sites $\gamma_{\nu+1}\cdots\gamma_\mu$. The evaluation of this sum

$$\sum_{(g_{\nu+1}\cdots g_{m+1})}\left(L^{-1/2}e^{-ikg}\left|\begin{matrix}lk_1\cdots k_m\\ g_1\cdots g_{m+1}\end{matrix}\right.\right)\left(L^{-1/2}e^{ikg}\left|\begin{matrix}kk_1\cdots k_m\\ g_1\cdots g_{m+1}\end{matrix}\right.\right)$$

$$\times\sum_{(\gamma_{\nu+1}\cdots\gamma_\mu)}\left(L^{-1/2}e^{-i\kappa\gamma}\left|\begin{matrix}\kappa_1\cdots\kappa_\mu\\ g_1\cdots g_\nu\gamma_{\nu+1}\cdots\gamma_\mu\end{matrix}\right.\right)\left(L^{-1/2}e^{i\kappa\gamma}\left|\begin{matrix}\kappa_1\cdots\kappa_\mu\\ g_1\cdots g_\nu\gamma_{\nu+1}\cdots\gamma_\mu\end{matrix}\right.\right) \tag{A3}$$

is a simple exercise in determinant manipulation. After the summation over k and l, one obtains

$$(\Phi|a_{h\uparrow}n_{g_1\uparrow}n_{g_1\downarrow}\cdots n_{g_\nu\downarrow}a_{f\uparrow}^\dagger|\Phi)=\begin{vmatrix}\chi(h-f) & -\chi(h-g_1) & -\chi(h-h_2)\cdots\\ \chi(g_1-f) & w(g_1-g_1) & w(g_1-g_2)\cdots\\ \chi(g_2-f) & w(g_2-g_1) & w(g_2-g_2)\cdots\\ \vdots & \vdots & \vdots\end{vmatrix}\begin{vmatrix}\omega(g_1-g_1) & \omega(g_1-g_2)\cdots\\ \omega(g_2-g_1) & \omega(g_1-g_2)\cdots\\ \vdots & \vdots\end{vmatrix}, \tag{A4}$$

where

$$\chi(h-f)=\delta_{hf}-w(h-f)=L^{-1}\sum_{k>k_F}\exp(ikh-ikf).$$

The formula (A4) is correct even for $h=f$. But in that case it is not always useful, because

$$(\Psi|a_{f\uparrow}a_{f\uparrow}^{\dagger}|\Psi)=(\Phi|a_{f\uparrow}\prod_{g\neq f}(1-n_{g\uparrow}n_{g\downarrow})a_{f\uparrow}^{\dagger}|\Phi), \tag{A5}$$

so that only sets $(g_1\cdots g_\nu)$ occur in the summation which do not contain the site f. There is no such restriction on the sets $(g_1\cdots g_\nu)$ if $f\neq h$. Therefore, we obtain a term in addition to the sum over the expressions (A4), but this term is restricted to $f=h$. In this manner, one finds that

$$(\Psi|a_{f\uparrow}^{\dagger}a_{h\uparrow}|\Psi)-\delta_{fh}(\Psi|\Psi)=-\chi(h-f)+\sum_{\nu=1}\frac{(-1)^\nu}{\nu!}\sum_{g_1\cdots g_\nu}\begin{vmatrix}-\chi(h-f) & \chi(h-g_1)\cdots\\ \chi(g_1-f) & w(g_1-g_1)\cdots\\ \vdots & \vdots\end{vmatrix}\begin{vmatrix}\omega(g_1-g_1) & \omega(g_1-g_2)\cdots\\ \omega(g_2-g_1) & \omega(g_2-g_2)\cdots\\ \vdots & \vdots\end{vmatrix}$$

$$+\delta_{hf}\sum_{\nu=0}\frac{(-1)^\nu}{\nu!}\sum_{g_1\cdots g_\nu}\begin{vmatrix}\chi(h-f) & w(h-g_1)\cdots\\ \chi(g_1-f) & w(g_1-g_1)\cdots\\ \vdots & \vdots\end{vmatrix}\begin{vmatrix}\omega(h-f) & \omega(h-g_1)\cdots\\ \omega(g_1-f) & \omega(g_1-g_1)\cdots\\ \vdots & \vdots\end{vmatrix}. \tag{A6}$$

This expression can be further simplified, first, by separating out of the first sum everything that is multiplied with the $-\chi(h-f)$ in the left-hand upper corner of the determinant, and remembering that

$$(\Psi|\Psi)=1+\sum_{\nu=1}\frac{(-1)^\nu}{\nu!}\sum_{g_1\cdots g_\nu}\begin{vmatrix}w(g_1-g_1) & w(g_1-g_2)\cdots\\ w(g_2-g_1) & w(g_2-g_2)\cdots\\ \vdots & \vdots\end{vmatrix}\begin{vmatrix}\omega(g_1-g_1) & \omega(g_1-g_2)\cdots\\ \omega(g_2-g_2) & \omega(g_2-g_2)\cdots\\ \vdots & \vdots\end{vmatrix}; \tag{A7}$$

second, by inserting the definition of χ into the second sum and then, instead of putting $h=f$, by writing $L^{-1}\sum(f=h)$; third, by carrying out the summation over h as far as possible with the help of the simple relation

$$\sum_g \omega(h-g)\omega(g-f)=\omega(h-f). \tag{A8}$$

This gives the final expression

$$(\Psi|a_{f\uparrow}^{\dagger}a_{h\uparrow}|\psi)=[w(h-f)-\delta_{hf}\mu/L](\Psi|\Psi)$$

$$+\sum_{\nu=1}\frac{(-1)^\nu}{\nu!}\sum_{g_1\cdots g_\nu}\begin{vmatrix}0 & \chi(h-g_1) & \chi(h-g_2)\cdots\\ \chi(g_1-f) & w(g_1-g_1) & w(g_1-g_2)\cdots\\ \chi(g_2-f) & w(g_2-g_1) & w(g_2-g_2)\cdots\\ \vdots & \vdots & \vdots\end{vmatrix}\begin{vmatrix}\omega(g_1-g_1) & \omega(g_1-g_2)\cdots\\ \omega(g_2-g_1) & \omega(g_1-g_1)\cdots\\ \vdots & \vdots\end{vmatrix}, \tag{A9}$$

where we have used the fact that $\omega(g-g)=\mu/L$.

Now, we can write the occupation probability in reciprocal space as

$$(\psi|n_{k\uparrow}|\Psi)=(1/L)\sum_{fh}e^{ik(f-h)}(\Psi|a_{f\uparrow}^{\dagger}a_{h\uparrow}|\Psi) \tag{A10}$$

and insert (A9). The first term on the right-hand side of (A9) is trivial, since the summation over f and h applied to $w(h-f)$ gives 1 for k inside and 0 for k outside the Fermi surface of the free spin-up electrons, whereas applied to δ_{hf} this summation gives 1 for all k. On the other hand, if k is outside the Fermi surface of the free spin-up electrons, the first row in the first determinant is modified to $(0,L^{-1/2}e^{ikg_1},\cdots,L^{-1/2}e^{-ikg_\nu})$ and the first column to $(0,L^{-1/2}e^{ikg_1},\cdots,L^{-1/2}e^{ikg_\nu})$ by the summation over h and f, whereas both the first row and the first column vanish, if k is inside the Fermi surface. Therefore, the occupation probability $n_{k\uparrow}$ for k inside the Fermi surface follows entirely from the first term on the right-hand side of (A9), and it is given by the value established in (17), q.e.d.

APPENDIX B

All the hypergeometric series of interest are of the form

$$F=(\alpha-m,\beta-\mu;L-m-\mu+\delta;\eta^2)=\sum_{\nu=0}^{\infty}\frac{(\alpha-m)\cdots(\alpha-m+\nu-1)(\beta-\mu)\cdots(\beta-\mu+\nu-1)}{(L-m-\mu+\delta)\cdots(L-m-\mu+\delta+\nu-1)\nu!}\eta^{2\nu}. \tag{B1}$$

Since m and μ are of the order of the number L of lattice sites, it is sufficient to evaluate the series by considering only the largest terms. The biggest terms occur for the values of ν, where

$$[(\alpha-m+\nu)(\beta-\mu+\nu)/(\nu+1)(L-m-\mu+\delta+\nu)]\eta^2\cong 1. \tag{B2}$$

With the help of Stirling's formula, it follows in the standard manner that the sum is essentially equal to its largest term. We can then approximate $F(\alpha-m,\beta-\mu;L-m-\mu+\delta;\eta^2)$ by

$$\frac{\Gamma(m-\alpha+1)\Gamma(\mu-\beta-1)\Gamma(L-m-\mu+\delta)\eta^{2\nu}}{\Gamma(m-\alpha-\nu+1)\Gamma(\mu-\beta-\nu+1)\Gamma(L-m-\mu+\delta+\nu)\nu!}, \tag{B3}$$

provided α, β, and δ are of order 1, and ν is obtained from the condition

$$[(m-\nu)(\mu-\nu)/\nu(L-m-\mu+\nu)]\eta^2=1. \tag{B4}$$

The result for the hypergeometric function is correct within a factor $[1+0(L^{-1})]$, provided one considers always ratios of two hypergeometric functions. Since ν is independent of α, β, and δ, the quotient of two hypergeometric functions becomes a quotient of Γ functions, which is a rational function, if α, β, and δ are integers.

As a first simple example we evaluate (30), and find immediately that

$$\rho(g\uparrow,g\downarrow;g\uparrow;g\downarrow)=\nu, \tag{B5}$$

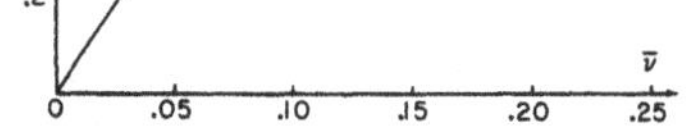

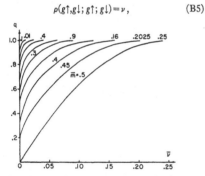

FIG. 4. Plot of q, the occupation probability inside the Fermi surface, versus $\bar{\nu}$, the density of doubly occupied sites.

with ν given by (B4). The quantity η^2 plays, therefore, the same role as the Boltzmann factor in the law of mass action. Indeed, $m-\nu$,and $\mu-\nu$ are the average numbers of dissociated spin-up and spin-down particles, whereas ν is the average number of "bound" spin pairs and $L-m-\mu+\nu$ is the average number of empty lattice sites. The method presented in this paper can be compared to the "quasichemical" method in the theory of mixtures.[6] As η varies from 0 to 1, ν varies from 0 to $m\mu/L$, which is the number of crowded sites without correlation.

In order to obtain the expectation value of the energy (36), we calculate with the help of (B3) the quotient

$$q=\frac{(m-\nu)(L-m-\mu+\nu)}{m(L-m)}\left(1+\frac{(\mu-\nu)\eta}{L-m-\mu+\nu}\right)^2. \tag{B6}$$

In the important special case $m=\mu$, we have

$$\eta=[\nu(L-2m+\nu)]^{1/2}/(m-\nu). \tag{B7}$$

This leads to the expression

$$q=(m-\nu)[(L-2m-\nu)^{1/2}+(\nu)^{1/2}]^2/m(L-m). \tag{B8}$$

The left-hand side is plotted as function of $\bar{\nu}=\nu/L$ for various values of $\bar{m}=m/L<\frac{1}{2}$ in Fig. 4. Note the infinite slope at $\bar{\nu}=0$ and the vanishing slope at $\bar{\nu}=\bar{m}^2$. A reduction in $\bar{\nu}$ can, therefore, be achieved without losing much of the kinetic and crystal potential energy.

[6] E. A. Guggenheim, *Mixtures* (Oxford University Press, New York, 1952), p. 38.

VOLUME 57, NUMBER 11 PHYSICAL REVIEW LETTERS 15 SEPTEMBER 1986

New Functional Integral Approach to Strongly Correlated Fermi Systems: The Gutzwiller Approximation as a Saddle Point

Gabriel Kotliar[(1)] and Andrei E. Ruckenstein[(2)]

[(1)] *Department of Physics, Massachusetts Institute of Technology, Cambridge, Massachusetts 02139*

[(2)] *Department of Physics, University of California at San Diego, La Jolla, California 92093*

(Received 21 April 1986)

We propose a new functional integral representation of the Hubbard and Anderson models of lattice fermions. The simplest saddle-point approximation leads, at zero temperature, to the results derived from the Gutzwiller variational wave function. This approach uncovers the limitations of the Gutzwiller approximation and clarifies its connection to the "auxiliary-boson" mean-field theory of the Anderson model. This formulation leads to a novel strong-coupling mean-field theory which allows for a unified treatment of antiferromagnetism and ferromagnetism, metal-to-insulator transition, and Kondo compensation effects.

PACS numbers: 71.10.+x, 71.30.+h, 75.10.Lp

Recently there has been an upsurge of interest in strongly correlated Fermi systems, mainly triggered by the remarkable properties of the newly discovered "heavy-electron" materials.[1] Interest in the problem was also stimulated by speculations concerning the interplay between disorder and correlation effects near the metal-to-insulator transition in doped semiconductors.[2] A theoretical understanding of these systems is still lacking and further progress appears to require new techniques, outside the scope of conventional weak-coupling approximations.

Apart from Monte Carlo simulations,[3] two analytical approaches to strongly correlated fermions have received a great deal of attention: the Gutzwiller variational approach, which originated in the context of the Hubbard model,[4] and the so-called "slave-boson" or auxiliary-boson formulation first proposed by Barnes[5] and rediscovered and extended by Coleman[6] and Read and Newns[7] in their work on the mixed-valence problem. The first is an appealing but uncontrolled approximation scheme for calculating the ground-state energy of a variational trial wave function. The second has so far been mainly used to treat the infinite-U Anderson model and consists of replacing the infinite correlations by a local constraint which is then handled by standard field-theoretical methods (see, however, Ref. 5). In this Letter we present a new functional integral formulation which (i) extends the collective boson approach to any value of the correlation[8]; (ii) reproduces for the first time the results of the Gutzwiller approach in a saddle-point approximation, and thus uncovers the limitations of the Gutzwiller approximation and suggests, at least in principle, systematic ways of improving it; and (iii) allows for a unified treatment of ferromagnetism, antiferromagnetism, and metal-insulator transitions in a mean-field theory which in the weak-coupling case agrees with Hartree-Fock theory while for strong coupling incorporates the qualitative physics expected from the few available exact results. (iv) In the Anderson model the resulting saddle point builds in the collective quenching of the local moments (Kondo effect) and will ultimately allow us to study the competition between the Kondo effect and magnetic order. To our knowledge this is the first method by which such a large number of phenomena become easily accessible within the same framework.

Qualitatively our approach is based on the idea that, in a strongly correlated system, in the process of hopping the electron is accompanied by a "backflow" of spin and density excitations of the medium. (In a quasiparticle picture this shows up as a renormalization of the hopping amplitude and simply leads to a change of the effective mass.) Formally, this qualitative idea can be realized by rewriting the original Hamiltonian in terms of the original fermions and a set of four projection operators which keep track of the environment by measuring the occupation numbers in each of the four possible states available for hopping.

To be explicit, we first concentrate on the Hubbard model[9] which is expected to capture the main features of the physics of lattice fermions in a narrow energy band. The corresponding Hamiltonian includes a nearest-neighbor hopping, t_{ij}, and an on-site repulsion between electrons of different spins, U:

$$H=\sum_{ij,\sigma} t_{ij} f_{i\sigma}^{\dagger} f_{j\sigma}+U\sum_{i} f_{i\sigma}^{\dagger} f_{i\sigma} f_{i-\sigma}^{\dagger} f_{i-\sigma}, \tag{1}$$

where $f_{i\sigma}^{\dagger}$ ($f_{i\sigma}$) are creation (annihilation) operators for an electron of spin σ ($=\pm 1$) at site i. In analogy with the "slave boson" approach we enlarge the Fock space at each site, to contain in addition to the original fermions a set of four bosons represented by the creation (annihilation) operators $e_i^{\dagger}$ (e_i), $p_{i\sigma}^{\dagger}$ ($p_{i\sigma}$), $d_i^{\dagger}$ (d_i). This enlarged space contains unphysical states which can be eliminated by imposing the set of constraints

$$\textstyle\sum_{\sigma} p_{i\sigma}^{\dagger} p_{i\sigma}+e_i^{\dagger} e_i+d_i^{\dagger} d_i=1, \tag{2a}$$

$$f_{i\sigma}^{\dagger} f_{i\sigma}=p_{i\sigma}^{\dagger} p_{i\sigma}+d_i^{\dagger} d_i, \quad \sigma \pm 1. \tag{2b}$$

When restricted by (2) the Bose fields e_i, $p_{i\sigma}$

VOLUME 57, NUMBER 11 PHYSICAL REVIEW LETTERS 15 SEPTEMBER 1986

$(\sigma=-1)$, and d_i act respectively as projection operators onto the empty, singly occupied (with spin up and down), and doubly occupied electronic states at each site. Equation (2a) can then be interpreted as a completeness relation and reflects the fact that no more and no less than one of the four possible states must be occupied at each site; the second constraint equates the two ways of counting the fermion occupancy of a given spin. It is easy to check that in the physical subspace defined by Eqs. (2) the Hamiltonian

$$\tilde{H}=\sum_{ij,\sigma} t_{ij} f_{i\sigma}^{\dagger} f_{j\sigma} z_{i\sigma}^{\dagger} z_{j\sigma}+U\sum_i d_i^{\dagger} d_i, \tag{3a}$$

$$z_{i\sigma}=e_i^{\dagger} p_{i\sigma}+p_{i-\sigma}^{\dagger} d_i, \tag{3b}$$

has the same matrix elements as those calculated for (1) in the original Hilbert space.

To calculate observable quantities we write down the partition function Z of model (3) as a functional integral over coherent states of Fermi and Bose fields.[6,7] We note that the constraints (2) commute with the Hamiltonian (3) and thus the physical Hilbert space is preserved under time evolution. The constraints (2a) and (2b) are thus enforced at each site by time-independent Lagrange multipliers, which we symbolize below by $\lambda_i^{(1)}$ and $\lambda_{i\sigma}^{(2)}$ $(\sigma=\pm1)$, respectively. We integrate out the Fermi fields (by using standard rules for integration over Grassmann variables) to reexpress Z in terms of the effective action for the bosons, $\tilde{S}$, as

$$Z=\int[De][Dp_{\pm\sigma}][Dd]\prod_{i\sigma}[d\lambda_i^{(1)}][d\lambda_{i\sigma}^{(2)}]\exp\left[-\int_0^{\beta} d\tau\,\tilde{S}(\tau)\right], \tag{4a}$$

$$\tilde{S}(\tau)=\sum_i\Big[e_i^{\dagger}(\partial_\tau+\lambda_i^{(1)})e_i+\sum_\sigma p_{i\sigma}^{\dagger}(\partial_\tau+\lambda_i^{(1)}-\lambda_{i\sigma}^{(2)})p_{i\sigma}+d_i^{\dagger}(\partial_\tau+U+\lambda_i^{(1)}-\lambda_{i\sigma}^{(2)})d_i\Big]$$
$$-\lambda_i^{(1)}+\mathrm{tr}\ln[\delta_{ij}(\partial_\tau-\mu-\sigma h+\lambda_{i\sigma}^{(2)})+t_{ij}z_{i\sigma}^{\dagger}z_{j\sigma}], \tag{4b}$$

where μ is the chemical potential which is adjusted to fix the average occupation of the site, $n=1-\delta$, and h is an external magnetic field. (Since the physics is symmetric about $n=1$ [10] we restrict ourselves to the case $\delta\geqslant 0$.)

We note that in the atomic limit $(t_{ij}=0)$ the functional integral (4) can be calculated exactly and leads to the known results.[9] For $t_{ij}\neq 0$ the simplest approach to (4) is the saddle-point approximation in which all Bose fields and Lagrange multipliers are taken to be independent of space and time. Unfortunately, the resulting saddle-point equations lead to the incorrect result in the noninteracting limit (which occurs either for $U=0$ or in the case of fully polarized spins). This is because in this approximation the constraints are only satisfied on the average, and not explicitly at each site of the lattice. For example, when $U=0$ and $\delta=0$, $e^2=d^2=p_\sigma^2=\frac{1}{4}$ and thus

$$\langle z_{i\sigma}^{\dagger} z_{j\sigma}\rangle=e^2p_\sigma^2+d^2p_{-\sigma}^2+2edp_\sigma p_{-\sigma}=\tfrac{1}{4},$$

rather than unity as it should be for the noninteracting system.

In order to resolve this problem we make use of the fact that the procedure described above is not unique; there are many different Hamiltonians, $\tilde{H}$, with different properties in the enlarged Hilbert space which lead to the same spectrum as (1) when restricted to the physical subspace defined by (2). Clearly this arbitrariness presents no difficulty as long as the constraints are handled exactly. However, any approximation which relaxes the constraints is sensitive to the precise choice of $\tilde{H}$. In any practical calculation this ambiguity can be used to our advantage, and the form of $\tilde{H}$ can be determined by requiring that the approximation scheme leads to physically sensible results in known limiting cases. In particular, in this case we replace $z_{i\sigma}$ in (3b) by another operator, $\tilde{z}_{i\sigma}$:

$$\tilde{z}_{i\sigma}=(1-d_i^{\dagger}d_i-p_{i\sigma}^{\dagger}p_{i\sigma})^{-1/2}z_{i\sigma}(1-e_i^{\dagger}e_i-p_{i-\sigma}^{\dagger}p_{i-\sigma})^{-1/2}, \tag{5}$$

which has the same eigenvalues and eigenvectors as z_i in the physical subspace but also leads to the correct $U=0$ limit in the saddle-point approximation.[11] The resulting saddle-point free-energy functional $f=-k_BT\ln Z/N$ can then be written as

$$f=Ud^2-k_BT\sum_\sigma\int_{-\infty}^{+\infty}d\xi\,\rho(\xi)\ln[1+e^{-\beta(q_\sigma\xi-\mu-\sigma h+\lambda_\sigma^{(2)})}]+\lambda^{(1)}\Big(\sum_\sigma p_\sigma^2+e^2+d^2-1\Big)-\sum_\sigma\lambda_\sigma^{(2)}(p_\sigma^2+d^2) \tag{6}$$

where $q_\sigma\equiv\langle\tilde{z}_{i\sigma}^{\dagger}\tilde{z}_{j\sigma}\rangle$, and T is the temperature.

As the simplest illustration of our approach we consider (6) in the paramagnetic phase of the half-filled-band Hubbard model (i.e., $n=1$, $\delta=0$). From general arguments as well as by a direct inspection of the saddle-point equations, $\mu=U/2$; also we set $h=0$, and we assume a symmetric density of states, $\rho(\xi)$. The Lagrange multipliers can then be easily eliminated to arrive at a (free energy) functional of d alone, $f=\bar{\epsilon}-Ts$. Here, $\bar{\epsilon}=2\int_{-\infty}^{+\infty}d\xi\,\rho(\xi)q\xi f(q\xi)+Ud^2$, and $q=8d^2(1-2d^2)$; s is the entropy per particle calculated for a lattice gas of free fermions with an effective hopping amplitude $\tilde{t}_{ij}=qt_{ij}$, and f is the Fermi function $f(\xi)=(e^{\beta\xi}+1)^{-1}$. At $T=0$, $\bar{\epsilon}$ is minimized by $d^2=1/4(1-U/U_c)$ with $U_c=16\int_0^{\infty}d\xi\,\rho(\xi)\xi$. Within our approximation this corresponds to the vanishing of the number of doubly occupied sites and indicates that the system is undergoing a metal-insulator transition at a finite critical value of U. The same result was derived by Brinkman and Rice[12] by using the variational wave function and the approximation scheme proposed by Gutzwiller.[4] The Gutzwiller approach to the half-filled Hubbard model has recently received a great deal of attention as a model for liquid ^{3}He close to the solidification curve.[13]

At sufficiently high temperatures we expect on physical grounds that the correlation effects become unimportant and q should approach unity. However, in the rigid-band picture the entropy favors $q=0$ and the system undergoes a first-order transition at a temperature of order W^2/U. We stress that none of the "slave boson" mean-field or Gutzwiller-type theories proposed to date[14] give a sensible crossover between low and high temperatures and we expect that new techniques treating fluctuations in fermion and boson degrees of freedom on an equal footing will be required in order to remedy this problem. Even so, it is likely that the behavior predicted from the finite-temperature saddle point to (4) is qualitatively correct at the lowest temperatures. In particular, it is appealing to interpret the initial decrease of q with temperature in terms of an increase of the coherent low-frequency fluctuations accompanying the hopping particle.

Our approach also provides a natural framework for studying magnetic properties of the Hubbard model. We have investigated the stability of the paramagnetic ground state with respect to both ferromagnetism and antiferromagnetism. At the ferromagnetic saddle point $p_\uparrow^2 - p_\downarrow^2 = m$ (m is the magnetization), while antiferromagnetism can be parametrized as usual by dividing the system into two sublattices, A and B, with the sublattice Bose fields satisfying the relations $e_A = e_B$, $d_A = d_B$, $p_{A\uparrow} = p_{B\downarrow}$, $p_{A\downarrow} = p_{B\uparrow}$, and $p_{A\uparrow}^2 - p_{A\downarrow}^2 = m_s$, the staggered magnetization. The ferromagnetic and antiferromagnetic phase boundaries (as determined by the vanishing of the inverse magnetic and staggered susceptibilities) are shown in Fig. 1.[15] It is intriguing that in this mean-field theory the possibility for ferromagnetism is restricted to very large values of U in contrast to the predictions of Stoner-type weak-coupling theories.[16] The tendency towards ferromagnetism for infinite U is in accordance with Nagaoka's theorem[10] which asserts that the ground state of a system with $d^2=0$ and $\delta = 1/N$ is ferromagnetic (N is the number of sites), while the disappearance of ferromagnetism at large U for $\delta \approx 0.38$ agrees qualitatively with the corresponding result of Kanamori.[17] Also, we stress that, as expected in lattices with perfect nesting, and in contrast with a previous calculation based on the Gutzwiller wave function due to Ogawa *et al.*,[18] the ground state of the half-filled Hubbard model is antiferromagnetic for an infinitesimal interaction U; for values of U up to $U \approx 40\,W$ the corresponding energy is lower than that given by Hartree-Fock theory.[16] The competition between ferromagnetism and antiferromagnetism, and the full phase diagram of the Hubbard model will be discussed elsewhere.[19]

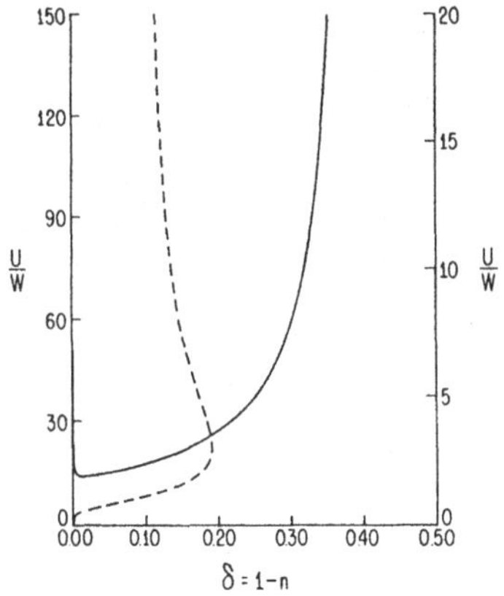

FIG. 1. The boundaries of stability of the paramagnetic phase of the Hubbard model with respect to ferromagnetism (solid line, left axis) and antiferromagnetism (dashed line, right axis) as determined from the vanishing of the inverse magnetic and staggered susceptibilities, respectively.

Finally, the approach outlined above can also be used to study the Anderson-lattice Hamiltonian[6,7,20] for all values of U. Here we make two comments about our results[19]: (i) In the paramagnetic phase and for infinite U ($d=0$) our calculation exactly agrees with that of Rice and Ueda,[20] who applied the Gutzwiller variational approach to the Anderson lattice. Their intriguing result that, for infinite U and in the Kondo regime, the spin-$\frac{1}{2}$ Anderson lattice has an instability with respect to ferromagnetism appears in our formulation as a consequence of the delocalization of "holes" in the f band, in analogy with the Nagaoka limit of the Hubbard model. (ii) We note that the boson introduced by Coleman[6] for the infinite-U case corresponds, strictly speaking, to the e_i ($e_i^\dagger$) of our treatment. It can then be easily seen that the resulting mean-field theories can only agree in the paramagnetic regime and in the limit of infinite spin degeneracy.

In this Letter we presented a new approach to strongly correlated Fermi systems, in the context of which we derive a new strong-coupling mean-field theory. While in the simplest physical situations the mean-field results are equivalent to those of the Gutzwiller variational approach, our formulation is

VOLUME 57, NUMBER 11 PHYSICAL REVIEW LETTERS 15 SEPTEMBER 1986

systematic and is applicable with equal ease to both magnetic and nonmagnetic phases of these systems. In the case of the Anderson model this mean-field theory incorporates the competition between magnetism and the Kondo effect. This work can be extended in several directions: One can study the effect of Gaussian fluctuations about the saddle point, and one can apply this mean-field approach to richer physical situations (disorder, several bands, deformable lattice). Work in these directions is currently in progress.

We would like to thank the Aspen Center for Physics and its participants for providing the stimulating environment where this collaboration began. We would especially like to thank Piers Coleman who played a major role in the early stages of this work, and Nick Read for his inspiring explanation of his treatment of the f^1-f^2 problem. The work of one of us (A.E.R.) was partially supported by a University of California, San Diego, Academic Senate research grant and a U.S. Office of Naval Research Young Investigator Award. The work of the other (G.K.) was supported by National Science Foundation Grant No. DMR-8521377.

[1]For a recent review see P. A. Lee, T. M. Rice, J. W. Serene, L. J. Sham, and J. W. Wilkins, Comments Solid State Phys. (to be published).

[2]P. W. Anderson, in *Proceedings of the International Conference on Localization, Interactions and Transport Properties in Impure Metals, Braunschweig, 1984,* edited by Y. Bruynseraede (Springer-Verlag, Berlin, 1985); A. E. Ruckenstein, M. A. Paalanen, and G. A. Thomas, in *Proceedings of the Seventeenth International Conference on the Physics of Semiconductors, San Francisco, 1984,* edited by D. J. Chadi (Springer, New York, 1985); E. Abrahams, P. W. Anderson, and G. Kotliar, unpublished.

[3]J. E. Hirsch, Phys. Rev. Lett. **51**, 1900 (1983); also in Proceedings of the Conference on the Frontiers of Quantum Monte-Carlo, Los Alamos, 1985, J. Stat. Phys. (to be published).

[4]M. Gutzwiller, Phys. Rev. Lett. **10**, 159 (1963), and Phys. Rev. **137**, A1726 (1965).

[5]S. E. Barnes, J. Phys. F **6**, 1375 (1976), and **7**, 2637 (1977).

[6]P. Coleman, Phys. Rev. B **29**, 3035 (1984).

[7]N. Read and D. Newns, J. Phys. C **16**, 3273 (1983); N. Read, J. Phys. C **18**, 2651 (1985).

[8]A completely different "slave boson" scheme for studying the finite-U Anderson model (restricted to paramagnetic phases) was discussed by N. Read, Ph.D. thesis, Imperial College, London, 1985 (unpublished). We also note that our approach is different from Barnes's two-boson theory in that it leads to a *simple* saddle point which is qualitatively correct in both weak- *and* strong-U limits, for all fillings (see text).

[9]J. Hubbard, Proc. Roy. Soc. London, Ser. A **296**, 100 (1966).

[10]Y. Nagaoka, Phys. Rev. **147**, 392 (1966).

[11]In the physical space, because of the projection-operator property of the bosons, the normal ordering of the square roots in (5) is not an issue; in defining the functional integral these factors can be translated directly into c-number fields.

[12]W. F. Brinkman and T. M. Rice, Phys. Rev. B **2**, 4302 (1970).

[13]D. Vollhardt, Rev. Mod. Phys. **56**, 99 (1984).

[14]K. Seiler, C. Gross, T. M. Rice, K. Ueda, and D. Vollhardt, to be published.

[15]We have assumed densities of states $\rho(\xi)=2[1-(\xi/W)^2]/\pi W$ and $\rho(\xi)=1/2W$ for the ferromagnetic and antiferromagnetic cases, respectively.

[16]D. R. Penn, Phys. Rev. **142**, 350 (1966).

[17]J. Kanamori, Prog. Theor. Phys. **30**, 257 (1964).

[18]T. Ogawa, K. Kanda, and T. Matsubara, Prog. Theor. Phys. **53**, 614 (1975).

[19]G. Kotliar and A. E. Ruckenstein, to be published.

[20]T. M. Rice and K. Ueda, Phys. Rev. Lett. **55**, 995 (1985); C. Varma, W. Weber, and L. J. Randall, Phys. Rev. **33**, 1015 (1985).

PHYSICAL REVIEW B VOLUME 14, NUMBER 7 1 OCTOBER 1976

Theory of the quasi-one-dimensional electron gas with strong "on-site" interactions*

V. J. Emery
Brookhaven National Laboratory, Upton, New York 11973
(Received 16 December 1975)

It is shown that a quasi-one-dimensional system with strong attractive or repulsive coupling between electrons on the same site is equivalent to a Bose gas with hard core and longer-range interactions and to a Heisenberg-Ising magnet. Interchain and intrachain hopping and coupling are included and the equivalences are derived by elementary degenerate perturbation theory. Electronic correlation functions are derived from known pseudomagnetic ones and are used to discuss the circumstances in which superconductivity, charge- or spin-density wave transitions occur in the coupled chains.

I. INTRODUCTION

Quasi-one-dimensional systems typically consist of well-spaced chains of molecules for which the coupling of the electronic motion from one chain to another is relatively weak, and most of the interesting behavior takes place at temperatures T which are small compared to the scale temperature T_S for motions along the chain. Quantum-mechanical effects are important and, frequently, there are two types of correlation (e.g., superconductive and charge-density wave for metallic systems) which are coupled in an essential way. Any theory must, therefore, give an accurate account of the one-dimensional motion and the use of mean-field theory or classical fluctuations is likely to give misleading results. It *is* frequently possible to use mean-field theory for coupling *between* the chains because the long-range correlations which build up along the chains tend to suppress fluctuations,[1] but this still requires a good solution of the one-dimensional problem.

Of the methods proposed to deal with these systems, the renormalization group has been of limited use because, even when there is a phase transition and a fixed point of the Gell-Mann-Low equations, it occurs at an intermediate value of the coupling constant for which it is difficult to carry out accurate analytic calculations.[2] Using a different method Luther and Emery[3] have found an exact solution of a quite general model for a particular value of one of the coupling constants and, with some imagination, the renormalization group may be used as a qualitative way of scaling onto other values. However, it is not always easy to calculate correlation functions and the inclusion of phonons in a dynamical way presents some problems. Moreover, the mathematical method, though not difficult, is not in common use, and this has led to expressions of doubt about some of the results. In particular, a question[4] about the existence of an energy gap in the spin-density wave spectrum for attractive interactions does not appear to be justified.[5,6] Accordingly it would be useful to have an approach which uses relatively elementary methods and is easier to generalize. The purpose of this paper is to describe such a method. The ultimate objective is to include dynamical coupling to phonons, but for the present, in order to give a simple description and to make contact with previous results, a general static coupling between electrons will be adopted.

The main assumption is that the largest energy in the problem is the coupling U between electrons on the same molecule. This may be repulsive if Coulomb forces dominate or attractive if the molecule is very polarizable or, what amounts to the same thing, if there is strong coupling to intramolecular vibrations. It is quite likely that one or other of these situations is found in real physical systems, although this may only be realized *a posteriori*, when the results of calculations are compared to experiment.

The method is described in Sec. II. Elementary degenerate perturbation theory is used to take account of intrachain and interchain hopping of electrons and the coupling between electrons on different sites. It is shown that the effective Hamiltonian can be represented as either a set of bosons with hard cores and longer-ranged interactions or, more usefully, as a Heisenberg pseudospin system. For $U<0$, the bosons are bound pairs of electrons and superconductivity occurs when they condense. In the spin picture, the pseudospin waves correspond to density waves and antiferromagnetic ordering in the z direction gives a charge-density wave state, whereas ferromagnetic ordering in the x-y plane is equivalent to a superconducting transition. The electron spin waves decouple from the charge-density waves and there is a gap in their energy spectrum as a consequence of pair breaking. This is analogous to the gap found by Luther and Emery.[3] The pseudospin Hamiltonian has an anisotropic coupling stemming partly from hopping and partly from direct electron coupling. It could be ferromagnetic or antiferromagnetic.

When the sign of U is reversed, the roles of charge- and spin-density waves are interchanged and there is a gap in the charge-density wave spectrum for the special case of a half-filled band. The pseudospins form an isotropically coupled antiferromagnet and ordering corresponds to a spin-density wave state of the original electron gas. These general conclusions agree with the results of Emery, Luther, and Peschel[7] for an electron gas on a lattice, with intermediate repulsive coupling between electrons. In the absence of an interaction between electrons on different sites, the Hamiltonian is a generalization of the Hubbard model and, for $U>0$, the equivalence to a Heisenberg antiferromagnet has been obtained previously by Anderson.[8]

In Sec. III, the electronic correlation functions are derived from those of the Heisenberg-Ising model and used to discuss long-range order in the ground state of a single chain and finite-temperature transitions in coupled chains. It is shown that hopping between chains can lead to any kind of transition but a density-density coupling leads only to a charge-density wave state. Superconductivity has the highest transition temperature only if the attractive electron-phonon coupling outweighs the Coulomb interaction between sites. Finally, the correlation functions and the method of calculating them are compared to those of the model of Luther and Emery.[3,7]

II. LARGE ON-SITE COUPLING

It is assumed that the system consists of a set of chains of molecules providing a total of M sites for N electrons. The Hamiltonian is

$$H=H_0+H_1+H_2\,, \tag{2.1}$$

with

$$H_0=U\sum_{i,\lambda} n_{\lambda i+}n_{\lambda i-}\,, \tag{2.2}$$

$$H_1=\sum_{\substack{\lambda,\lambda'\\ i,j,\sigma}} t_{\lambda,i;\lambda',j}\,c^{\dagger}_{\lambda i\sigma}c_{\lambda' j\sigma}\,, \tag{2.3}$$

$$H_2=\sum_{\substack{\lambda,\lambda'\\ i,j,\sigma,\sigma'}} V_{\lambda,i;\lambda',j}\,n_{\lambda i\sigma}\,n_{\lambda' j\sigma'}\,. \tag{2.4}$$

Here, λ,λ' identify the chains, i,j the sites along each chain, and $\sigma=\pm$ specifies the spins of the electrons. The $c^{\dagger}_{\lambda i\sigma}$ create electrons in states which are localized on the molecules and $n_{\lambda i\sigma}$ are the corresponding number operators. The hopping integrals $t_{\lambda,i;\lambda',j}$ and the coupling $V_{\lambda,i;\lambda',j}$ are assumed to be zero for $\lambda=\lambda'$ and $i=j$, otherwise they are left unspecified. It is desirable to keep this general form since the geometric arrangement of molecules may allow the hopping and coupling between second and third neighbors along a chain to be comparable to that between electrons on molecules of different chains. The coupling in H_2 is a competition between Coulomb and electron-phonon interactions and its sign can vary from system to system. The largest energy is assumed to be U. If Coulomb interactions dominate $U>0$; but very polarizable molecules, which also allow the electrons to be reasonably far apart, can have a reduced Coulomb interaction and a strong attractive attraction from intramolecular vibrations which, together, can lead to $U<0$. Accordingly it will be sufficient to regard H_1 and H_2 as perturbations and to work to the lowest nontrivial order in H_1/U and H_2/U. It is simpler to separate the discussion of the two cases:

(a) $U<0$: Here, the unperturbed ground state has $N/2$ of the sites occupied by pairs of opposite-spin electrons. These are real bound pairs which replace Cooper pairs[9] in the strong coupling limit, although they are still responsible for superconductivity. When $N/2<M$, the ground state is degenerate because the energy $NU/2$ is independent of which sites are occupied. The addition of H_0 and H_1 splits the degeneracy to form a band of charge-density wave states in which the pairs move from site to site. On the other hand, a spin-density excitation turns over a spin to break a pair at a cost $|U|$ in energy which (by assumption) is much larger than the bandwidth of the low-lying states. This energy gap causes the spin-density waves to be frozen out at low temperatures.

(b) $U>0$: In this case the ground state has only one electron per site and it is degenerate because the energy does not depend upon which sites are occupied or upon the spins of the electrons. Once again, H_1 and H_2 split the degeneracy but there is no energy gap unless $N=M$ (half-filled band), when the excitation of charge-density waves is inevitably accompanied by double occupancy of sites at a cost in energy of at least $|U|$.

A more quantitative expression of this picture may be obtained from elementary degenerate perturbation theory. Attractive interactions, $U<0$, will be considered first and it will then be shown how the results may be used for $U>0$ without further calculation.

A. Attractive on-site interactions

The discussion will be restricted to the lowest band, in which all electrons remain paired. This is sufficient for a calculation of the low-temperature properties of the system. In first-order degenerate perturbation theory, $H_1 + H_2$ has to be diagonalized in the space of paired states and, within that space, it is an effective Hamiltonian from which the thermodynamic properties may be obtained. But H_1 breaks pairs and so has no first-order matrix elements. It must therefore be calculated to second order, allowing virtual transitions into the next band of states which have one pair broken. Suppose $H_2 = 0$ for the moment, and let the various degenerate ground states of H_0 be denoted by $|\alpha\rangle$ with energy $E_0 = NU/2$. The Schrödinger equation is

$$(E - H_0)\,|\psi\rangle = H_1\,|\psi\rangle \tag{2.5}$$

and, dividing both sides by $E - H_0$ and rearranging, gives

$$|\psi\rangle = \sum_{\alpha} |\alpha\rangle \frac{\langle\alpha|H_1|\psi\rangle}{E - E_0} + \frac{P}{E - H_0} H_1 |\psi\rangle , \tag{2.6}$$

where $P = 1 - \sum_\alpha |\alpha\rangle\langle\alpha|$ projects out of the unperturbed ground states. By substitution, it can be seen that Eq. (2.6) is equivalent to

$$|\psi\rangle = \sum_{\alpha} a_\alpha |\psi_\alpha\rangle , \tag{2.7}$$

where

$$|\psi_\alpha\rangle = |\alpha\rangle + [P/(E - H_0)]\, H_1 |\psi_\alpha\rangle \tag{2.8}$$

and

$$a_\alpha = \langle\alpha|H_1|\psi\rangle/(E - E_0) . \tag{2.9}$$

To first order, Eq. (2.8) is

$$\begin{aligned} |\psi_\alpha\rangle &= |\alpha\rangle + [P/(E - H_0)] H_1 |\alpha\rangle \\ &= |\alpha\rangle + (1/U) H_1 |\alpha\rangle . \end{aligned} \tag{2.10}$$

The last line follows because H_1 breaks exactly one pair to give an excitation energy $-U$ and P is irrelevant because $H_1|\alpha\rangle$ has no component in the ground states. Then, substituting Eqs. (2.7) and (2.10) into Eq. (2.9) gives

$$(E - E_0) a_\alpha = \frac{1}{U} \sum_{\alpha'} \langle\alpha|H_1^2|\alpha'\rangle a_{\alpha'} , \tag{2.11}$$

which is a Schrödinger equation in the $|\alpha\rangle$ subspace with effective Hamiltonian H_1^2/U. To ensure that H_1 acts between the ground states, it is necessary that if the first application of H_1 transfers an electron with spin σ from site j to site i, then the second application of H_1 either returns the electron to its original site or transfers an electron with spin $-\sigma$ from j to i. Thus, the effective Hamiltonian is

$$\begin{aligned} H_1' = -\sum_{\substack{\lambda,\lambda' \\ i,j,\sigma}} \frac{t^2_{\lambda,i;\lambda',j}}{|U|} \big(& c^\dagger_{\lambda i\sigma} c_{\lambda' j\sigma} c^\dagger_{\lambda' j\sigma} c_{\lambda i\sigma} \\ & + c^\dagger_{\lambda i,-\sigma} c_{\lambda' j,-\sigma} c^\dagger_{\lambda i\sigma} c_{\lambda' j\sigma} \big) . \end{aligned} \tag{2.12}$$

Since it is sufficient to work to first order in H_2,

$$H' = H_1' + H_2 \tag{2.13}$$

is the total effective Hamiltonian which may be used to obtain the thermodynamic properties at temperatures $T \ll |U|$.

There are several equivalent ways of rewriting H' which relate it to more familiar systems. Define

$$n_{\lambda i} = \tfrac{1}{2}(n_{\lambda i+} + n_{\lambda i-} - 1) , \tag{2.14}$$

$$b_{\lambda i} = c_{\lambda i+} c_{\lambda i-} , \tag{2.15}$$

$$\sigma_{\lambda i} = \tfrac{1}{2}(n_{\lambda i+} - n_{\lambda i-}) . \tag{2.16}$$

Then

$$\begin{aligned} H' = \sum_{\substack{\lambda,\lambda' \\ i,j,\sigma}} \bigg(& \frac{2t^2_{\lambda,i;\lambda',j}}{|U|} (-b^\dagger_{\lambda i} b_{\lambda' j} + n_{\lambda i} n_{\lambda' j} + \sigma_{\lambda i}\sigma_{\lambda' j} - \tfrac{1}{2}) \\ & + V_{\lambda,i;\lambda',j} n_{\lambda i} n_{\lambda' j} \bigg) . \end{aligned} \tag{2.17}$$

The operators $\sigma_{\lambda i}$ commute with the $b_{\lambda' j}$ and $n_{\lambda' j}$ and give zero acting on every doubly occupied site, so they are dynamically insignificant and will be ignored. The n_i and b_i satisfy the commutation relations

$$\begin{aligned} &[b_{\lambda i}, b^\dagger_{\lambda' j}] = [b_{\lambda i}, b_{\lambda' j}] = 0 , \quad i \neq j \\ &b^2_{\lambda i} = 0 , \\ &[n_{\lambda i}, b^\dagger_{\lambda' j}] = b^\dagger_{\lambda' j}\delta_{ij}\delta_{\lambda\lambda'} . \end{aligned} \tag{2.18}$$

It is then possible to interpret the $b_{\lambda i}$ as boson operators and the $n_{\lambda i}$ as the corresponding number operators, provided a hard-core interaction is added to ensure that at most one boson occupies each site. For a single chain, such a model has been considered by Schultz.[10] The question of the existence of superconductivity in the original system is now rephrased as the existence of a condensation of the bosons. In the BCS theory of superconductivity,[11] the pairs of fermions condense into a macroscopic occupation of a zero-momentum state as they form whereas, in the strong-coupling limit, the formation of pairs and the condensation of their center-of-mass motion occurs independently. Indeed, for a single chain with only hard-core interactions, the ground state is composed entirely of pairs but there is no superconductivity because the bosons do not condense.[10]

A more useful representation of the system is obtained by noticing that Eqs. (2.18) are the commutation relations of spin-$\frac{1}{2}$ operators with the identification

$$b_{\lambda i}=s^-_{\lambda i},\quad b^\dagger_{\lambda i}=s^+_{\lambda i},\quad n_{\lambda i}=s^z_{\lambda i}\,, \qquad (2.19)$$

where $s^\pm_{\lambda i}=s^x_{\lambda i}\pm s^y_{\lambda i}$. Comparison of Eqs. (2.14) and (2.19) shows that $s^z_{\lambda i}$ is +1 for a site occupied by a pair and −1 for an unoccupied site. As in the case of the Bose gas, it is much more convenient to carry out calculations in the grand canonical distribution, and since the number operator is essentially $\sum_{\lambda,i}s^z_{\lambda i}$, the chemical potential is an effective magnetic field and will be denoted by h. Then, in this representation,

$$H'=\sum_{\substack{\lambda,\lambda'\\ i,j,\sigma}}\left(\frac{2t^2_{\lambda,i;\lambda',j}}{|U|}(s^z_{\lambda i}s^z_{\lambda' j}-s^x_{\lambda i}s^x_{\lambda' j}-s^y_{\lambda i}s^y_{\lambda' j})\right.$$
$$\left.+V_{\lambda,i;\lambda',j}s^z_{\lambda i}s^z_{\lambda' j}-h\sum_{\lambda i}s^z_{\lambda i}\right) \qquad (2.20)$$

and "spin-wave" excitations correspond to charge-density waves for the original electron system. In Sec. II B it will be shown that antiferromagnetic order in the z-components is equivalent to charge-density wave order in the electron gas and ferromagnetic order in the x-y direction represents condensation of the bosons or superconductivity of the electrons. The connection between the boson and the spin pictures is analogous to the pseudospin model of liquid helium.[12]

B. Repulsive on-site interactions

Turning now to the case $U>0$, it will be assumed that $N=M$ (half-filled band), otherwise the hopping term H_1 breaks the degeneracy in first order by transferring an electron from a singly occupied site to an unoccupied site. This case is also of considerable physical interest. The effective Hamiltonian may be obtained directly from Eqs. (2.14)–(2.17) by making the canonical transformation

$$c_{\lambda i-}=\bar{c}^\dagger_{\lambda i-},\quad c_{\lambda i+}=\bar{c}_{\lambda i+}\,, \qquad (2.21)$$

for which

$$\bar{H}_0=-U\sum_{i,\lambda}\bar{n}_{\lambda i+}\bar{n}_{\lambda i-}+U\sum_{i,\lambda}\bar{n}_{\lambda i+}\,, \qquad (2.22)$$

$$\bar{H}_1=\sum_{\lambda,\lambda',i,j}t_{\lambda,i;\lambda',j}(\bar{c}^\dagger_{\lambda i+}\bar{c}_{\lambda' j+}-\bar{c}^\dagger_{\lambda i-}\bar{c}_{\lambda' j-})\,. \qquad (2.23)$$

In the unperturbed states, sites which were occupied by an electron with $\sigma=-1$ are now empty whereas other sites are doubly occupied. The "on-site" coupling becomes attractive and the number of pairs is equal to the original number of spins with $\sigma=+1$. Second-order degenerate perturbation theory may now be used exactly as before and, on reversing the transformation (2.21), the effective Hamiltonian is given by Eq. (2.17) with the sign of the $b^\dagger_{\lambda i}b_{\lambda' j}$ term changed.

The variables have a rather different interpretation, since $b_{\lambda i}$ becomes $c_{\lambda i+}c^\dagger_{\lambda i-}$ and $\sigma_{\lambda i}$ replaces $n_{\lambda i}$ in the commutation relations (2.18) and in the relation to the spin operators in Eq. (2.19). Since every site is singly occupied, all of the $n_{\lambda i}$ give unity when applied to the ground states and so are dynamically insignificant. They will therefore be omitted and the effective spin Hamiltonian takes the antiferromagnetic, spin-isotropic, Heisenberg form:

$$H''=\sum_{\lambda,\lambda',i,j}\frac{2t^2_{\lambda,i;\lambda',j}}{|U|}\,\vec{S}_{\lambda i}\cdot\vec{S}_{\lambda' j}\,. \qquad (2.24)$$

There is no magnetic field because the number of electrons is fixed. The pseudospin-wave excitations are now true spin waves and changing the sign of U has interchanged the roles of charge- and spin-density excitations. This feature was found for a single chain by Emery, Luther, and Peschel[7] when they obtained an exact solution for intermediate coupling. Because the original Hamiltonian had no direct spin-spin interaction, the exchange constants in Eq. (2.24) come entirely from the hopping term. For the repulsive Hubbard model, $U>0$ and $H_2=0$, Eq. (2.24) has previously been obtained by Harris and Lange[8] using a canonical transformation method.

III. CORRELATION FUNCTIONS

To discuss the existence of the various kinds of phase transition mentioned earlier, it is necessary to evaluate the correlation functions generated by $c^\dagger_{\lambda i\sigma}c_{\lambda' j\sigma'}$ or $c_{\lambda i\sigma}c_{\lambda' j\sigma'}$. When $U<0$, the wave functions for the low-lying states are linear combinations of the $|\psi_\alpha\rangle$ of Eq. (2.10) which, in turn, are linear combinations of states with all sites occupied by pairs of electrons of opposite spin or with pairs broken without spin flip, and only $c^\dagger_{\lambda i\pm}c_{\lambda' j\pm}$ or $c^\dagger_{\lambda i+}c^\dagger_{\lambda' j-}$ have matrix elements within this space. The other combinations connect to states in which there are unpaired electrons with parallel spins and are separted from the ground state by an energy gap $|U|$ which prevents divergences at zero frequency. This agrees with the conclusion of Lee.[5,6]

To lowest order in $|t_{\lambda,i;\lambda',j}/U|$ it is sufficient to take $|\psi_\alpha\rangle\approx|\alpha\rangle$ and merely use H_1 and H_2 to determine the coefficients a_α for Eq. (2.7). Then only the operators

$$\tfrac{1}{2}(c^\dagger_{\lambda i+}c_{\lambda i+}+c^\dagger_{\lambda i-}c_{\lambda i-})\equiv s^z_{\lambda i}+\tfrac{1}{2}\,, \qquad (3.1)$$

$$c^\dagger_{\lambda i+}c^\dagger_{\lambda i-}\equiv s^+_{\lambda i} \qquad (3.2)$$

are relevant. In writing Eqs. (3.1) and (3.2) use has been made of Eqs. (2.14), (2.15), and (2.19). Since $\sigma_{\lambda i}$ commutes with the other operators, the

$c^\dagger_{\lambda i\pm}c_{\lambda i\pm}$ and $c^\dagger_{\lambda i+}c^\dagger_{\lambda i-}$ correlation functions are equivalent to the $s^z_{\lambda i}$ and $s^+_{\lambda i}$ correlation functions of the spin representation of H' as asserted earlier. A charge-density wave then corresponds to antiferromagnetic ordering of the $s^z_{\lambda i}$, with the wave vector of the condensation determined by the applied field which represents the chemical potential. Similarly, superconductivity corresponds to ferromagnetic ordering in the $s^x_{i\lambda}$ and $s^y_{i\lambda}$ variables.

A similar argument may be made for $U>0$. Starting from a half-filled band, $c^\dagger_{\lambda i+}c^\dagger_{\lambda' j-}$ create states in which sites are doubly occupied at a cost of energy $2|U|$ relative to the ground state. The ordering variables are

$$s^+_{\lambda i}\equiv c_{\lambda i-}c^\dagger_{\lambda i+} \tag{3.3}$$

and

$$s^z_{\lambda i}\equiv\sigma_{\lambda i}=\tfrac{1}{2}(c^\dagger_{\lambda i+}c_{\lambda i+}-c^\dagger_{\lambda i-}c_{\lambda i-}), \tag{3.4}$$

which correspond to transverse and longitudinal spin-density waves, consistent with the spin isotropy of the Hamiltonian.

Given these representations, it is possible to make use of what is known about the Heisenberg-Ising model to calculate the properties of the electron system. The discussion will be restricted to the physically interesting case of near-neighbor coupling. First consider a single chain and choose units so that $2t^2_{\lambda i;\lambda,i+1}/|U|=1$. Dropping the subscript λ, the contribution to the Hamiltonian from a single chain is, for $h=0$,

$$H_1=-\sum_i(s^x_is^x_{i+1}+s^y_is^y_{i+1}+J_zs^z_is^z_{i+1})\,. \tag{3.5}$$

Here, for $U>0$, $J_z=-1$ gives the contribution to H'' in Eq. (2.24) after rotation of axes on alternate sites about the z direction and, for $U<0$, the contribution to H' in Eq. (2.20) is obtained if

$$-J_z=V_{\lambda i;\lambda,i+1}+1\,. \tag{3.6}$$

The case of greatest interest is $0\ge J_z\ge-1$, for which the asymptotic forms of the correlation functions have been obtained by Luther and Peschel.[13] It will be seen that this range is important for the discussion of the circumstances favorable to superconductivity. It is physically less realistic to have $J_z>0$ because it requires a large attractive $V_{\lambda i;\lambda,i+1}$ and the pairs would form clusters in which adjacent sites were occupied. The properties for $J_z<-1$ may be obtained from the numerical calculations of Bonner and Fisher.[14]

The route followed by Luther and Peschel[13] is to use a Jordan-Wigner transformation[15] to rewrite H_1 as a fermion Hamiltonian and then to replace the kinetic energy by a linear spectrum to obtain a Luttinger model, for which it is relatively easy to obtain the correlation functions.[16] If the site label i is replaced by the distance r along the chain, then the asymptotic forms of the s^z and $s^\pm$ correlation functions are given by[13]

$$\langle s^z(r,t)s^z\rangle=(2\pi^2\alpha^2)^{-1}\cos 2k_Fr\left(\frac{\alpha^2}{r^2-c^2t^2}\right)^{1/2\theta} \tag{3.7}$$

and

$$\langle s^+(r,t)s^-\rangle+\langle s^-(r,t)s^+\rangle=(2\pi^2\alpha^2)^{-1}\left(\frac{\alpha^2}{r^2-c^2t^2}\right)^{\theta/2} \tag{3.8}$$

for $T=0$. Here, c is the Fermi velocity, t the time, α a cutoff, and

$$\theta=\tfrac{1}{2}-\pi^{-1}\arcsin J_z\,. \tag{3.9}$$

For large r, the correlation functions in Eqs. (3.7) and (3.8) fall off as $r^{-\theta^{-1}}$ and $r^{-\theta}$, respectively, and since, according to Eq. (3.9), $\frac{1}{2}\le\theta\le1$ when $0\ge J_z\ge-1$, there is no long-range order in the ground state. This conclusion has been reached previously by Schultz[10] for the special case[17] $J_z=0$, $\theta=\frac{1}{2}$, which corresponds to the Bose gas with hard-core interactions.

On the other hand, in a three-dimensional system, interchain coupling may produce long-range order at a finite temperature T_c. If T_c is small (in units of the exchange integral), it is possible to use mean-field theory for the interchain coupling provided the motion along the chains is treated accurately. Using Eqs. (3.7) and (3.8) together with the results of Ref. 1, Eq. (9),

$$T_c\sim|2t^2_{\lambda i;\lambda+1,i}/|U|+V_{\lambda i;\lambda+1,i}|^{1/(2-\theta^{-1})} \tag{3.10}$$

for a charge-density wave transition ($U<0$), and

$$T_c\sim(2t^2_{\lambda i;\lambda+1,i}/|U|)^{1/(2-\theta)} \tag{3.11}$$

for superconductivity ($U<0$) or a spin-density wave state ($U>0$, $\theta=1$). Note that the arguments in Eqs. (3.10) and (3.11) are small because they are interchain couplings in units of the intrachain exchange integral. It is clear that hopping between chains can drive any of the transitions but the density-density coupling $V_{\lambda i;\lambda+1,i}$ can give rise to a charge-density wave instability only.[18] Equations (3.6) and (3.9)–(3.11) also show that if superconductivity is to occur at a higher temperature than the charge-density wave instability, the electron-phonon interaction must outweigh the Coulomb force to make $V_{\lambda i,\lambda' j}$ attractive. This would either decrease the argument of Eq. (3.10) or, more effectively, decrease θ.

It is interesting to compare these conclusions with previous calculations.[3,7] The Fourier transforms of the correlation functions in Eqs. (3.7) and (3.8) are proportional to ω^μ, with μ equal to $\theta^{-1}-2$ and $\theta-2$, respectively. When $V_{\lambda i;\lambda,i+1}=0$

as in Refs. 3 and 7, it follows from Eqs. (3.6) and (3.9) that $\mu = -1$ in both cases. This is in agreement with the results of Luther and Emery[3] and Emery, Luther, and Peschel[7] for a lattice model provided there is a half-filled band. In that case, in the low-temperature limit, when $U<0$ backward scattering produces a gap in the spin-density wave spectrum and umklapp scattering renormalizes to zero the charge-density wave coupling v' which appears in Table I of Ref. 3. For $U>0$, the roles of backward scattering and umklapp scattering and of charge-density and spin-density waves are interchanged. This does not mean that the special values of the coupling constants considered in Refs. 3 and 7 are effectively in the large-$|U|$ limit but, rather, for a half-filled band, the exponents are independent of U. It is necessary to keep this in mind in making comparisons with the phase diagrams for coupled chains obtained by Klemm and Gutfreund,[18] which are rather different from those obtained here.

It is clear that the evaluation of the correlation functions is considerably simpler in the strong-coupling limit than in the case considered by Luther and Emery[3] and Emery, Luther, and Peschel.[7] It is also possible to work with more general Hamiltonians. The reason for this is that H_1 in Eq. (3.5) is equivalent to a set of *spinless* fermions for which umklapp scattering and backward scattering do not play a particularly crucial role. In the model of Luther and Emery[3,7] it is necessary to use a renormalization-group argument to deal with one or the other of these processes in calculating the exponents for an electron gas on a lattice, when there is a half-filled band. Also, numerical factors in the correlation functions involve boson representations of powers of fermion field operators, which are difficult to evaluate. Finally, Eqs. (3.7) and (3.8) give the correlation functions for the more general Hamiltonian (3.5), which includes the effects of direct near-neighbor coupling $V_{\lambda i;\lambda' j}$ as well as hopping.

It is hoped that the general approach described in this paper is simple enough that it can give a description of the ordered states and be extended to include the dynamical effects of phonons. These topics are under investigation.

ACKNOWLEDGMENT

I wish to acknowledge valuable discussions with S. Krinsky on various aspects of this work.

---

*Work supported by Energy Research and Development Administration.

[1]D. J. Scalapino, Y. Imry, and P. Pincus, Phys. Rev. B 5, 2042 (1975); Y. Imry, P. Pincus, and D. J. Scalapino, *ibid.* 12, 1978 (1975).

[2]L. Mihaly, and J. Solyom (unpublished).

[3]A. Luther and V. J. Emery, Phys. Rev. Lett. 33, 589 (1974).

[4]H. U. Everts and H. Schulz (unpublished). Their premise that an energy gap implies the existence of long-range order is clearly shown to be false in the limit discussed in subsequent sections of the present paper.

[5]P. A. Lee, Phys. Rev. Lett. 34, 1247 (1975).

[6]This does not refer to the point made by Lee (see Ref. 5), who showed that an alternative calculation of two out of eight correlation functions disagreed with the result of Ref. 3. The question is difficult to resolve for infinite bandwidth, where exact calculations may be made, but Lee's result appears to be correct for finite bandwidth.

[7]V. J. Emery, A. Luther, and I. Peschel, Phys. Rev B 13, 1272 (1976); see also H. Gutfreund and R. A. Klemm, Phys. Rev. B (to be published).

[8]P. W. Anderson, in *Solid State Physics*, edited by F. Seitz and D. Turnbull (Academic, New York, 1963), Vol. 14, p. 99.

[9]V. J. Emery, Nucl. Phys. 12, 69 (1959).

[10]T. D. Schultz, J. Math. Phys. 4, 666 (1963).

[11]J. Bardeen L. N. Cooper, and J. R. Schrieffer, Phys. Rev. 108, 1175 (1957).

[12]T. Matsubara and H. Matsuda, Prog. Theor. Phys. 16, 569 (1956); 17, 19 (1957).

[13]A. Luther and I. Peschel, Phys. Rev. B 12, 3908 (1975).

[14]J. C. Bonner and M. E. Fisher, Phys. Rev. 135, A640 (1964).

[15]P. Jordon and E. Wigner, Z. Phys. 47, 631 (1928).

[16]A. Theumann, J. Math. Phys. 8, 2460 (1967); C. B. Dover, Ann. Phys. (N.Y.) 50, 500 (1968); A. Luther and I. Peschel, Phys. Rev. B 9, 2911 (1974).

[17]After this paper was submitted for publication, Dr. A. Zawadowski showed me a paper by K. V. Efetov and A. I. Larkin, Zh. Eksp. Teor. Fiz. 69, 764 (1975) [English translation not yet available], who also considered the large-U limit and used the boson picture to obtain correlation functions for $J_z = 0$. The discussion of sound-wave excitations given in this paper is physically similar to that of Luther and Emery (Ref. 3) but omits umklapp scattering (Ref. 7), which has a significant effect on the conclusions.

[18]This has also been shown for intermediate coupling by R. A. Klemm and H. Gutfreund [Phys. Rev. B (to be published)], who also discussed the use of mean-field theory for coupling between the chains.

PHYSICAL REVIEW B VOLUME 2, NUMBER 10 15 NOVEMBER 1970

Application of Gutzwiller's Variational Method to the Metal-Insulator Transition

W. F. Brinkman and T. M. Rice
Bell Telephone Laboratories, Murray Hill, New Jersey 07974
(Received 16 April 1970)

It is shown that the approximate variational calculation of Gutzwiller predicts a metal-insulator transition as the intra-atomic Coulomb interaction is increased for the case of one electron per atom. The susceptibility and effective mass are calculated in the metallic phase and are found to be enhanced by a common factor which diverges at the critical value of the interaction.

Several years ago, Gutzwiller[1] performed an approximate variational calculation of the ground-state wave function for a model Hamiltonian with a single tight-binding band and with only intra-atomic Coulomb interactions between the electrons. This model Hamiltonian, introduced earlier by Hubbard,[2] Gutzwiller,[3] and Kanamori,[4] is generally known as the Hubbard model and has been studied by many authors. Using Gutzwiller's[1] notation, as we shall in this paper, the model Hamiltonian has the form

$$H=\sum_{\vec{k}}\epsilon_{\vec{k}}(a^{\dagger}_{\vec{k}\uparrow}a_{\vec{k}\uparrow}+a^{\dagger}_{\vec{k}\downarrow}a_{\vec{k}\downarrow})+C\sum_{\vec{g}}a^{\dagger}_{\vec{g}\uparrow}a^{\dagger}_{\vec{g}\downarrow}a_{\vec{g}\downarrow}a_{\vec{g}\uparrow}\ , \quad (1)$$

where $a^{\dagger}_{\vec{k}}$ and $a^{\dagger}_{\vec{g}}$ are the creation operators for electrons in the Bloch state $\{\vec{k}\}$ and the Wannier state $\{\vec{g}\}$, respectively, C is the intra-atomic Coulomb repulsion, and $\epsilon_{\vec{k}}$ is the kinetic energy, with the zero of energy chosen so that $\sum_{\vec{k}} \epsilon_{\vec{k}} = 0$.

Gutzwiller[1] constructed a trial wave function by starting with the conventional Bloch state for noninteracting electrons and reducing the amplitude of all components in which ν atoms are doubly occupied by an amount η^{ν}, where $0<\eta<1$. He calculated the needed matrix elements by neglecting the kinetic energy of the down-spin electrons, arguing that this procedure should be a good approximation to an optimally chosen generalization of his wave function. This led to an explicit and spin-symmetric expression for the energy as a function of η. This expression was then minimized with respect to η and the ground-state energy obtained. Gutzwiller used his results to obtain a criterion for itinerant ferromagnetism. In this paper we wish to apply his calculation to the problem of the metal-insulator transition.

We consider only the case in which there is one electron per atom. (For any other number of electrons per atom Gutzwiller's variational state is always metallic.) With one electron per atom, Eqs. (B7) and (B8) of Gutzwiller's give η and q in terms of $\bar{\nu}(\equiv \langle n_{i\uparrow} n_{i\downarrow}\rangle_\eta)$:

$$\eta = \bar{\nu}/(\tfrac{1}{2}-\bar{\nu}) , \tag{2}$$

$$q = 16\bar{\nu}\,(\tfrac{1}{2}-\bar{\nu}) . \tag{3}$$

Here q is the discontinuity in the single-particle occupation number $\langle n_{\vec{k}}\rangle$ at the Fermi surface. Substituting these results into Eq. (36) for the energy and minimizing with respect to $\bar{\nu}$, we find for the lowest-energy state

$$\bar{\nu} = \tfrac{1}{4}\,(1 - C/C_0) , \tag{4}$$

$$q = 1-(C/C_0)^2 , \tag{5}$$

and the expectation value of the energy in the (paramagnetic) ground state

$$\langle H\rangle_N = \bar{\epsilon}\,(1-C/C_0)^2 . \tag{6}$$

Here

$$\bar{\epsilon} = 2\sum_{k<k_F} \epsilon_{\vec{k}} < 0$$

is the average energy without correlation and $C_0 = -8\bar{\epsilon}$. Thus, at a critical value of the interaction strength $C = C_0$ the number of doubly occupied sites and the discontinuity in the single-particle occupation number at the Fermi surface go to zero. The value of the energy (6) also approaches zero, the expectation value of the energy of a paramagnetic localized insulating state. However, it is clear that some magnetically ordered insulating ground state will have a lower energy than the paramagnetic insulating state and a transition to an insulating magnetically ordered ground state will occur for a value of C less than C_0.

Nevertheless, it is interesting to calculate the properties of this trial wave function in the metallic state. If we assume that the effective-mass renormalization m^*/m is due to the frequency dependence of the self-energy only, as for example, in the electron-phonon and paramagnon problems,[5-7] then m^*/m equals the reciprocal of the discontinuity at the Fermi surface in the single-particle occupation number,

$$m^*/m = q^{-1} = [1-(C/C_0)^2]^{-1} . \tag{7}$$

The effective mass, therefore, diverges as C approaches C_0. Gutzwiller also calculated the minimum energy for states with differing numbers of up- and down-spin electrons and the static susceptibility χ_s can be obtained by expanding his results to second order in the magnetization. Defining the magnetization $2\zeta = (\langle N_\uparrow\rangle - \langle N_\downarrow\rangle)/N$, we find

$$\bar{\epsilon}_\zeta = \bar{\epsilon} + \zeta^2/\rho(\epsilon_F) , \tag{8}$$

$$q_\zeta = 16\bar{\nu}\,(\tfrac{1}{2}-\bar{\nu})\,\{1+\zeta^2[4-\tfrac{1}{4}(\tfrac{1}{2}-\bar{\nu})^{-2}]\} , \tag{9}$$

where $\rho(\epsilon_F)$ is the noninteracting one electron density of states at the Fermi energy. Upon substituting (8) and (9) into the expression from the ground-state energy we find

$$\chi_s^{-1} = \frac{1-(C/C_0)^2}{\rho(\epsilon_F)}\left[1-\rho(\epsilon_F)C\left(\frac{1+(C/2C_0)}{[1+(C/C_0)]^2}\right)\right] . \tag{10}$$

Therefore, as C approaches C_0 both the susceptibility and the effective mass diverge in proportion to $[1-(C/C_0)^2]^{-1}$. This result is quite different from the type of result obtained from paramagnon theory[6,7] near a ferromagnetic instability. In that theory the mass is proportional to the logarithm of the susceptibility, and although both χ_0 and m^* diverge, the ratio of the two goes to infinity. In the present case this ratio goes to a finite value. We note that if range effects are ignored in paramagnon theory, corresponding to a uniform enhancement of the static wave-vector-dependent susceptibility, then χ_0 and m^* would scale as we find.

Examining Eqs. (7) and (10), it is clear that the susceptibility enhancement is not coming from the usual Stoner enhancement factor $[1-\rho(\epsilon_F)C]^{-1}$, but rather from the effective mass. The Stoner factor has been replaced by the expression in the brackets in (10). This expression becomes small only if $\rho(\epsilon_F)$ is considerably larger than the average density of states in the band. Therefore, the possibility of itinerant ferromagnetism prior to the metal-insulator transition is greatly reduced. We have not been able to calculate $\chi(\vec{Q})$, the static wave-vector-dependent susceptibility. However, if we accept that $\chi(\vec{Q})$ is roughly independent of Q, as one naively expects for a localized instability, then it is interesting to speculate that the correlation effects could possibly also suppress itinerant antiferromagnetism and lead to a first-order transition between a paramagnetic metallic state and an antiferromagnetic insulating state.

The results obtained from Gutzwiller's method are to be contrasted with those found by Hubbard[2] using a Green's-function decoupling approximation. While Hubbard's approximation is reasonable for the insulating phase, it certainly is incorrect for the metallic phase since it does not properly describe the Fermi surface as emphasized by Herring[8] and by Edwards and Hewson.[9] Further, in the Hubbard approximation the density of states at the Fermi surface approaches zero as $C \to C_0$. The Gutzwiller calculation, on the other hand, builds in the Fermi surface from the start and gives an appealing description of a metallic state in which the discontinuity in the single-particle occupation number at the Fermi surface becomes small as the system becomes closer to the metal-insulator transition.

In conclusion, it is interesting to compare the above results with the experimental properties of the metallic state of V_2O_3.[10-12] This type of comparison may be meaningless since V_2O_3 is surely a complicated many-band situation for which the simple model studied by Gutzwiller is not applicable. Nevertheless, the Gutzwiller results are not strongly dependent on the density of states, and it is interesting that the specific heat and the susceptibility appear to be enhanced by roughly the same amount. In V_2O_3 the susceptibility and specific-heat density of states of the metallic phase both appear to be quite large. An extrapolation of the susceptibility to 0 °K gives a value for χ_s, expressed as a density of states, of 35 states/eV molecule. A rough estimate of the specific-heat density of states $N(\epsilon_F)$, can be obtained as follows. If we assume that the difference between the metallic and insulating specific heats is of the form $\Delta C_v = \gamma T + \beta T^3$, then the parameters γ and β can be estimated by setting (a) $\int_0^{T_N} \Delta C_v dT/T = \Delta S$, the change in entropy at the metal-to-antiferromagnetic-insulating phase transition in pure V_2O_3 at 1 atm, and (b) $\Delta C_v(T_N = 170\,°K) = 0$, in agreement with Anderson's[13] experimental results. This gives $N(\epsilon_F) = 20$ states/eV molecule, which is quite large. However, the ratio $\chi_s/N(\epsilon_F)$ is only 1.75, so that the two quantities appear to be roughly equally enhanced.

The authors would like to thank D. B. McWhan and D. R. Hamann for several useful comments.

[1]M. C. Gutzwiller, Phys. Rev. 137, A1726 (1965).

[2]J. Hubbard, Proc. Roy. Soc. (London) A276, 238 (1963); 277, 238 (1964).

[3]M. C. Gutzwiller, Phys. Rev. Letters 10, 159 (1963); Phys. Rev. 134, A923 (1964).

[4]J. Kanamori, Progr. Theoret. Phys. (Kyoto) 30, 275 (1963).

[5]J. R. Schrieffer, *Theory of Superconductivity* (Benjamin, New York, 1964).

[6]N. F. Berk and J. R. Schrieffer, Phys. Rev. Letters 17, 433 (1966).

[7]S. Doniach and S. Engelsberg, Phys. Rev. Letters 17, 750 (1969).

[8]C. Herring, in *Magnetism*, edited by G. T. Rado and H. Suhl (Academic, New York, 1966), Vol. 4.

[9]D. M. Edwards and A. C. Hewson, Rev. Mod. Phys. 40, 810 (1968).

[10]D. B. McWhan, T. M. Rice, and J. P. Remeika, Phys. Rev. Letters 23, 1384 (1969).

[11]T. M. Rice and D. B. McWhan, IBM J. Res. Develop. 14, 251 (1970).

[12]D. B. McWhan and J. P. Remaika, Phys. Rev. (to be published).

[13]C. T. Anderson, J. Am. Chem. Soc. 58, 564 (1936).

VOLUME 54, NUMBER 12 PHYSICAL REVIEW LETTERS 25 MARCH 1985

Attractive Interaction and Pairing in Fermion Systems with Strong On-Site Repulsion

J. E. Hirsch
Department of Physics, University of California, San Diego, La Jolla, California 92093
(Received 19 November 1984)

It is shown that an effective attractive interaction between nearest-neighbor antiparallel spins arises in the Hubbard and Anderson lattice models in the limit of large on-site electron-electron repulsion. Results of Monte Carlo simulations of the Hubbard model show enhancement of anisotropic singlet-pairing correlations and suppression of triplet-pairing correlations. It is proposed that this interaction leads to an anisotropic singlet superconducting state in the heavy-fermion superconductors.

PACS numbers: 75.10.Hk, 74.20.-z

The discovery of superconductivity in heavy-fermion systems[1] has posed many interesting questions, in particular whether it is the first example of non-phonon-induced superconductivity. It has been suggested that these systems exhibit triplet superconductivity,[2,3] with the attractive interaction due to paramagnon exchange.[3] Other theoretical treatments assume the usual BCS state induced by electron-phonon interactions.[4]

Two models of interacting fermions, the Hubbard and Anderson models, have been used to describe these systems.[5-7] While the Anderson model is closer to the real materials, the Hubbard model is simpler and probably shares some of its features. Here, we study these models in the strong-coupling limit. The Hubbard model is defined by

$$H=-t\sum_{\substack{\langle i,j\rangle\\ \sigma}}(c_{i\sigma}^{\dagger}c_{j\sigma}+\text{H.c.})+U\sum_i n_{i\uparrow}n_{i\downarrow}. \qquad (1)$$

In the limit of large U, H reduces to an antiferromagnetic Heisenberg (AFH) model in the half-filled band case (one electron per site).[8] Klein and Seitz[9] showed that an AFH model still describes the *spin* degrees of freedom for other than half-filled cases in one dimension in strong coupling. Here, we derive an effective Hamilonian for the spin *and charge* degrees of freedom. We take as zeroth-order Hamiltonian the interaction term, and as perturbation the hopping. The ground state to zeroth order is highly degenerate, consisting of an arbitrary arrangement of *singly occupied* sites. To first order, the kinetic-energy term forms a band described by the Hamiltonian

$$H_1=-t\sum_{\substack{\langle i,j\rangle\\ \sigma}}(h_{i\sigma}^{\dagger}h_{j\sigma}+\text{H.c.}). \qquad (2)$$

The operators $h_{i\sigma}$ hop single electrons from site to site but are not true fermion operators since they do not allow double occupancy, i.e., $h_{i\uparrow}^{\dagger}h_{i\downarrow}^{\dagger}=0$. In one dimension, it is easy to see that Eq. (2) describes a noninteracting gas of *spinless fermions*. In higher dimensions the properties of the Hamiltonian of Eq. (2) are unknown, but it is plausible to assume that the system can be described by a Fermi liquid of spinless fermions as far as the charge degrees of freedom are concerned, by analogy with one dimension. Such a system should become unstable in the presence of a small attractive interaction.

To next order in t/U, we allow for virtual transitions into states with doubly occupied sits. The effective Hamiltonian, acting only on the states with single occupied sites, is

$$H_{\text{eff}}=-t'\sum_{\substack{\langle i,j\rangle\\ \sigma}}(h_{i\sigma}^{\dagger}h_{j\sigma}+\text{H.c.})-V'\sum_{\langle i,j\rangle}n_in_j+V'\sum_{\langle i,j\rangle}\sigma_i\cdot\sigma_j+\frac{V'}{2}\sum_{\substack{\langle i_1,i_2,i_3\rangle\\ \sigma}}(h_{i_1,\sigma}^{\dagger}n_{i_2,-\sigma}h_{i_3,\sigma}+\text{H.c.})$$
$$-\frac{V'}{2}\sum_{\substack{\langle i_1,i_2,i_3\rangle\\ \sigma}}(h_{i_1,\sigma}^{\dagger}h_{i_2,-\sigma}^{\dagger}h_{i_2,-\sigma}h_{i_2,\sigma}h_{i_3,-\sigma}+\text{H.c.}), \qquad (3)$$

where i_1, i_2, and i_3 are neighboring sites, $n_{i\sigma}=h_{i\sigma}^{\dagger}h_{i\sigma}$, $n_i=n_{i\uparrow}+n_{i\downarrow}$, $t'=t$, and $V'=2t^2/U$. Note that an effective attraction between nearest-neighbor electrons occurs, as a reuslt of virtual transitions to the excluded doubly occupied states. Actually, the attraction is only between antiparallel spins; the one between parallel spins is canceled by the third term in the Hamiltonian, a nearest-neighbor antiferromagnetic exchange. The last two terms in H describe the hopping of pairs of nearest-neighbor electrons with opposite spin, with and without spin flip.

Figure 1 shows the nearest-neighbor charge-density correlation function $\langle n_i n_{i+1}\rangle$ in the ground state of a four-site Hubbard chain with two electrons of opposite spin. Note that it *increases* as U decreases, because of the nearest-neighbor attraction discussed above, and peaks around $U\sim 4$ (in units of t). The results obtained from the effective Hamiltonian of Eq. (3) are shown as the dotted line. Exact diagonalizations up to

VOLUME 54, NUMBER 12 PHYSICAL REVIEW LETTERS 25 MARCH 1985

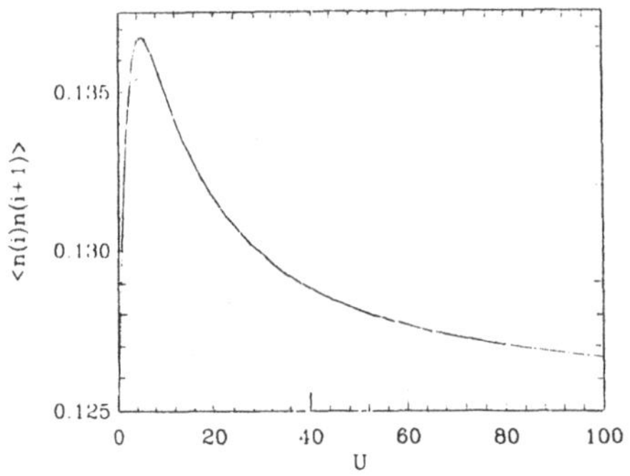

FIG. 1. Nearest-neighbor charge-density correlations in a four-site Hubbard model with two electrons of opposite spin. The dotted line shows the results obtained from the effective Hamiltonian, Eq. (3).

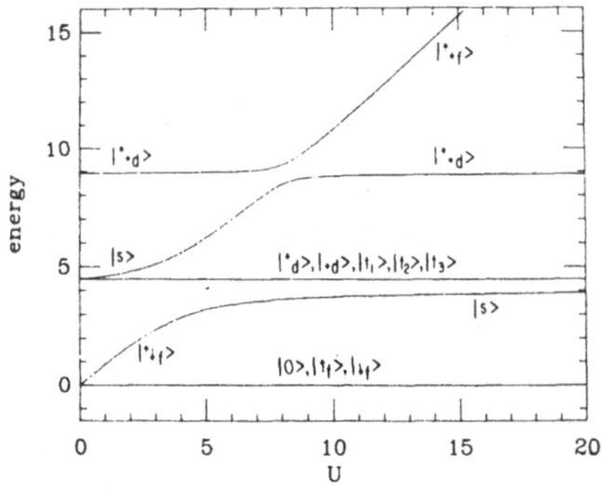

FIG. 2. Lowest-energy level vs U for a single-site Anderson model with $\epsilon_d - \epsilon_f = 4$, $V = 1$. The subindex (d or f) in a state indicates that the state is predominantly of that character, with a small admixture of states of the other band. s and t refer to singlet and triplet states. ϵ_f is chosen so that the states $|0\rangle$, $|\uparrow_f\rangle$, and $|\downarrow_f\rangle$ are degenerate.

eight sites and Monte Carlo simulations for systems up to 64 sites in one and two dimensions show that the qualitative behavior shown in Fig. 1 always occurs for band fillings not too far from $\frac{1}{4}$.[10]

For systems where the intersite hopping occurs through direct overlap of the atomic wave functions, like transition metals, we do not expect this effective attractive interaction to play a role, since it will be overwhelmed by the larger nearest-neighbor direct Coulomb repulsion. However, as is shown below, we obtain the same effective Hamiltonian to describe the f electrons in the Anderson model. For the heavy-fermion superconductors, there is probably no direct f-f overlap[1] so that this cancellation would not occur.

We consider the Anderson lattice Hamiltonian

$$H = -t\sum_{\substack{\langle i,j\rangle \\ \sigma}} (d_{i\sigma}^\dagger d_{j\sigma} + \text{H.c.}) + V\sum_i (d_{i\sigma}^\dagger f_{i\sigma} + \text{H.c.}) + \epsilon_d \sum_{i,\sigma} n_{di,\sigma} + \epsilon_f \sum_{i,\sigma} n_{fi,\sigma} + U\sum_i n_{fi\uparrow} n_{ni\downarrow}, \tag{4}$$

and start by solving the site problem, i.e., $t = 0$. The lowest-energy levels as a function of U are shown in Fig. 2 for a particular case. We will keep only the states labled $|0\rangle$, $|\uparrow_f\rangle$, and $|\downarrow_f\rangle$. These states are essentially f-electron states, with a small admixture [of order $V/(\epsilon_d - \epsilon_f)$] of d-electron states. We take into account the influence of the higher states in perturbation theory in t,[11] and keep only the dominant contributions to second order, involving transitions to the states labeled $|\uparrow_d\rangle$, $|\downarrow_d\rangle$, $|t_1\rangle$, $|t_2\rangle$, $|t_3\rangle$, and $|s\rangle$ in Fig. 2 (this effectively amounts to our taking $U = \infty$). Details of the calculation will be given elsewhere. The result is an effective Hamiltonian identical to Eq. (3) for the "renormalized" f electrons, with

$$t' = [V^2/(\epsilon_d - \epsilon_f)^2]t, \tag{5a}$$

$$V' = 2V^4 t^2/(\epsilon_d - \epsilon_f)^5, \tag{5b}$$

to lowest order in $V/(\epsilon_d - \epsilon_f)$ (the general expressions will be given elsewhere). Here, the effective attraction has a different origin than in the Hubbard case: It arises from the fact that the energy of the singlet state for $U \to \infty$ is lowered from its energy at $U = 0$ (see Fig. 2) by the Kondo coupling

$$\Delta E = 4J_{\text{eff}} = 2V^2/(\epsilon_d - \epsilon_f), \tag{6}$$

to lowest order in $V/(\epsilon_d - \epsilon_f)$, as one would expect from the Schrieffer-Wolff transformation.[12] A similar expansion for $U = 0$ gives attractive and repulsive interactions that exactly cancel because of the degeneracy of the singlet and triplet states.

As mentioned earlier, we do not expect a direct nearest-neighbor repulsion between f electrons in the heavy-fermion systems because of the large separation between f atoms.[1] Still, a nearest-neighbor repulsion V_c between the d electrons in the Hamiltonian of Eq. (4) would contribute an effective repulsion between the renormalized f electrons. Because of screening by the conduction electrons V_c is likely to be quite small, so that its contribution to the effective interaction, which is further reduced by f-d overlap matrix elements, is probably negligible. A detailed quantitative estimate, however, has not been performed.

Our effective Hamiltonian for the Anderson model describes "heavy" f electrons, since the effective hop-

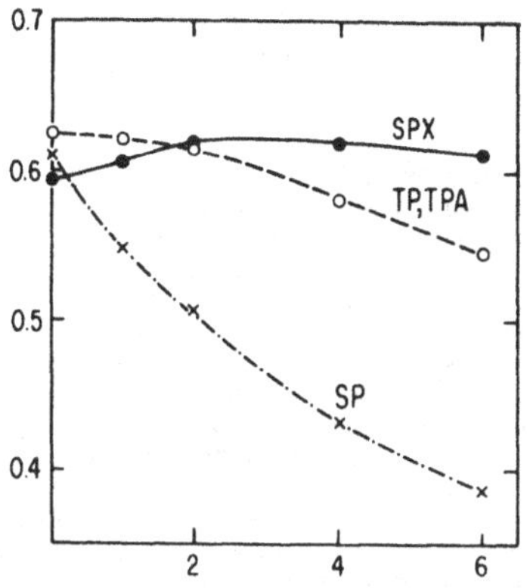

FIG. 3. Pairing-correlation functions at $q=0$ vs U for a 6×6 Hubbard model at $\beta=3$, band filling $\rho=0.67$, from Monte Carlo simulations. The statistical error is smaller than the points.

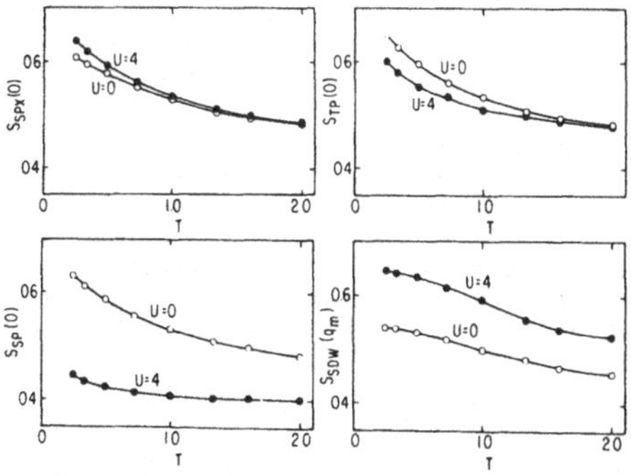

FIG. 4. Temperature dependence of pairing and spin-density-wave (SPW) correlations for a 6×6 Hubbard model, $\rho=0.67$.

ing t' [Eq. (5a)] is probably quite small in these systems. Because of the effective attraction, one might expect pairing correlations to develop at low temperatures. One potential use of this effective Hamiltonian is for exact diagonalization studies: Since the Hilbert space has only two states per site, much larger systems can be studied than for the full Anderson or Hubbard models. Here, we use the effective Hamiltonian only as a guide, which suggests that one might have a tendency towards pairing of electrons on neighboring sites and antiferromagnetism. We have studied static correlation functions of the form

$$S_0(q)=N^{-1}\sum \exp[iq(R_i-R_j)]\langle O_i O_j^\dagger\rangle \qquad (7)$$

in the Hubbard model by Monte Carlo simulations,[13] with O_i pairing operators of the form

$$\begin{aligned} O_i &= c_{i\uparrow}c_{i\downarrow} \quad \text{(SP)},\\ O_i &= c_{i\uparrow}c_{i+\hat{x}\downarrow}-c_{i\downarrow}c_{i+\hat{x}\uparrow} \quad \text{(SPX)},\\ O_i &= c_{i\uparrow}c_{i+\hat{x}\uparrow} \quad \text{(TP)},\\ O_i &= c_{i\uparrow}c_{i+\hat{x}\downarrow}+c_{i\downarrow}c_{i+\hat{x}\uparrow} \quad \text{(TPA)}, \end{aligned} \qquad (8)$$

for singlet pairing (SP), extended singlet pairing (SPX), and triplet pairing of parallel (TP) and of antiparallel (TPA) spins. Figure 3 shows the dependence on U on a 6×6 two-dimensional lattice at temperature $\beta=3$ and band filling $\rho=0.67$. Qualitatively similar results were obtained for a three-dimensional $4\times 4\times 4$ lattice. It can be seen that $S_{\rm SPX}(q=0)$ is enhanced by the Hubbard interaction, while the others are suppressed (the results for $S_{\rm TP}$ and $S_{\rm TPA}$ are identical within statistical error). Since we derived the same effective Hamiltonian in strong coupling for the Anderson and Hubbard models, we expect these results for the Hubbard model to be relevant to the Anderson model in the appropriate regime. For other band fillings, the results are qualitatively similar to Fig. 3, with the effect of U becoming smaller as the band filling decreases. In particular, we never found enhancement of triplet-pairing correlations.

Figure 4 shows the temperature dependence of the pairing correlations and of the spin-density correlations for $U=4$, $\rho=0.67$. The q value for which $S_{\rm SDW}$ is maximum, q_m, is close to (π,π) and somewhat temperature dependent. $S_{\rm SDW}$ is strongly enhanced, but the ratio to the $U=0$ values is approximately constant as T decreases. $S_{\rm SPX}$ instead is further enhanced as T is lowered. If $S_{\rm SPX}(q=0)$ were to diverge at a critical temperature, it would signal a transition to a singlet superconducting state with an anisotropic energy gap of the form

$$\Delta(k)=\Delta(\cos k_x+\cos k_y+\cos k_z) \qquad (9)$$

in a simple cubic lattice. Such a state was discussed by Ohkawa and Fukuyama.[14] Because the gap can vanish along a line on the Fermi surface, the system can exhibit power-law temperature dependence of the specific heat and ultrasonic attenuation instead of the usual exponential dependence, as is found in UBe_{13} and UPt_3.[1]

In summary, we have shown that an effective nearest-neighbor attraction arises in fermion systems with strong on-site repulsion. This attractive interaction leads to enhancement of extended singlet-pairing correlations, as demonstrated by our Monte Carlo simulation results on the Hubbard model. We derived an effective Hamiltonian for the Anderson model which describes a narrow band of heavy electrons with an interaction that can lead to an anisotropic singlet superconducting state at low temperatures with features analogous to those found in the heavy fermion systems, induced solely by the strong on-site electron-electron repulsion.

The computations reported here were performed on the Cray-1 computer at the University of Minnesota, funded through the National Science Foundation program for access to supercomputers. In addition, this work was supported by the National Science Foundation under Grant No. DMR-82-17881 and by the Alfred P. Sloan Foundation through a research fellowship.

[1]G. R. Stewart, Rev. Mod. Phys. **56**, 755 (1984), and references therein.

[2]C. M. Varma, to be published.

[3]P. W. Anderson, Phys. Rev. B **30**, 1549 (1984).

[4]H. Razafimandimky, P. Fulde, and J. Keller, Z. Phys. **54**, 111 (1984).

[5]M. Taichiki and S. Maekawa, Phys. Rev. B **29**, 2497 (1984).

[6]N. Grewe, Z. Phys B **56**, 111 (1984).

[7]K. Ueda and T. M. Rice, to be published.

[8]V. J. Emery, Phys. Rev. B **14**, 2989 (1976).

[9]D. J. Klein and W. A. Seitz, Phys. Rev. B **10**, 3217 (1974).

[10]J. E. Hirsch, unpublished.

[11]We have recently studied a Kondo lattice model in perturbation theory in t (J. E. Hirsch, to be published), for a different regime, however, than the one considered here.

[12]J. R. Schrieffer and P. A. Wolff, Phys. Rev. **149**, 491 (1966).

[13]J. E. Hirsch, Phys. Rev. Lett. **51**, 1900 (1983); J. E. Hirsch and D. J. Scalapino, Phys. Rev. Lett. **53**, 706 (1984).

[14]F. Ohkawa and H. Fukuyama, to be published.

PHYSICAL REVIEW VOLUME 147, NUMBER 1 8 JULY 1966

Ferromagnetism in a Narrow, Almost Half-Filled s Band*

YOSUKE NAGAOKA†
Department of Physics, University of California, San Diego, La Jolla, California
(Received 17 January 1966)

We consider conduction electrons in a narrow s band with a strong repulsive potential which acts when two electrons are at the same atomic site. It is assumed that the electron-transfer matrix elements are nonvanishing only between nearest-neighbor sites, and that the band is almost half-filled, or $|N-N_e|\ll N$, N and N_e being, respectively, the number of atoms and electrons in the crystal. Then it is proved quite rigorously that, if the repulsive potential is sufficiently strong, the ferromagnetic state with the maximum total spin is the ground state for simple cubic and body centered cubic structures as well as for face centered cubic and hexagonal closed packed structures with $N_e>N$, and is not the ground state for face centered cubic and hexagonal closed packed structures with $N_e<N$. For the former case, it is also shown that it is not the ground state if the repulsive potential is weaker than some critical value which is of the order (bandwidth)$\times N/|N-N_e|$.

1. INTRODUCTION

THE problem of metallic ferromagnetism has been long investigated by many authors,[1–5] and various models and approximations have been proposed for it. It should be said, however, that so far we have no definite answer to the question what is essential for the appearance of ferromagnetism in metals. Therefore, it still has some meaning to study a simple model in detail and to examine whether it can be ferromagnetic or not.

Hubbard,[2] Kanamori,[3] and some other authors[4,5] studied a simple model of a ferromagnetic metal such that conduction electrons in a narrow s band interact with one another by a strong repulsive potential of the δ-function type. The model may be best explained by writing down the Hamiltonian:

$$H=\sum_{ij}\sum_{\sigma} t_{ij}c_{i\sigma}^{\dagger}c_{j\sigma}+I\sum_{i} n_{i\uparrow}n_{i\downarrow}, \qquad n_{i\sigma}=c_{i\sigma}^{\dagger}c_{i\sigma}, \tag{1.1}$$

where $c_{i\sigma}^{\dagger}$ and $c_{i\sigma}$ are, respectively, creation and annihilation operators for an electron of spin σ in the atomic state at the ith lattice site, t_{ij} is the matrix element of the electron transfer between the states at the ith and jth sites, and I is the repulsive potential which acts only when two electrons are at the same site. The model is interesting in some aspects, one of which is its simplicity. Because of its simplicity we can treat it in a rather rigorous way for some limiting cases.

To study the model, Hubbard[2] used the Green's-function method and introduced an approximation which is exact in the limit $t_{ij}\rightarrow 0$. Though he obtained some reasonable results, the validity of his approximation is quite obscure for finite t_{ij}. On the other hand, Kanamori[3] used the Brueckner approximation[6] for nuclear matter which is valid in the low-density limit. Therefore his theory is applicable to the case where the band is almost vacant or almost filled.

There is another interesting limiting case to which Kanamori's theory is inapplicable, i.e., the case where the band is almost half-filled. The purpose of this paper is to study this case in some detail, and to see whether the ground state of the system is ferromagnetic or not.[7,8] It is easily shown that, when the band is exactly half-filled, the ferromagnetic state with the maximum total spin, $S_{\max}$, can never be the ground state. What is interesting now is the role of holes or excess electrons in an almost half-filled band.

We introduce here the following assumptions:

(a) The band is almost half-filled, or

$$|n|\ll N, \tag{1.2}$$

where

$$n=N-N_e,$$

N_e and N being, respectively, the number of electrons and atoms in the crystal.

(b) t_{ij} is nonvanishing only between nearest-neighbor sites, which is denoted by $t(<0)$.

(c) The bandwidth is narrow enough compared with the repulsive potential, or

$$|t|\ll I. \tag{1.3}$$

* Research supported by U. S. Air Force Grant No. AF-AFOSR-610-64, Theory of Solids.

† On leave of absence from Research Institute for Fundamental Physics, Kyoto University, Kyoto, Japan. Present address: Department of Physics, Nagoya University, Nagoya, Japan.

[1] Some of the recent works are: D. C. Mattis, Physics 1, 183 (1964); T. Moriya, Progr. Theoret. Phys. (Kyoto) 33, 157 (1965). Other references are cited in these papers.

[2] J. Hubbard, Proc. Roy. Soc. (London) A276, 238 (1963); 281, 401 (1964).

[3] J. Kanamori, Progr. Theoret. Phys. (Kyoto) 30, 275 (1963).

[4] M. C. Gutzwiller, Phys. Rev. Letters 10, 159 (1963); Phys. Rev. 134, A923 (1964); 137, A1726 (1965).

[5] J. Callaway and D. M. Edwards, Phys. Rev. 136, A1333 (1964); J. Callaway, *ibid.* 140, A618 (1965).

[6] K. A. Brueckner and C. A. Levinson, Phys. Rev. 97, 1344 (1955).

[7] Thouless [D. J. Thouless (private communication); Proc. Phys. Soc. (London) 86, 893 (1965)] studied the same case quite independently by a different method, and obtained the same results as obtained here. Though his method is simpler than ours on some points, his proof seems to be incomplete as far as the present author knows it.

[8] The preliminary report of this work was given by Y. Nagaoka, Solid State Commun. 3, 409 (1965).

(d) The crystal structure is simple cubic (sc), body centered cubic (bcc), face centered cubic (fcc), or hexagonal closed packed (hcp).

Though assumptions (b) and (d) are not necessary in some of the following arguments, we shall adopt them througbout the paper, for simplicity.

As will be shown later, the following two cases should be discussed separately:

Case A. sc and bcc; fcc and hcp with $N_e>N$;

Case B. fcc and hcp with $N_e<N$.

It is worthwhile to remark that the case $N_e>N$ can be discussed in a parallel way to the case $N_e<N$ by considering holes, in the former case, instead of electrons, in the latter case, and by changing the sign of t. It is also noticed here that for sc and bcc the lattice can be divided into two sublattices such that all the nearest neighbors of a lattice point on one sublattice belong to the other sublattice. Then, if we introduce an extra phase factor (-1) to the atomic wave functions at the lattice points on one sublattice, t changes its sign. Thus, without loss of generality, it can be assumed that $N_e<N$ and that

$$t>0, \quad \text{for Case A}, \qquad t<0, \quad \text{for Case B}. \tag{1.4}$$

We first consider a simple case where $n=1$ and $I=\infty$. Then our conclusion is summarized as follows:

(I) *For Case A with $n=1$ and $I=\infty$, the ferromagnetic state with the maximum total spin is the ground state of the system.*

(II) *For Case B with $n=1$ and $I=\infty$, the ferromagnetic state with the maximum total spin is not the ground state.*

When $n \ll N$, it is quite reasonable to expect that the total energy is expanded with n as

$$E=E_0+E_1(S)n+\cdots, \tag{1.5}$$

where S denotes the total spin of the system. As the first term E_0 is just the energy when the band is exactly half-filled, it is independent of the spin configuration if $I=\infty$. We put $E_0=0$ by measuring the energy from E_0. The second term depends on the magnitude of the total spin. As far as this term is concerned, it is sufficient to consider the case $n=1$. Therefore the above theorems I and II hold even for finite n, if it is sufficiently small. To see how small it should be, we have to study higher terms in the expansion (1.5). We shall discuss this problem later for a special case.

As can be seen easily, if I if finite, it tends to destroy the ferromagnetic state. Then for Case A with finite n and I, there arises a question how large I should be for the ferromagnetic state to be the ground state. In this paper we only give a necessary condition for it. Our conclusion is as follows:

(III) *For Case A with finite n and I, the ferromagnetic state with the maximum total spin is not the ground state if*

$$\alpha n/N<t/I, \tag{1.6}$$

where α is a numerical constant of the order unity, which is different for different lattice structures.

In the following sections we shall give the proofs of the above theorems I, II, and III with some generalization of them and some additional arguments.

2. CASE A: PROOF OF (I)

In this section we consider Case A with $n=1$ and $I=\infty$. Then the problem is to see what the lowest energy state is for arbitrarily given numbers of up- and down-spin electrons, $N_\uparrow$ and $N_\downarrow$, in the manifold spanned by

$$\Psi_{i\alpha_i}=(-1)^i c_{1\sigma_1}{}^\dagger c_{2\sigma_2}{}^\dagger\cdots c_{i-1\sigma_{i-1}}{}^\dagger c_{i+1\sigma_{i+1}}{}^\dagger\cdots c_{N\sigma_N}{}^\dagger|0\rangle, \tag{2.1}$$

where α_i denotes the set $(\sigma_1,\sigma_2,\cdots,\sigma_{i-1},\sigma_{i+1},\cdots,\sigma_N)$ and $|0\rangle$ is the vacuum state, when the Hamiltonian is given by

$$H'=t\sum_{ij}{}'\sum_\sigma c_{i\sigma}{}^\dagger c_{j\sigma}, \tag{2.2}$$

where $t>0$ and in the summation $\sum_{ij}'$, i and j should be the nearest neighbors.

For the case where $N_\uparrow=N_e$ and $N_\downarrow=0$ (or $N_\downarrow=N_e$ and $N_\uparrow=0$), it is quite a trivial task to find the lowest energy state. It is given by filling up the band for up-spin electrons (or down-spin electrons) except one level at the band top. It is the state with

$$S=S_{\max}=\tfrac{1}{2}N_e, \quad S_z=\pm S, \tag{2.3}$$

$$E=-zt, \tag{2.4}$$

z being the number of nearest neighbors. Now we prove (I) by two steps: i.e., first it is shown that there is no state with $E<-zt$, and secondly that there is no state with $E=-zt$ and $S<S_{\max}$. Then it immediately follows that the state with $E=-zt$ and $S=S_{\max}$ is the ground state of the system.

For later convenience, we introduce here the notation of a *superlattice* which consists of the set $\{(i\alpha_i)\}$. In this superlattice each $(i\alpha_i)$ represents a lattice point and $(i\alpha_i)$ and $(j\beta_j)$ are called nearest neighbors if $\langle\Psi_{i\alpha_i}|H'|\Psi_{j\beta_j}\rangle\neq 0$. That is, we have

$$\begin{aligned}\langle\Psi_{i\alpha_i}|H'|\Psi_{j\beta_j}\rangle&=-t, \quad \text{for} \quad (j\beta_j)=n[(i\alpha_i)]\\ &=0, \qquad \text{otherwise},\end{aligned} \tag{2.5}$$

where $n[(i\alpha_i)]$ denotes nearest neighbors of $(i\alpha_i)$. It is clear that each lattice point in the superlattice has z nearest neighbors as in the real lattice, and that the superlattice is equivalent to the real lattice for $N_\uparrow=N_e$ and $N_\downarrow=0$ (or $N_\downarrow=N_e$ and $N_\uparrow=0$).

Let

$$(j\beta_j|i\alpha_i)_\omega=\left\langle\Psi_{j\beta_j}\left|\frac{1}{\omega-H'}\right|\Psi_{i\alpha_i}\right\rangle, \tag{2.6}$$

where ω is a complex number in general, though in the following we only consider $(j\beta_j|i\alpha_i)_\omega$ on the real axis. The poles of $(i\alpha_i|i\alpha_i)_\omega$ lie on the real axis and give the energy eigenvalues of the states with $S \geq |N_\uparrow - N_\downarrow|/2$. Using the identity

$$\frac{1}{\omega - H'} = \frac{1}{\omega} + \frac{1}{\omega} H' \frac{1}{\omega - H'}, \tag{2.7}$$

we can derive the equation for $(j\beta_j|i\alpha_i)_\omega$ as

$$\begin{aligned} \omega(j\beta_j|i\alpha_i)_\omega &= \delta_{ij}\delta_{\alpha_i\beta_j} + \langle \Psi_{j\beta_j}|H'(\omega - H')^{-1}|\Psi_{i\alpha_i}\rangle \\ &= \delta_{ij}\delta_{\alpha_i\beta_j} - t \sum_{(k\gamma_k) = n[(j\beta_j)]} (k\gamma_k|i\alpha_i)\omega . \end{aligned} \tag{2.8}$$

Equation (2.8) can be solved by a successive iteration, and $(i\alpha_i|i\alpha_i)_\omega$ is found to be

$$(i\alpha_i|i\alpha_i)_\omega^{-1} = \omega[1 - f(\omega)], \tag{2.9}$$

$$f(\omega) = \sum_{p=2}^{\infty} A_p \left(\frac{-t}{\omega}\right)^p = \sum_{p=2}^{\infty} \frac{A_p}{z^p}\left(\frac{-zt}{\omega}\right)^p, \tag{2.10}$$

where A_p is the number of paths in which a particle starts from $(i\alpha_i)$ in the superlattice and comes back to the same site after p nearest-neighbor steps, without passing $(i\alpha_i)$ on the way. As z^p is just the number of paths in which a particle starts from $(i\alpha_i)$ and takes p nearest-neighbor steps without any restrictions, it is clear that

$$A_p < z^p . \tag{2.11}$$

Therefore $f(\omega)$ is absolutely convergent for real $\omega < -zt$.

For the case where $N_\uparrow = N_e$ and $N_\downarrow = 0$, Eqs. (2.9) and (2.10) become

$$(i|i)_\omega^{-1} = \omega[1 - f_0(\omega)] \tag{2.12}$$

$$f_0(\omega) = \sum_{p=2}^{\infty} (A_p^0/z^p)(-zt/\omega)^p , \tag{2.13}$$

where we dropped the index α_i, for the states are uniquely specified only by i in this case. Here, as the superlattice is equivalent to the real lattice, A_p^0 can be defined as the number of paths in the real lattice. Since the lowest energy for this case is given by $-zt$, $(i|i)_\omega$ has no pole for $\omega < -zt$. This means that

$$f_0(\omega) \neq 1 \quad \text{for} \quad \omega < -zt. \tag{2.14}$$

We again consider general cases of arbitrary numbers of $N_\uparrow$ and $N_\downarrow$. Let us introduce

$$v_{\alpha\beta}(\omega) = \sum_{p=2}^{\infty} (C_p^{\alpha\beta}/z^p)(-zt/\omega)^p , \tag{2.15}$$

where $C_p^{\alpha\beta}$ is the number of paths in which a particle starts from $(i\alpha_i)$ and arrives at $(i\beta_i)$ after p nearest-neighbor steps, without passing any $(i\gamma_i)$ with the same i on the way. $v_{\alpha\beta}(\omega)$ is also absolutely convergent for $\omega < -zt$. Furthermore, it is shown that

$$v_{\alpha\beta}(\omega) > 0, \quad \text{for} \quad \omega < -zt, \tag{2.16}$$

$$\textstyle\sum_\beta v_{\alpha\beta}(\omega) = f_0(\omega) . \tag{2.17}$$

Equation (2.16) is obvious, while Eq. (2.17) follows from $\sum_\beta C_p^{\alpha\beta} = A_p^0$, which is proved by watching the motion of i in each path contributing to $C_p^{\alpha\beta}$.

Using $v_{\alpha\beta}(\omega)$, we rewrite $f(\omega)$ as

$$\begin{aligned} f(\omega) = v_{\alpha\alpha}(\omega) &+ \sum_{\beta(\neq\alpha)} v_{\alpha\beta}(\omega) v_{\beta\alpha}(\omega) \\ &\sum_{\beta(\neq\alpha)} \sum_{\gamma(\neq\alpha)} v_{\alpha\beta}(\omega) v_{\beta\gamma}(\omega) v_{\gamma\alpha}(\omega) + \cdots , \end{aligned} \tag{2.18}$$

where the nth term is the contribution from the paths in which the particle passes the lattice points with the same i $(n-1)$ times on the way. Using Eq. (2.17) repeatedly, we can further rewrite it as follows:

$$\begin{aligned} f(\omega) &= f_0(\omega) + \sum_{\beta(\neq\alpha)} v_{\alpha\beta}(\omega)[v_{\beta\alpha}(\omega) - 1] + \cdots \\ &= f_0(\omega) + [f_0(\omega) - 1] \sum_{\beta(\neq\alpha)} v_{\alpha\beta}(\omega) \\ &\quad + \sum_{\beta(\neq\alpha)} \sum_{\gamma(\neq\alpha)} v_{\alpha\beta}(\omega) v_{\beta\gamma}(\omega)[v_{\gamma\alpha}(\omega) - 1] + \cdots . \end{aligned}$$

Finally, we get

$$\begin{aligned} f(\omega) - 1 = [f_0(\omega) - 1][1 &+ \sum_{\beta(\neq\alpha)} v_{\alpha\beta}(\omega) \\ &\sum_{\beta(\neq\alpha)} \sum_{\gamma(\neq\alpha)} v_{\alpha\beta}(\omega) v_{\beta\gamma}(\omega) + \cdots] . \end{aligned} \tag{2.19}$$

Because of Eq. (2.16) the second factor of the right-hand side is positive for $\omega < -zt$, while because of Eq. (2.14) the first factor is nonvanishing for $\omega < -zt$. Thus it turns out that

$$f(\omega) \neq 1, \quad \text{for} \quad \omega < -zt. \tag{2.20}$$

This means that $(i\alpha_i|i\alpha_i)_\omega$ for arbitrary numbers of $N_\uparrow$ and $N_\downarrow$ has no pole for $\omega < -zt$, and hence that there is no energy eigenvalue lower than $-zt$.

It has been proved that the ground-state energy should be $-zt$, while it is already known that there is a state with $E = -zt$, i.e., the state with $S = S_{max}$. Then there arises a question whether there is another state with $E = -zt$ and $S < S_{max}$. The answer is no.

To prove it, we have to know the structure of the superlattice. To clarify the situation, let us first consider the case of a one-dimensional lattice. We show three spin configurations for $N = 5$ and $N_\uparrow = N_\downarrow = 2$ in Fig. 1. It is clear that one can arrive at (b) starting from (a) by moving spins to their nearest-neighbor vacant site one by one, but that one can never arrive at (c) starting from (a) or (b). This means that the lattice point in the superlattice corresponding to configuration (a) is connected with the lattice point corresponding to (b), but disconnected from the lattice

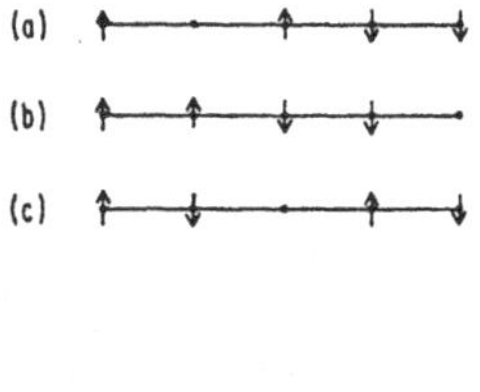

FIG. 1. Some examples of the configuration of spins in the one-dimensional lattice with $N = 5$ and $N_\uparrow = N_\downarrow = 2$.

point corresponding to (c). Thus it turns out that the superlattice for the one-dimensional case is decomposed into some disconnected parts.

The situation is, however, quite different for the two- and three-dimensional cases. It is shown in Appendix A that for these cases all lattice points in the superlattice are connected directly or indirectly. This property of the superlattice is essential for the following proof.

Let us denote the wave function of the state with $E=-zt$ and arbitrarily given $N_\uparrow$ and $N_\downarrow$ by Ψ, and expand it with $\Psi_{i\alpha_i}$ as

$$\Psi=\sum_{(i\alpha_i)}\Gamma(i\alpha_i)\Psi_{i\alpha_i}. \tag{2.21}$$

Then from

$$H'\Psi=-zt\Psi, \tag{2.22}$$

we get the equation for Γ as

$$\Gamma(i\alpha_i)=z^{-1}\sum_{(j\beta_j)=n[(i\alpha_i)]}\Gamma(j\beta_j). \tag{2.23}$$

Equation (2.23) is a kind of Laplace equation in a discrete space of the superlattice. It is clear from Eq. (2.23) that both the real and the imaginary parts of $\Gamma(i\alpha_i)$ should have no extremum. Then, if we take the periodic boundary condition,

$$\Gamma(i\alpha_i)=\text{const} \tag{2.24}$$

should be the unique solution of Eq. (2.23). The solution (2.24) corresponds to the state with $S=S_{\max}$ and $S_z=\frac{1}{2}(N_\uparrow-N_\downarrow)$, and there is no state with $E=-zt$ and $S<S_{\max}$.

3. GENERAL FORMULATION FOR ONE-SPIN FLIPPED STATES

Before proceeding to the proofs of (II) and (III), some general formulas are derived in this section for a special case where $N_\uparrow=N_e-1$, $N_\downarrow=1$ and I and n are finite. The formulas will be applied to Case B in Sec. 4 to prove (II), and to Case A in Sec. 5 to prove (III). The assumptions (b) and (d) are not required in the discussion of this section.

For the case where $N_\uparrow=N_e-1$ and $N_\downarrow=1$, the wave function of the system can be written in the form

$$\begin{aligned}\Psi=\{&\sum\Phi_0(j;i_1\cdots i_n)c_{j\downarrow}{}^\dagger c_{j\uparrow}c_{i_n\uparrow}\cdots c_{i_1\uparrow}\\&+\sum\Phi_1(j;i_1\cdots i_{n+1})c_{j\downarrow}{}^\dagger c_{i_{n+1}\uparrow}\cdots c_{i_1\uparrow}\}\Psi_f\\ \equiv{}& T\Psi_f,\end{aligned}\tag{3.1}$$

where

$$\Psi_f=\prod_{i=1}^{N}c_{i\uparrow}{}^\dagger|0\rangle. \tag{3.2}$$

Here we assume that $\Phi_0(j;i_1\cdots i_n)$ and $\Phi_1(j;i_1\cdots i_{n+1})$ are antisymmetric with respect to the permutation of i's and that

$$\begin{aligned}\Phi_0(j;i_1\cdots j\cdots i_n)&=0,\\ \Phi_1(j;i_1\cdots j\cdots i_{n+1})&=0.\end{aligned}\tag{3.3}$$

The equations for Φ_0 and Φ_1 are derived from

$$\{[H,T]-ET\}\Psi_f=0. \tag{3.4}$$

Calculating the commutator, we get

$$\begin{aligned}[\sum\{&-\sum_{\lambda=1}^{n}\sum_{i'}t_{i'i_\lambda}\Phi_0(j;i_1\cdots i'^\lambda\cdots i_n)-\sum_{\lambda=1}^{n}t_{ji_\lambda}\Phi_0(i_\lambda;i_1\cdots j^\lambda\cdots i_n)-(n+1)\sum_{i'}t_{i'j}\Phi_1(j;i_1\cdots i'^{n+1})\\&+(n+1)\sum_{i'}t_{ji'}\Phi_1(i';i_1\cdots j^{n+1})-E\Phi_0(j;i_1\cdots i_n)\}c_{j\downarrow}{}^\dagger c_{j\uparrow}c_{i_n\uparrow}\cdots c_{i_1\uparrow}\\&+\sum{}'\{-t_{ji_{n+1}}\Phi_0(j;i_1\cdots i_n)+t_{ji_{n+1}}\Phi_0(i_{n+1};i_1\cdots i_n)-\sum_{\lambda=1}^{n+1}\sum_{i'}t_{i'i_\lambda}\Phi_1(j;i_1\cdots i'^\lambda\cdots i_{n+1})\\&\qquad+\sum_{i'}t_{ji'}\Phi_1(i';i_1\cdots i_{n+1})+(I-E)\Phi_1(j;i_1\cdots i_{n+1})\}c_{j\downarrow}{}^\dagger c_{i_{n+1}\uparrow}\cdots c_{i_1\uparrow}]\Psi_f=0,\end{aligned}$$

where the term $j=i_\mu(\mu=1,\ldots,n+1)$ should be excluded in the summation $\sum'$, and i'^λ means that i_λ is replaced by i'. From the first term, assuming $j\neq i_\mu(\mu=1,\ldots,n)$, we get

$$\begin{aligned}E\Phi_0(j;i_1\cdots i_n)=&-\sum_{\lambda=1}^{n}\sum_{i'}t_{i'i_\lambda}\Phi_0(j;i_1\cdots i'^\lambda\cdots i_n)-\sum_{\lambda=1}^{n}t_{ji_\lambda}\Phi_0(i_\lambda;i_1\cdots j^\lambda\cdots i_n)\\&-(n+1)\sum_{j'}\{t_{j'j}\Phi_1(j;i_1\cdots j'^{n+1})-t_{jj'}\Phi_1(j';i_1\cdots j^{n+1})\}.\end{aligned}\tag{3.5}$$

From the second term, antisymmetrizing the coefficient with respect to the permutation of i's and assuming $j\neq i_\mu(\mu=1,\ldots,n+1)$, we get

$$\begin{aligned}(E-I)\Phi_1(j;i_1\cdots i_{n+1})=&-\sum_{\lambda=1}^{n+1}\sum_{i'}t_{i'i_\lambda}\Phi_1(j;i_1\cdots i'^\lambda\cdots i_{n+1})\\&+\sum_{j'}t_{jj'}\Phi_j(j';i_1\cdots i_{n+1})-\frac{1}{n+1}\{t_{ji_{n+1}}[\Phi_0(j;i_1\cdots i_n)-\Phi_0(i_{n+1};i_1\cdots i_n)]\\&-\sum_{\lambda=1}^{n}t_{ji_\lambda}[\Phi_0(j;i_1\cdots i_{n+1}{}^\lambda\cdots i_n)-\Phi_0(i_\lambda;i_1\cdots i_{n+1}{}^\lambda\cdots i_n)]\}.\end{aligned}\tag{3.6}$$

So far, Eqs. (3.5) and (3.6) are exact. Now we introduce the assumptions (a) and (c) and calculate the energy to the order nt and t^2/I. From Eq. (3.6) we obtain to the first order of t

$$\Phi_1(j;i_1\cdots j'^{n+1})=((n+1)I)^{-1}\{t_{jj'}[\Phi_0(j;i_1\cdots i_n)-\Phi_0(j';i_1\cdots i_n)]$$
$$-\sum_{\lambda=1}^{n} t_{ji_\lambda}[\Phi_0(j;i_1\cdots j'^{\lambda}\cdots i_n)-\Phi_0(i_\lambda;i_1\cdots j'^{\lambda}\cdots i_n)]\}. \quad (3.7)$$

Substitution of Eq. (3.7) in Eq. (3.5) gives

$$E\Phi_0(j;i_1\cdots i_n)=-\sum_{\lambda=1}^{n}\sum_{i'} t_{i'i_\lambda}\Phi_0(j;i_1\cdots i'^{\lambda}\cdots i_n)-\sum_{\lambda=1}^{n} t_{ji_\lambda}\Phi_0(i_\lambda;i_1\cdots j^{\lambda}\cdots i_n)$$
$$+\frac{2}{I}\sum_{j'} t_{jj'}t_{j'j}[\Phi_0(j';i_1\cdots i_n)-\Phi_0(j;i_1\cdots i_n)]-\frac{1}{I}\sum_{\lambda=1}^{n}\sum_{j'}\{t_{jj'}t_{j'i_\lambda}[\Phi_0(j';i_1\cdots j^{\lambda}\cdots i_n)$$
$$-\Phi_0(i_\lambda;i_1\cdots j^{\lambda}\cdots i_n)]-t_{ji_\lambda}t_{j'j}[\Phi_0(j;i_1\cdots j'^{\lambda}\cdots i_n)-\Phi_0(i_\lambda;i_1\cdots j'^{\lambda}\cdots i_n)]\}. \quad (3.8)$$

Equation (3.8) is valid only for $j\neq i_\mu(\mu=1,\ldots,n)$. If we add a term

$$\sum_{\lambda=1}^{n}\delta_{ji_\lambda}\sum_{i'} t_{i'i_\lambda}\Phi_0(j;i_1\cdots i'^{\lambda}\cdots i_n)$$

to the right-hand side of the equation, it becomes consistent with Eq. (3.3) for $j=i_\mu$. Further, we neglect the last two lines of the right-hand side, because their contribution is of the order nt^2/I. Thus we finally obtain

$$E\Phi_0(j;i_1\cdots i_n)=-\sum_{\lambda=1}^{n}(1-\delta_{ji_\lambda})\sum_{i'} t_{i'i_\lambda}\Phi_0(j;i_1\cdots i'^{\lambda}\cdots i_n)$$
$$-\sum_{\lambda=1}^{n} t_{ji_\lambda}\Phi_0(i_\lambda;i_1\cdots j^{\lambda}\cdots i_n)+(2/I)\sum_{j'} t_{jj'}t_{j'j}[\Phi_0(j';i_1\cdots i_n)-\Phi_0(j;i_1\cdots i_n)]. \quad (3.9)$$

In solving Eq. (3.9), it is convenient to introduce Fourier transforms. Then we have to discuss the case of sc, bcc, and fcc and the case of hcp separately, for in the latter lattice structure a unit cell contains two atoms. In this section we only consider the former case, while the latter will be discussed in Appendix C.

Let us introduce Fourier transforms by

$$\Phi_0(j;i_1\cdots i_n)=\sum_{\mathbf{q}\mathbf{k}_1\ldots\mathbf{k}_n}\Phi_0(\mathbf{q};\mathbf{k}_1\cdots\mathbf{k}_n)\exp[-i\mathbf{q}\cdot\mathbf{r}_j-i\sum_{\lambda=1}^{n}\mathbf{k}_\lambda\cdot(\mathbf{r}_{i_\lambda}-\mathbf{r}_j)] \quad (3.10)$$

$$t_{ij}=N^{-1}\sum_{\mathbf{k}} t(\mathbf{k})e^{i\mathbf{k}\cdot(\mathbf{r}_i-\mathbf{r}_j)}, \quad (3.11)$$

where $\Phi_0(\mathbf{q};\mathbf{k}_1\ldots\mathbf{k}_n)$ is antisymmetric with respect to the permutation of $\mathbf{k}$'s and satisfies

$$\sum_{\mathbf{k}_\lambda}\Phi_0(\mathbf{q};\mathbf{k}_1\cdots\mathbf{k}_n)=0, \quad (3.12)$$

which comes from Eq. (3.3). Then from Eq. (3.9) we have

$$\{E+\sum_{\lambda=1}^{n} t(\mathbf{k}_\lambda)+(2/NI)\sum_{\mathbf{k}'} t(\mathbf{k}')[t(\mathbf{k}')-t(\mathbf{k}'+\sum_{\lambda=1}^{n}\mathbf{k}_\lambda-\mathbf{q})]\}\Phi_0(\mathbf{q};\mathbf{k}_1\cdots\mathbf{k}_n)$$
$$=N^{-1}\sum_{\lambda=1}^{n}\sum_{\mathbf{k}'}[t(\mathbf{k}')-t(\mathbf{k}'+\sum_{\mu=1}^{n}\mathbf{k}_\mu-\mathbf{q})]\Phi_0(\mathbf{q};\mathbf{k}_1\cdots\mathbf{k}'^{\lambda}\cdots\mathbf{k}_n). \quad (3.13)$$

It is worthwhile to point out here that Eq. (3.13) has a solution given by

$$\mathbf{q}=\sum_{\lambda=1}^{n}\mathbf{k}_\lambda,\quad E=-\sum_{\lambda=1}^{n} t(\mathbf{k}_\lambda). \quad (3.14)$$

This solution corresponds to the state with $S=S_{\max}$ and $S_z=S-1$. Especially, if the levels with the wave vectors $\mathbf{k}_1\cdots\mathbf{k}_n$ are n highest levels at the band top, we get the state with the lowest energy. In the following sections we shall look for the solution of Eq. (3.13) other than Eq. (3.14) for some cases.

4. CASE B: PROOF OF (II)

We first apply the formula obtained in the last section to Case B with $n=1$ and $I=\infty$ to prove (II). In this section, starting from Eq. (3.13), we only consider the case of fcc. The case of hcp will be discussed in Appendix C.

Under the assumption (b), $t(\mathbf{k})$ for fcc is given by

$$t(\mathbf{k})=4t[\cos k_x a\cos k_y a+\cos k_y a\cos k_z a+\cos k_z a\cos k_x a] \quad (4.1)$$
$$\equiv -4|t|\theta_{\mathbf{k}},$$

where a is the lattice constant. In this case, the lowest energy of the $S=S_{\max}$ states is given by $-4|t|$. We have to see whether there is any state with $E<-4|t|$.

We consider the case $\mathbf{q}=0$ in Eq. (3.13). Putting $n=1$, $I=\infty$, and $\mathbf{q}=0$ in Eq. (3.13), we get

$$\{E+t(\mathbf{k})\}\Phi(\mathbf{k})=\frac{1}{N}\sum_{\mathbf{k}'}[t(\mathbf{k}')-t(\mathbf{k}+\mathbf{k}')]\Phi(\mathbf{k}'), \quad (4.2)$$

where we put $\Phi_0(0;\mathbf{k})=\Phi(\mathbf{k})$. Substitution of Eq. (4.1) in Eq. (4.2) gives

$$(\epsilon-\theta_{\mathbf{k}})\Phi(\mathbf{k})=\sum_{(\lambda,\mu)}\{-(1-\cos k_\lambda a\cos k_\mu a)\Phi_{c\lambda\mu}+\sin k_\lambda a\sin k_\mu a\Phi_{s\lambda\mu}-\cos k_\lambda a\sin k_\mu a\Phi_{1\lambda\mu}-\sin k_\lambda a\cos k_\mu a\Phi_{1\mu\lambda}\}, \quad (4.3)$$

where $\sum_{(\lambda,\mu)}$ is the summation over $(\lambda,\mu)=(x,y)$, (y,z), and (z,x) and

$$\epsilon=E/4|t|,$$
$$\Phi_{c\lambda\mu}=N^{-1}\sum_{\mathbf{k}}\cos k_\lambda a\cos k_\mu a\Phi(\mathbf{k}),$$
$$\Phi_{s\lambda\mu}=N^{-1}\sum_{\mathbf{k}}\sin k_\lambda a\sin k_\mu a\Phi(\mathbf{k}), \quad (4.4)$$
$$\Phi_{1\lambda\mu}=N^{-1}\sum_{\mathbf{k}}\cos k_\lambda a\sin k_\mu a\Phi(\mathbf{k}).$$

From Eq. (4.3) the equations for $\Phi_{c\lambda\mu}$, $\Phi_{s\lambda\mu}$, and $\Phi_{1\lambda\mu}$ are obtained as

$$[1+F_1(\epsilon)]\Phi_{c\lambda\mu}+F_2(\epsilon)(\Phi_{c\mu\nu}+\Phi_{c\nu\lambda})=0, \quad (4.5)$$
$$[1-G(\epsilon)]\Phi_{s\lambda\mu}=0, \quad (4.6)$$
$$[1+H_1(\epsilon)]\Phi_{1\lambda\mu}+H_2(\epsilon)\Phi_{1\nu\mu}=0, \quad (4.7)$$

where (λ,μ,ν) denotes a permutation of (x,y,z) and

$$F_1(\epsilon)=\frac{1}{N}\sum_{\mathbf{k}}\frac{(1-\cos k_x a\cos k_y a)\cos k_x a\cos k_y a}{\epsilon-\theta_{\mathbf{k}}}, \quad (4.8a)$$

$$F_2(\epsilon)=\frac{1}{N}\sum_{\mathbf{k}}\frac{(1-\cos k_y a\cos k_z a)\cos k_x a\cos k_y a}{\epsilon-\theta_{\mathbf{k}}}, \quad (4.8b)$$

$$G(\epsilon)=\frac{1}{N}\sum_{\mathbf{k}}\frac{\sin^2 k_x a\sin^2 k_y a}{\epsilon-\theta_{\mathbf{k}}}, \quad (4.9)$$

$$H_1(\epsilon)=\frac{1}{N}\sum_{\mathbf{k}}\frac{\sin^2 k_x a\cos^2 k_y a}{\epsilon-\theta_{\mathbf{k}}}, \quad (4.10a)$$

$$H_2(\epsilon)=\frac{1}{N}\sum_{\mathbf{k}}\frac{\sin^2 k_x a\cos k_y a\cos k_z a}{\epsilon-\theta_{\mathbf{k}}}. \quad (4.10b)$$

It is easily verified that Eqs. (4.5) and (4.6) have no nontrivial solution for $\epsilon<-1$ (i.e., $E<-4|t|$). Therefore we have to examine Eq. (4.7) in some detail.

In order that Eq. (4.7) has some nontrivial solutions, it should be that $[1+H_1(\epsilon)]^2-H_2(\epsilon)^2=0$, or

$$H_\pm(\epsilon)=-1, \quad (4.11)$$

where

$$H_\pm(\epsilon)=H_1(\epsilon)\pm H_2(\epsilon)$$
$$=\frac{1}{N}\sum_{\mathbf{k}}\frac{\sin^2 k_x a\cos k_y a(\cos k_y a\pm\cos k_z a)}{\epsilon-\theta_{\mathbf{k}}}. \quad (4.12)$$

Let us examine the behavior of $H_-(\epsilon)$ at $\epsilon\lesssim-1$. The main contribution to $H_-(\epsilon)$ at $\epsilon\lesssim-1$ comes from $\mathbf{k}$'s at the band top where $\theta_{\mathbf{k}}\simeq-1$. Especially we have

$$\lim_{\epsilon\to-1-0}H_-(\epsilon)=\langle\sin^2 k_x a\cos k_y a(\cos k_y a-\cos k_z a)\rangle_{\text{band top}}\times\lim_{\epsilon\to-1-0}N^{-1}\sum_{\mathbf{k}}(\epsilon-\theta_{\mathbf{k}})^{-1}, \quad (4.13)$$

where $\langle\cdots\rangle_{\text{band top}}$ means the average at the band top given by

$$\cos k_x a\cos k_y a=-1,\ \cos k_y a\cos k_z a=-1$$
$$\text{or} \quad (4.14)$$
$$\cos k_z a\cos k_x a=-1.$$

As

$$\langle\sin^2 k_x a\cos k_y a(\cos k_y a-\cos k_z a)\rangle_{\text{band top}}=\tfrac{1}{3}, \quad (4.15)$$

we obtain

$$\lim_{\epsilon\to-1-0}H_-(\epsilon)=-\infty. \quad (4.16)$$

From Eq. (4.16) it is clear that $H_-(\epsilon)=-1$ has at least one solution for $\epsilon<-1$. This means that there exists at least one state with $E<-4|t|$, and that the state with $S=S_{\max}$ and $E=-4|t|$ is not the ground state.

Thus we have proved (II) for the case of fcc. It is also proved for the case of hcp in a similar way, as will be shown in Appendix C.

5. SPIN-WAVE SPECTRUM FOR CASE A: PROOF OF (III)

Now let us apply Eq. (3.13) to Case A with finite n and I, and look for a spin-wave solution of it to prove (III). Although we can take a much simpler order-of-magnitude argument if we are only concerned with the proof of (III), the following argument is useful to generalize it to some extent as well as to calculate α explicitly. These problems will also be discussed in this section.

It is already known that for Case A the ground state is the state with $S=S_{\max}$, if n is sufficiently small and I is sufficiently large. The ground state is given by

$$\Phi_0(\mathbf{q};\mathbf{k}_1\cdots\mathbf{k}_n)=\delta_{q0}\delta_{\{\mathbf{k}_\lambda\},\{\mathbf{p}_\lambda\}}\Phi_0+\Phi' \quad (5.1)$$

$$E=-\sum_{\lambda=1}^{n}t(\mathbf{p}_\lambda)\equiv E_0, \quad (5.2)$$

where $\{\mathbf{p}_\lambda\}$ is the set of wave vectors of n highest levels in the band, and Φ' is at most of the order Φ_0/N. $\{\mathbf{p}_\lambda\}$ satisfies $\sum_\lambda \mathbf{p}_\lambda=0$.

It is expected that the wave functions of the spin-wave states are not much different from that of the ground state. Therefore we look for the solution of Eq. (3.13) by putting

$$\begin{aligned}\Phi_0(\mathbf{q};\mathbf{p}_1\cdots\mathbf{p}_n)&=\Phi_\mathbf{q},\\ \Phi_0(\mathbf{q};\mathbf{p}_1\cdots\mathbf{k}^\lambda\cdots\mathbf{p}_n)&=\Phi_\mathbf{q}(\mathbf{p}_\lambda|\mathbf{k}),\\ \Phi_0(\mathbf{q};\mathbf{p}_1\cdots\mathbf{k}^\lambda\cdots\mathbf{k}^\mu\cdots\mathbf{p}_n)&=\Phi_\mathbf{q}(\mathbf{p}_\lambda\mathbf{p}_\mu|\mathbf{k}\mathbf{k}'),\end{aligned}\tag{5.3}$$

and so on. From Eq. (3.12) they should satisfy

$$\begin{aligned}&\Phi_\mathbf{q}+\sum_{\mathbf{k}<}\Phi_\mathbf{q}(\mathbf{p}|\mathbf{k})=0,\\ &\Phi_\mathbf{q}(\mathbf{p}|\mathbf{k})+\sum_{\mathbf{k}'<}\Phi_\mathbf{q}(\mathbf{p}\mathbf{p}'|\mathbf{k}\mathbf{k}')=0,\cdots,\end{aligned}\tag{5.4}$$

where $\mathbf{p}$ or $\mathbf{p}'$ denotes one of $\{\mathbf{p}_\lambda\}$ and $\sum_{\mathbf{k}<}$ is the summation over all levels other than $\{\mathbf{p}_\lambda\}$. The summation over $\{\mathbf{p}_\lambda\}$ will be denoted by $\sum_{\mathbf{p}>}$.

By the use of Eq. (5.3), Eq. (3.13) becomes

$$\{E_\mathbf{q}+(2/NI)\sum_\mathbf{k} t(\mathbf{k})[t(\mathbf{k})-t(\mathbf{k}-\mathbf{q})]-N^{-1}\sum_{\mathbf{p}>}[t(\mathbf{p})-t(\mathbf{p}-\mathbf{q})]\}\Phi_\mathbf{q}=N^{-1}\sum_{\mathbf{p}>}\sum_{\mathbf{k}<}[t(\mathbf{k})-t(\mathbf{p}-\mathbf{q})]\Phi_\mathbf{q}(\mathbf{p}|\mathbf{k}),\tag{5.5}$$

$$\begin{aligned}\{E_\mathbf{q}+t(\mathbf{p})-t(\mathbf{k})+(2/NI)\sum_{\mathbf{k}'} t(\mathbf{k}')[t(\mathbf{k}')-t(\mathbf{k}'+\mathbf{k}-\mathbf{p}-\mathbf{q})]\}\Phi_\mathbf{q}(\mathbf{p}|\mathbf{k})=N^{-1}[t(\mathbf{p})-t(\mathbf{k}-\mathbf{q})]\Phi_\mathbf{q}\\ +N^{-1}\sum_{\mathbf{k}'<}[t(\mathbf{k}')-t(\mathbf{k}'+\mathbf{k}-\mathbf{p}-\mathbf{q})]\Phi_\mathbf{q}(\mathbf{p}|\mathbf{k}')+N^{-1}\sum_{\mathbf{p}'>}\sum_{\mathbf{k}'<}[t(\mathbf{k}')-t(\mathbf{k}'+\mathbf{k}-\mathbf{p}-\mathbf{q})]\Phi_\mathbf{q}(\mathbf{p}\mathbf{p}'|\mathbf{k}\mathbf{k}')\end{aligned}\tag{5.6}$$

and so on, where

$$E_\mathbf{q}=E-E_0\tag{5.7}$$

is the spin-wave energy to be determined in the following calculation.

Under the assumptions (a) and (c), Eqs. (5.5) and (5.6) can be solved by neglecting higher order terms. In Eq. (5.5) $\Phi_\mathbf{q}(\mathbf{p}|\mathbf{k})$ only appears in the term which contributes to $E_\mathbf{q}$ in the order nt. Therefore we have only to determine it to the lowest order. Neglecting the terms of the order nt and t^2/I in Eq. (5.6), we have

$$[E_\mathbf{q}+t(\mathbf{p})-t(\mathbf{k})]\Phi_\mathbf{q}(\mathbf{p}|\mathbf{k})-N^{-1}\sum_{\mathbf{k}'<}[t(\mathbf{k}')-t(\mathbf{k}'+\mathbf{k}+\mathbf{p}-\mathbf{q})]\Phi_\mathbf{q}(\mathbf{p}|\mathbf{k}')=N^{-1}[t(\mathbf{p})-t(\mathbf{k}-\mathbf{q})]\Phi_\mathbf{q}.\tag{5.8}$$

Now it is rather easy to prove (III) without any further calculations. From Eq. (5.5) it turns out that $E_\mathbf{q}$ consists of two contributions; one is of the order t^2/I, which comes from the second term in the wavy brackets of the left-hand side, and the other is of the order nt, which comes from the rest terms. It is clear that the second contribution is positive, for, if $I=\infty$, the $S=S_{\max}$ state is the ground state for Case A, and hence the spin-wave energy should be positive. On the other hand, the first contribution is negative, because

$$N^{-1}\sum_\mathbf{k} t(\mathbf{k})[t(\mathbf{k})-t(\mathbf{k}-\mathbf{q})]=\sum_j |t_{ij}|^2[1-\cos\mathbf{q}\cdot(\mathbf{r}_i-\mathbf{r}_j)]>0.$$

Then, if the magnitude of the first contribution is larger than the second one, or if the condition (1.6) holds, the spin-wave energy becomes negative and the state with $S=S_{\max}$ cannot be the ground state. This is the proof of (III). Though Eqs. (5.5) and (5.6) are valid only for sc, bcc, and fcc, the situation is exactly the same for hcp, and we can get the same conclusion for it.

The remaining problem is to determine the numerical constant α for each lattice structure. In the following we calculate it only for sc, the simplest case among the four. Though there is no essential difficulty in calculating it for other cases, the calculation involved there will be much more complicated.

For sc, $t(\mathbf{k})$ is given by

$$t(\mathbf{k})=2t\sum_{\lambda=x,y,z}\cos(k_\lambda a)\equiv 2t\theta_\mathbf{k},\tag{5.9}$$

where a is the lattice constant. Because of the assumption $t>0$, the band top is located at $\mathbf{k}=0$. Substitution of Eq. (5.9) in Eq. (5.5) gives

$$\begin{aligned}&[\epsilon_\mathbf{q}+(2t/I)\sum_\lambda(1-\cos q_\lambda a)-N^{-1}\sum_{\mathbf{p}>}\sum_\lambda(1-\cos q_\lambda a)\cos p_\lambda a]\Phi_\mathbf{q}\\ &\qquad=\sum_{\mathbf{p}>}\sum_\lambda[(1-\cos q_\lambda a)\Phi_\mathbf{q}^{c\lambda}(\mathbf{p})-(\sin q_\lambda a)\Phi_\mathbf{q}^{s\lambda}(\mathbf{p})],\end{aligned}\tag{5.10}$$

where

$$\epsilon_q = E_q/2t,$$
$$\Phi_q{}^{c\lambda}(\mathbf{p}) = N^{-1}\sum_{\mathbf{k}<}(\cos k_\lambda a)\Phi_q(\mathbf{p}|\mathbf{k}), \tag{5.11}$$
$$\Phi_q{}^{s\lambda}(\mathbf{p}) = N^{-1}\sum_{\mathbf{k}<}(\sin k_\lambda a)\Phi_q(\mathbf{p}|\mathbf{k}).$$

The equations for $\Phi_q{}^{c\lambda}$ and $\Phi_q{}^{s\lambda}$ are derived from Eq. (5.8) as

$$\{1+F(\theta_p)-[\cos(p_\lambda+q_\lambda)a]G_1(\theta_p)\}\Phi_q{}^{c\lambda}(\mathbf{p})+\sum_{\mu\neq\lambda}\{F(\theta_p)-[\cos(p_\mu+q_\mu)a]G_2(\theta_p)\}\Phi_q{}^{c\mu}(\mathbf{p})$$
$$-[\sin(p_\lambda+q_\lambda)a]G_1(\theta_p)\Phi_q{}^{s\lambda}(\mathbf{p})-\sum_{\mu\neq\lambda}[\sin(p_\mu+q_\mu)a]G_2(\theta_p)\Phi_q{}^{s\mu}(\mathbf{p})$$
$$=-[(1-\cos q_\lambda a)G_1(\theta_p)+\sum_{\mu\neq\lambda}(1-\cos q_\mu a)G_2(\theta_p)]\Phi_q/N \tag{5.12}$$

$$\{1+[\cos(p_\lambda+q_\lambda)a]H(\theta_p)\}\Phi_q{}^{s\lambda}(\mathbf{p})-[\sin(p_\lambda+q_\lambda)a]H(\theta_p)\Phi_q{}^{c\lambda}(\mathbf{p})=[\sin q_\lambda a]H(\theta_p)\Phi_q/N, \tag{5.13}$$

where

$$F(\theta)=\frac{1}{N}\sum_{\mathbf{k}}P\frac{\cos k_x a}{\theta-\theta_{\mathbf{k}}},$$
$$G_1(\theta)=\frac{1}{N}\sum_{\mathbf{k}}P\frac{\cos^2 k_x a}{\theta-\theta_{\mathbf{k}}},$$
$$G_2(\theta)=\frac{1}{N}\sum_{\mathbf{k}}P\frac{\cos k_x a\,\cos k_y a}{\theta-\theta_{\mathbf{k}}}, \tag{5.14}$$
$$H(\theta)=\frac{1}{N}\sum_{\mathbf{k}}P\frac{\sin^2 k_x a}{\theta-\theta_{\mathbf{k}}}.$$

In the derivation of Eqs. (5.12) and (5.13), there first appear functions of the type

$$N^{-1}\sum_{\mathbf{k}<}f(\mathbf{k})/(\theta-\theta_{\mathbf{k}}) \tag{5.15}$$

rather than those defined by Eq. (5.14). If we rewrite them as

$$\frac{1}{N}\sum_{\mathbf{k}<}\frac{f(\mathbf{k})}{\theta-\theta_{\mathbf{k}}}=\frac{1}{N}\sum_{\mathbf{k}}P\frac{f(\mathbf{k})}{\theta-\theta_{\mathbf{k}}}-\frac{1}{N}\sum_{\mathbf{p}>}P\frac{f(\mathbf{p})}{\theta-\theta_{\mathbf{p}}}, \tag{5.16}$$

then the first term is a continuous function of θ, while the second term, which is of the order n/N, has a logarithmic singularity at $\theta=\theta_m$, θ_m being the maximum value of θ_{p_λ}. However, what we need in calculating ϵ_q is not the function itself but its integration over θ which is convergent. The second term in Eq. (5.16) only contributes to ϵ_q in the order $(n/N)^2$ which is neglected in the present calculation. Thus in Eqs. (5.12) and (5.13) we replaced the functions of the type (5.15) by those defined by Eq. (5.14).

If we confine the calculation to the terms linear to n, we further can simplify the equation by putting

$$\sum_{\mathbf{p}>}f(\mathbf{p})\cong nf(0). \tag{5.17}$$

Then, eliminating $\Phi_q{}^{s\lambda}$ from Eqs. (5.12) and (5.13), we have, to the lowest order of n,

$$[\epsilon_q+(2t/I)\sum_\lambda(1-\cos q_\lambda a)-(n/N)\sum_\lambda A_q{}^\lambda]\Phi_q=n\sum_\lambda A_q{}^\lambda\Phi_q{}^{c\lambda}(0), \tag{5.18}$$

$$(1+F-G_1+G_1A_q{}^\lambda)\Phi_q{}^{c\lambda}(0)+\sum_{\mu\neq\lambda}(F-G_2+G_2A_q{}^\mu)\Phi_q{}^{c\mu}(0)=-(G_1A^\lambda{}_q+G_2\sum_{\mu\neq\lambda}A_q{}^\mu)\Phi_q/N, \tag{5.19}$$

where

$$A_{\mathbf{q}}^{\lambda}=1-\frac{H+\cos q_{\lambda}a}{1+H\cos q_{\lambda}a} \tag{5.20}$$

and F, G_1, G_2, and H denote, respectively, the values of the corresponding functions at $\theta=3$.

Solving Eq. (5.19) for $\Phi_{\mathbf{q}}^{c\lambda}(0)(\lambda=x,y,z)$, and inserting them in Eq. (5.18), we finally obtain the spin-wave spectrum as

$$E_{\mathbf{q}}=(2nt/N)[\mathfrak{F}(\mathbf{q})-\kappa\mathfrak{G}(\mathbf{q})], \tag{5.21}$$

where

$$\kappa=\left(\frac{2t}{I}\right)\Big/\left(\frac{n}{N}\right), \tag{5.22}$$

$$\mathfrak{G}(\mathbf{q})=3-\sum_{\lambda}\cos q_{\lambda}a, \tag{5.23}$$

$$\mathfrak{F}(\mathbf{q})=\frac{\sum_{\lambda}A_{\mathbf{q}}^{\lambda}+a\sum_{(\lambda,\mu)}A_{\mathbf{q}}^{\lambda}A_{\mathbf{q}}^{\mu}+bA_{\mathbf{q}}^{x}A_{\mathbf{q}}^{y}A_{\mathbf{q}}^{z}}{1+c\sum_{\lambda}A_{\mathbf{q}}^{\lambda}+d\sum_{(\lambda,\mu)}A_{\mathbf{q}}^{\lambda}A_{\mathbf{q}}^{\mu}+eA_{\mathbf{q}}^{x}A_{\mathbf{q}}^{y}A_{\mathbf{q}}^{z}}, \tag{5.24}$$

$$\begin{aligned}
a&=\frac{2(G_1-G_2)}{1-G_1+G_2},\\
b&=3\left(\frac{G_1-G_2}{1-G_1+G_2}\right)^2,\\
c&=\frac{(1+2F-G_1-G_2)G_1-2(F-G_2)G_2}{(1-G_1+G_2)(1+3F-G_1-2G_2)},\\
d&=\frac{(G_1-G_2)[(1+F-G_1)(G_1+G_2)-2(F-G_2)G_2]}{(1-G_1+G_2)^2(1+3F-G_1-2G_2)},\\
e&=\frac{(G_1-G_2)^2(G_1+2G_2)}{(1-G_1+G_2)^2(1+3F-G_1-2G_2)}.
\end{aligned} \tag{5.25}$$

The numerical calculation of the constants F, G_1, G_2, and H will be given in Appendix C. Using the results obtained there, we have

$$\begin{aligned}
&H=0.208,\\
&a=0.463,\quad b=0.161,\\
&c=0.326,\quad d=0.0974,\quad e=0.0276.
\end{aligned} \tag{5.26}$$

The magnitudes of $\mathfrak{F}(\mathbf{q})$ and $\mathfrak{G}(\mathbf{q})$ for some special values of $\mathbf{q}$ are given in Table I. $E_{\mathbf{q}}$ as a function of $|\mathbf{q}|$ is shown in Fig. 2 for $\mathbf{q}$ parallel to (1.1.1) when κ takes several values. When κ increases, $E_{\mathbf{q}}$ first becomes zero at $\mathbf{q}=1/a(\pi,\pi,\pi)$ for $\kappa=0.492$. If $\kappa>0.492$, there appear spin-wave modes with negative energy and the ferromagnetic state becomes unstable. Then it follows that if

$$0.246n/N<t/I, \tag{5.27}$$

the ferromagnetic state with $S=S_{\max}$ is not the ground state.

We can also calculate the spin-wave spectrum to the next order of n from Eqs. (5.10), (5.12), and (5.13) simply by taking into account the $\mathbf{p}$ dependence of the functions. For the purpose of examining the stability of the ferromagnetic state it is sufficient to calculate it at $\mathbf{q}=\mathbf{q}_0\equiv1/a(\pi,\pi,\pi)$. Putting $\mathbf{q}=\mathbf{q}_0$ in Eqs. (5.10), (5.12), and (5.13), and eliminating $\Phi_{\mathbf{q}_0}^{s\lambda}(\mathbf{p})$ from Eqs. (5.12) and (5.13), we have

$$\left[\epsilon_{\mathbf{q}_0}+\frac{12t}{I}-\frac{2}{N}\sum_{\mathbf{p}>}\sum_{\lambda}\cos p_{\lambda}a\right]\Phi_{\mathbf{q}_0}=2\sum_{\mathbf{p}>}\sum_{\lambda}\Phi_{\mathbf{q}_0}^{c\lambda}(\mathbf{p}), \tag{5.28}$$

$$\left[1+F(\theta_p)+\frac{\cos p_\lambda a-H(\theta_p)}{1-(\cos p_\lambda a)H(\theta_p)}G_1(\theta_p)\right]\Phi_{q_0}{}^{c\lambda}(\mathbf{p})+\sum_{\mu\neq\lambda}\left[F(\theta_p)+\frac{\cos p_\mu a-H(\theta_p)}{1-(\cos p_\mu a)H(\theta_p)}G_2(\theta_p)\right]\Phi_{q_0}{}^{c\mu}(\mathbf{p})$$
$$=-2[G_1(\theta_p)+2G_2(\theta_p)]\Phi_{q_0}/N. \quad (5.29)$$

From Eq. (5.29) we obtain, to the order of p^2,

$$\sum_\lambda \Phi_{q_0}{}^{c\lambda}(p)=-\left\{\frac{18F}{1+6F}+\left[\frac{1+H}{1-H}\left(\frac{3F}{1+6F}\right)^2-\frac{6(3\beta+F+3F^2)}{(1+6F)^2}\right]\times\tfrac{1}{2}(pa)^2\right\}2\Phi_{q_0}\Big/N, \quad (5.30)$$

where use has been made of Eqs. (C4) and (C14). Then the spin-wave energy is found from Eqs. (5.28) and (5.30) to be

$$E_{q_0}=2t\left\{-\frac{12t}{I}+\frac{6}{1+6F}\frac{n}{N}-\left[1+\frac{1+H}{1-H}\left(\frac{3F}{1+6F}\right)^2-\frac{6(3\beta+F+3F^2)}{(1+6F)^2}\right]\frac{1}{N}\sum_{p>}(pa)^2\right\}. \quad (5.31)$$

Using

$$N^{-1}\sum_{p>}(pa)^2\cong\tfrac{3}{5}(6\pi^2)^{2/3}(n/N)^{5/3}$$

and substituting the numerical values calculated in Appendix C, we finally obtain

$$E_{q_0}=12t[-2t/I+0.492(n/N)-0.404(n/N)^{5/3}]. \quad (5.32)$$

The boundary in $(2t/I)$-(n/N) plane at which the ferromagnetic state becomes unstable is given by

$$2t/I=0.492(n/N)-0.404(n/N)^{5/3}, \quad (5.33)$$

which is shown in Fig. 3. It should be noticed here that, as n increases, the curve deviates downwards from the linear one given by the first term.

TABLE I. The magnitudes of $\mathfrak{F}(\mathbf{q})$ and $\mathcal{G}(\mathbf{q})$ for some special values of $\mathbf{q}$.

$\mathbf{q}a$	$\mathfrak{F}(\mathbf{q})$	$\mathcal{G}(\mathbf{q})$
$\simeq 0$	$0.656\times\frac{1}{2}(qa)^2$	$\frac{1}{2}(qa)^2$
$(\pi,0,0)$	1.21	2
$(\pi,\pi,0)$	2.17	4
$\simeq(\pi,\pi,\pi)$[a]	$2.96-0.370\times\frac{1}{2}(q'a)^2$	$6-\frac{1}{2}(q'a)^2$

[a] $\mathbf{q}'=\mathbf{q}-\mathbf{q}_0$, $\mathbf{q}_0a=(\pi,\pi,\pi)$.

6. DISCUSSION

In the preceding sections we gave the proof of the theorems (I), (II), and (III) stated in Sec. 1. For Case B (fcc and hcp with $N_e>N$) it was found that when $n=1$ and $I=\infty$ the ferromagnetic state with $S=S_{\max}$ is not the ground state. This seems to be true for finite n and I as long as n/N and t/I are sufficiently small, for there is no interaction which favors the ferromagnetic ordering. On the other hand, it was shown for Case A (sc and bcc; fcc and hcp with $N_e<N$) that if the ratio $\kappa=(2t/I)/(n/N)$ as well as t/I, and n/N is sufficiently small, the ferromagnetic state with $S=S_{\max}$ *is* the ground state, and that if κ is larger than some constant it *is not* the ground state. In the following we shall discuss this case in some detail.

When $n=0$ and I is finite, i.e., on the $(2t/I)$ axis in the (n/N)-$(2t/I)$ plane, it is very likely that the ground state is antiferromagnetic, because we have an effective antiferromagnetic interaction of the order t^2/I if t/I is small. This seems to be true even for finite n if κ is sufficiently large. Then there arises the question how the transition occurs from the ferromagnetic state with $S=S_{\max}(F)$ to the antiferromagnetic state (A) when κ increases. There are various possibilities, which are shown schematically in Fig. 4 as a phase diagram in the (n/N)-$(2t/I)$ plane. In Fig. 4(a) the transition $F\rightarrow A$ occurs discontinuously. What we can say for this case is that the transition should occur at the value of κ small than κ_0 predicted by the calculation of the spin-wave energy in Sec. 5. In Fig. 4(b), where F' denotes the ferromagnetic state with $0<S<S_{\max}$, there are four possibilities according that the transitions $F\rightarrow F'$ and $F'\rightarrow A$ occur continuously or discontinuously. In this case, the transition $F\rightarrow F'$ should occur just at $\kappa=\kappa_0$ if it is continuous, and at $\kappa<\kappa_0$ if it is discontinuous. Within the argument of this paper we cannot determine which is the case for each lattice structure among these possibilities.

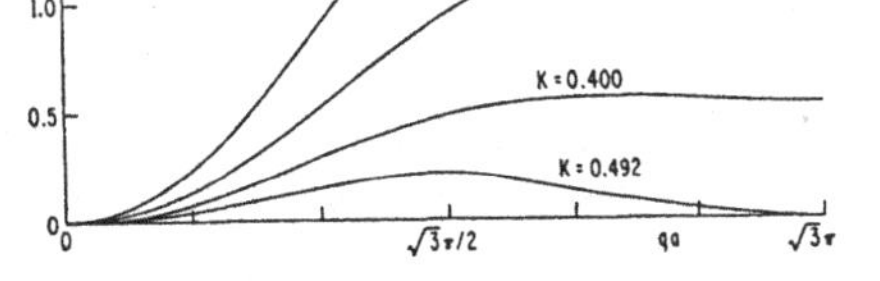

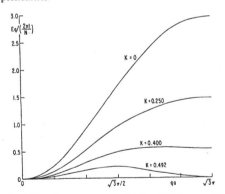

FIG. 2. The spin-wave energy as a function of $|\mathbf{q}|$ for $\mathbf{q}$ parallel to (1,1,1) when κ takes four values, 0, 0.250, 0.400, and 0.492.

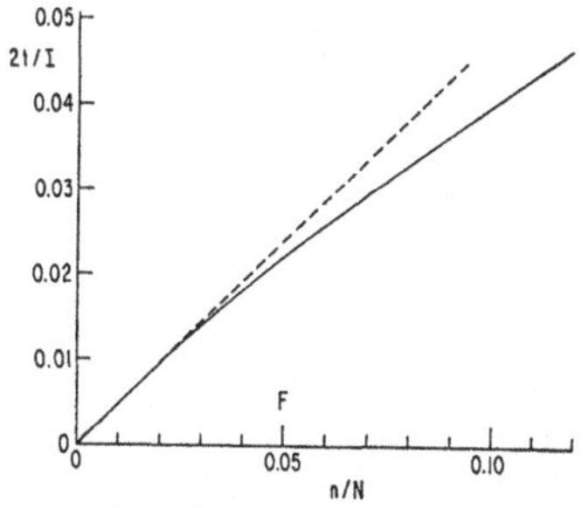

FIG. 3. The boundary curve for sc in the $(2t/I)$-(n/N) plane at which the ferromagnetic with $S=S_{max}$ becomes unstable for spin-wave excitations. F denotes the region where the ferromagnetic state with $S=S_{max}$ is the ground state.

Next we consider what happens when n increases with given I. For small n, the ferromagnetic ordering of the spins are favored energetically by the motion of the holes. As n increases, the volume in which the holes can move decreases. Then it is plausible that the energy gain per hole due to the ferromagnetic ordering decreases with increasing n. The boundary curve of the ferromagnetic region F is therefore expected to bend downwards and to reach finally the (n/N) axis at some value of n. The calculation for sc in Sec. 5 shows that this behavior is seen in the next order of n. According to Kanamori's theory,[3,5] the ground state is not ferromagnetic for sc and bcc if $n\simeq N$. This is consistent with the above prediction. For fcc and hcp it is very likely that the ground state is ferromagnetic when $n\simeq N$, for the density of states diverges at the band edge. Then it seems that the (n/N) axis is divided into three (or more) regions, among which the regions $0<n<n_1$ and $n_2<n<1$ are those of the ferromagnetic ground state.

There remain some problems even for the limiting case $n/N\ll1$ and $t/I\ll1$. In this paper we first considered the case $t/I=0$ and $n/N\neq0$, and then proceeded to the case $t/I\neq0$ and $n/N\neq0$. The next problem is to start from the case $n/N=0$ and $t/I\neq0$. This seems to be more difficult than the present problem, since we have no exact ground state even at the starting point $n/N=0$ and $t/I\neq0$. The problem will be left for future investigations.

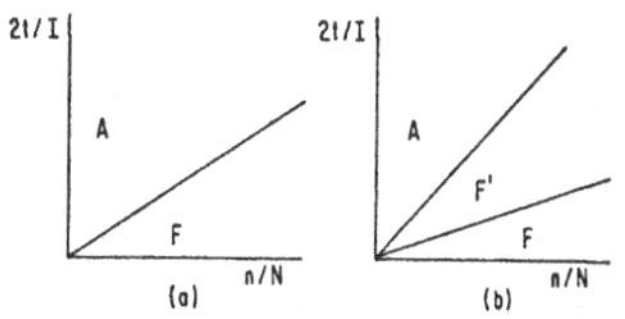

FIG. 4. Possible phase diagrams in the $(2t/I)$-(n/N) plane for Case A: F, ferromagnetic region, $S=S_{max}$; F', ferromagnetic region, $0<S<S_{max}$; A, antiferromagnetic region, $S=0$.

Note added in proof. Recently Penn[8a] studied the same model for sc, using the self-consistent field approximation. His result for small n/N, t/I and κ seems to be consistent with our rigorous conclusion.

ACKNOWLEDGMENTS

I should like to thank Professor Harry Suhl and Professor Kazumi Maki for valuable discussions, R. M. More for reading the manuscript, and Dr. D. J. Thouless for informing me of his unpublished work.

APPENDIX A

In this Appendix we show briefly that for two- and three-dimensional lattices the corresponding superlattice defined in Sec. 2 cannot be decomposed into two or more disconnected parts, or in other words, that it is always possible to rearrange spins in the lattice from one configuration to another.

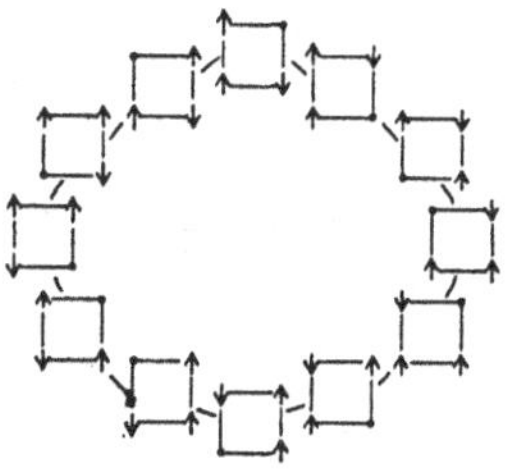

FIG. 5. The superlattice for the case of 2×2 square lattice ($N=4$) with $N_\uparrow=2$ and $N_\downarrow=1$.

Let us first consider a two-dimensional square lattice by means of mathematical induction. For a 2×2 lattice ($N=4$, $N_e=3$), it can be shown directly for any numbers of $N_\uparrow$ and $N_\downarrow$. When $N_\uparrow=3$ (or $N_\downarrow=3$), it is obvious. When $N_\uparrow=2$ and $N_\downarrow=1$ (or $N_\downarrow=2$ and $N_\uparrow=1$), we have 12 lattice points in the superlattice, which are shown in Fig. 5. They form one connected ring. Next, let us assume that for an $N\times M$ lattice all lattice points in the corresponding superlattice are connected, and construct an $N\times(M+1)$ lattice by adding one column at the edge. In rearranging the spins from one configuration to another, we first consider spins on the added column. Then it is always possible to replace any unwanted spin on the column by the wanted one by moving the vacant site on and near the column. This can easily be verified by considering some examples. If we have succeeded in getting the spin configuration we want on the column, there remains only the problem of the rearrangement in an $N\times M$ lattice, which is always possible by the assumption. Thus it is proved for a two-dimensional square lattice that all lattice points in the superlattice are connected. It can also be proved for a two-dimensional triangle lattice in a similar way.

Now it is easy to prove it for three-dimensional

[8a] D. R. Penn, Phys. Rev. 142, 350 (1966).

lattices. We have only to divide the lattice into many two-dimensional sheets, and to make the rearrangement for each sheet one by one.

APPENDIX B

In this Appendix we prove (II) for hcp in the same way as we did for fcc in Sec. 4.

Putting $n=1$ and $I=\infty$ in Eq. (3.9), we have

$$E\Phi_0(j;i)=-(1-\delta_{ij})\sum_{i'}t_{i'i}\Phi_0(j;i')-t_{ji}\Phi_0(i;j). \quad \text{(B1)}$$

In taking the Fourier transform of Eq. (B1), we should be careful because a unit cell contains two atoms in a hcp structure. Let us denote two atomic sites in the ith cell by i_1 and i_2, and define four $\Phi_{\alpha\beta}(j;i)$ by

$$\Phi_{\alpha\beta}(j;i)=\Phi_0(j_\alpha;i_\beta), \quad \alpha,\beta=1,2. \quad \text{(B2)}$$

Then, introducing Fourier transforms by

$$\Phi_{\alpha\beta}(j;i)=\sum_{\mathbf{qk}}\Phi_{\alpha\beta}(\mathbf{q};\mathbf{k})e^{-i\mathbf{q}\cdot\mathbf{r}_j-i\mathbf{k}\cdot(\mathbf{r}_i-\mathbf{r}_j)}, \quad \text{(B3)}$$

$$t_{a,b}(\mathbf{k})=\sum_j t_{j1,2i1}e^{-i\mathbf{k}\cdot(\mathbf{r}_j-\mathbf{r}_i)}, \quad \text{(B4)}$$

where $\mathbf{r}_i$ is the position vector of the representative point of the ith cell, we obtain the equation for $\Phi_{\alpha\beta}(\mathbf{q};\mathbf{k})$ as

$$\begin{aligned}
E\Phi_{11}(\mathbf{q};\mathbf{k})&=-t_a(\mathbf{k})\Phi_{11}(\mathbf{q};\mathbf{k})-t_b(\mathbf{k})\Phi_{12}(\mathbf{q};\mathbf{k})\\
&\quad+(2/N)\sum_{\mathbf{k}'}[t(\mathbf{k}')-t(\mathbf{k}+\mathbf{k}'-\mathbf{q})]\Phi_{11}(\mathbf{q};\mathbf{k}')+(2/N)\sum_{\mathbf{k}'}t_b(\mathbf{k}')\Phi_{12}(\mathbf{q};\mathbf{k}'),\\
E\Phi_{12}(\mathbf{q};\mathbf{k})&=-t_a(\mathbf{k})\Phi_{12}(\mathbf{q};\mathbf{k})-t_b^*(\mathbf{k})\Phi_{11}(\mathbf{q};\mathbf{k})-(2/N)\sum_{\mathbf{k}'}t_b^*(\mathbf{k}+\mathbf{k}'-\mathbf{q})\Phi_{21}(\mathbf{q};\mathbf{k}'),\\
E\Phi_{21}(\mathbf{q};\mathbf{k})&=-t_a(\mathbf{k})\Phi_{21}(\mathbf{q};\mathbf{k})-t_b(\mathbf{k})\Phi_{22}(\mathbf{q};\mathbf{k})-(2/N)\sum_{\mathbf{k}'}t_b(\mathbf{k}+\mathbf{k}'-\mathbf{q})\Phi_{12}(\mathbf{q};\mathbf{k}'),\\
E\Phi_{22}(\mathbf{q};\mathbf{k})&=-t_a(\mathbf{k})\Phi_{22}(\mathbf{q};\mathbf{k})-t_b^*(\mathbf{k})\Phi_{21}(\mathbf{q};\mathbf{k})\\
&\quad+(2/N)\sum_{\mathbf{k}'}[t_a(\mathbf{k}')-t_a(\mathbf{k}+\mathbf{k}'-\mathbf{q})]\Phi_{22}(\mathbf{q};\mathbf{k}')+(2/N)\sum_{\mathbf{k}'}t_b^*(\mathbf{k}')\Phi_{21}(\mathbf{q};\mathbf{k}').
\end{aligned} \quad \text{(B5)}$$

The solution of Eq. (B5) which corresponds to the solution (3.14) of Eq. (3.13) is given by

$$\begin{aligned}
\Phi_{11}(\mathbf{q};\mathbf{k})&=A(\delta_{\mathbf{kq}}-2/N), & \Phi_{12}(\mathbf{q};\mathbf{k})&=B\delta_{\mathbf{kq}},\\
\Phi_{21}(\mathbf{q};\mathbf{k})&=A\delta_{\mathbf{kq}}, & \Phi_{22}(\mathbf{q};\mathbf{k})&=B(\delta_{\mathbf{kq}}-2/N),
\end{aligned} \quad \text{(B6)}$$

with

$$\begin{aligned}
EA&=-t_a(\mathbf{q})A-t_b(\mathbf{q})B,\\
EB&=-t_a(\mathbf{q})B-t_b^*(\mathbf{q})A.
\end{aligned} \quad \text{(B7)}$$

So far we have assumed nothing about the lattice structure and the transfer matrix elements t_{ij} except that a unit cell contains two atoms. For a special case of hcp with assumption (b) and $t<0$, we have

$$\begin{aligned}
t_a(\mathbf{k})&=-|t|\theta_a(\mathbf{k}), \quad \theta_a(\mathbf{k})=\sum_{i=1}^{3}(e^{i\mathbf{k}\cdot\mathbf{a}_i}+e^{-i\mathbf{k}\cdot\mathbf{a}_i}),\\
t_b(\mathbf{k})&=-|t|\theta_b(\mathbf{k}), \quad \theta_b(\mathbf{k})=\sum_{j=0}^{2}(e^{i\mathbf{k}\cdot\mathbf{a}_j}+e^{i\mathbf{k}\cdot\mathbf{a}_{j+4}}),
\end{aligned} \quad \text{(B8)}$$

where

$$\begin{aligned}
\mathbf{a}_0&=0,\\
\mathbf{a}_1&=a(\tfrac{1}{2},-\sqrt{3}/2,0),\\
\mathbf{a}_2&=a(1,0,0),\\
\mathbf{a}_3&=a(\tfrac{1}{2},\sqrt{3}/2,0),\\
\mathbf{a}_{j+4}&=\mathbf{a}_j+a(0,0,-2\sqrt{\tfrac{2}{3}}),
\end{aligned} \quad \text{(B9)}$$

a denoting the lattice constant. The energy spectrum of the band is given by

$$\begin{aligned}
t_\pm(\mathbf{k})&=-|t|\theta_\pm(\mathbf{k}),\\
\theta_\pm(\mathbf{k})&=\theta_a(\mathbf{k})\pm|\theta_b(\mathbf{k})|.
\end{aligned} \quad \text{(B10)}$$

The band top is located at

$$k_z=0, \quad \tfrac{1}{2}k_xa=\pm(\pi/2) \quad \text{or} \quad \tfrac{1}{2}k_xa\pm(\sqrt{3}/2)k_ya=\pm\pi, \quad \text{(B11)}$$

where the energy is given by $4|t|$. Then the lowest energy of the $S=S_{\max}$ states is given by $-4|t|$.

For $\mathbf{q}=0$, substitution of Eq. (B8) in Eq. (B5) gives

$$
\begin{aligned}
&[\epsilon-\theta_a(\mathbf{k})]\Phi_{11}(\mathbf{k})-\theta_b(\mathbf{k})\Phi_{12}(\mathbf{k})=-\sum_{i=1}^{3}[(1-e^{i\mathbf{k}\cdot\mathbf{a}_i})\Phi_{11}{}^i+(1-e^{-i\mathbf{k}\cdot\mathbf{a}_i})\bar{\Phi}_{11}{}^i]-\sum_{j=0}^{2}(\Phi_{12}{}^j+\bar{\Phi}_{12}{}^j)\,,\\
&[\epsilon-\theta_a(\mathbf{k})]\Phi_{12}(\mathbf{k})-\theta_b{}^*(\mathbf{k})\Phi_{11}(\mathbf{k})=\sum_{j=0}^{2}[e^{-i\mathbf{k}\cdot\mathbf{a}_j}\Phi_{21}{}^j+e^{-i\mathbf{k}\cdot\mathbf{a}_{j+1}}\bar{\Phi}_{21}{}^j]\,,\\
&[\epsilon-\theta_a(\mathbf{k})]\Phi_{21}(\mathbf{k})-\theta_b(\mathbf{k})\Phi_{22}(\mathbf{k})=\sum_{j=0}^{2}[e^{i\mathbf{k}\cdot\mathbf{a}_j}\Phi_{12}{}^j+e^{i\mathbf{k}\cdot\mathbf{a}_{j+1}}\bar{\Phi}_{12}{}^j]\,,\\
&[\epsilon-\theta_a(\mathbf{k})]\Phi_{22}(\mathbf{k})-\theta_b{}^*(\mathbf{k})\Phi_{21}(\mathbf{k})=-\sum_{i=1}^{3}[(1-e^{i\mathbf{k}\cdot\mathbf{a}_i})\Phi_{22}{}^i+(1-e^{-i\mathbf{k}\cdot\mathbf{a}_i})\bar{\Phi}_{22}{}^i]-\sum_{j=0}^{2}(\Phi_{21}{}^j+\bar{\Phi}_{21}{}^j)\,,
\end{aligned}
\tag{B12}
$$

where

$$
\begin{aligned}
\Phi_{\alpha\beta}(\mathbf{k})&=\Phi_{\alpha\beta}(0;\mathbf{k})\,,\\
\epsilon&=E/|t|\,,\\
\Phi_{\alpha\alpha}{}^i&=(2/N)\sum_{\mathbf{k}}e^{i\mathbf{k}\cdot\mathbf{a}_i}\Phi_{\alpha\alpha}(\mathbf{k})\,,\\
\bar{\Phi}_{\alpha\alpha}{}^i&=(2/N)\sum_{\mathbf{k}}e^{-i\mathbf{k}\cdot\mathbf{a}_i}\Phi_{\alpha\alpha}(\mathbf{k})\,,\\
\Phi_{12}{}^j&=(2/N)\sum_{\mathbf{k}}e^{i\mathbf{k}\cdot\mathbf{a}_j}\Phi_{12}(\mathbf{k})\,,\\
\bar{\Phi}_{12}{}^j&=(2/N)\sum_{\mathbf{k}}e^{i\mathbf{k}\cdot\mathbf{a}_{j+1}}\Phi_{12}(\mathbf{k})\,,\\
\Phi_{21}{}^j&=(2/N)\sum_{\mathbf{k}}e^{-i\mathbf{k}\cdot\mathbf{a}_j}\Phi_{21}(\mathbf{k})\,,\\
\bar{\Phi}_{21}{}^j&=(2/N)\sum_{\mathbf{k}}e^{-i\mathbf{k}\cdot\mathbf{a}_{j+1}}\Phi_{21}(\mathbf{k})\,.
\end{aligned}
\tag{B13}
$$

Now our problem is to see whether Eq. (B12) has any nontrivial solution for $\epsilon<-4$ (i.e., $E<-4|t|$). From Eq. (B12) we can derive a set of simultaneous equations for 24 components, $\Phi_{\alpha\alpha}{}^i$, $\bar{\Phi}_{\alpha\alpha}{}^i$, $\Phi_{\alpha\beta}{}^j$, and $\bar{\Phi}_{\alpha\beta}{}^j$. Though these equations can be diagonalized easily, for our purpose it is sufficient to show, if any, the equation of one component which has a nontrivial solution for $\epsilon<-4$.

Putting

$$\Phi_\alpha=(\Phi_{\alpha\alpha}{}^1-\bar{\Phi}_{\alpha\alpha}{}^1)-(\Phi_{\alpha\alpha}{}^2-\bar{\Phi}_{\alpha\alpha}{}^2)+(\Phi_{\alpha\alpha}{}^3-\bar{\Phi}_{\alpha\alpha}{}^3)\,,\quad \alpha=1,2, \tag{B14}$$

we have the equation for Φ_α as

$$[1-J(\epsilon)]\Phi_\alpha=0\,, \tag{B15}$$

where

$$
\begin{aligned}
J(\epsilon)&=\frac{1}{N}\sum_{\mathbf{k}}j(\mathbf{k})\left\{\frac{1}{\epsilon-\theta_+(\mathbf{k})}+\frac{1}{\epsilon-\theta_-(\mathbf{k})}\right\}\\
j(\mathbf{k})&=-2[\sin^2\mathbf{k}\cdot\mathbf{a}_2+(\cos\mathbf{k}\cdot\mathbf{a}_1-1)\\
&\qquad\times\cos\mathbf{k}\cdot\mathbf{a}_2-\sin\mathbf{k}\cdot\mathbf{a}_1\sin\mathbf{k}\cdot\mathbf{a}_2]\,.
\end{aligned}
\tag{B16}
$$

The behavior of $J(\epsilon)$ for $\epsilon\to-4-0$ can be seen in a similar way to the case of $H_-(\epsilon)$ in Sec. 4. As we have

$$\langle j(\mathbf{k})\rangle_{\text{band top}}=-\tfrac{4}{3}\,, \tag{B17}$$

it turns out that

$$
\begin{aligned}
\lim_{\epsilon\to-4-0}J(\epsilon)&=\langle j(\mathbf{k})\rangle_{\text{band top}}\lim_{\epsilon\to-4-0}\frac{1}{N}\\
&\qquad\times\sum_{\mathbf{k}}\left\{\frac{1}{\epsilon-\theta_+(\mathbf{k})}+\frac{1}{\epsilon-\theta_-(\mathbf{k})}\right\}\\
&=\infty\,.
\end{aligned}
\tag{B18}
$$

It is clear from Eq. (B18) that Eq. (B15) has at least one nontrivial solution for $\epsilon<-4$, which means the $S=S_{\max}$ state is not the ground state. This is the proof of (II) for hcp.

APPENDIX C

In this Appendix we calculate some numerical constants which appeared in the calculation of Sec. 5. Introducing the function

$$I(\theta)=N^{-1}\sum P(\theta-\theta_{\mathbf{k}})^{-1}\,, \tag{C1}$$

we can prove the following relations:

$$G_1(\theta)+H(\theta)=I(\theta)\,, \tag{C2}$$

$$F(\theta)=\tfrac{1}{3}[\theta I(\theta)-1]\,, \tag{C3}$$

$$G_1(\theta)+2G_2(\theta)=\theta F(\theta)\,. \tag{C4}$$

Equation (C2) is obvious, while Eq. (C3) is derived as

$$\begin{aligned} F(\theta)&=\frac{1}{3N}\sum_{\mathbf{k}}\frac{\sum_\lambda \cos k_\lambda a}{\theta-\sum_\lambda \cos k_\lambda a}\\ &=\frac{1}{3N}\sum_{\mathbf{k}}\left(\frac{\theta}{\theta-\sum_\lambda \cos k_\lambda a}-1\right)\\ &=\tfrac{1}{3}[\theta I(\theta)-1]\,. \end{aligned}$$

Equation (C4) can also be proved in a similar way.

$I(3)$ is one of the so-called Watson integrals,[9] which is given by

$$I(3)=0.50546\,. \tag{C5}$$

For $\theta\geq 3$, $H(\theta)$ can be rewritten as

$$\begin{aligned} H(\theta)&=\frac{1}{\pi^3}\int\!\!\int\!\!\int_0^\pi \frac{\sin^2\alpha_1\prod_i^3 d\alpha_i}{\theta-\sum_i^3\cos\alpha_i}\\ &=\int_0^\infty dt\frac{1}{\pi^3}\int\!\!\int\!\!\int_0^\pi \sin^2\alpha_1 \times\exp[-(\theta-\sum_i^3\cos\alpha_i)t]\prod_i^3 d\alpha_i\\ &=\int_0^\infty dt e^{-\theta t}\frac{1}{t}I_1(t)[I_0(t)]^2\,, \end{aligned} \tag{C6}$$

[9] G. N. Watson, Quart. J. Math. (Oxford) 10, 266 (1939).

where $I_0(t)$ and $I_1(t)$ are modified Bessel functions of the first kind. Using Eq. (C6), we calculate $H(3)$ numerically.[10] The result is

$$H(3)=0.208\,. \tag{C7}$$

From Eqs. (C2)-(C5) and (C7) we obtain

$$G_1(3)=0.297\,, \tag{C8}$$

$$G_2(3)=0.109\,, \tag{C9}$$

$$F(3)=0.172\,. \tag{C10}$$

Next we consider the θ dependence of $I(\theta)$ near $\theta=3$. For $\theta\gtrsim 3$, it is expanded as

$$I(\theta)=I(3)-(\sqrt{2}\pi)^{-1}(\theta-3)^{1/2}-\alpha(\theta-3)+\cdots\,. \tag{C11}$$

If we consider $z=\theta-3$ as a complex number, and make an analytic continuation of the function from the positive real axis to the negative real axis, then the $\sqrt{z}$ term only contributes to the imaginary part, and we have for $\theta\lesssim 3$

$$I(\theta)=I(3)+\alpha(3-\theta)+\cdots\,. \tag{C12}$$

The coefficient α can be estimated numerically from the table given by Mannari and Kawabata.[11] We get

$$\alpha=0.054\,. \tag{C13}$$

From Eqs. (C3), (C5), (C12) and (C13) we obtain for $\theta\lesssim 3$

$$F(\theta)=F(3)-\beta(3-\theta)+\cdots\,, \tag{C14}$$

where

$$\beta=0.114\,. \tag{C15}$$

[10] It seems to the author that there is an analytical method to calculate it, though he could not find it.

[11] I. Mannari and C. Kawabata, Research Notes of the Department of Physics, Okayama University, Japan, No. 15, 1964 (unpublished).

PHYSICAL REVIEW B VOLUME 41, NUMBER 4 1 FEBRUARY 1990

Instability of the Nagaoka ferromagnetic state of the $U=\infty$ Hubbard model

B. S. Shastry*
AT&T Bell Laboratories, Murray Hill, New Jersey 07974

H. R. Krishnamurthy† and P. W. Anderson
Joseph Henry Laboratories of Physics, Princeton University, Princeton, New Jersey 08544
(Received 24 April 1989)

We identify a "k_F instability" of the Nagaoka ferromagnetic state of the $U=\infty$ Hubbard model. We show rigorously that for a large enough hole concentration the ferromagnet possesses an instability with respect to overturning an up-spin electron at the Fermi surface and placing it at the bottom of a down-spin band made very narrow by correlation effects. We find a low-energy scale for spin waves in this strong-coupling limit, in the form of a spin-wave stiffness that is much smaller than its random-phase-approximation value.

In this paper we present some variational and exact results concerning the Hubbard model in the limit of large U—pertaining mainly to the stability or otherwise of the Nagaoka ferromagnet. The result of Nagaoka[1] (see also Thouless[2]) is of considerable importance since it is a nonperturbative and exact statement about the Hubbard model for strong coupling, i.e., $U=\infty$. The recent revival of interest in the Hubbard model for large (but not infinite) U, following Anderson's suggestion[3] of its relevance in high-T_c superconductivity, has focused mainly on the so-called Heisenberg-Hubbard model, which in fact contains the $U=\infty$ Hubbard model kinetic energy as one of its two pieces. In addition, the theory of itinerant electron ferromagnetism has traditionally relied upon the Nagaoka ferromagnet as a clearly demonstrable case of the existence of ferromagnetism in a one-band Hubbard model.[4]

Given the importance of the Nagaoka ferromagnet, the "thermodynamic frailty" of the methods used to prove it have been a source of concern to several workers over the years. Nagaoka shows that the fully saturated ferromagnet is a ground state in the case of one hole (measured from half-filling) for $U=\infty$ and on appropriate lattices. This method fails to prove ferromagnetism for a few as two holes. In fact in the case of two holes we can readily show, by essentially a Peierls construction, that a singlet state must exist with an energy only $O(1/L^2)$ above that of the ferromagnet (we could cut the lattice into two equal domains and confine one hole into each, and further form the largest spin state for each domain and couple these two domain ferromagnets into a singlet—the energy cost is only a boundary effect). For a thermodynamic concentration of holes, such considerations really do not serve as proper guides. However, the one-dimensional Hubbard model, with $U=\infty$, has a separation of charge and spin, and so it is impossible to find a state with lower energy than the Nagaoka state at any concentration of holes. This leads to a possible scenario in which the ferromagnet could survive (at $U=\infty$) in two and three dimensions for *any* hole concentration.

It is the purpose of this paper to show that the preceding scenario is false—we present a variational wave function with one spin down with a finite wave vector, k_F, which has a lower energy than the Nagaoka state in two and three dimensions for a sufficiently large concentration of holes. Our "excitation energy" is a strict upper bound to the true excitation energy, and becomes negative for large enough δ (density of holes) but it remains non-negative in one dimension. In fact, our criterion for the instability of the ferromagnet (namely the "excitation energy" going negative) captures the subtleties of the Nagaoka theorem (related to the signs of the hopping matrix element on nonbipartite lattices). We also present variational estimates on how large the Coulomb interaction U must be in order to stabilize the ferromagnet. These are, however, not optimal for all δ. We believe that this is the first published demonstration of the instability of the Nagaoka state for any hole concentration (at $U=\infty$) which has a variational (and hence rigorous) basis, and which is thermodynamically relevant.

In order to motivate our wave functions, we would like to review, briefly, the work of Richmond and Rickayzen,[5] who performed an interesting calculation with a similar objective to ours. These authors also consider the problem of $N_\uparrow=N_s(1-\delta)$ up (-spin) electrons and one down (-spin) electron N_s being the number of lattice sites) and construct a variational wave function obtained by freezing the motion of the down electron and solving exactly for the up electron gas which now sees a simple potential scatterer (strength U) at one site. The up-electron energies are shifted by $O(1/N_s)$ each, and the net cost is $O(1)$, whereas the possibility of virtually admixing the doubly occupied site gains an exchange energy. The final conclusion of this study is that the Nagaoka ferromagnet is unstable with respect to reducing U from infinity—however, they find that at $U=\infty$ the Nagaoka ferromagnet is always stable for any δ.

It appears that the preceding stability of the Nagaoka ferromagnet arises in their calculation by the inability of the wave function to allow the down-spin electron to hop around. The overturned electron would prefer to be in a highly delocalized state. For, if we imagine the fully spin-polarized Stoner state (which is just the Nagaoka state at $U=\infty$) and switch off U, then the leading instability would correspond to destroying an up electron at the (Stoner-Nagaoka) Fermi surface and creating a down electron at the band bottom. This picture immediately suggests that an appropriate strategy for the large-U problem would be to take such a "Fermi surface restoring" excitation and to correct for strong Coulomb repulsion by a variational projection.

Explicitly we write the Hamiltonian

$$H=-\sum_{ij} t_{ij} C_{i\sigma}^{\dagger} C_{j\sigma}+U\sum_{i} n_{i\uparrow}n_{i\downarrow} . \tag{1}$$

In the $U=\infty$ limit, the above considerations lead us to the variational wave function[6]

$$|\chi_v\rangle=(N_s)^{-1/2}\sum_{m} e^{iqr_m} C_{m\downarrow}^{\dagger}(1-n_{m\uparrow})C_{k_F\uparrow}|F\rangle , \tag{2}$$

where $|F\rangle$ is the ferromagnetic Nagaoka-Stoner state $\prod_{0\le|k|\le k_F} C_{k\uparrow}^{\dagger}|\text{vac}\rangle$, and k_F refers to one of the Fermi surface vectors. A straightforward calculation gives the excitation energy

$$\langle\chi_v|(H-E_0)|\chi_v\rangle/\langle\chi_v|\chi_v\rangle ,$$

where E_0 is the energy of the Nagaoka-Stoner state, as

$$\lambda_v(q)=(\hat{\mu}-\varepsilon_F)+\varepsilon_q\delta(1-\hat{\mu}^2/z^2t^2) . \tag{3}$$

Here $\rho\equiv1-\delta$, $\hat{\mu}\equiv-E_0/(N\delta)$, and $\varepsilon_F\equiv\varepsilon_{k_F}$, with

$$\varepsilon_k=-(1/N_s)\sum_{ij} t_{ij}e^{i\mathbf{k}\cdot(\mathbf{r}_i-\mathbf{r}_j)} ,$$

z = coordination number and further we assume that $t_{ij}=t$ for i,j nearest neighbor and zero otherwise. In terms of the density of states $\rho(\varepsilon)$ per site and per spin we have

$$\hat{\mu}\delta=\int_{\varepsilon_F}^{W_{\text{top}}}\varepsilon\rho(\varepsilon)d\varepsilon \quad\text{and}\quad \delta=\int_{\varepsilon_F}^{W_{\text{top}}}\rho(\varepsilon)d\varepsilon , \tag{4}$$

where W_{top} is the band top energy, whence $\varepsilon_F\le\hat{\mu}\le W_{\text{top}}$. In Eq. (3), the two terms are, respectively, the energy loss of the up spins brought about by the up electrons having to avoid the inserted down spin (it is a net loss since $\hat{\mu}\ge\varepsilon_F$) and the energy gain of the down electron that can move around on the vacant sites left behind in the ferromagnet. The coefficient of ε_q in (3) represents the effective "band width" reduction factor of the down electron—which is, in fact, the hole-density–hole-density correlation function at nearest-neighbor separation in the Nagaoka ferromagnet divided by δ. The physics of this term is simply that given a hole at a site, a down spin is inserted there, and its hopping requires a hole at a neighboring site—thus we need the conditional probability of finding a second hole at a nearest-neighbor site given one hole at a site.

Clearly the lowest value of $\lambda_v(q)$ is obtained by setting ε_q as the band bottom energy $-|W_{\text{bot}}|$. We distinguish two cases here depending on whether (a) $W_{\text{top}}=z|t|$ or (b) $W_{\text{top}}<z|t|$. Case (a) applies to the square lattice, the triangular lattice with $t<0$, the simple cubic, the bcc lattice, and the fcc lattice with $t<0$. Case (b) corresponds to the triangular and fcc lattices with $t>0$. We assert for all the lattices in case (b) that the ferromagnet is unstable for arbitrarily small δ; the instability is of course exactly what Nagaoka's theorem would predict for a single hole— it arises in (3) because the first term is a positive number of $O(\delta)$ and the second is also of $O(\delta)$ but negative, with a larger coefficient. Case (a) is, however, more subtle. The fact that $W_{\text{top}}=z|t|$ and Eq. (4) imply that $(\hat{\mu}/zt)^2$ tends to 1 as $\delta\to0$, whence the second term in (3) is of $O(\delta^2)$. This guarantees that there must exist a nonzero region around $\delta=0$ where λ_v is non-negative—this robustness of the Nagaoka ferromagnet in this case stems from the rather curious fact that the hole-density–hole-density correlation function of the ferromagnet, at nearest-neighbor separation, is of $O(\delta^3)$ rather than $O(\delta^2)$ as one might naively expect. In effect holes repel very strongly in this case thereby enhancing the stability of the Nagaoka state. Table I lists the critical values of δ above which λ_v goes negative in various cases. It is seen that there are surprisingly stable cases—the case (a) triangular lattice and fcc lattice, which appear to be good candidates for itinerant ferromagnetism.

Having found an excitation with possibly vanishing energy, we observe that the preceding instability has a wave vector corresponding to the Stoner-Nagaoka Fermi momentum relative to the band bottom state's momentum. This is a generalized spin wave with a (fixed) nonzero wave vector $q=k_F$; and brings us to the question of the (Goldstone) long-wavelength spin waves, which must on symmetry grounds possess a vanishing energy.[7] We therefore construct a variational wave function which contains long-wavelength spin waves and also interpolates to contain the leading instability already discussed as

$$|\phi\rangle=\frac{1}{\sqrt{\beta(N_s-N)}}\sum_{m,k} e^{i(k+q)r_m}\psi_k C_{m\downarrow}^{\dagger}(1-n_{m\uparrow})C_{k\uparrow}|F\rangle, \tag{5}$$

where ψ_k is an unspecified amplitude for the wave vector k. The wave function[8] $|\phi\rangle$ is characterized by the wave vector q, and is motivated by the RPA (Ref. 9) which can be recovered by neglecting the (projection) factor $(1-n_{m\uparrow})$. If we choose ψ_k to be a Kronecker δ function at $k=k_F$, this reduces to our wave function $|\chi_v\rangle$ in Eq. (2). If we set $q=0$ and let ψ_k be independent of k, then $|\phi\rangle$ is simply the state obtained by acting on $|F\rangle$ with the total spin lowering operator, and hence is degenerate with $|F\rangle$.

The constant β in (5) is determined by normalization as

$$\beta=\sum_{k}|\psi_k|^2 f_k+\frac{1}{N_s\delta}\sum_{p,k}\psi_p^*\psi_k f_p f_k , \tag{6}$$

where $f_k=\Theta(\varepsilon_F-\varepsilon_k)$ restricts the sum to the Stoner-Nagaoka Fermi sea. We calculate the "spin-wave" excitation energy [i.e., $\langle\phi|(H-E_0)|\phi\rangle$] to be

TABLE I. The spin-wave stiffness in our scheme and in the RPA, in units of zt where z is the coordination number for different lattices, and the critical hole concentration δ_{cr} where Eq. (3) is zero.

		Square	Triangle	sc	bcc	fcc
$\delta=0.1$	D_{RPA}	0.023	0.03	0.012	0.01	0.006
	D	0.009	0.016	0.006	0.005	0.004
$\delta=0.2$	D_{RPA}	0.044	0.056	0.024	0.015	0.01
	D	0.014	0.026	0.007	0.006	0.006
δ_{cr}		0.49	1	0.32	0.32	0.62

$$\varepsilon(q)=\frac{1}{\beta}\sum_k \lambda_k(q)\psi_k^*\psi_k f_k + \frac{1}{\beta N_s}\sum_{kp} f_k f_p \psi_p^*\psi_k K(k,p;q)\ , \tag{7}$$

where

$$\lambda_k(q)=\tilde{\mu}-\varepsilon_k+\delta(1-\tilde{\mu}^2/z^2t^2)\varepsilon_{k+q} \tag{8}$$

and

$$K(k,p;q)\equiv(\varepsilon_{k+q}+\varepsilon_{p+q})+(\tilde{\mu}/zt)(\varepsilon_q+\varepsilon_{k+p+q})\ . \tag{9}$$

Varying with respect to ψ_k^*, with the preceding normalization constraint, we find the eigenvalue equation

$$[\varepsilon(q)-\lambda_k(q)]\psi_k=\frac{1}{N_s}\sum_p f_p\left[K(k,p;q)-\frac{1}{\delta}\varepsilon(q)\right]\psi_p\ . \tag{10}$$

This integral equation has two classes of solutions, the continuum of scattering states obtained by omitting the right-hand side [which shifts the energies only by $O(1/N)$] and a bound state, the (Goldstone) spin wave, starting at zero energy at $q=0$. The scattering continuum is analogous to the Stoner particle-hole continuum in the weak-coupling ferromagnet, and is bounded from below by the minimum of $\lambda_k(q)$ for a given q. Its value at $q=0$ is the effective exchange splitting in this strong-coupling theory. The instability discussed in the previous sections is precisely contained in this scheme, since $\lambda_k(q)$ at $\mathbf{q}=\mathbf{k}_f$ has a minimum at $\mathbf{k}=-\mathbf{k}_F$, thus our previous discussion is tantamount to the statement that the lower edge of the scattering continuum has a minimum at $q=k_F$, and that this minimum comes down with increasing δ, until at some critical value, it hits the abscissa, signifying the instability of the ferromagnet.[10] This scheme also contains the Goldstone mode, since a calculation shows that Eq. (7), in the limit of $q=0$, has a zero eigenvalue with the eigenfunction ψ_k independent of k. (This phenomenon is a statement of the rotational invariance of our scheme.)

The small-q spin-wave spectrum can be extracted from the (separable kernel) integral equation (10). We find the eigenfunction

$$\psi_k=1+q_x\phi_k+O(q_x^2)\ ,$$

and ϕ_k is obtained as

$$\phi_k=\frac{v_{k_x}}{\lambda_k(0)}\frac{1}{1+z\tilde{\mu}I/(2\theta)}\ , \tag{11}$$

where

$$v_{k_x}=(\partial/\partial k_x)\varepsilon_k\ ,$$

and

$$I=(1/N_s)\sum_k f_k v_{k_x}^2/\lambda_k(0)\ .$$

These are appropriate in all the lattices of case (a), with energy measured in units of W_{top}. In the remainder of the paper, we use the same units. The constant $\Theta=1$ in all cases except the triangular lattice where $\Theta=2$, and we treat this lattice as a square lattice (with lattice constant a) and all diagonal bonds running, say, northeast. The spin-wave energy goes as

$$\varepsilon_{SW}(q)=Dq_x^2a^2+O(q^4)\ ,$$

with the stiffness given (in units of zt) by

$$D=D_{RPA}\left[1-\frac{Iz}{\tilde{\mu}\Theta(1+\tilde{\mu}Iz/2\Theta)}\right]\ , \tag{12}$$

where $D_{RPA}=\Theta\tilde{\mu}\delta/z$. Note that I is always positive, and hence the spin waves are always softer than what RPA suggests. In Table I we list the stiffness for two typical values of δ (0.1 and 0.2) and also the RPA values for comparison.

It is remarkable that the stiffness is much smaller than D_{RPA}, and 2 orders of magnitude less than zt is almost all the cases considered. This low-energy scale of the long-wavelength excitations would lead to a transition temperature that is considerably lower than the Stoner-Hartree-Fock estimates, and has a bearing on the question of why the T_c of itinerant ferromagnets is so low[11] (spin-wave theory[12] for the simple cubic lattice for $\delta=0.2$ would give a $T_c=0.029\ z|t|$). In any case our calculation gives an upper bound on the spin-wave stiffness. A finite exchange energy (t^2/U) would reduce this further.

For general nonzero q, the integral equation (10) reduces to a set of algebraic equations by using the separability of the kernel, and was solved numerically. In Fig. 1 we sketch the bound-state spectrum and the scattering continuum for the square lattice in two cases; one case

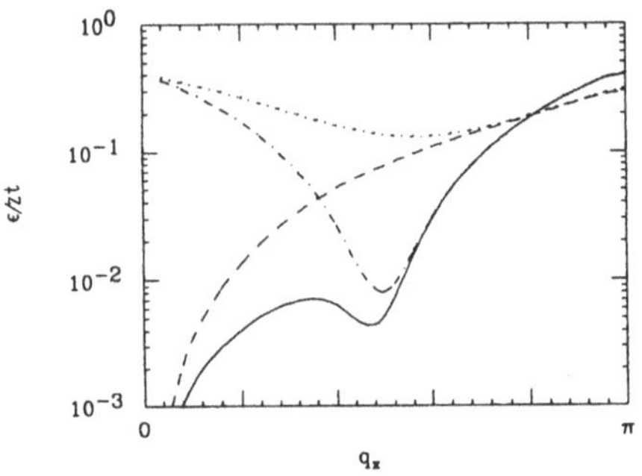

FIG. 1. The spin-wave energy (solid and dashed curves) and the bottom of the scattering continuum (dot-dashed and dotted curves) for $\delta \lesssim \delta_{cr}$ and $\delta \sim 0.34$, respectively, on a logarithmic scale, against q [along (1,1) direction].

corresponds to a small enough δ for which we only have positive-definite excitations, and the second corresponds to $\delta \sim 0.49$ at which the Nagaoka ferromagnet is close to being unstable.

From Fig. 1, it is seen that the spin-wave bound state is at a much lower scale than the single-particle continuum, and unlike in the usual weak-coupling case, does not enter the continuum for any q. The effect of increasing the hole density is to bring down the entire continuum rapidly, and the dip at $q = k_F$ precipitously. This situation is somewhat reminiscent of soft modes in ferroelectrics, but is a more severe instability since essentially an infinite number of states are going soft. The spin waves respond by going soft slightly before $q = k_F$, and hug the bottom of the continuum for larger q.

We have also used a simple generalization of the wave function of Eq. (2) to determine the critical value of U below which the Nagaoka state is definitely unstable. The variational function is chosen as a Gutzwiller incomplete projected version of Eq. (2) and written as

$$|\Phi_G\rangle = \prod_m [1+(g-1)n_{m\uparrow}n_{m\downarrow}]C_{q\downarrow}^{\dagger}C_{k_{F_1}}|F\rangle , \tag{13}$$

where g is the usual Gutzwiller parameter. The variational energy now reads

$$\lambda_v(g,q,U) = C^{-1}\{(g-1)^2\hat{\mu}\delta - C\varepsilon_F + g^2\rho U + \varepsilon_q[(\delta+\rho g)^2 - \delta^2(g-1)^2\hat{\mu}^2]\} , \tag{14}$$

where $C = \delta + \rho g^2$, $\rho = 1-\delta$, and the various terms are recognizable in analogy with Eq. (3) (obtained by $g \to 0$), except the third which is the Coulomb interaction energy. For a fixed δ and U, we can minimize Eq. (14) with respect to g; and U_{cr} is defined by $\lambda_v = 0$. We believe that this estimate of U_{cr} is a reasonably good guide for $\delta \to \delta_{cr}$, where U_{cr} diverges [like $(\delta_{cr}-\delta)^{-1}$]; but is far from optimal for $\delta \to 0$. In the limit of small δ and large U, the main contribution to the excitation energy comes from the first, second, and last term [i.e., by ignoring the kinetic energy of the down spin which is of $O(\delta^2)$], and we find the leading behavior goes as $(\hat{\mu}-\varepsilon_F) - \delta\hat{\mu}^2/U$. Thus, both terms are of order δ and hence we find that in order to stabilize the ferromagnet we must have $U > U_1$ with U_1 an $O(1)$ energy $\{=\lim_{\delta\to 0}[\delta\hat{\mu}^2/(\hat{\mu}-\varepsilon_F)]\}$. This is not as good as the result of Richmond and Rickayzen, who find the exchange contribution [i.e., term of $O(t^2/U)$] to be independent of δ, whereby $U_{cr} \to \infty$ as $\delta \to 0$. Their result is of course more reliable in this limit since their calculation is exact whenever the down-electron kinetic energy is neglected. In Fig. 2, we juxtapose for the square lattice our result for the stability regime with that of Ref. 4 for the case of the square lattice, to get a rigorous limit on the regime of stability of the Nagaoka ferromagnet.

In summary, we have shown in this paper that the Nagaoka ferromagnet is unstable for a sufficiently large concentration of holes at $U = \infty$ by identifying a soft Fermi-surface restoring excitation. The Nagaoka-Stoner Fermi surface forces the kinetic energy of the up electron to be much greater than that in the Luttinger, or normal (Fermi liquid) Fermi surface, and the instability corresponds to the rectification of this state of affairs—for large enough δ, the down-electron band width becomes large enough to benefit from this collapse. Our estimate of the down-electron band width as $O(\delta^2)$, rather than $O(\delta)$ is specific to the use of our variational wave function—if this is true in general then we have a generic argument for the stability of the Nagaoka ferromagnet for small enough δ. If a state can be found that has a band width for down electrons of $O(\delta)$ while costing only an energy of $O(\delta)$ for the up spins, then it would be possible to destabilize the ferromagnet for any $\delta > 0$. Although we cannot rule out this possibility categorically, we feel it is unlikely since our wave function can only be improved upon by an admixture of particle-hole excitations in the up-electron Fermi surface, which should be quite small [to keep the up-electron energy cost low, of $O(\delta)$].

The instability, with respect to reducing U from

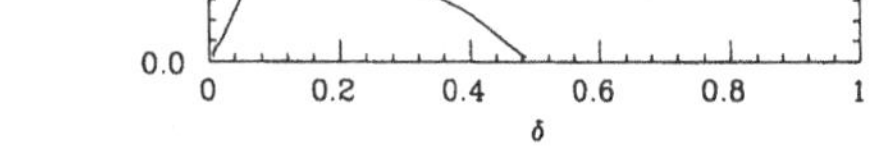

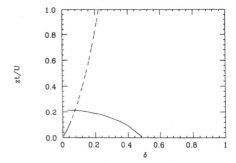

FIG. 2. zt/U_{cr} vs δ for the square lattice as found from our Eq. (14) (solid line) and from Ref. (4) (dashed line). The Nagaoka ferromagnetic is then definitely unstable outside the area bounded by the two curves and the abscissa.

infinity, occurs in two physically distinct ways. One is that at low δ the exchange energy of $O(t^2/U)$ competes with the ferromagnet and leads, presumably through a first-order transition, to an antiferromagnetically correlated state. In the other regime of $\delta \lesssim \delta_{cr}$, we find that terms of $O(1/U)$ bring down the strong coupling continuum to a lower energy, and the k_F instability becomes more pronounced—leading again, we suspect, via a first-order transition—to a metal with a normal Luttinger-like Fermi surface and strong antiferromagnetic correlations.

ACKNOWLEDGMENTS

We would like to thank T. V. Ramakrishnan for several interesting discussions about the Nagaoka ferromagnet. We thank D. M. Edwards for bringing to our attention the work of Roth (see Refs. 6 and 10) and of the unpublished thesis of Tan, (see Ref. 10), of which we were unaware.

---

*On leave from Tata Institute of Fundamental Research, Bombay, India.

†On leave from Indian Institute of Science, Bangalore, India.

[1]Y. Nagaoka, Phys. Rev. **147**, 392 (1966); Solid State Commun. 3, 409 (1965).

[2]D. J. Thouless, Proc. Phys. Soc. London **86**, 893 (1965).

[3]P. W. Anderson, Science **235**, 1196 (1987).

[4]Ferromagnetism in the Hubbard model has been studied using a variety of approximate methods over the years. Some early references, apart from Hubbard's own paper, J. Hubbard, Proc. R. Soc. London A **276**, 238 (1963), are J. Kanamori, Prog. Theor. Phys. **30**, 276 (1963); A. B. Harris and R. V. Lange, Phys. Rev. **157**, 295 (1967). For some recent reviews see *Metallic Magnetism*, edited by H. Capellmann (Springer-Verlag, Berlin, 1987). In what follows, we do not attempt to review the vast literature, and cite only a few papers that are *directly* related to our work.

[5]P. Richmond and G. Rickayzen, J. Phys. C **2**, 528 (1969).

[6]This wave function turns out to have a long history, and was first written down by Laura M. Roth, J. Phys. Chem. Solids **28**, 1549 (1967), who, however, seems to have missed the instability contained in it, which we discuss in what follows. The same wave function has also been rediscovered by S. Schmitt-Rink and A. Ruckenstein (unpublished).

[7]In the case of the electron gas the ferromagnetic state, found within the Hartree-Fock approximation by Bloch, was shown to be unstable with respect to long-wavelength spin-wave excitations in a small range of r_s by Herring [C. Herring, *Magnetism*, edited by G. T. Rado and H. Suhl (Academic, New York, 1966), Vol. IV, p. 104]. It is therefore necessary to examine the long-wavelength spin waves to make sure of the stability of the ferromagnet.

[8]This wave function has also been discussed by L. M. Roth (Ref. 5) who used it only to study the long-wavelength spin waves.

[9]Reviewed by C. Herring in Ref. 7, Chap. XIV. (By RPA we refer to the limit $U=\infty$ of the RPA results noted in this review).

[10]A similar instability of the Nagaoka-Stoner state has also been discussed by Laura M. Roth, Phys. Rev. **186**, 428 (1969), using a Green's-function decoupling scheme. Even though her considerations did not have a variational significance, B. W. Tan, in his unpublished thesis (Imperial College, London, 1974) has shown that Roth's decoupling scheme results can be obtained using a variational wave function for one-spin-down excitations. See also D. M. Edwards and J. A. Hertz, J. Phys. F **3**, 2191 (1973); S. R. Allan and D. M. Edwards, J. Phys. F **12**, 1203 (1982).

[11]D. M. Edwards, Magn. Magn. Mater. **15-18**, 262 (1980).

[12]We are using a crude estimate taken from a simple minded spin-wave calculation, which gives $\Delta M/M(0)=(0.0587/S)(k_BT/D)^3/2$, [see, e.g., C. Kittel, in *Introduction to Solid State Physics*, Sixth Ed. (Wiley, New York, 1986), p. 435], and use $S=\frac{1}{2}$ with $\Delta M/M(0)=1$ in order to find the estimate quoted in the text.

PHYSICAL REVIEW B VOLUME 42, NUMBER 10 1 OCTOBER 1990

Single spin flip in the Nagaoka state: Problems with the Gutzwiller wave function

Thilo Kopp and Andrei E. Ruckenstein
Serin Physics Laboratory, Rutgers University, Piscataway, New Jersey 08855

Stefan Schmitt-Rink
AT&T Bell Laboratories, Murray Hill, New Jersey 07974
(Received 14 June 1990)

We reconsider the problem of a single spin flip in the saturated ferromagnetic state (the Nagaoka state) of the infinite-U Hubbard model for a small (but finite) hole concentration. By studying the ground-state energy and the spin-flip spectral function, we argue that recently proposed Gutzwiller-like variational wave functions are not appropriate for studying this problem in one and two dimensions.

With the recent upsurge of interest in the single-band Hubbard model, the old question of the possibility of a ferromagnetic phase has received renewed attention. For infinite Coulomb repulsion U, saturated ferromagnetism (the Nagaoka state) is known to occur in the limit of a single hole in an otherwise half-filled band.[1] For large hole concentrations, on the other hand, the ground state may be shown to be always paramagnetic.[2,3] If ferromagnetism exists at all, it must therefore arise as a continuation of the Nagaoka state to a small density of holes and large, but finite values of U.

This line of argument has led to two kinds of approaches for infinite U: (i) exact studies of a few holes in finite systems[4-7] and (ii) Gutzwiller-like variational treatments of finite hole concentrations.[8-10] It is the latter that we are addressing in the present paper. (Although the exact studies have shown that for two holes the Nagaoka state is unstable with respect to a single spin flip,[4-7] the physical relevance of these resuts is not clear at present.)

In the Gutzwiller-like approach and for $U=\infty$, the wave function describing a single overturned (down) spin moving in an up-spin Fermi sea is written as[8-11]

$$|\Psi_{\mathbf{k}}\rangle = P\frac{1}{\sqrt{N}}\sum_j e^{i\mathbf{k}\cdot R_j}c_{j\downarrow}^{\dagger}|F\rangle\,, \tag{1}$$

where $P=\prod_j(1-n_{j\downarrow}n_{j\uparrow})$ projects out doubly occupied sites and $|F\rangle=\prod_{k\le k_F}c_{k\uparrow}^{\dagger}|\mathrm{vac}\rangle$ represents the saturated ferromagnetic state. The normalized expectation value of the Hubbard Hamiltonian taken with this wave function is

$$E_{\mathbf{k}}=E_{\uparrow}-\tilde{t}_{\mathbf{k}\downarrow}+\Delta\,, \tag{2}$$

where $E_{\uparrow}=-\sum_{\mathbf{k}}t_{\mathbf{k}}n_{\mathbf{k}\uparrow}$ is the kinetic energy of the up spins,

$$\Delta=-\frac{1}{N}\frac{1}{1-n_{\uparrow}}E_{\uparrow} \tag{3}$$

is the change in this energy due to projecting out the down-spin site, and

$$\tilde{t}_{\mathbf{k}\downarrow}=\frac{1}{N}\frac{1}{1-n_{\uparrow}}\times\sum_{i,j}t_{ij}e^{-i\mathbf{k}\cdot(\mathbf{R}_i-\mathbf{R}_j)}\langle F|(1-n_{i\uparrow})(1-n_{j\uparrow})|F\rangle \tag{4}$$

is the renormalized down-spin hopping matrix element, proportional to the up-spin hole-hole correlation function. From Eq. (2) we obtain the minimum excitation energy for a single spin flip, $-\tilde{t}_{0\downarrow}+\Delta-\mu_{\uparrow}$, which is shown in Fig. 1 as a function of hole concentration for the two-dimensional square lattice. In agreement with earlier studies,[9,10] we find that this excitation energy becomes negative and thus the Nagaoka state becomes unstable at a finite critical hole concentration of $1-n_{\uparrow}=0.49$. We

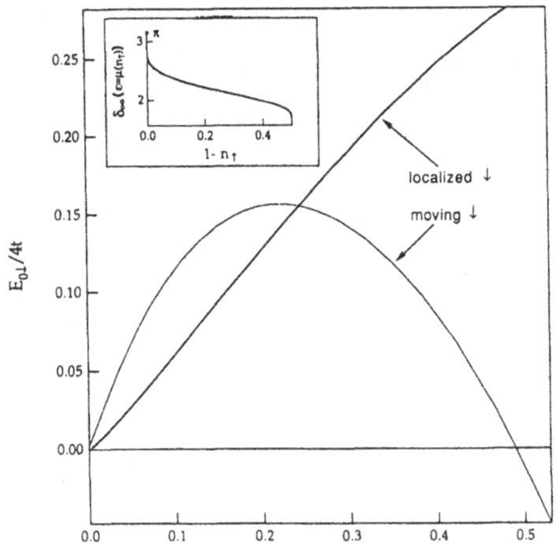

FIG. 1. Excitation energy for a single spin flip $E_{0\downarrow}$ as a function of hole concentration for the two-dimensional square lattice. The inset shows the scattering phase shift at the Fermi energy as a function of hole concentration.

note that since this instability corresponds to the softening of the bottom of the particle-hole continuum, it also implies a softening of the spin wave.[9,10]

The wave function Eq. (1) and the respective spectral function[8]

$$A_\downarrow(\mathbf{k},\omega)=2\pi(1-n_\uparrow)\delta(\omega+\tilde{t}_{\mathbf{k}\downarrow}-\Delta) \quad (5)$$

correspond to a down-spin quasiparticle carrying the full spectral weight of the lower Hubbard band and thus ignore any incoherent excitations. One might then wonder about the effect of the latter: We recall that in the one-dimensional Hubbard model the exact spin-flip spectrum has no quasiparticle part; i.e., it is purely incoherent. Indeed, even in two dimensions one expects that, sufficiently close to half filling, shakeup processes could modify the spectrum considerably.[8]

To study this question we start by considering the limit of an infinitely heavy down spin, a problem first discussed in Ref. 12. In that case, the change in the up-spin kinetic energy is given by Fumi's theorem in terms of the $U=\infty$ scattering phase shift,

$$\Delta=\int_{-4t}^{\mu_\uparrow} d\varepsilon \frac{\delta_\infty(\varepsilon)}{\pi}, \quad (6)$$

where

$$\delta_\infty(\varepsilon)=-\tan^{-1}\left[\pi\sum_{\mathbf{k}}\delta(\varepsilon+t_{\mathbf{k}})\Big/\sum_{\mathbf{k}}\frac{1}{\varepsilon+t_{\mathbf{k}}}\right]. \quad (7)$$

The resulting excitation energy is also shown in Fig. 1. As in one dimension,[8] *the localized spin flip leads to a lower variational bound than the Gutzwiller wave function close to half filling.* In other words, the lowering in the up-spin kinetic energy (which is now treated exactly) dominates the increase in the down-spin energy for $n_\uparrow\to 1$. The down-spin spectral function is now purely incoherent and, at low frequencies, approximately given by the power law

$$A_\downarrow(\omega)\propto\Theta(\omega-\Delta)(\omega-\Delta)^{g-1}, \quad (8)$$

where $g=[\delta_\infty(\mu_\uparrow)/\pi]^2$.[13]

Lowering the energy further by constructing delocalized solutions out of the above localized down-spin states is a subtle matter: In the adiabatic limit, in which the up-spin Fermi sea relaxes about the instantaneous position of the down spin, the down-spin kinetic energy scales to zero due to Anderson's orthogonality catastrophe.[14,15] Correspondingly, the exchange splitting reduces to Eq. (6). In order to study the effects of nonadiabatic corrections we use the lowest-order cluster expansion for a large but finite value of U. In the localized case this is known to give the exact form of Eq. (8), but with the Born-approximation phase shift replacing the exact one in the expression for g. The main features of the down-spin spectral function are expected to remain unaffected by taking the limit $U\to\infty$.[16,17]

The leading-order cluster-expansion result for the $\mathbf{k}=0$ time-dependent down-spin Green's function reads

$$G_\downarrow(0,t)=-i\Theta(t)\exp\left[-i(-t_0+Un_\uparrow)t-\left(\frac{U}{N}\right)^2\sum_{\mathbf{k},\mathbf{k}'}n_{\mathbf{k}\uparrow}(1-n_{\mathbf{k}'\uparrow})\frac{1-e^{i\Delta_{\mathbf{k}\mathbf{k}'}t}+i\Delta_{\mathbf{k}\mathbf{k}'}t}{\Delta_{\mathbf{k}\mathbf{k}'}^2}\right], \quad (9)$$

where

$$\Delta_{\mathbf{k}\mathbf{k}'}=-t_0+t_{\mathbf{k}-\mathbf{k}'}-t_{\mathbf{k}}+t_{\mathbf{k}'}. \quad (10)$$

The corresponding spectral function is of the form

$$A_\downarrow(0,\omega)=2\pi z_{0\downarrow}\delta(\omega-\varepsilon_{0\downarrow})+A_\downarrow^{\mathrm{inc}}(0,\omega). \quad (11)$$

Here

$$\varepsilon_{0\downarrow}=-t_0+Un_\uparrow+\left(\frac{U}{N}\right)^2\sum_{\mathbf{k},\mathbf{k}'}\frac{n_{\mathbf{k}\uparrow}(1-n_{\mathbf{k}'\uparrow})}{\Delta_{\mathbf{k}\mathbf{k}'}} \quad (12)$$

is the renormalized down-spin energy and

$$z_{0\downarrow}=\exp\left[-\left(\frac{U}{N}\right)^2\sum_{\mathbf{k},\mathbf{k}'}\frac{n_{\mathbf{k}\uparrow}(1-n_{\mathbf{k}'\uparrow})}{\Delta_{\mathbf{k}\mathbf{k}'}^2}\right] \quad (13)$$

is the down-spin renormalization constant which is of the form of a Debye-Waller factor. As in the electron-phonon problem, $z_{0\downarrow}$ can be written in terms of the density of states, $S(\omega)$, of virtual up-spin particle-hole excitations accompanying the down spin,

$$z_{0\downarrow}=\exp\left[-\int_0^\infty \frac{d\omega}{\omega^2}S(\omega)\right], \quad (14)$$

where

$$S(\omega)=\left(\frac{U}{N}\right)^2\sum_{\mathbf{k},\mathbf{k}'}n_{\mathbf{k}\uparrow}(1-n_{\mathbf{k}'\uparrow})\delta(\omega+\Delta_{\mathbf{k}\mathbf{k}'}). \quad (15)$$

The integral in Eq. (14) is dominated by the low-frequency behavior of $S(\omega)$ which may be shown to take the form

$$S(\omega)\sim g\xi(\omega/\xi)^\alpha,\ \omega<\xi. \quad (16)$$

Here, the cutoff ξ is proportional to the hopping amplitude and the exponent $\alpha=(d+1)/2$ for a mobile down spin and $\alpha=1$ for a localized one.[18] For the low-frequency limit of the spectral function and $\alpha>1$ we thus obtain

$$\ln(z_{0\downarrow})\propto -g/(\alpha-1) \quad (17)$$

and

$$A_\downarrow^{\mathrm{inc}}(0,\omega)\propto\Theta(\omega-\varepsilon_{0\downarrow})(\omega-\varepsilon_{0\downarrow})^{\alpha-2}, \quad (18)$$

while for $\alpha=1$, $z_{0\downarrow}=0$, and $A_\downarrow^{\mathrm{inc}}(0,\omega)$ is given by Eq. (8).

In one dimension we thus find no quasiparticle part, in agreement with the exact results. This is a consequence of the linear low-frequency behavior of $S(\omega)$ and reflects the fact that the motion of the down spin induces an *infinite* number of particle-hole pairs. Note that this result does not depend on the magnitude of U (or, equivalently, that of the phase shift at the Fermi energy): It only relies on *all* the up-spin electrons being phase shifted, as emphasized by Anderson.[19] In two and three dimensions this is not the case: The quasiparticle weight is always *finite.*

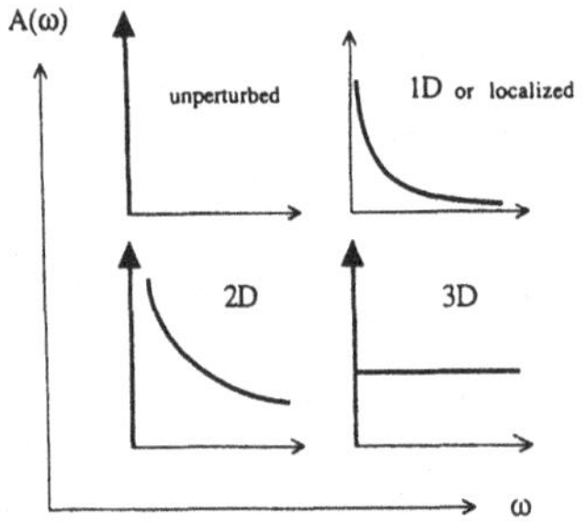

FIG. 2. Sketch of the $\mathbf{k}=0$ spectral function for a localized down spin and a mobile down spin in one, two, and three dimensions.

This is because the motion of the down spin is now a nonadiabatic process, i.e, not all up-spin electrons in the system are affected. Put another way, the orthogonality catastrophe is now suppressed by the recoil of the down spin.

In Fig. 2 we have sketched the complete low-frequency behavior of the down-spin spectral function. The notable feature in two dimensions is that, in spite of the finite quasiparticle weight, the incoherent part exhibits a square-root singularity, implying a large but finite number of low-energy states arbitrarily close to the bottom of the down-spin band. This should be contrasted with the Gutzwiller spectral function in Eq. (5) which consists only of a quasiparticle part.

The main message of this paper should be clear: Especially close to half filling, the Gutzwiller wave function is a poor representation of the physics of a single spin flip, since it completely ignores the incoherent excitations mentioned above. This is particularly serious in two dimensions, where the respective incoherent spectral function still shows a power-law behavior (qualitatively similar to that found in one dimension). We note that for a small but finite density of down spins we expect that Fermi statistics will push weight away from the Fermi level to higher energies, although in two dimensions a logarithmic divergence still remains (this can be trivially seen, for example, from second-order perturbation theory). Unfortunately, from the above discussion nothing can be concluded concerning the stability of the Nagaoka state for a small hole concentration.

A.E.R. would like to thank Fred Zawadowski for discussions, and also acknowledges the Sloan Foundation for support.

[1]Y. Nagaoka, Phys. Rev. **147**, 392 (1966).

[2]J. Kanamori, Prog. Theor. Phys. **30**, 275 (1963).

[3]D. C. Mattis, *The Theory of Magnetism* (Springer-Verlag, Berlin, 1981).

[4]M. Takahashi, J. Phys. Soc. Jpn. **51**, 3475 (1982).

[5]Yan Fang, Andrei E. Ruckenstein, Elbio Dagotto, and Stefan Schmitt-Rink, Phys. Rev. B **40**, 7406 (1989).

[6]J. A. Riera and A. P. Young, Phys. Rev. B **40**, 5285 (1989).

[7]A. Barbieri, J. A. Riera, and A. P. Young (unpublished).

[8]A. E. Ruckenstein and S. Schmitt-Rink, Int. J. Mod. Phys. B 3, 1809 (1989).

[9]B. S. Shastry, H. R. Krishnamurthy, and P. W. Anderson, Phys. Rev. B **41**, 2375 (1990).

[10]Anthony G. Basile and Veit Elser, Phys. Rev. B **41**, 4842 (1990).

[11]L. M. Roth, J. Phys. Chem. Solids **28**, 1549 (1967).

[12]P. Richmond and G. Rickayzen, J. Phys. C **2**, 528 (1969).

[13]P. Nozières and C. T. de Dominicis, Phys Rev. **178**, 1097 (1969).

[14]P. W. Anderson, Phys. Rev. Lett. **18**, 1049 (1967).

[15]K. Yamada, A. Sakurai, and M. Takeshige, Prog. Theor. Phys. **70**, 73 (1983).

[16]S. Doniach, Phys. Rev. B **2**, 3898 (1970).

[17]E. Mueller-Hartmann, T. V. Ramakrishnan, and G. Toulouse, Phys. Rev. B **3**, 1102 (1971).

[18]The result of Eq. (16) is most easily understood by considering the calculation of $S(\omega)$ in the continuum limit, $t_{\mathbf{k}}\rightarrow \mathbf{k}^2/2m$. For example, in 2D, after performing the angular integration one obtains

$$S(\omega)\propto\int_{\mu}^{\infty}d\varepsilon\int_{0}^{\mu+\omega-\varepsilon}d\varepsilon'\Theta(\varepsilon'-\omega^2/4\varepsilon)(4\varepsilon\varepsilon'-\omega^2)^{-1/2}$$
$$=\int_{\mu}^{\infty}(d\varepsilon/\sqrt{\varepsilon})\Theta(\mu+\omega-\varepsilon-\omega^2/4\varepsilon)\times(\mu+\omega-\varepsilon-\omega^2/4\varepsilon)^{1/2}.$$

In the limit $\omega\ll\mu$ this reduces to $S(\omega)\sim g\int_0^{\omega}(d\varepsilon/\sqrt{\mu})\times(\omega-\varepsilon)^{1/2}=g\omega^{3/2}/\sqrt{\mu}$. The same result is obtained in the case $m_{\uparrow}\neq m_{\downarrow}$, in the regime $\omega\ll m_{\uparrow}\mu/m_{\downarrow}$. It should be noted that our nonadiabatic limit is not accessed by a simple hopping expansion: This only incorporates particle-hole excitations with energies above the down-spin band width, in which case $S(\omega)\sim\omega$, independent of dimensionality [see, for example, G. T. Zimanyi, K. Vladar, and A. Zawadowski, Phys. Rev. B **36**, 3186 (1987), and references therein].

[19]P. W. Anderson, Phys. Rev. Lett. **64**, 1839 (1990).

PHYSICAL REVIEW B VOLUME 37, NUMBER 13 1 MAY 1988

Solution of the one-dimensional Hubbard model for arbitrary electron density and large U

José Carmelo* and Dionys Baeriswyl†
International Centre for Theoretical Physics, 34100 Trieste, Italy
(Received 28 October 1987)

The one-dimensional Hubbard model is studied for general electron densities and large U. On the basis of the coupled integral equations of Lieb and Wu exact asymptotic expansions are derived to fifth order in U^{-1} for the ground-state energy, the kinetic energy, and the number of doubly occupied sites. A closed-form expression for the distribution functions of charge and spin degrees of freedom is proposed which yields the same large-U expansion as that derived directly from the Lieb-Wu equations. Our analytical results are in excellent agreement with previous numerical studies.

I. INTRODUCTION

Electron correlations in narrow energy bands play an important role in the theory of itinerant magnetism and in the description of insulator-metal transitions. The simplest model capable of covering both the bandlike and the atomiclike behavior has been introduced by Hubbard.[1] The Hamiltonian consists of a kinetic term describing the electronic motion between atomic sites (transfer integrals t_{ij}) and an on-site Coulomb repulsion (Hubbard parameter U). Often the interatomic transfer is restricted to nearest-neighbor sites ($t_{ij}=-t$ for neighboring atoms and zero otherwise). In this case the model contains the three parameters t, U, and $n=N/N_a$, where N is the number of electrons and N_a the number of sites.

For $U=0$ the Hamiltonian corresponds to a system of noninteracting electrons moving in a band of width $W=2zt$, where z is the coordination number. In the opposite limit $U \gg t$ the free motion is increasingly suppressed due to the large energy of configurations with doubly occupied sites. In particular for $n=1$ the electrons tend to become localized (one at every site) with their spins ordered antiferromagnetically.[2] Therefore in the intermediate regime one expects a metal-insulator transition, as already pointed out by Hubbard.[1] Using a variational procedure proposed by Gutzwiller,[3] Brinkman and Rice[4] found such a transition at a critical value $U_c=(32/\pi)t$. They considered only the paramagnetic phase and, therefore, could not answer the question at which side of the metal-insulator transition antiferromagnetism will appear. This problem was studied by Ogawa, Kanda, and Matsubara[5] who found that antiferromagnetism already sets in within the metallic phase; however, the complicated interplay between localization and antiferromagnetism appears to be far from being understood.[6]

In 1968 following Yang's solution of the one-dimensional Fermi gas with δ-function interaction,[7] Lieb and Wu succeeded in solving the one-dimensional Hubbard model using the Bethe-ansatz technique.[8] They were able to reduce the many-particle Schrödinger equation for the ground state to a set of four simple integral equations which were solved explicitly for $n=1$. Lieb and Wu found that the ground state (for $n=1$) is insulating for all positive values of U, which implies that in one dimension the Mott-Hubbard transition occurs at $U=0$. The excitation spectra for spin and charge degrees of freedom were studied by Ovchinnikov,[9] who concluded that there is no gap for magnetic excitations whereas for $U>0$ a finite energy is required for creating an electron-hole pair (optical gap).[9,10] The magnetic susceptibility has been calculated by Takahashi.[11] It increases monotonically with U from the initial Pauli susceptibility for $U=0$ to the susceptibility of the one-dimensional Heisenberg model[12] with $J=4t^2/U$ for $U\rightarrow\infty$. From the explicit form of the ground-state energy both the local spin fluctuation $\langle S_i^2\rangle$ (Ref. 13) and the kinetic energy[14] can be derived by simple differentiation.

The integral equations of Lieb and Wu have not been solved analytically for $n\neq 1$. Approximating them in terms of a set of linear algebraic equations, Shiba calculated numerically both the ground-state energy and the magnetic susceptibility.[13] The magnetic susceptibility increases with increasing U and decreasing carrier density n ($n<1$). The excitation spectrum was analyzed numerically by Coll for various values of n.[15] In contrast to the case $n=1$, where there is an optical gap for all positive values of U, there is no gap in the charge-excitation spectrum for $n\neq 1$, however large U is. In this paper we derive an analytic large-U expansion for the ground-state energy for general n. (We can restrict ourselves to the case $n\leq 1$ by virtue of electron-hole symmetry.) Explicit expressions are presented up to fifth order in t/U. It turns out that our expansion agrees very well with the numerical results of Shiba[13] for $U \gtrsim 4t$.

The exact solution of the Hubbard model in one dimension provides a crucial test for approximate theories. Besides being a cornerstone of many-body theory, the one-dimensional Hubbard Hamiltonian represents an important model system in its own right. There exist in fact a number of quasi-one-dimensional materials with partially filled bands where effects due to electronic correlation have been observed, for instance in terms of an enhanced magnetic susceptibility[16] or through a reduced oscillator strength for intraband transitions.[17]

The paper is organized as follows. The Lieb-Wu equations are presented in Sec. II. The expansion in powers of t/U is derived in Sec. III. A closed-form expression for the distribution functions of charge and spin degrees of freedom is proposed. Our large-U expansion for the ground-state energy agrees very well with Shiba's numerical calculations[13] in the region $U \gtrsim 4t$. The kinetic energy and the number of doubly occupied sites are derived in Sec. IV. Section V gives a brief summary and a few remarks on the role of dimensionality.

II. THE LIEB-WU EQUATIONS

We consider the one-dimensional Hubbard Hamiltonian

$$H=-t\sum_{i\sigma}(c_{i\sigma}^{\dagger}c_{i+1\sigma}+c_{i+1\sigma}^{\dagger}c_{i\sigma})+U\sum_{i}n_{i\uparrow}n_{i\downarrow}\ , \qquad (1)$$

where $c_{i\sigma}^{\dagger}$ and $c_{i\sigma}$, $i=1,\ldots,N_a$, are, respectively, creation and annihilation operators for an electron with spin σ at site i. We use periodic boundary conditions $c_{N+1}\equiv c_1$. Following the work of Gaudin[18] and Yang[7] for the one-dimensional Fermi gas on a continuous line with δ-function interactions, Lieb and Wu were able to reduce the eigenvalue problem for the Hubbard chain to a set of integral equations involving two types of distribution functions, $\rho(k)$, $-Q<k\le Q$ and $\sigma(\Lambda)$, $-B<\Lambda<B$, where the cutoffs Q and B are determined self-consistently.[8] The function $\rho(k)$ can be considered as the momentum distribution function between collisions which produce phase shifts and modify the density of states. Indeed the ground-state energy is simply given by

$$E=-2N_a t\int_{-Q}^{Q}dk\,\rho(k)\cos k\ , \qquad (2)$$

which is formally identical to the noninteracting limit, where $\rho(k)$ is replaced by the Fermi distribution function and Q by the Fermi wave vector $k_F=\pi n/2$, $n=N/N_a$ being the electron density per site. The distribution function $\sigma(\Lambda)$ is associated with the spin degrees of freedom. The total spin is zero in the ground state which corresponds to $B=\infty$. The Lieb-Wu equations are

$$2\pi\rho(k)=1+2u\cos k\int_{-\infty}^{\infty}d\Lambda\,\sigma(\Lambda)/[u^2+(\sin k-\Lambda)^2]\ , \qquad (3a)$$

$$2\pi\sigma(\Lambda)=2u\int_{-Q}^{Q}dk\,\rho(k)/[u^2+(\sin k-\Lambda)^2] -u\int_{-\infty}^{\infty}d\Lambda'\,\sigma(\Lambda')/[u^2+\tfrac{1}{4}(\Lambda-\Lambda')^2]\ , \qquad (3b)$$

where the distribution functions are normalized according to

$$\int_{-Q}^{Q}dk\,\rho(k)=n\ , \qquad (3c)$$

$$\int_{-\infty}^{\infty}d\Lambda\,\sigma(\Lambda)=\tfrac{1}{2}n\ , \qquad (3d)$$

and we have introduced the dimensionless parameter $u=U/(4t)$. One can show that both $\rho(k)$ and $\sigma(\Lambda)$ are even functions with maximum at $k=0$ and $\Lambda=0$, respectively.[15] Equations (3a)–(3d) have been solved in closed form for $n=1$.[8] Numerical solutions for general n have been presented by Shiba,[13] who found also a large-U expansion to first order in u^{-1} and by Coll.[15] Equation (3b) can be solved in terms of the Fourier transform of $\sigma(\Lambda)$,[8,15] giving

$$\sigma(\Lambda)=(4u)^{-1}\int_{-Q}^{Q}dk\,\rho(k)\mathrm{sech}[(\pi/2u)(\Lambda-\sin k)]\ . \qquad (4)$$

Inserting this expression into Eq. (3a) results in an integral equation for $\rho(k)$

$$2\pi\rho(k)=1+\cos k\int_{-Q}^{Q}dk'\,\Gamma(k,k')\rho(k') \qquad (5)$$

with a kernel

$$\Gamma(k,k')=\tfrac{1}{2}\int_{-\infty}^{\infty}d\Lambda(u^2+\Lambda^2)^{-1} \times\mathrm{sech}[(\pi/2u)(\Lambda+\sin k-\sin k')]\ . \qquad (6)$$

It is easy to convince oneself using Eq. (4) that the normalization condition (3d) is automatically satisfied if Eq. (3c) holds. Equations (3c), (4), and (5) are equivalent to the Lieb-Wu equations (3a)–(3d).

III. LARGE-U EXPANSION

In order to derive the large-U expansion for the ground-state energy we first transform the expression for the kernel $\Gamma(k,k')$, Eq. (6), into a more convenient form using the relation

$$\mathrm{sech}(\pi x/2u)=\pi^{-1}\int_{0}^{\infty}dy\,\cos(xy/2u)\mathrm{sech}(y/2) \qquad (7)$$

for $x=\Lambda+\sin k-\sin k'$. The kernel is transformed to

$$\Gamma(k,k')=u^{-1}\int_{0}^{\infty}dy(1+e^{y})^{-1} \times\cos[(\sin k-\sin k')y/(2u)]\ , \qquad (8)$$

a suitable form for expanding in powers of u^{-1}. We find

$$\Gamma(k,k')=u^{-1}\left[\ln 2+\sum_{l=1}^{\infty}(-1)^l[(\sin k-\sin k')/(2u)]^{2l}(1-2^{-2l})\zeta(2l+1)\right]\ . \qquad (9)$$

This expression shows that the iteration of Eq. (5) leads to a straightforward expansion in powers of u^{-1}. We find

$$2\pi\rho(k)=1+u^{-1}\cos k\left[n\ln 2+\sum_{l=1}^{\infty}(-1)^l(1-2^{-2l})\zeta(2l+1)(2u)^{-2l}\sum_{m=0}^{l}\binom{2l}{2m}M_m(\sin k)^{2(l-m)}\right]\ , \qquad (10)$$

where we have introduced the moments

$$M_m=\int_{-Q}^{Q}dk\,\rho(k)(\sin k)^{2m}\ . \qquad (11)$$

The large-U expansion for these moments is readily obtained by multiplying Eq. (5) by $(\sin k)^{2m}$ and integrating. We find

$$M_m=(2\pi)^{-1}\int_{-Q}^{Q}dk(\sin k)^{2m}+(\pi u)^{-1}\Bigg[n\ln 2(\sin Q)^{2m+1}/(2m+1)+\sum_{l=1}^{\infty}(-1)^l(1-2^{-2l})\zeta(2l+1)(2u)^{-2l}\times\sum_{\mu=0}^{l}\binom{2l}{2\mu}M_\mu(\sin Q)^{2(l+m-\mu)+1}/[2(l+m-\mu)+1]\Bigg]. \tag{12}$$

We have solved Eqs. (10) and (12) iteratively and used the normalization condition (3c) for replacing the cutoff parameter Q by the density n. We obtain the following expansion for the ground-state energy Eq. (2),

$$E/N_a t=-(2/\pi)\sin(\pi n)-\sum_{l=1}^{\infty}\kappa_l(n)(t/U)^l, \tag{13}$$

where the coefficients $\kappa_l(n)$ are independent of t and U. The first five terms are given in Table I. The corresponding result for the ground-state energy is shown in Fig. 1 in comparison with the numerical results of Shiba.[13] Our expansion to fifth order in u^{-1} is in agreement with the numerical results for $U \gtrsim 4t$.

The corresponding expansion for the distribution function $\rho(k)$ is found to be

$$2\pi\rho(k)=1+\sum_{l=1}^{\infty}F_l(k,n)(t/U)^l. \tag{14}$$

The first five coefficients are given in Table II. A similar equation for $\sigma(\Lambda)$ is obtained using Eqs. (4) and (14). The distribution function $\rho(k)$ is shown in Figs. 2(a) and 2(b) for various values of U and n. In the interval $-Q<k<Q$ it represents the density of k states of the interacting system. One notices that the cutoff wave vector Q increases from k_F for $U=0$ to $2k_F$ for $U\to\infty$. Our result agrees with the numerical calculations of Coll[15] who computed this function for the quarter-filled band ($n=\frac{1}{2}$).

It is instructive to compare the distribution function $2\pi\rho(k)$ with the occupation numbers

$$n(k)=\sum_\sigma\langle c_{k\sigma}^\dagger c_{k\sigma}\rangle, \tag{15}$$

where $c_{k\sigma}$ is the Fourier transform of $c_{i\sigma}$. For $U=0$ the two functions are identical $2\pi\rho(k)=n(k)=2$ for $|k|<k_F$. For finite U they are expected to be different. Consider the particular case $n=1$ and $U\gg t$. The ground state assumes the form[19-21]

$$|\psi\rangle=|\phi\rangle-U^{-1}H_0|\phi\rangle+O(t^2/U^2), \tag{16}$$

where $|\phi\rangle$ is the ground state of the one-dimensional Heisenberg Hamiltonian (with exchange constant $J=4t^2/U$) and H_0 is the kinetic part of Eq. (1). We find

$$\begin{aligned}n(k)&=1-\cos k\langle\psi|H_0/(N_a t)|\psi\rangle/\langle\psi|\psi\rangle+O(t^2/U^2)\\&=1+2\cos k\langle\phi|H_0^2/(N_a tU)|\phi\rangle/\langle\phi|\phi\rangle+O(t^2/U^2)\\&=1+8\ln 2(t/U)\cos k+O(t^2/U^2)\end{aligned} \tag{17}$$

and

$$\begin{aligned}2\pi\rho(k)&=1-\cos k\langle\psi|H/(N_a t)|\psi\rangle/\langle\psi|\psi\rangle+O(t^2/U^2)\\&=1+\cos k\langle\phi|H_0^2/(N_a tU)|\phi\rangle/\langle\phi|\phi\rangle+O(t^2/U^2)\\&=1+4\ln 2(t/U)\cos k+O(t^2/U^2).\end{aligned} \tag{18}$$

One notices that $n(k)$ is related to the kinetic energy whereas $2\pi\rho(k)$ depends on the total energy. Therefore the two functions are different for general values of U and become equal only in the limits $U\to 0$ and $U\to\infty$, as exemplified in Figs. 3(a) and 3(b). We want to emphasize that neither $n(k)$ nor $\rho(k)$ are discontinuous at the Fermi surface $k_F=\pi/2$ for finite U. We will come back to this point in Sec. V.

For the special case of a half-filled band ($n=1$) one easily verifies that the functions

$$2\pi\sigma(\Lambda)=(4u)^{-1}\int_{-\pi n}^{\pi n}dk\,\mathrm{sech}\{(\pi/2u)[\Lambda-f(k)]\}, \tag{19}$$

$$2\pi\rho(k)=1+\pi^{-1}\cos k\int_{-\pi n}^{\pi n}dk'\int_0^\infty d\omega\cos(\omega\sin k)\cos[\omega f(k')]/(1+e^{2\omega u}) \tag{20}$$

satisfy the Lieb-Wu equations (3a)–(3d) provided that

$$Q=\begin{cases}\frac{1}{2}\pi, & u=0,\\ \pi, & u>0,\end{cases} \tag{21}$$

and

$$f(k)=\sin(k-\phi_k), \tag{22}$$

where

$$\phi_k=\begin{cases}\frac{1}{2}k, & u=0,\\ 0, & u>0.\end{cases} \tag{23}$$

TABLE I. First five coefficients $\kappa_l(n)$ of the expansion (13). Notation: $A=4\ln 2$, $s=\sin(\pi n)$, $f=\sin(2\pi n)/2\pi n$.

l	$\kappa_l(n)$
1	$An^2(1-f)$
2	$-\pi^{-1}A^2n^2s^3$
3	$\frac{4}{3}A^3n^4s^2f+\zeta(3)n^2[3(1-f)(2f-3)+2s^2f]$
4	$\pi^{-1}An^2s^3\{4\zeta(3)[5(1-f)+2s^2]-\frac{5}{12}A^3n^2(4-5s^2)\}$
5	$\frac{2}{3}A^5n^6s^2f(5-9s^2)$ $-4\zeta(3)A^2n^4s^2\{3f[3(1-f)+2s^2]+4s^4/(\pi n)^2\}$ $+\zeta(5)n^2\{15(1-f)[5-\frac{3}{2}(3+2s^2)f]-(5+4s^2)s^2f\}$

It is also easy to convince oneself that these solutions agree with the expressions of Lieb and Wu.[8] It is interesting to note that the function $|f(k)|$ is proportional to the spin-wave dispersion both for $u=0$ and $u\to\infty$.

We have taken Eqs. (19) and (20) as an ansatz for general band fillings. Indeed it exactly solves Eqs. (3a)–(3d) for $u=0$ and $u\to\infty$ if $f(k)$ is again chosen as in Eq. (22) with ϕ_k given by

$$\phi_k=\begin{cases}\frac{1}{2}k, & u=0,\\ 0, & u\to\infty.\end{cases} \tag{24}$$

We did not succeed in finding an explicit form for ϕ_k which solves the problem in general. Instead we found the following large-U expansion (details can be found in the Appendix)

$$\phi_k=(n/u)\ln 2\sin k-\tfrac{1}{2}[(n/u)\ln 2]^2\sin 2k+O(u^{-3}) . \tag{25}$$

Inserting this expression into Eqs. (19), (20), and (22) one recovers the expression (14) up to fifth order in u^{-1}. We have extended this analysis and verified that an expansion of ϕ_k to fourth order in u^{-1} leads to the exact expansion of $\rho(k)$ to seventh order in u^{-1}. This suggests that our ansatz defined by Eqs. (19) and (20) is valid for a large region of parameter space. In this region the cutoff parameter Q is simply related to the density n by

$$Q=\pi n-\phi_{\pi n} . \tag{26}$$

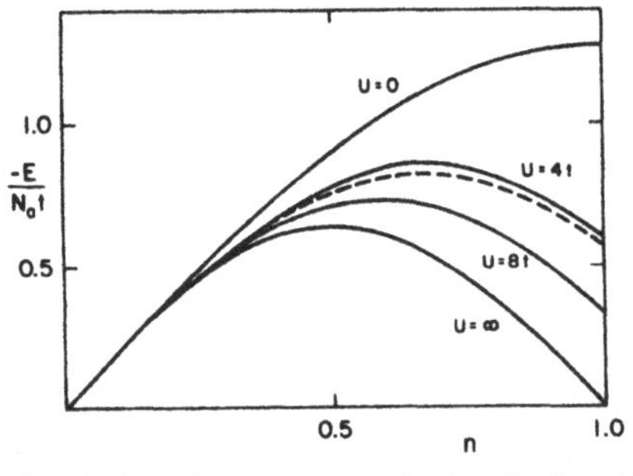

FIG. 1. Ground-state energy as a function of n for various values of U. Solid lines represent the present analytical expansion, the dashed line Shiba's numerical results for $U=4t$ (Ref. 13). For $U=8t$ the difference between our expansion and the numerical results is not visible on the scale of this figure.

We have used Eqs. (19) and (25) for calculating the distribution function $\sigma(\Lambda)$ for $U\gtrsim 4t$. For $U\to\infty$, $\sigma(\Lambda)$ assumes the asymptotic form

$$\sigma(\Lambda)\sim(n/4u)\operatorname{sech}(\pi\Lambda/2u),\quad U\to\infty . \tag{27}$$

It is interesting to compare the large-U behavior with the $U=0$ limit. For $U=0$ $\sigma(\Lambda)$ is given by Eqs. (19) and (24) and found to vanish for $|\Lambda|>\sin(\pi n/2)$. For $|\Lambda|<\sin(\pi n/2)$ we find

$$\sigma(\Lambda)=(4\pi)^{-1}\int_{-\pi n}^{\pi n}dk\,\delta[\Lambda-\sin(k/2)]$$
$$=(2\pi)^{-1}(1-\Lambda^2)^{-1/2},\quad U=0 . \tag{28}$$

Figure 4(a) shows $\sigma(\Lambda)$ both for $U=0$ [Eq. (28)] and for $U>0$ [Eqs. (19) and (25)]. One notices that the $U=0$ behavior is qualitatively different from that found in the large-U limit. The dependence on the density n is illustrated in Fig. 4(b) for $U=4t$.

IV. KINETIC ENERGY AND LOCAL SPIN FLUCTUATIONS

The kinetic energy is simply obtained by differentiating the total energy with respect to t.[14] Using the expansion of Sec. III we find

$$E_{\rm kin}=t\partial_t E=E^0_{\rm kin}\cos(n\pi/2)$$
$$-N_a t\sum_{l=1}^{\infty}(l+1)\kappa_l(n)(t/U)^l , \tag{29}$$

where

$$E^0_{\rm kin}=-(4N_a t/\pi)\sin(n\pi/2)$$

TABLE II. First five coefficients $F_l(k,n)$ of the expansion (14). Notation as in Table I.

l	$F_l(k,n)$
1	$An\cos k$
2	0
3	$-6\zeta(3)n(1-f+2\sin^2 k)\cos k$
4	$8\pi^{-1}\zeta(3)Ans^3\cos k$
5	$3n\{5\zeta(5)[4\sin^4 k+\frac{3}{2}(1-f)(1+8\sin^2 k)-s^2f]$ $-4\zeta(3)A^2n^2s^2f\}\cos k$

is the kinetic energy for $U=0$. The first term of Eq. (29) represents the free motion of N_a-N holes in a ground state without doubly occupied sites. The following term, proportional to t^2/U, corresponds to a second-order process where initial and final states have only singly occupied sites, but the intermediate state has one doubly occupied site. For $n=1$ this term is identical to $t\partial_t E_H$, where E_H is the ground-state energy of the Heisenberg chain with $J=4t^2/U$. The kinetic energy is shown in Fig. 5 where we have used the expansion (29) to fifth order in (t/U) together with the exact result[14] for $n=1$. The absolute value of the kinetic energy decreases with U for all values of n. Asymptotically it tends to zero only for $n=1$ and approaches a finite limit for $n<1$.

The number of doubly occupied sites

$$D=\sum_i \langle n_{i\uparrow} n_{i\downarrow}\rangle \tag{30}$$

can also be obtained by simple differentiation. For $n=1$ we use the exact expression for the energy[8] and find

$$D=\partial_U E = N_a \int_0^\infty d\omega \frac{J_0(\omega)J_1(\omega)}{1+\cosh(2\omega u)} . \tag{31}$$

For general n, D assumes its maximum value $D_0=N_a(n/2)^2$ for $U=0$. It decreases monotonically with increasing U and vanishes for $U\to\infty$. For large U we find using Eq. (13),

$$D=N_a \sum_{l=1}^{\infty} l\kappa_l(n)(t/U)^{l+1} , \tag{32}$$

D determines the local spin-spin correlation function $\langle \mathbf{S}_i^2\rangle$ through the relation[13]

$$\langle \mathbf{S}_i^2\rangle = 3(N-2D)/(4N_a) . \tag{33}$$

The number of doubly occupied sites, as obtained from the fifth order expansion of the energy is shown in Fig. 6 as a function of U for different values of n together with the exact result (31) for $n=1$. For $\frac{1}{2}<n<1$ D/D_0 depends only weakly on n. For smaller values of n the reduction of the number of doubly occupied sites becomes stronger.

Figures 5 and 6 confirm that for $n=1$ the large-U expansion breaks down for $U\lesssim 4t$. On the other hand, Eq. (25) suggests that the expansion parameter is n/u. This agrees with the observation that the coefficients $\kappa_l(n)$ tend to zero with $n\to 0$ as n^{l+3} (see Table I).

V. DISCUSSION

In this paper we have solved the Lieb-Wu equations[8] for general electron densities n and large U. We have derived analytic expansions in powers of (t/U) for the ground-state energy, the kinetic energy, and the number of doubly occupied sites. They agree very well with numerical computations[13] for $U \gtrsim 4t$. The smaller n the faster the convergence of the power series. We have also found closed-form expressions for the distribution functions $\rho(k)$ and $\sigma(\Lambda)$ which agree with the Lieb-Wu solution for $n=1$ and also are exact in the limiting cases

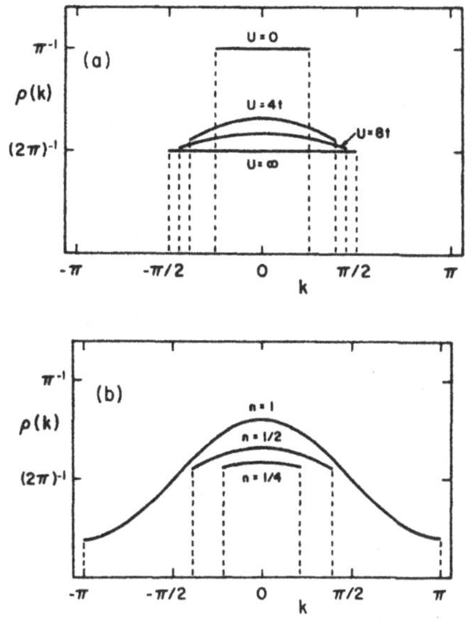

FIG. 2. Distribution function $\rho(k)$: (a) for $n=\frac{1}{2}$ and different values of U; (b) for $U=4t$ and different values of n.

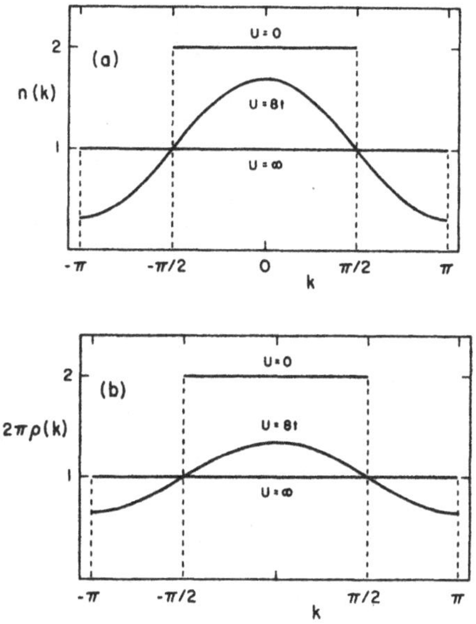

FIG. 3. Comparison between (a) the occupation number $n(k)$ and (b) the distribution function $2\pi\rho(k)$ for $n=1$ and different values of U.

$U=0$ and $U\to\infty$ for general n. We have verified that for $n<1$ these expressions reproduce the large-U expansion of the energy at least to seventh order in U^{-1}.

For the particular case $n=1$ we have compared [to first order in (t/U)] the distribution function $2\pi\rho(k)$ with the occupation number $n(k)$. We have found that, although they are not identical, they are quite similar. In particular both functions are continuous at the Fermi points $\pm\pi/2$. This is in agreement with the general understanding that in one dimension, electron-electron interactions eliminate the discontinuity of $n(k)$ at the Fermi surface (this has been demonstrated for the Tomonaga-Luttinger model[22]).

Recent theoretical discussions of high-temperature superconductivity have been frequently based on the two-dimensional Hubbard model.[23] In two dimensions several properties are expected to differ from those of the Hubbard chain. For instance, it has been proposed that for large U the ground state has antiferromagnetic long-range order in two dimensions,[24] whereas the spin correlations decay with a power law in one dimension.[25] Furthermore Nagaoka's theorem tells us that for sufficiently large U and n close to 1 the ground state will be ferromagnetic in certain two- and three-dimensional lattices.[26] However the theorem does not apply to one dimension due to topological reasons, and indeed it has been proven that the ground state of the one-dimensional system is always a spin singlet.[27] Recent experiments on $Ba_2YCu_3O_7$ compounds (which become superconducting around 93 K) indicate that Cu-O chains play an important role.[28,29] Therefore the one-dimensional Hubbard model may not be entirely academic in this context.

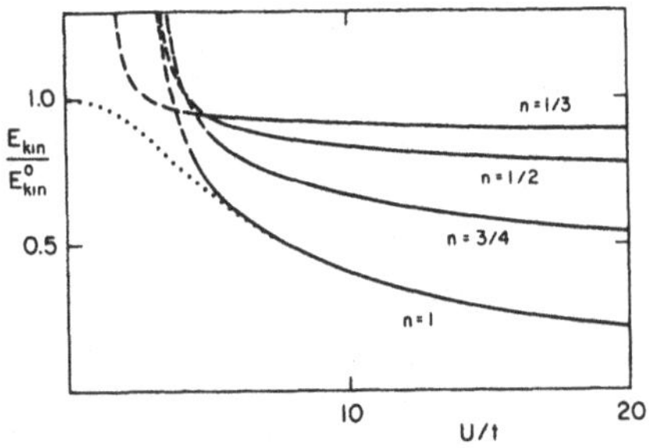

FIG. 5. Kinetic energy as a function of U for various values of n. Solid and dashed lines represent the large-U expansion, dots the exact solution for $n=1$.

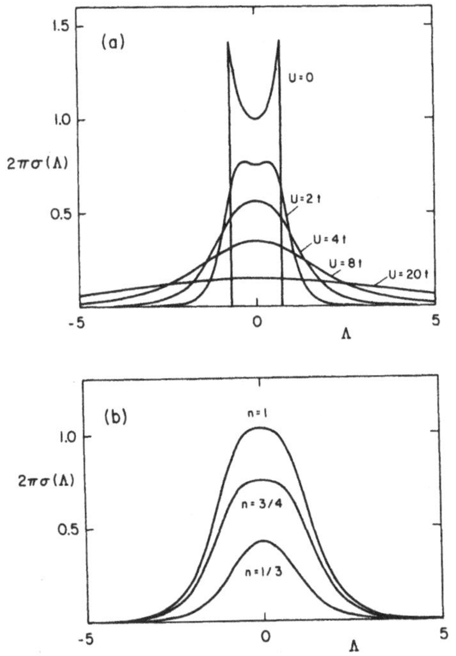

FIG. 4. Distribution function $\sigma(\Lambda)$: (a) for $n=\frac{1}{2}$ and different values of U; (b) for $U=4t$ and different values of n.

ACKNOWLEDGMENTS

We have profitted from discussions with John Ipsen in an early stage of this work. We also thank Professor Shiba for providing us copies of his original figures. One of us (J.C.) has been supported by the Portuguese National Science Foundation (INIC).

APPENDIX

We want to show that Eqs. (19) and (20) satisfy the Lieb-Wu equations (3a)–(3d) for large U. Equations (3a) and (3d) are indeed satisfied for general u and n. Using Eq. (19) for eliminating $\sigma(\Lambda)$ in Eq. (3b) and making the Fourier transform we find

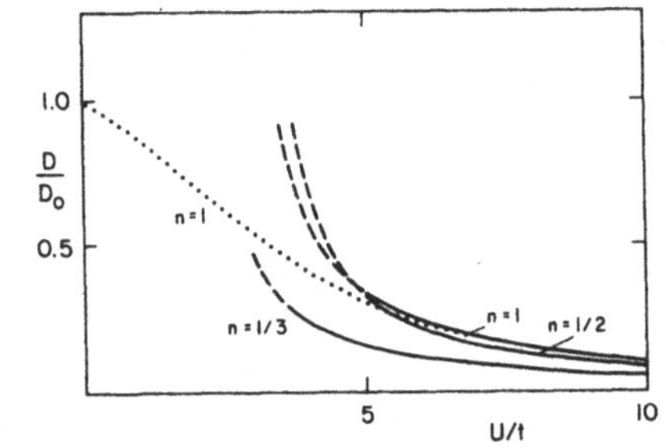

FIG. 6. Number of doubly occupied sites as a function of U for various values of n. Solid and dashed lines correspond to the large-U expansion, dots to the exact solution for $n=1$.

TABLE III. Large-U expansion of coefficients D_{2l}. Notation as in Table I. The expansion parameter is $x=4(\ln 2)n\sin(\pi n)(t/U)$.

l	D_{2l}
0	πn
1	$(\pi/2)n(1-f)-\frac{2}{3}s^2x+\pi nfx^2+O(x^3)$
2	$(\pi/8)n[3(1-f)-2s^2f]-\frac{4}{5}s^4x+2\pi ns^2fx^2+O(x^3)$
3	$(5\pi/16)n[1-f-\frac{2}{15}(5+4s^2)s^2f]+O(x)$

$$2\pi\int_0^Q dk\,\rho(k)\cos(\omega\sin k)=\int_0^{\pi n}dk\,\cos[\omega f(k)]\ . \tag{A1}$$

This equation not only replaces Eq. (3b) in the region of parameter space, where the ansatz (19) is valid, but also reproduces Eq. (3c) [one has simply to put $\omega=0$ in Eq. (A1) and to remember that $\rho(k)$ is an even function of k]. Inserting Eq. (20) into Eq. (A1) we obtain an integral equation for $f(k)$. It assumes a simple form in terms of the function

$$D(\Lambda)=\int_0^{n\pi}dk\,\cos[\Lambda f(k)]\ , \tag{A2}$$

namely,

$$D(\Lambda)=D_0(\Lambda)+\int_{-\infty}^{\infty}d\omega\,K(\omega,\Lambda)D(\omega)\ , \tag{A3}$$

where

$$D_0(\Lambda)=\int_0^Q dk\,\cos(\Lambda\sin k) \tag{A4}$$

and

$$K(\omega,\Lambda)=\frac{\sin[(\Lambda-\omega)\sin Q]}{\{\pi(\Lambda-\omega)[1+\exp(2\,|\,\omega\,|\,u)]\}}\ . \tag{A5}$$

The following relations hold for the kernel $K(\omega,\Lambda)$:

$$\int_{-\infty}^{\infty}d\Lambda\int_{-\infty}^{\infty}d\omega\,|\,K(\omega,\Lambda)\,|^2=(\ln 2-\tfrac{1}{2})\sin Q/(2\pi u) \tag{A6}$$

and

$$\int_{-\infty}^{\infty}d\Lambda\,|\,K(\omega,\Lambda)\,|^2\le(1/8\pi)\sin Q\ . \tag{A7}$$

Furthermore $D_0(\Lambda)$ is square summable. Therefore Eq. (A3) is a Fredholm integral equation of the second kind. For

$$u>(\ln 2-\tfrac{1}{2})\sin Q/(2\pi) \tag{A8}$$

the iterative solution of Eq. (A3) converges and is unique. It remains to be shown that a function $f(k)$ exists such that the solution of the integral equation can be expressed by Eq. (A2). We show now that this is indeed possible for large U and that $f(k)$ is given by Eqs. (22) and (25). In view of Eq. (A2), $D(\Lambda)$ has to be an even function of Λ. We introduce the Taylor expansion

$$D(\Lambda)=\sum_{l=0}^{\infty}[(2l)!]^{-1}(-1)^l D_{2l}\Lambda^{2l}\ , \tag{A9}$$

where

$$D_{2l}=(-1)^l d^{2l}D/d\Lambda^{2l}\,|_{\Lambda=0} \tag{A10}$$

into the integral equation (A3). Expanding the denominator of the kernel (A5), performing the integrals, and solving the resulting algebraic equations order by order we find the system of linear equations

$$D_{2l}=D_{2l}^0+\sum_{ij=0}^{\infty}\frac{C_{ij}}{u^{2(i+j)+1}}\frac{(\sin Q)^{2(l+j)+1}}{2(l+j)+1}D_{2i} \tag{A11}$$

where

$$D_{2l}^0=\int_0^Q dk\,\sin^{2l}k \tag{A12}$$

and

$$C_{ij}=\begin{cases}\pi^{-1}\ln 2, & \text{for } i=j=0,\\ \pi^{-1}(-1)^{i+j}\binom{2i+2j}{2i}\dfrac{2^{2(i+j)}-1}{2^{4(i+j)}}\zeta[2(i+j)+1], & \text{otherwise}\ .\end{cases} \tag{A13}$$

Similarly the Taylor expansion of Eq. (A2) leads to the equation

$$D_{2l}=\int_0^{n\pi}dk\,f^{2l}(k)\ . \tag{A14}$$

We have calculated the first few coefficients D_{2l} from Eq. (A11) to low order in u^{-1}. This yields the coefficients D_{2l} as functions of Q and n. The cutoff parameter Q is eliminated by solving the normalization condition Eq. (3c) or, equivalently, Eq. (A11) for $l=0$ with $D_0=n\pi$. We have verified that, indeed, the result is consistent with Eq. (A14) with $f(k)$ given by Eqs. (22) and (25) and Q given by Eq. (26). The result is shown in Table III.

The distribution function $\rho(k)$, given by Eq. (20), is related to $D(\omega)$ by

$$2\pi\rho(k)=1+(2/\pi)\cos k\int_0^{\infty}d\omega\frac{\cos(\omega\sin k)}{1+\exp(2\omega u)}D(\omega)\ . \tag{A15}$$

The expansion (A9) thus allows to derive an expansion for $\rho(k)$ which coincides with Eq. (14), at least for the first few terms and to low order in u^{-1}. The ground-state energy is obtained as

$$E/N_a t = -4\int_0^Q dk\, \rho(k)\cos k$$
$$= -(2/\pi)\sin Q - 4\sum_{i,j=0}^{\infty} C_{ij} u^{-[2(i+j)+1]} D_{2i} \int_0^Q dk[\cos k \sin^j(k)]^2 \,. \tag{A16}$$

Using this expansion with the coefficients of Table III (and again expressing Q by n) yields the ground-state energy to fifth order in u^{-1}, as in Eq. (13) with the coefficients of Table I.

---

*Permanent address: Instituto Nacional de Investigação Científica, Centro de Física da Matéria Condensada, Av. Prof. Gama Pinto, 2 1699 Lisboa Codex, Portugal and Departamento de Física, Universidad de Évora, Apartado 94 Postal No. 7001 Évora Codex, Portugal.

†Permanent address: Institut für Theoretische Physik, ETH-Hönggerberg, 8093 Zürich, Switzerland.

[1]J. Hubbard, Proc. R. Soc. London, Sec. A **276**, 238 (1963); **277**, 237 (1964).

[2]P. W. Anderson, Solid State Phys. **14**, 99 (1963).

[3]M. C. Gutzwiller, Phys. Rev. Lett. **10**, 159 (1963); Phys. Rev. **134**, A923 (1964); **137**, A1726 (1965).

[4]W. F. Brinkman and T. M. Rice, Phys. Rev. B **2**, 4302 (1970).

[5]T. Ogawa, K. Kanda, and T. Matsubara, Prog. Theor. Phys. **53**, 614 (1975).

[6]M. Cyrot, Physica B **91**, 141 (1977).

[7]C. N. Yang, Phys. Rev. Lett. **19**, 1312 (1967).

[8]E. H. Lieb and F. Y. Wu, Phys. Rev. Lett. **20**, 1445 (1968).

[9]A. A. Ovchinnikov, Zh. Eksp. Teor. Fiz. **57**, 2137 (1969) [Sov. Phys. JETP **30**, 1160 (1970)].

[10]A. A. Ovchinnikov, I. I. Ukrainskii, and G. V. Kventsel, Usp. Fiz. Nauk. **108**, 81 (1972) [Sov. Phys. Usp. **15**, 575 (1973)].

[11]M. Takahashi, Prog. Theor. Phys. **42**, 1098 (1969); **43**, 1619 (1970).

[12]R. Griffiths, Phys. Rev. **133**, A768 (1964).

[13]H. Shiba, Phys. Rev. B **6**, 930 (1972).

[14]D. Baeriswyl, J. Carmelo, and A. Luther, Phys. Rev. B **33**, 7247 (1986); **34**, 8976(E) (1986).

[15]C. F. Coll, III, Phys. Rev. B **9**, 2150 (1974).

[16]D. Jérome and H. J. Schulz, Adv. Phys. **31**, 299 (1982).

[17]C. S. Jacobsen, Ib Johannsen, and K. Bechgaard, Phys. Rev. Lett. **53**, 194 (1984).

[18]M. Gaudin, Phys. Lett. **24A**, 55 (1967).

[19]V. J. Emery, Phys. Rev. B **14**, 2989 (1976).

[20]D. Baeriswyl, in *Proceedings of the International Conference on Nonlinearity in Condensed Matter*, Vol. 69 of *Solid-State Sciences*, edited by A. R. Biship, D. K. Campbell, P. Kumar, and S. E. Trullinger (Springer, Berlin, 1987), p. 183.

[21]J. Carmelo, Ph.D. thesis, Copenhagen, 1986.

[22]D. C. Mattis and E. H. Lieb, J. Math. Phys. **6**, 304 (1965).

[23]For a recent review, see, T. M. Rice, Z. Phys. B **67**, 141 (1987).

[24]J. Oitmaa and D. D. Betts, Can. J. Phys. **56**, 897 (1978).

[25]N. M. Bogoliubov, A. G. Izergin, and V. E. Korepin, Nucl. Phys. B **275** [FS17], 687 (1986).

[26]Y. Nagaoka, Phys. Rev. **147**, 392 (1966).

[27]E. H. Lieb and D. C. Mattis, Phys. Rev. **125**, 164 (1962).

[28]F. Beech, S. Miraglia, A. Santoro, and R. S. Roth, Phys. Rev. B **35**, 8778 (1987).

[29]J. D. Jorgensen, B. W. Veal, W. K. Kwok, G. W. Crabtree, A. Umezawa, L. J. Nowicki, and A. P. Paulikas, Phys. Rev. B **36**, 5731 (1987).

PHYSICAL REVIEW B VOLUME 8, NUMBER 1 1 JULY 1973

Strong-Coupling Hubbard Chain*

G. Beni, T. Holstein, and P. Pincus
Department of Physics, University of California, Los Angeles, California 90024
(Received 17 November 1972)

An exact expression is obtained for the partition function of the arbitrary-electron-density Hubbard chain in the infinite-coupling ($U\to\infty$) limit. It is shown that the magnetic susceptibility obeys a Curie law, while the orbital contribution to the specific heat corresponds to that of a noninteracting band of spinless fermions. The electronic mobilities in this limit are shown to be infinite.

I. INTRODUCTION

One-dimensional systems[1] have received considerable theoretical attention for several reasons. Often problems exactly solvable in one dimension may only be approximately treated in higher dimensionalities. Thus one expects that the insight gained in the exact solution may lead to a better understanding of the underlying physics applicable in other situations. On the other hand, there exist special features of one-dimensional problems which often lead to drastically different results from similar problems, say, in three dimensions. A well-known example is the absence of long-range order in one dimension. Over recent years the discovery of a variety of compounds possessing quasi-one-dimensional structures has lent further impetus to the theoretical investigation of lower-dimensionality models. Some examples of such systems include (1) magnetic insulators such as (TMMC) tetramethylmanganese chloride[2] and various hydrated salts[3]; (2) one-dimensional "metallic" compounds such as the square planar platinum salts[4] and the organic charge-transfer salts based on tetracyanoquinodimethan (TCNQ).[4] It is this latter situation which interests us here. The salt *N*-methylphenizinium (NMP)-TCNQ has been studied in detail by the University of Pennsylvania group[5] and its conducting-insulating transition analyzed in terms of a half-filled-band Hubbard model. Other TCNQ salts may be described as more or less than half-filled bands, e.g., in quinolinium (Q) $TCNQ_2$, there presumably exists one electron per two TCNQ molecules and thus a one-quarter-filled band. Therefore the one-dimensional Hubbard model with variable electron density is of interest.

The one-dimensional Hubbard-model[6] Hamiltonian which is to be discussed here is

$$\mathcal{H} = -t\sum_{i,\sigma}\left(c^\dagger_{(i+1)\sigma}c_{i\sigma} + c^\dagger_{i\sigma}c_{(i+1)\sigma}\right) + U\sum_i n_{i\uparrow}n_{i\downarrow} \quad , \tag{1.1}$$

where t is the hopping integral which is the matrix element for the transfer of an electron to a nearest-neighbor site; $c_{i\sigma}$, $c^\dagger_{i\sigma}$ are, respectively, annihilation and creation operators for an electron with spin σ at the ith site; U is the local Coulomb repulsion which operates when two electrons occupy the same orbital; $n_{i\sigma}=c^\dagger_{i\sigma}c_{i\sigma}$ is the number operator for electrons on the ith site. We consider only one nondegenerate orbital at each site. Lieb and Wu[7] have solved for the ground state exactly for arbitrary t, U, and electron density ρ. Takahashi[8] has calculated the magnetic susceptibility at $T=0\,^\circ$K for the half-filled band ($\rho=1$) and Shiba[9] has extended this result to arbitrary density. Ovchinnikov[10] has computed elementary excitations (at $T=0\,^\circ$K) of both magnon and single-particle character; Coll[11] has extended his results to arbitrary density. Thus we have a fairly complete description of the $T=0\,^\circ$K properties for all values of the parameters. At finite temperatures, there exist fewer results. For finite chains, of up to six sites with $\rho=1$, the partition function and thermodynamic properties (specific heat, spin susceptibility) have been computed[12] in both the canonical and grand canonical ensembles for arbitrary U and t. For infinite chains, the statistical mechanics of the noninteracting limit ($U\to 0$) is of course simply a problem of Fermi statistics on a tight-binding chain. The purely local limit ($t\to 0$) is also completely solvable.[13]

The purpose of this paper is to discuss another situation which leads to exact thermodynamics and for which correlation functions associated with dynamical response functions can be calculated. This is the strong-coupling limit, i.e., t and ρ arbitrary but $U\to\infty$. In this case, Sokoloff[14] has already derived an expression for the partition function and shown that the susceptibility obeys a Curie law. By a rather different method, we rederive these results showing that there is a complete decoupling of the orbital and spin degrees of freedom, i.e., (1) the spin susceptibility is a Curie law corresponding to an electron density of $(2-\rho)$ (for $\rho>1$) and ρ (for $\rho<1$); and (2) the specific heat is that for a noninteracting tight-binding band of spinless fermions of density $|\rho-1|$. Finally we demonstrate that correlation functions associated with dynamical properties can be calculated. As an example, we explicitly compute the velocity-velocity correlation function and show that the mobility is infinite.

In Sec. II, we consider the case of a half-filled band with one extra electron (or hole). This situation has previously been considered by Brinkman and Rice[15] who showed that the mobility is infinite. We shall reconsider their calculation in detail and then generalize the result to arbitrary densities in Sec. III.

II. HALF-FILLED BAND PLUS ONE CARRIER

In this section, we consider a uniform-Hubbard-model chain described by the Hamiltonian (1.1) in the limit $U\rightarrow\infty$ containing one electron per site plus one additional carrier. In particular, we take the extra carrier to be an electron, but by electron-hole symmetry, this is equivalent to adding a hole (i.e., subtracting one electron from the half-filled band). The infinite local repulsion described by $U\rightarrow\infty$ ensures that states of the chain with more than one doubly occupied site are excluded. By projecting out such states explicitly, Brinkman and Rice[15] point out that the resulting Hamiltonian commutes (in one dimension only) with the velocity operator,

$$v=(ita/\hbar)\sum_{i,\sigma}(c^{\dagger}_{(i+1)\sigma}c_{i\sigma}-c^{\dagger}_{i\sigma}c_{(i+1)\sigma}) \ , \qquad (2.1)$$

where a is the lattice constant. Thus, the total velocity is a constant of the motion and the mobility is infinite. These authors also explicitly computed the velocity correlation function

$$\phi(\tau)=\langle v(\tau)v(0)+v(0)v(\tau)\rangle \ , \qquad (2.2)$$

where $v(\tau)$ is the velocity at time τ and the brackets indicate an ensemble average, and showed that indeed the correlation function $\phi(\tau)$ is independent of time, τ. In this section, we shall reconsider this problem in a slightly different formalism and derive the partition function, yielding the thermodynamic properties, as well as the velocity correlation function. In the following section, the method is easily extended to arbitrary density.

Since we are interested in the strong-coupling situation, it is appropriate to solve the interaction term in the Hamiltonian (proportional to U) exactly and treat the transfer term (proportional to t) by perturbation theory. For $t=0$, each site is singly occupied except for the one doubly occupied site. The energy of the system is then U, which we now take as our zero of energy. Each singly occupied site has spin $\frac{1}{2}$ and thus the application of a magnetic field removes the up-down degeneracy by $\pm\mu_B H\equiv\pm h$ (μ_B is the Bohr magneton and H the external field), where + and − refer, respectively, to spin parallel and antiparallel to the field. For the zero-transfer case, each site is independent and the partition function is then simply

$$Z_0=(2\cosh\beta h)^{N-1} \ , \qquad (2.3)$$

where N is the number of sites on the chain. The doubly occupied site is necessarily in a singlet state involving no spin degeneracy and thus does not contribute to the partition function; hence, the exponent $N-1$. The extreme simplicity of Z_0 is a direct consequence of the fact that we need not take into account multiply-occupied sites because $U=\infty$. Denoting the transfer term [the first term of (1.1)] by V, standard formal thermodynamic perturbation theory[16] for the thermodynamic potential Ω gives

$$e^{-\beta(\Omega-\Omega_0)}=1+\sum_{n=1}^{\infty}(-1)^n\int_0^{\beta}\cdots\int_0^{x_{n-2}}\int_0^{x_{n-1}}dx_n\ldots dx_1 \times\langle V(x_1)V(x_2)\cdots V(x_n)\rangle_0 \ , \qquad (2.4)$$

where

$$V(x)=e^{x\mathcal{H}_0}Ve^{-x\mathcal{H}_0} \qquad (2.5)$$

and $\mathcal{H}_0$ is the unperturbed Hamiltonian, in this case the interaction term [second term of (1.1)]; Ω_0 is the unperturbed thermodynamic potential ($\Omega_0=-\beta^{-1}\ln Z_0$); and the ensemble average in (2.4), $\langle\ \rangle_0$, is with respect to the unperturbed density matrix. The $U=\infty$ limit ensures that the transfer term V has nonvanishing matrix elements only when operating on one of the electrons of the doubly occupied site. Thus the nth-order term of the sum in (2.4) involves the doubly occupied site moving n steps, and because we are calculating a trace the final step must be a return to the initial site. Such a round trip must be independent of the spin configuration of all singly occupied sites because (1) no matter whether a given singly occupied site has spin up or down an electron of the appropriate spin may move onto it, and (2) in one dimension all paths contributing to a diagonal matrix element must be retraced leaving the original spin configuration unaltered. Furthermore, since any motion of the doubly occupied site leaves the interaction energy (or $\mathcal{H}_0$) unchanged, $V(x)$ is independent of x. Only the even terms in V contribute to the sum; the $(2n)$th term is then simply

$$[(\beta t)^{2n}/(2n)!]\,p(2n) \ , \qquad (2.6)$$

where $p(2n)$ is the number of distinct paths by which the doubly occupied site may move and return to its original site in $2n$ steps. This must involve n steps to the right and n steps to the left; thus $p(2n)$ is the number of ways one may make n right steps out of $2n$ steps, i.e.,

$$p(2n)=(2n)!/(n!)^2 \ . \qquad (2.7)$$

Substituting, (2.6) and (2.7) into (2.4), we obtain an exact expression for the partition function,

$$e^{-\beta(\Omega-\Omega_0)}=1+\sum_{n=1}^{\infty}(\beta t)^{2n}/(n!)^2=I_0(2\beta t) \ , \qquad (2.8)$$

where $I_0(x)$ is the hyperbolic Bessel function of zero order.[17] The thermodynamic properties are now easily computed from the thermodynamic potential Ω. In particular, since $\Omega - \Omega_0$ is independent of magnetic field, the spin susceptibility χ is completely determined by Ω_0 and is easily shown to be a Curie law corresponding to $(N-1)$ noninteracting spins,

$$\chi = (N-1)\mu_B^2 \beta \ . \tag{2.9}$$

In a finite magnetic field, there exist two contributions to the specific heat: (1) the Schottky-type anomaly associated with the $N-1$ localized spins whose Zeeman levels are split by the field; (2) the kinetic heat capacity arising from the motion of doubly occupied site. The Schottky term is well known and we shall not discuss it further. The kinetic contribution is given by

$$C_v = -\frac{d^2\Omega}{dT^2} = k_B x^2 \frac{d}{dx}\frac{I_1(x)}{I_0(x)} \ , \tag{2.10}$$

where $x = 2\beta t$. For $k_B T \gg t$, $C_v \approx \frac{1}{2} k_B x^2$ and for $k_B T \ll t$, $C_v \approx \frac{1}{2} k_B$. The low-temperature result is just the familiar Dulong-Petit value of $\frac{1}{2} k_B$ per degree of freedom (which is one for a free particle in one dimension). This result obtains because at low temperature the band-structure effects on a single carrier are negligible; the zone boundaries play no role.

Let us now turn our attention to the velocity correlation function $\phi(\tau)$ which should be explicitly independent of time, according to Brinkman and Rice.[15] This quantity is explicitly

$$\phi(\tau) = \langle v(\tau)v(0) + v(0)v(\tau)\rangle$$
$$= Z^{-1}\, \mathrm{Tr}\{e^{-\beta\mathcal{H}}[e^{i\mathcal{H}(\tau/\hbar)} v e^{-i\mathcal{H}(\tau/\hbar)} v + v e^{i\mathcal{H}(\tau/\hbar)} v e^{-i\mathcal{H}(\tau/\hbar)}]\} \ , \tag{2.11}$$

where the velocity operator v is given by (2.1). Again rewriting the Hubbard Hamiltonian (1.1) as $H_0 + V$, where the transfer term V is treated by a perturbation expansion to all orders, the correlation function becomes

$$\phi(\tau) = Z^{-1}\, \mathrm{Tr}\{Q(\beta, i\tau/\hbar) v Q(i\tau/\hbar, 0) v + Q(\beta + i\tau/\hbar, 0) v Q(0, i\tau/\hbar) v\} \ , \tag{2.12}$$

where

$$Q(x,y) = \sum_{n=0}^{\infty} Q_n(x,y) \tag{2.13}$$

and

$$Q_n(x,y) = (-1)^n \int_y^x dx_1 \int_y^{x_1} dx_2 \cdots \int_y^{x_{n-1}} dx_n \times \langle V(x_1)V(x_2)\cdots V(x_{n-1})\rangle \ . \tag{2.14}$$

The expression (2.12) for $\phi(\tau)$ can be easily interpreted in the following way: Suppose that the carrier (i.e., the doubly occupied site) is initially at some site which we may take as the origin. The velocity operator causes the carrier to move one step either to the right or left, i.e., $0 \to \pm 1$. The final result must be independent of the sign of the initial step, so we can arbitrarily choose to consider the first step to be to the right $(0 \to 1)$. The development operator Q, because it contains an arbitrary number of transfer operations, allows the carrier to move an arbitrary number of steps to some site q, i.e., $1 \to q$. The velocity operator again generates a shift by one site, i.e., $q \to q \pm 1$. Finally the operator Q must return the carrier to its original location, i.e., $q \pm 1 \to 0$. Thus in the site representation,

$$\phi(\tau) = 2Z^{-1} \sum_{q=-\infty}^{\infty} \sum_{\eta=\pm 1} [\langle 0|Q(\beta, i\tau/\hbar)|q+\eta\rangle\langle q+\eta|v|q\rangle \times \langle q|Q(i\tau/\hbar, 0)|1\rangle\langle 1|v|0\rangle + \langle 0|Q(\beta + i\tau/\hbar, 0)|q+\eta\rangle\langle q+\eta|v|q\rangle \times \langle q|Q(0, i\tau/\hbar)|1\rangle\langle 1|v|0\rangle] \ , \tag{2.15}$$

where the factor of 2 arises from the fact that the first step may be either to the right or left, the $U=\infty$ condition again forces $V(x)$ in the Q integrals to be independent of x; also the second term in the square brackets of (2.15) is easily seen to be the complex conjugate of the first. Then performing the integrals in (2.14) and using the velocity matrix elements,

$$\langle q+\eta|v|q\rangle = (ita\eta/\hbar) \ , \quad \eta = \pm 1 \tag{2.16}$$

leads us to

$$\phi(\tau) = 4\left(\frac{ta}{\hbar}\right)^2\left(\frac{Z_0}{Z}\right)\mathrm{Re}\left\{\sum_{q=-\infty}^{\infty}\left[\sum_{n=0}^{\infty}\left(\frac{-it\tau/\hbar}{n!}\right)^n P_n(1,q)\right] \times\left[\sum_{m=0}^{\infty}\frac{[t(i\tau/\hbar - \beta)]^m}{m!}[P_m(q-1,0) - P_m(q+1,0)]\right]\right\} , \tag{2.17}$$

where $P_n(a,b)$ is the number of ways one can move from site b to site a in n steps; the factor Z_0 appears because the motion of the carrier is independent of the over-all spin configuration of the chain. The combinatorial function $P_n(a,b)$ is

$$P_n(a,b) = \frac{n!}{[\frac{1}{2}(n+b-a)]!\,[\frac{1}{2}(n+a-b)]!} \tag{2.18}$$

for $n > |b-a|$ and zero otherwise. Using (2.18), the typical series that occurs in (2.17) is easily summed[17]:

$$\sum_{n=0}^{\infty} \frac{Z^n}{n!} P_n(a,b) = J_{|b-a|}(2iZ) i^{|b-a|} \ , \tag{2.19}$$

where $J_\alpha(x)$ is the Bessel function of order α. Inserting (2.19) into (2.17), and after some manipulations,

$$\phi(\tau) = 4\left(\frac{Z_0}{Z}\right)\left(\frac{ta}{\hbar}\right)^2 \mathrm{Re}\left\{\sum_{q=0}^{\infty} (-1)^q \alpha_q J_q(-x)\right.$$

$$\times[J_{q+2}(y)+2J_q(y)+J_{q-2}(y)]\Big\} \quad , \qquad (2.20)$$

where $\alpha_q=\frac{1}{2}$ for $q=0$, 1 otherwise, and where

$$x=2t\tau/\hbar \; , \quad y=x+2it\beta \quad . \qquad (2.21)$$

Finally, the q sum is performed with the aid of the Bessel-function-addition theorem,[18] which yields the result that $\phi(\tau)$ is only a function of $y-x$; i.e., independent of the time τ. Using the partition function, (2.8), the explicit expression for the correlation function is

$$\phi(\tau)=4(ta/\hbar)^2\{1-[I_2(2t\beta)/I_0(2t\beta)]\} \quad . \qquad (2.22)$$

Thus we indeed find that the mobility is infinite in agreement with Brinkman and Rice.[14] The function ϕ is then simply $2\langle v^2\rangle$, i.e., twice the average square velocity of the carrier which according to (2.22) tends to zero as $T\rightarrow 0$ and for $\beta t\ll 1$, tends to $4(ta/\hbar)^2$; i.e., $\langle v^2\rangle \rightarrow 2(ta/\hbar)^2$. In the next section we generalize the experience gained in this problem to the case of arbitrary density.

III. ARBITRARY DENSITY

Let us now consider the $U\rightarrow\infty$ limit in the situation where we add an arbitrary number of electrons ($2>\rho>1$) to a half-filled band. Electron-hole symmetry dictates that the identical results should obtain for the physical quantities of interest when $1>\rho>0$. Sokoloff[14] derived an expression for the partition function for this situation by generalizing the method of the Lieb-Wu[7] calculation of the ground-state energy to finite temperature. Our method is to generalize the technique utilized in the preceding section. In particular, we begin with the general expression for the partition function (2.4). In that case where there existed only one carrier (doubly occupied site), the nth term in the perturbation expansion was determined by summing up the contributions from all possible excursions of the carrier such that it returned to its initial site in exactly n steps. For the present situation, we must carry out the same type of analysis with the additional constraint imposed by the Pauli principle that two doubly occupied sites cannot pass through one another. This clearly limits the number of available excursions and greatly complicates the combinatorial problem. However, the path counting can be avoided by noting that it is identical to an already solved problem. Consider a noninteracting tight-binding chain ($U=0$) which contains a density Γ of electrons with all spins parallel. In principle, we may compute the partition function for this problem by using (2.4) and summing to all orders. Notice that Fermi statistics imposes the same paths to count for this problem as for the $U=\infty$ problem with a density $\rho=1+\Gamma$. Thus, we are immediately led to the result

$$Z=Z_0Z' \quad , \qquad (3.1)$$

where Z_0 is the partition function for the singly occupied sites:

$$Z_0=|2\cosh(\beta h)|^{N(2-\rho)} \qquad (3.2)$$

and Z' is the partition function for $n=N(\rho-1)$ spinless fermions in a tight-binding band. The magnetic field dependence of the partition function is entirely in Z_0 leading immediately to the Curie susceptibility

$$\chi=N(2-\rho)\mu_B^2\beta \quad . \qquad (3.3)$$

Note that $\chi\rightarrow\infty$ as $T\rightarrow 0$ in agreement with Shiba's[9] calculation that $\chi(0)\rightarrow\infty$ as $U\rightarrow\infty$. On the other hand, the properties based on the orbital motion are completely associated with the spinless doubly occupied sites. The fermion partition function in the grand canonical ensemble is given by

$$Z'=\mathrm{Tr}\, e^{-\beta(\mathcal{H}-\mu\mathfrak{N})} \quad , \qquad (3.4)$$

where the chemical potential μ is determined by

$$\mathfrak{N}=N(\rho-1)=\sum_k f(\epsilon_k) \quad . \qquad (3.5)$$

Here $f(x)$ is the usual fermion function

$$f(x)=1/(e^{\beta(x-\mu)}+1) \qquad (3.6)$$

and ϵ_k denotes the energy spectrum of the noninteracting tight-binding band,

$$\epsilon_k=2t\cos k \; , \quad -\pi<k<\pi \quad . \qquad (3.7)$$

The internal energy W is calculated in the usual way,

$$W=\sum_k \epsilon_k f(\epsilon_k) \quad , \qquad (3.8)$$

which using (3.6) and (3.7) is

$$W=\frac{2tN}{\pi}\int_{-1}^{1}\frac{yf(y)}{(1-y^2)^{1/2}}\,dy \quad . \qquad (3.9)$$

The specific heat $C_v=\partial W/\partial T$ is easily computed numerically and is shown in Fig. 1 for several densities. Let us compute the ground-state energy W_g explicitly. First the Fermi momentum k_F at $T=0\,°\mathrm{K}$ is found from (3.5),

$$\mathfrak{N}=\sum_k f(\epsilon_k)=\frac{N}{\pi}\int_0^{k_F}dk=\frac{Nk_F}{\pi} \qquad (3.10)$$

or $k_F=\pi(\rho-1)$. The ground-state energy is then

$$W_g=\sum_k \epsilon_k f(\epsilon_k)=-(2tN/\pi)\int_0^{k_f}\cos k\,dk$$
$$=-(2tN/\pi)\sin k_F=(2tN/\pi)\sin\pi\rho \quad , \qquad (3.11)$$

which may be written as $W_g=-(2tN/\pi)|\sin\pi\rho|$ and is identical to the $U=\infty$ limit of Shiba's calculation.[9] At infinite temperature, $f(\epsilon_k)$ must be in-

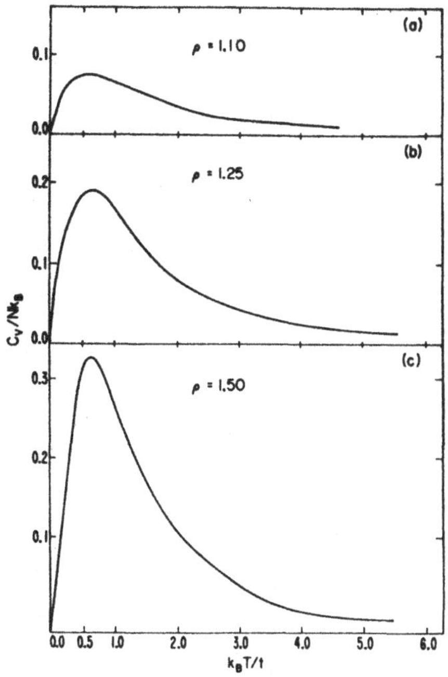

FIG. 1. The orbital contribution to the specific heat in zero external field for various values of the electron density ρ.

dependent of k, leading to $W(\infty)=0$.

As in the case of one carrier considered previously, the mobility is infinite. This follows directly from the argument given by Brinkman and Rice[15] that the velocity operator commutes with the Hamiltonian when $U=\infty$. The mean square velocity is

$$\langle v^2\rangle = \left(\frac{a}{\hbar}\right)^2 \sum_k \left(\frac{\partial\epsilon_k}{\partial k}\right)^2 f(\epsilon_k)$$

$$= \left(\frac{2ta}{\hbar}\right)^2 \sum_k \sin^2 k f(\epsilon_k) \quad . \tag{3.12}$$

At absolute zero, using the Fermi momentum given in (3.10),

$$\langle v^2\rangle_0 = (2ta/\hbar)^2(N/2)[\rho - 1 - (2\pi)^{-1}\sin 2\pi\rho] \quad , \tag{3.13}$$

which for $\rho = 1 + N^{-1}$ gives zero, in agreement with the result of Sec. II for one extra carrier. In the infinite-temperature limit all $f(\epsilon_k)$'s are equal to $\rho - 1$, leading to

$$\langle v^2\rangle_\infty = (2ta/\hbar)^2(N/2)(\rho - 1) \quad , \tag{3.14}$$

which for $\rho - 1 + N^{-1}$ again agrees with the result of Sec. II.

As we have seen the $U=\infty$ limit of the arbitrary-density Hubbard chain is exactly soluble, at least for its thermodynamic quantities. The basic result is that the magnetic and orbital properties are completely decoupled. The magnetic behavior is that of a set of localized spins while the kinetic effects correspond to the appropriate density of spinless noninteracting fermions. Future studies will consider the use of these exact results as a guide to obtain good approximations for finite but large U situations.

ACKNOWLEDGMENTS

We are happy to acknowledge useful discussions with J. Kanamori, P. Chaikin, and C. Coll, III.

*Supported in part by the National Science Foundation and the Office of Naval Research.
[1]For example, L. Hi Lieb and D. C. Mattis, *Mathematical Physics in One Dimension* (Academic, New York, 1966).
[2]R. J. Birgeneau, R. Dingle, M. T. Hutchings, G. Shirane, and S. L. Holt, Phys. Rev. Lett. **26**, 718 (1971).
[3]See, for example, R. N. Rogers, F. Carboni, and P. M. Richards, Phys. Rev. Lett. **19**, 1016 (1967).
[4]See the review article by I. F. Shchegolev, Phys. Status Solidi **12**, 9 (1972).
[5]A. J. Epstein, S. Etemad, A. F. Garito, and A. J. Heeger, Phys. Rev. B **5**, 952 (1972).
[6]J. Hubbard, Proc. R. Soc. A **276**, 238 (1963); Proc. R. Soc. A **281**, 401 (1964).
[7]L. H. Lieb and F. Y. Wu, Phys. Rev. Lett. **20**, 1445 (1968).
[8]M. Takahashi, Prog. Theor. Phys. **42**, 1098 (1969); Prog. Theor. Phys. **43**, 1619 (1970).
[9]H. Shiba, Phys. Rev. B **6**, 930 (1972).
[10]A. A. Ovchinnikov, Zh. Eksp. Teor. Fiz. **57**, 2137 (1969) [Sov. Phys.-JETP **30**, 1160 (1970)].
[11]C. Coll (unpublished).
[12]H. Shiba and P. Pincus, Phys. Rev. B **5**, 1966 (1972); H. Shiba, Prog. Theor. Phys. **48**, 2171 (1972).
[13]T. A. Kaplan and P. N. Argyres, Phys. Rev. B **1**, 2457 (1970).
[14]J. B. Sokoloff, Phys. Rev. B **2**, 779 (1970).
[15]W. F. Brinkman and T. M. Rice, Phys. Rev. B **2**, 1324 (1970).
[16]See, for example, A. A. Abrikosov, L. P. Gorkov, and I. E. Dzyaloshinsky, *Methods of Quantum Field Theory in Statistical Physics* (Prentice-Hall, Englewood Cliffs, N.J., 1963), Chap. 3.
[17]P. M. Morse and H. Feshbach, *Methods of Theoretical Physics II* (McGraw-Hill, New York, 1963), Chap. 10.
[18]G. N. Watson, *A Treatise on the Theory of Bessel Functions* (Cambridge U. P., New York, 1945), Chap. XI.

Chapter 4

RESULTS IN DIFFERENT PHYSICAL LIMITS

Opposite to the strong-coupling regime, the weak-coupling region has also been investigated in the literature[r4.1],[r4.2],[4.1],[4.2]. Even this limit, due to the antiferromagnetic correlations which exist at any $U \neq 0$, is nontrivial. Nevertheless, the ground-state energy in low dimensions can be obtained by standard diagrammatic perturbation techniques[r4.1]. A general approach to weak-coupling expansions for correlated electrons has been developed in [4.1], by deriving and partially solving a self-consistent approximation scheme which satisfies microscopic conservation laws. The tremendous numerical effort underlying this kind of calculations can be possibly avoided by resorting first to the study of the model for $D = \infty$[4.2],[r4.3], and hopefully connecting then the solution, by perturbative techniques, to the case of finite dimension.

For intermediate values of the coupling constant, one must resort to mean-field type calculations. The Hartree-Fock approach to the Hubbard hamiltonian[r4.4],[5.1], allows one to draw a temptative $T = 0$ phase diagram, which, in two dimensions[5.1], is quite rich, providing, besides the paramagnetic phase, both a ferromagnetic and an antiferromagnetic ground state, depending on the filling of the band. In particular, it is found that, for large U, the ferromagnetic state has the lowest energy also for δ very close to $\frac{1}{2}$, in agreement with Nagaoka's result.

In classical statistical physics, as well as for localized quantum spin systems, mean-field type solutions are equivalent to solving the model in the limit $D = \infty$. Remarkably enough, this seems no longer true for itinerant electron systems. Indeed, it has been shown[4.2] that, for the Hubbard model, in $D = \infty$ and in the weak-coupling limit, the correlations among fermions remain non-trivial, leading nevertheless to dramatic simplifications in the diagrammatic expansion of the solution. A "collapse of diagrams" occurs, i.e. different vertices can be identified with a single site, and many results coming from the Gutzwiller approximation of different variational wave functions, are recovered, by exact explicit evaluation of the wave functions.

In the framework of mean-field type solutions, an alternative approach was also developed in [4.3], where, instead of approximating the Coulomb interaction term, which is left unchanged, it is the hopping term to be treated within a *fermi-linearization* scheme. The scheme was successively implemented[r4.5], so as to deal with clusters of sites in an exact way, while only the hopping between neighbouring clusters is approximated. In the latter case, a ground state with non-zero pairing was found for a two-band Hubbard model for sufficiently large U.

The intermediate coupling region has been also explored in the limit of low concentration of electrons. In this limit ($\delta \ll \frac{1}{2}$) Kanamori[4.4] found that the ground state is paramagnetic. This implies that the Nagaoka's ferromagnetic state, if still stable at some finite U, is unstable for large enough concentration of holes.

Finally, the alternative approach of the renormalization group has been used[4.5] as well, in the study of the model behavior in a half-filled band. The method reproduces quite well in one dimension most of the features of the exact solution, and, applied to the case of two and three dimensions, predicts a Mott-transition at finite interaction strength, providing the first example of the possible relevance of renormalization group approach even for itinerant electron systems.

In conclusion, let's remark how the different approximate results presented give, especially for the two-dimensional case, a reasonably coherent picture of the zero-temperature phase space. Nevertheless, such a picture is far from complete, and in particular there is no evidence of superconducting properties of the model, at least for the one-band case.

[r4.1] W. Metzner and D. Vollhardt, *Phys. Rev.* **B39**, 4462 (1989).
[r4.2] J.R. Schrieffer, X.G. Wen, and S.C. Zhang, *Phys. Rev.* **B39**, 11663 (1989).
[r4.3] E. Mueller Hartmann, *Z. Phys.* **B74**, 507 (1989).
[r4.4] D. Penn, *Phys. Rev.* **142**, 350 (1966).
[r4.5] A. Montorsi and M. Rasetti, *Mod. Phys. Lett.* **B4**, 613 (1990).

VOLUME 62, NUMBER 8 PHYSICAL REVIEW LETTERS 20 FEBRUARY 1989

Conserving Approximations for Strongly Correlated Electron Systems: Bethe-Salpeter Equation and Dynamics for the Two-Dimensional Hubbard Model

N. E. Bickers[(a)]
Institute for Theoretical Physics, University of California at Santa Barbara, Santa Barbara, California 93106

D. J. Scalapino and S. R. White
Department of Physics, University of California at Santa Barbara, Santa Barbara, California 93106
(Received 18 July 1988)

In this Letter we describe a new technique for investigating phase transitions and dynamics in interacting electron systems. This technique is based on the derivation and self-consistent solution of infinite-order conserving approximations. It provides a new approach to the study of two-particle correlations with strong frequency and momentum dependence. We use this technique to derive a low-temperature phase diagram and dynamic correlation functions for the two-dimensional Hubbard model.

PACS numbers: 71.10.+x, 74.20.−z, 74.65.+n, 75.10.Jm

Experimental studies of heavy-electron compounds,[1] the organic Bechgaard salts,[2] the recently reported bis-(ethylenedithiolo)-tetrathiafulvalene superconductors,[3] and the oxide superconductors[4] have emphasized the need for new approaches to strongly correlated electronic systems. While quantum Monte Carlo (QMC) techniques can, in principle, provide exact results for electronic correlation functions, such techniques are at present limited to relatively small systems and high temperatures. Furthermore, QMC furnishes direct information on imaginary-time, rather than real-time correlations; the treatment of real-time dynamics remains problematic. In this Letter, we describe a semianalytical approach to strongly correlated systems which satisfies microscopic conservation laws, treats strong frequency and momentum dependences, and provides information on both static and dynamic properties. This approach may be used to treat large systems and temperatures lower than those currently accessible to finite-temperature QMC. Furthermore, this approach naturally incorporates realistic features such as Fermi surface nonsphericity, which may explain property variations within a series of otherwise similar compounds.

In this Letter, we restrict attention to the two-dimensional Hubbard model. This lattice model is an interesting test for approximate many-body approaches since it involves a potential competition between antiferromagnetic and superconducting ground states. The simplest Hubbard Hamiltonian for electrons moving on a square lattice is characterized by a hopping integral t, which connects neighboring sites with separation a, an on-site Coulomb repulsion U, and a temperature-dependent chemical potential μ, which must be adjusted to fix the electronic filling n. A "conserving approximation" for this model, i.e., an approximation consistent with microscopic conservation laws for particle number, energy, and momentum, may be generated following an approach introduced by Baym[5] to treat the Coulomb gas: (a) First, write down a "free-energy" functional Φ of the dressed single-particle Green's function G and the interaction U. The diagrammatic representation of this functional formally resembles the expansion for the ground-state energy. (b) Generate an approximation for the single-particle self-energy Σ by functional differentiation of Φ with respect to G:

$$\Sigma(1,2)=\delta\Phi/\delta G(2,1)\,.$$

(Our notation is that of Ref. 5.) Compute G self-consistently using this self-energy. (c) Generate an approximation for particle-hole (or particle-particle) response functions L by computing G in the presence of an external potential $\mathcal{U}$, then performing a functional differentiation with respect to $\mathcal{U}$:

$$L(1,2;1',2')=\delta G(1,1')/\delta\mathcal{U}(2',2)\,|_{\mathcal{U}=0}\,.$$

The correlation functions generated in this way automatically obey microscopic conservation laws, are consistent with Luttinger's theorem,[6] and lead to Fermi-liquid behavior at low temperatures in the normal state. The Φ diagram describing the repulsion of opposite-spin electrons in the absence of fluctuations is shown in Fig. 1(a). The conserving approximation generated by this diagram is just Hartree-Fock, or mean-field, theory. The diagrams describing the interaction of electrons with spin, density, and two-particle fluctuations are shown in Figs. 1(b)–1(d). These diagrams lead to a "higher order" Hartree-Fock theory which consistently incorporates the effect of the simplest fluctuations. We shall hereafter refer to this approach as the "fluctuation-exchange approximation."[7,8] In the past, technical limitations have generally prevented the self-consistent solution of conserving approximations beyond mean-field level. We describe below a general approach for solving approximations which incorporate strong collective fluctuations, and use this technique to obtain a fully self-consistent solution of the fluctuation-exchange approximation for

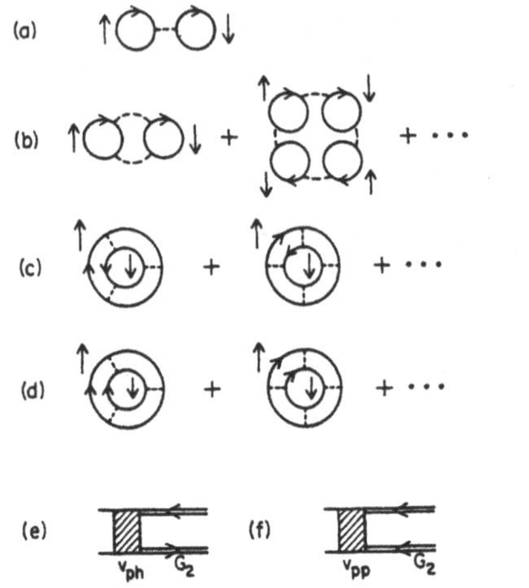

FIG. 1. Hubbard model diagrams for the fluctuation-exchange approximation. The interaction U is represented by a dashed line. (a) Lowest order Φ diagram (Hartree-Fock theory), (b) Φ ring diagrams representing the interaction of longitudinal spin and density fluctuations, (c) transverse spin fluctuations, (d) particle-particle fluctuations, (e) Bethe-Salpeter kernels for the particle-hole, and (f) particle-particle channels. The double lines represent self-consistent Green's functions based on the Φ approximation in (a)-(d).

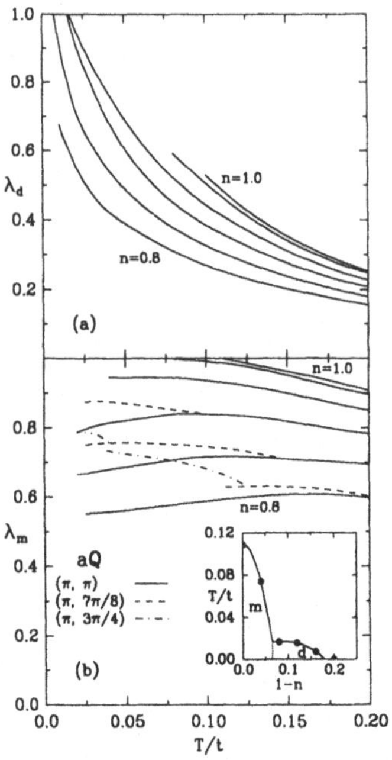

FIG. 2. (a) Leading d-wave singlet eigenvalues of the particle-particle kernel for $U/t=4$. These results are based on a 16×16 discretization of the Brillouin zone. Successive curves vary in n by increments of 0.04 from $n=1.0$ to 0.8. (b) Leading commensurate $[a\mathbf{Q}=(\pi,\pi)]$ and incommensurate magnetic eigenvalues of the particle-hole kernel. As in (a), values of the filling vary regularly by increments of 0.04. Incommensurate eigenvalues (for fillings $n=0.80$, 0.84, and 0.88) are shown only when they exceed the corresponding commensurate eigenvalues. Inset: Approximate phase diagram derived from the transitions (Ref. 10) in (a) and (b).

the Hubbard model.

In general, steps (b) and (c) above lead to singular integral equations in the Fourier transform variables $k=(\mathbf{k},\omega)$. These equations cannot be solved in closed form, but may be conveniently treated by (a) discretizing the Brillouin zone (or, equivalently, restricting the calculation to a finite lattice in real space), and (b) analytically continuing the energy variable ω to discrete Matsubara frequencies on the imaginary axis. Using the self-energy generated by Fig. 1, we have solved iteratively for the single-particle Green's function, taking into account its full momentum and frequency dependence. Use of symmetries in momentum and frequency space significantly reduces the computation time and storage requirements. Real-frequency correlation functions, including the one-electron density of states and the magnetic spectral density (or neutron-scattering function), may be recovered using N-point Padé approximants[9] to reverse the analytic continuation.

The competition between two or more ordering transitions[10] may be studied using the results of a conserving approximation and the Bethe-Salpeter equation. Singlet and triplet superconductivity correspond to the development of a scattering pole in the particle-particle channel, while spin and charge ordering correspond to a scattering pole in the particle-hole channel. In the nearly half-filled Hubbard model, an almost instantaneous attractive interaction occurs in the triplet particle-hole channel, corresponding to spin ordering with wave vector at or near the zone-boundary vector $a\mathbf{Q}=(\pi,\pi)$. On the other hand, it is believed from the random-phase-approximation[11,12] and from QMC[13] studies that a strongly retarded attractive interaction occurs in a particle-particle channel with singlet $d_{x^2-y^2}$ symmetry.[14] The Bethe-Salpeter equation, which treats time-dependent two-particle interactions, allows the study of both instabilities.

An instability occurs when an eigenvalue of the Bethe-Salpeter kernel K reaches unity. The imaginary-frequency kernel takes the form shown in Figs. 1(e) and 1(f). It is a product of two single-particle Green's functions and v, the effective particle-hole (or particle-

particle) interaction. We show in Figs. 2(a) and 2(b) the behavior of the leading magnetic and d-wave superconducting eigenvalues for calculations performed using the Φ approximation in Figs. 1(a)-1(d) with a 256-point discretization of the Brillouin zone and a frequency cutoff of $\pm 10t$. The Coulomb energy $U/t=4$ and results are shown for electronic fillings varying from 0.8 to 1.0. Leading eigenvalues and eigenvectors of K are computed by repeated projection on test vectors of the appropriate symmetry. The largest kernels we have treated with this technique have row dimension 256×322 in $(\mathbf{k},i\omega)$ space.[15] Finite-size studies with 16- and 64-point Brillouin zones suggest that discretization error in the calculated transition temperatures is on the order of 10% or less. For deviations of less than ~6% from half filling, an antiferromagnetic state is favored, while for deviations between 6% and 18% a d-wave superconducting state is favored. A phase diagram is shown in the inset of Fig. 2(b). Note that the maximum predicted superconducting transition temperature is of order $0.01t$. This is an order of magnitude smaller than the random-phase-approximation estimate which would follow neglecting self-energies altogether. Combined with recent QMC results[13] which explore the effect of the two-particle vertex on pairing susceptibilities, these results strongly suggest that the Hubbard model has a d-wave superconducting ground state for a small range of fillings. The predicted transition temperature for a bare bandwidth $8t=1$ eV is $T_d \sim 15$ K.

The accuracy of our conserving approximation for the Hubbard model may be gauged by direct comparison with QMC results.[16] We show in Fig. 3 results for the imaginary-time Green's function obtained from the fluctuation-exchange approximation and from QMC for fillings $n=0.88$ and 1.00. The approximate Green's function is particularly accurate at half filling; for smaller fillings, the two functions remain similar, though the fluctuation-exchange approximation underestimates G at intermediate values of τ. (This reduction is reflected in susceptibilities, which involve integrals of G.)

In order to illustrate the information on dynamics which may be obtained by this approach, we show in Fig. 4 a plot of the absorptive part of the magnetic susceptibility for $T/t=0.20$ at filling $n=0.96$. This plot shows the development of a sharp quasielastic peak at wave vector $a\mathbf{Q}^*=(\pi,\pi)$ and a softening of the paramagnon dispersion curve at the zone center, as well as the zone boundary. The velocity in the (1,1) direction is in this case approximately $0.35t/a^{-1}$, or $1.4t/\text{Å}^{-1}$ for a lattice constant of 4 Å. The one-electron density of states, scattering rate, and mass enhancement have also been obtained using the Padé technique[9] and will be discussed elsewhere.[17]

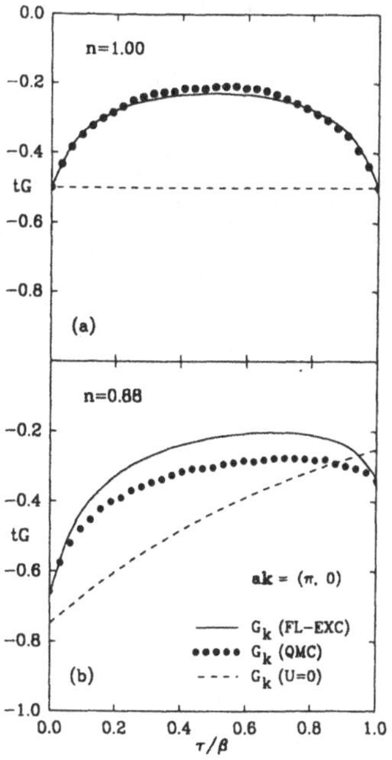

FIG. 3. Comparison of the imaginary-time Green's functions $G(\tau)$ derived from the fluctuation-exchange approximation and from QMC for $T/t=0.25$ with (a) $n=1.00$ and (b) $n=0.88$. Results are shown for the momentum-space Green's function $G_{\mathbf{k}}$ with $a\mathbf{k}=(\pi,0)$. The corresponding noninteracting functions are plotted as well to indicate the effect of interactions. All results shown are based on an 8×8 momentum-space discretization (or real-space lattice).

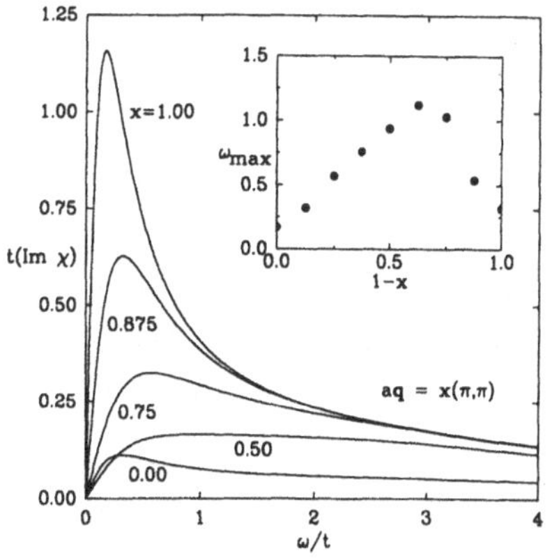

FIG. 4. Magnetic spectral density $\text{Im}\chi(\mathbf{q},\omega)$ for $n=0.96$ and $T/t=0.20$. Inset: Peak locations $\omega_{\max}$, i.e., the paramagnon dispersion relation, for $a\mathbf{q}=(1-x)(\pi,\pi)$.

The calculation described above may be extended to more elaborate Hubbard or electron-phonon models and to more complex lattice structures. In the future, the self-consistent solution of conserving lattice approximations, which treat frequency and momentum dependence on an equal footing, may prove useful in developing a hybrid many-body band-theory approach to strongly interacting systems.

This work has been supported in part by the National Science Foundation under Grants No. DMR86-15454 and No. PHY82-17853 and by funds from NASA. The numerical calculations reported in this paper were performed on the Cray-XMP at the San Diego Supercomputer Center. We wish to thank SDSC for its support.

[(a)]Present address: Department of Physics, University of Southern California, University Park, Los Angeles, CA 90089.

[1]G. R. Stewart, Rev. Mod. Phys. **56**, 755 (1984).

[2]D. Jerome, J. Phys. (Paris), Colloq. **44**, C3-755 (1983).

[3]K. Oshima, H. Urayama, H. Yamochi, and G. Saito, J. Phys. Soc. Jpn. **57**, 730 (1988).

[4]J. G. Bednorz and K. A. Müller, Z. Phys. B **64**, 18 (1986); M. K. Wu *et al.*, Phys. Rev. Lett. **58**, 908 (1987).

[5]G. Baym, Phys. Rev. **127**, 1391 (1962).

[6]J. M. Luttinger, Phys. Rev. **119**, 1153 (1960).

[7]We have intentionally included spin, density, and two-particle fluctuations on the same footing. As expected from random-phase-approximation estimates, magnetic fluctuations are by far the most important contribution.

[8]This approximation satisfies approximate crossing symmetry and may be viewed as a first step toward a full summation of crossing-symmetric parquet diagrams.

[9]See, e.g., H. J. Vidberg and J. W. Serene, J. Low. Temp. Phys. **29**, 179 (1977).

[10]No true finite-temperature magnetic transition and only a possible Kosterlitz-Thouless superconducting transition are expected in the two-dimensional Hubbard model. For this reason, finite-temperature "transitions" should be interpreted as crossovers signaling the appearance of a large, but finite, correlation length. Weak three-dimensional coupling is sufficient to induce true long-range order in such phases.

[11]D. J. Scalapino, E. Loh, and J. E. Hirsch, Phys. Rev. B **35**, 6694 (1987).

[12]N. E. Bickers, D. J. Scalapino, and R. T. Scalettar, Int. J. Mod. Phys. B **1**, 687 (1987).

[13]S. R. White, D. J. Scalapino, N. E. Bickers, and R. T. Scalettar (to be published).

[14]A number of approximate approaches have been previously applied to investigate *d*-wave superconductivity: J. Miyake, S. Schmitt-Rink, and C. M. Varma, Phys. Rev. B **34**, 6554 (1986); C. Gros, R. Joynt, and T. M. Rice, Z. Phys. B **68**, 425 (1987); A. J. Millis, S. Sachdev, and C. M. Varma, Phys. Rev. B **37**, 4975 (1988); G. Kotliar, *ibid.* **37**, 3664 (1988); H. Shimahara and S. Takada, J. Phys. Soc. Jpn. **57**, 1044 (1988).

[15]The only conserving diagrams omitted from the calculation of the vertex $v_{\text{p-h}}$ ($v_{\text{p-p}}$) describe the decay of a particle-hole (particle-particle) pair into two collective fluctuations. We have established, using studies with a sixteen-point Brillouin zone, that these contributions are small in comparison with the contributions retained.

[16]For a discussion of the QMC algorithm, see S. R. White, R. L. Sugar, and R. T. Scalettar (to be published).

[17]The one-electron self-energy is nontrivial and incorporates two opposing effects: The dressing of low-energy electrons by spin fluctuations tends to increase the electronic mass (in analogy with mass enhancement by phonons); on the other hand, the coupling of nearly degenerate electronic states in the presence of strong nesting repels electronic density from the Fermi level.

VOLUME 62, NUMBER 3 PHYSICAL REVIEW LETTERS 16 JANUARY 1989

Correlated Lattice Fermions in $d=\infty$ Dimensions

Walter Metzner and Dieter Vollhardt
Institut für Theoretische Physik C, Technische Hochschule Aachen, Sommerfeldstrasse 26/28, D-5100 Aachen, Federal Republic of Germany
(Received 28 September 1988)

We show that even in $d=\infty$ dimensions the Hubbard model, when scaled properly, describes nontrivial correlations among fermions. Diagrammatic treatments are found to be substantially simpler than in finite dimensions. The weak-coupling correlation energy is seen to be a good approximation for that in $d=3$. Recent approximations based on slave-boson techniques are recovered by the exact evaluation of explicit variational wave functions in $d=\infty$.

PACS numbers: 75.10.Jm, 71.10.+x, 71.28.+d

The spin-$\frac{1}{2}$, single-band Hubbard model[1] for interacting fermions on a lattice plays a particularly important role in condensed-matter physics. In spite of its apparent simplicity, exact solutions have so far only been possible in $d=1$ space dimension.[2] However, in view of the special properties of one-dimensional systems it is not clear how relevant these results are for higher-dimensional systems, e.g., for $d=3$. There is another limiting dimension, $d=\infty$, where exact solutions have been obtained for various, mostly classical, spin systems.[3] In this case there exist close relations to mean-field–type solutions. In the case of the Hubbard model, an exact solution in $d=\infty$ is not available so far.

In this Letter properties of the Hubbard model in $d=\infty$ will be discussed. The Hubbard Hamiltonian[1] has the form

$$\hat{H}=\hat{H}_{\text{kin}}+U\sum_i \hat{n}_{i\uparrow}\hat{n}_{i\downarrow}, \tag{1a}$$

$$\hat{H}_{\text{kin}}=\sum_\sigma\sum_{i,j} t_{ij}\hat{c}^+_{i\sigma}\hat{c}_{j\sigma}=\sum_{k,\sigma}\epsilon_k\hat{n}_{k\sigma}, \tag{1b}$$

where $\hat{H}_{\text{kin}}$ is the kinetic-energy operator expressed in position space and momentum space, respectively. For next-neighbor hopping on a d-dimensional simple cubic lattice with unit lattice spacing, we have

$$\epsilon_{\mathbf{k}}=-2t\sum_{j=1}^{d}\cos k_j, \tag{2}$$

where t is the hopping amplitude and $\mathbf{k}=(k_1,\ldots,k_d)$.

For $d\to\infty$, the on-site interaction is still well defined but the hopping rate in the kinetic energy has to be scaled properly to yield a nontrivial model. This is most easily seen from the corresponding density of states (DOS) for $U=0$, which in $d=\infty$ is determined by the central-limit theorem as

$$D(E)\underset{d\to\infty}{=}\frac{1}{2t(\pi d)^{1/2}}\exp[-(E/2t\sqrt{d})^2]. \tag{3}$$

Clearly, only the scaling $t=t^*/(2d)^{1/2}$ with fixed t^* (henceforth $t^*\equiv 1$) yields a finite DOS[4] and thereby leads to a finite average kinetic energy $\bar{\epsilon}_0(n_\uparrow,n_\downarrow)$ of the noninteracting particles for arbitrary densities $n_\uparrow,n_\downarrow$. Any other scaling makes $\bar{\epsilon}_0(n_\uparrow,n_\downarrow)$ either zero or infinite so that the model becomes immediately trivial.

We have recently shown that for weak coupling the correlation energy of the Hubbard model in $d=1,2,3$ can be calculated within ordinary Goldstone perturbation theory.[5] In particular, in the limit $d\to\infty$, the evaluation of the corresponding diagrams is greatly simplified. This may be demonstrated in the case of the second-order contribution given by

$$E_2=-\frac{LU^2}{(2\pi)^{3d}}\int d\mathbf{k}\,d\mathbf{k}'\,d\mathbf{q}\frac{n^0_{\mathbf{k}\uparrow}n^0_{\mathbf{k}'\downarrow}(1-n^0_{\mathbf{k}+\mathbf{q}\uparrow})(1-n_{\mathbf{k}'-\mathbf{q}\downarrow})}{\epsilon_{\mathbf{k}}+\epsilon_{\mathbf{k}'}-\epsilon_{\mathbf{k}+\mathbf{q}}-\epsilon_{\mathbf{k}'-\mathbf{q}}}, \tag{4}$$

where the integrations extend over a Brillouin zone; here $n^0_{\mathbf{k}\sigma}\equiv 1$ for $|\mathbf{k}|<k_{F\sigma}$ and zero elsewhere, and L is the number of lattice sites. If we write the energy denominator in (4) as an integral over an exponential factor, and express momentum conservation in (4) explicitly by an integral over a δ function, which is then converted into a lattice sum via $\delta(\mathbf{k})=(2\pi)^{-d}\sum_{\mathbf{f}}\exp(i\mathbf{k}\cdot\mathbf{f})$, (4) is written as

$$E_2/LU^2=-\int_0^\infty d\lambda\sum_{\mathbf{f}}F^+_\uparrow(\lambda;\mathbf{f})F^+_\downarrow(\lambda;\mathbf{f})F^-_\uparrow(\lambda;\mathbf{f})F^-_\downarrow(\lambda;\mathbf{f}). \tag{5}$$

Here $F^+_\sigma(\lambda;\mathbf{f})$, $F^-_\sigma(\lambda;\mathbf{f})$ are the Fourier transforms of $n^0_{\mathbf{k}\sigma}\exp(\lambda\epsilon_{\mathbf{k}})$ and $(1-n^0_{\mathbf{k}\sigma})\exp(-\lambda\epsilon_{\mathbf{k}})$, respectively. If we make use of the fact that $\sum_{\mathbf{f}}[F^\pm_\sigma(\lambda;\mathbf{f})]^2$ is finite and count the number of nearest neighbors, next-nearest neighbors, etc., of the site $\mathbf{f}=0$, it follows that $F^\pm_\sigma(\lambda;\mathbf{f}\neq 0)$ vanishes at least as $1/\sqrt{d}$ for $d\to\infty$. Consequently, the off-site ($f\neq 0$) contributions in (5) vanish as $1/d$, such that only the $\mathbf{f}=0$ term remains. Using the DOS (3) we thus find

$$E_2/LU^2=-\int_0^\infty d\lambda\, e^{2\lambda^2}\prod_\sigma P(E_{F\sigma}-\lambda)P(-E_{F\sigma}-\lambda). \tag{6}$$

Here $P(x)$ is the Gaussian probability function. Hence the evaluation of E_2 for $d \to \infty$ is seen to reduce to a *single* integral over probability functions and is therefore the simplest of all dimensions. The simplification is due to the fact that in $d=\infty$, and for an arbitrary choice of the momenta $\mathbf{k}$, $\mathbf{k}'$, and $\mathbf{q}$ in (4), the energies $\epsilon_{\mathbf{k}}$, $\epsilon_{\mathbf{k}'}$, $\epsilon_{\mathbf{k}+\mathbf{q}}$, and $\epsilon_{\mathbf{k}'-\mathbf{q}}$ are randomized by umklapp processes generated when the lattice momenta are added.[6] Hence the energies become mutually *independent*, which allows one to replace the momentum integrations in (4) by energy integrations over the DOS (3). The energy E_2, plotted as a function of density n $(n_\uparrow = n_\downarrow = n/2)$, is shown in Fig. 1 in comparison with the respective numerical evaluation for $d=1$ and 3.[5] The result for $d=\infty$ is seen to be very close to that for $d=3$ and can therefore be considered as an easily tractable, reliable approximation. The existence of a nonzero second-order contribution to the exact ground-state energy also shows that the antiferromagnetic Hartree-Fock approximation[7] is *not* the exact result even in $d=\infty$, since the asymptotic expansion of the latter terminates after the linear term in U.

We note that $F_\sigma^+(0;\mathbf{f}-\mathbf{h})$ is just the one-particle density matrix of the noninteracting system

$$F_\sigma^+(0;\mathbf{f}-\mathbf{h}) = P^0_{\sigma,\mathbf{fh}} \equiv \langle \Phi_0 | \hat{c}^+_{\mathbf{f}\sigma} \hat{c}_{\mathbf{h}\sigma} | \Phi_0 \rangle ,$$

where $|\Phi_0\rangle$ is the noninteracting ground state. Hence $P^0_{\sigma,\mathbf{fh}}$ vanishes at least as $1/\sqrt{d}$ for $d \to \infty$, if $\mathbf{f} \neq \mathbf{h}$. The result remains true even if $|\Phi_0\rangle$ is chosen to be a more general one-particle wave function, e.g., a Hartree-Fock spin-density wave. This property allows for an exact evaluation of expectation values $\langle \hat{O} \rangle = \langle \Psi | \hat{O} | \Psi \rangle / \langle \Psi | \Psi \rangle$ in terms of variational wave functions of the form

$$|\Psi\rangle = g^{\hat{D}} |\Phi_0\rangle , \qquad (7)$$

which are generalizations of the Gutzwiller wave function.[8] Here $|\Phi_0\rangle$ is an arbitrary, not necessarily translationally invariant, one-particle wave function, g is a variational parameter, $0 \le g \le 1$, and $\hat{D} = \sum_i \hat{D}_i = \sum_i \hat{n}_{i\uparrow}\hat{n}_{i\downarrow}$ is the number operator for doubly occupied sites. Generalizations of $|\Phi_0\rangle$ to BCS-type wave functions, leading to a resonating-valence-bond-type state in the limit $g=0$,[9] are only slightly more complicated. The case where $|\Phi_0\rangle$ is a simple Fermi sea has already been discussed earlier,[6] where it was shown that an analytic evaluation of ground-state properties of the Hubbard model is possible in $d=1$. This earlier analysis may be simplified and further generalized by employment of well known many-body techniques. To this end we recall that the diagrams which determine the one-particle density matrix $P_{\sigma,\mathbf{fh}} = \langle \hat{c}^+_{\mathbf{f}\sigma} \hat{c}_{\mathbf{h}\sigma} \rangle$ are identical in form to the ones for the two-point functions of a Φ^4 theory with lines corresponding to $P^0_{\sigma,\mathbf{fh}}$. In analogy to the diagrammatic representation of Green's functions, one may therefore define a "self-energy" $S_{\sigma,\mathbf{fh}}$, which is built up by the same diagrams as the self-energy for Green's functions, and a "proper" self-energy $S^*_{\sigma,\mathbf{fh}}$ being the sum over all one-particle irreducible diagrams. S_σ and S^*_σ are related by the usual Dyson equation. The one-particle density matrix then takes the form

$$P_{\sigma,\mathbf{fh}} = P^0_{\sigma,\mathbf{fh}} + 1_{\mathbf{fh}} \left(\frac{1-g}{1+g} P^0_\sigma S_\sigma - \frac{S_\sigma}{(1+g)^2} \right)_{\mathbf{ff}} + \left[\left(P^0_\sigma - \frac{1}{1+g} \right) S_\sigma \left(P^0_\sigma - \frac{1}{1+g} \right) \right]_{\mathbf{fh}} . \qquad (8)$$

Here P_σ, P^0_σ, and S_σ are taken as matrices, with $\mathbf{1}$ as the unit matrix, with elements $P_{\sigma,\mathbf{fh}}$, etc. The expectation value for the interaction term of the Hubbard Hamiltonian can be expressed in terms of S as

$$\bar{d} \equiv L^{-1} \langle \hat{D} \rangle = L^{-1} [g^2/(1-g^2)] \mathrm{Tr}(P^0_\sigma S_\sigma) , \qquad (9)$$

where $\mathrm{Tr}(\;) = \sum_{\mathbf{f}} (\;)_{\mathbf{ff}}$. To investigate the consequences of the fact that $P^0_{\sigma,\mathbf{fh}}$, $\mathbf{f} \neq \mathbf{h}$, vanishes at least as $1/\sqrt{d}$ for $d \to \infty$, we consider a diagram in which two vertices $\mathbf{f},\mathbf{h}$ are connected by three or more separate paths. The evaluation of the diagram involves the lattice sum over $\mathbf{f},\mathbf{h}$ and all other vertices. However, the contributions from $\mathbf{f} \neq \mathbf{h}$, are suppressed by factors of order $1/\sqrt{d}$. Consequently, in the limit $d \to \infty$, only the on-site terms $(\mathbf{f}=\mathbf{h})$ remain, i.e., the two vertices collapse into a single vertex. Consequently, since external vertices of proper diagrams are always connected by three separate paths, $S^*_{\sigma,\mathbf{fh}}$ is seen to be *diagonal*, i.e., $S^*_{\sigma,\mathbf{fh}} \equiv \delta_{\mathbf{fh}} S^*_{\sigma\mathbf{f}}$.

As a first application, we treat the simple Gutzwiller wave function[8] with $|\Phi_0\rangle$ in (7) given by the Fermi sea. Because of translational invariance $S^*_{\sigma\mathbf{f}}$ is independent of $\mathbf{f}$: $S^*_{\sigma\mathbf{f}} = S^*_\sigma$ for all $\mathbf{f}$. The evaluation of $\langle \hat{n}_{\mathbf{k}\sigma} \rangle$, the Fourier transform of $P_{\sigma,\mathbf{fh}}$ in (8), and of $\langle \hat{D} \rangle$ in (9) is thus reduced to the calculation of a single *number* S^*_σ.

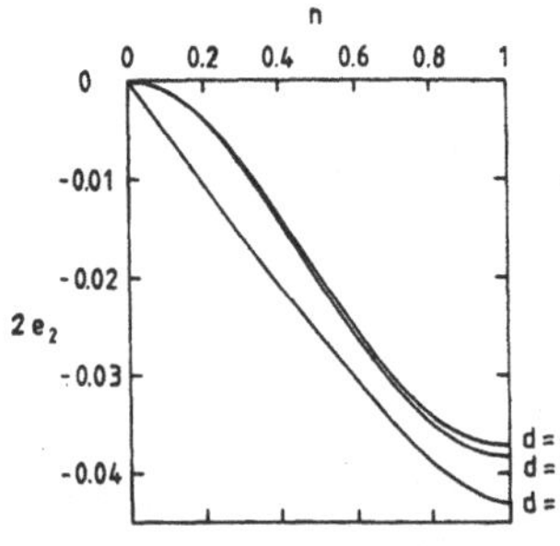

FIG. 1. Second-order correlation energy $e_2 \equiv E_2/[LU^2/|\bar{\epsilon}_0(\frac{1}{2},\frac{1}{2})|]$ vs density for lattice dimensions $d=1$, 3, and ∞.

With $L^{-1}\sum_k n_{k\sigma}=n_\sigma$ the latter is determined by a quadratic equation, yielding

$$S_\sigma^* = \{A_\sigma - [A_\sigma^2 - 4(1-g^2)(1-n_\sigma)n_{-\sigma}]^{1/2}\}/2(1-n_\sigma), \tag{10}$$

with $A_\sigma = 1-(1-g^2)(n_\sigma - n_{-\sigma})$. By Dyson's equation, the self-energy $S_\sigma(\mathbf{k})$ is then given by $S_\sigma = S_\sigma^*/(1-S_\sigma^*)$ for $k<k_{F\sigma}$ and by $S_\sigma = S_\sigma^*$ for $k>k_{F\sigma}$. The momentum distribution $\langle \hat{n}_{\mathbf{k}\sigma}\rangle$ is therefore a step function with a discontinuity

$$q_\sigma = 1-[S_\sigma^*/(1+g)^2][1-g^2/(1-S_\sigma^*)]$$

and

$$\langle \hat{D}\rangle = L[g^2 n_\sigma/(1-g^2)]S_\sigma^*/(1-S_\sigma^*).$$

These are precisely the results of the Gutzwiller *approximation*[8] (for a detailed discussion, see Ref. 10), which is therefore seen to become the exact result for the expectation value of $\hat{H}$ in terms of the Gutzwiller wave function in the limit $d=\infty$.[6]

To determine $S_{\sigma f}^*$ for general starting wave functions $|\Phi_0\rangle$ we introduce skeleton diagrams by defining dressed lines $\bar{P}_\sigma = P_\sigma^0 + P_\sigma^0 S_\sigma P_\sigma^0$. The first three diagrams appearing in a skeleton expansion of S^* are shown in Fig. 2. Solid (broken) loops correspond to factors $\bar{P}_{\uparrow ff}$ ($\bar{P}_{\downarrow ff}$). Because of their simple structure, the collapsed skeleton diagrams can be summed up exactly, yielding

$$S_{\sigma f}^* = -\{1-[1+4(1-g^2)\bar{P}_{\sigma,ff}\bar{P}_{-\sigma,ff}]^{1/2}\}/2\bar{P}_{\sigma,ff}, \tag{11}$$

which, together with the definition of $\bar{P}_\sigma$ and the Dyson equation, determines $S_{\sigma f}^*$ and $S_{\sigma,fh}$ for given $P_{\sigma,fh}^0$ and g. We note that the series of diagrams in Fig. 2 contains all diagrams, i.e., does not represent a particular subclass.

The above formalism enables one to evaluate the Hubbard Hamiltonian in terms of increasingly refined wave functions. For example, $|\Phi_0\rangle$ in (7) may allow for antiferromagnetic long-range order

$$|\Psi_{AF}\rangle = g^{\hat{D}}\prod_\sigma \prod_{|\mathbf{k}|<k_F}[\cos\theta_k \hat{a}_{\mathbf{k}\sigma}^+ + \sigma\sin\theta_k \hat{a}_{\mathbf{k}+\mathbf{Q},\sigma}^+]|0\rangle. \tag{12}$$

Here $\mathbf{Q}$ is half a reciprocal-lattice vector, for which the perfect nesting property $\epsilon_{\mathbf{k}+\mathbf{Q}} = -\epsilon_{\mathbf{k}}$ is supposed to be valid (*AB* lattice). For detailed numerical investigations of $|\Psi_{AF}\rangle$ in $d=1,2$, see Ref. 11. In the case of $|\Psi_{AF}\rangle$, $P_{\sigma,fh}^0$ is no longer translationally invariant, although translations among points of the A- and the B-type sublattice, respectively, are still allowed. This particular situation may be treated conveniently by the introduction of a matrix representation distinguishing between translations from A to A, A to B, B to A, and B to B.

The ground-state energy $E\{g;\theta_k\} = \langle \hat{H}\rangle$ is then obtained as a functional of g and θ_k. The minimization with respect to the function θ_k can be performed exactly yielding

$$|\sin 2\theta_k| = \frac{[\mu/(\mu^2-1)]\epsilon_k^2 + \Delta(\epsilon_k^2+\Delta^2)^{1/2}}{[\mu^2/(\mu^2-1)]\epsilon_k^2+\Delta^2}, \tag{13}$$

where μ is a parameter related to g and Δ parametrizes the sublattice magnetization

$$m \equiv |\langle n_{i\uparrow} - n_{i\downarrow}\rangle| = 2\Delta L^{-1}\sum_{|\mathbf{k}|<k_F}(\epsilon_k^2+\Delta^2)^{-1/2}.$$

We note that θ_k in (13) is, in general, different from the Hartree-Fock form, which has been used in numerical calculations,[11] and which is seen to be a special case ($g=1$, i.e., $\mu=-\infty$). If we use (13) and express $E\{g;\theta_k\}$ in terms of m and $\bar{d}$ (9), E may be cast into the form

$$E(m,\bar{d})/L = q\bar{\epsilon}_{HF}(m) + U\bar{d}, \tag{14a}$$

where $\bar{\epsilon}_{HF}(m)$ is the antiferromagnetic Hartree-Fock result for the kinetic energy and

$$q = \frac{4(n-2\bar{d})[\bar{d}(\bar{d}+1-n)]^{1/2} + 2(2\bar{d}+1-n)[(n-2\bar{d})^2-m^2]^{1/2}}{\{(n^2-m^2)[(2-n)^2-m^2]\}^{1/2}} \tag{14b}$$

is a renormalization factor. The minimization of E with respect to $m,\bar{d}$ has to be performed numerically. The result (14) is seen to be identical to that by Kotliar and Ruckenstein,[12] who obtained their results by a slave-boson technique. Here we have constructed the explicit *wave function* $|\Psi_{AF}\rangle$, for which this result is exact in the limit $d=\infty$.

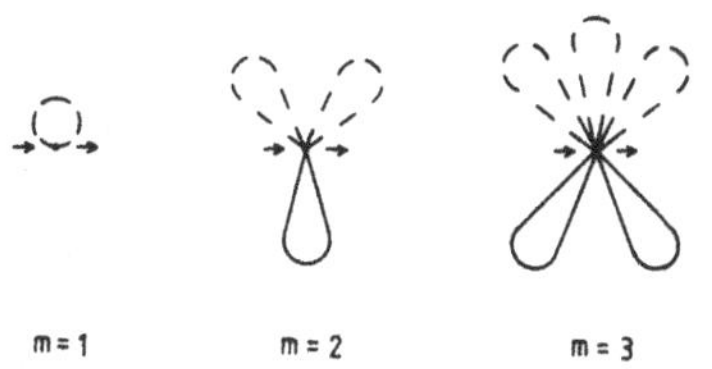

FIG. 2. Skeleton diagrams for the first three orders of the proper self-energy.

For $n=1$, the minimization shows that for all $U>0$ one has $m>0$ and $q<1$, i.e., there is a transition at $U=0$ to an antiferromagnetic insulator as is expected from the perfect nesting property. This is in contrast to earlier attempts to generalize the Gutzwiller approximation to the antiferromagnetic case.[13] The energy is always lower than that obtained by use of the Gutzwiller wave function or the Hartree-Fock results. For $U\rightarrow\infty$, $|\Psi_{AF}\rangle$ approaches the Néel state which is, in fact, the *exact* ground state of the half-filled Hubbard model in $d=\infty$ for large U.[14] [However, for smaller U neither $|\Psi_{AF}\rangle$ (12) nor the Gutzwiller wave function is the exact ground state in $d=\infty$, e.g., they yield only 85% of the exact second-order correlation energy E_2 (6).] For less than half-filling there exists a critical density $n_c \approx 0.85$ below which $m=0$ for all U, while for $n_c < n < 1$ there exists a bounded regime in U with $m > 0$. One finds $q < 1$ for all n and $U > 0$.

In summary, we have investigated the Hubbard model in the limit $d=\infty$ using perturbation theory and variational wave functions. Applying well known many-body techniques to the case $d=\infty$, we have shown that (i) in $d=\infty$ the diagrammatic calculation of ground-state properties is greatly simplified; (ii) at least in the weak-coupling limit, the correlation energy in $d=3$ is very close to that in $d=\infty$; (iii) the mean-field antiferromagnetic Hartree-Fock solution does *not* yield the exact results in $d=\infty$; (iv) calculations with variational wave functions of increasing refinement are analytically tractable. In this way the results of several well known approximations[8,12] used in finite-dimensional systems are recovered. The methods developed here for $d=\infty$ may equally be used to study correlation functions and other Hamiltonians such as the periodic Anderson model.[15] In particular, our results should also be valuable for the investigation of these models within a general Green's-function approach. Indeed, preliminary results for the Hubbard model in $d=\infty$, obtained by application of the above methods and ideas to the one-particle propagator, are very promising.[16]

We thank P. van Dongen and F. Gebhard for useful discussions.

[1]M. C. Gutzwiller, Phys. Rev. Lett. **10**, 159 (1963); J. Hubbard, Proc. Roy. Soc. London A **276**, 238 (1963); J. Kanamori, Prog. Theor. Phys. **30**, 275 (1963).

[2]E. H. Lieb and F. Y. Wu, Phys. Rev. Lett. **20**, 1445 (1968).

[3]See, for example, H. E. Stanley, in *Phase Transitions and Critical Phenomena*, edited by C. Domb and M. S. Green (Academic, London, 1974), Vol. 3, p. 485.

[4]U. Wolff, Nucl. Phys. **B225**, 391 (1983).

[5]W. Metzner and D. Vollhardt, Phys. Rev. B (to be published).

[6]W. Metzner and D. Vollhardt, Phys. Rev. Lett. **59**, 121 (1987), and Phys. Rev. B **37**, 7382 (1988).

[7]D. R. Penn, Phys. Rev. **142**, 350 (1966); M. Cyrot, J. Phys. (Paris) **33**, 125 (1972).

[8]M. C. Gutzwiller, Phys. Rev. **137A**, 1726 (1965).

[9]P. W. Anderson, Science **235**, 1196 (1987).

[10]D. Vollhardt, Rev. Mod. Phys. **56**, 99 (1984); D. Vollhardt, P. Wölfle, and P. W. Anderson, Phys. Rev. B **35**, 6703 (1987).

[11]H. Yokoyama and H. Shiba, J. Phys. Soc. Jpn. **56**, 3582 (1987).

[12]G. Kotliar and A. E. Ruckenstein, Phys. Rev. Lett. **57**, 1362 (1986).

[13]T. Ogawa, K. Kanda, and T. Matsubara, Prog. Theor. Phys. **53**, 614 (1975); F. Takano and M. Uchinami, Prog. Theor. Phys. **53**, 1267 (1975); J. Florencio and K. A. Chao, Phys. Rev. B **14**, 3121 (1976).

[14]T. Kennedy, E. H. Lieb, and B. S. Shastry, Phys. Rev. Lett. **61**, 2582 (1988).

[15]F. Gebhard, P. van Dongen, and D. Vollhardt, Rheinisch-Westfälische Technische Hochschule Report No. RWTH/ITP-C 12/88 (to be published); F. Gebhard and D. Vollhardt, Rheinisch-Westfälische Technische Hochschule Report No. RWTH/ITP-C 13/88 (to be published).

[16]E. Müller-Hartmann, private communication.

THE LINEARIZED HUBBARD MODEL: DYNAMICAL SUPERALGEBRA AND SUPERSYMMETRY

ARIANNA MONTORSI
Dipartimento di Fisica del Politecnico, 10129 Torino, Italy

MARIO RASETTI*
The Institute for Advanced Study, Princeton, NJ 08540, USA

and

ALLAN I. SOLOMON
Faculty of Mathematics, The Open University, Milton Keynes, UK

Received 20 September 1988

The dynamical superalgebras for the conventional Hubbard model and for its extension to a superlattice given by the union of two sublattices are considered. Self consistency equations are obtained for the relevant order parameters. Moreover the models are shown to be the linearized form of a class of supersymmetric models obtained from charges living in the fermionic sector of these superalgebras. The superconductive phase transition exhibited by these models is accounted for as a spontaneous breaking of supersymmetry.

1. Introduction

The recent discovery of a number of oxides unexpectedly exhibiting high superconductive transition temperatures[1] has revived theoretical interest in the Hubbard model, and variations or extensions thereof. Indeed both in the strong coupling to local spin configurations picture[2] and in the resonating valence bond state model,[3] superconductivity is inherent to a Mott insulator phase of the system,[4] and the partially filled Hubbard model[5] has been the natural candidate to consider with a view to finding a mechanism able to produce phase transition temperatures of the required magnitude.

The Hubbard model is in fact designed just to describe the main features of electron correlations in narrow energy bands in a tight-binding basis. In its conventional form it is defined by a hamiltonian of the form:

$$H_{\mathrm{Hub}} = \sum_{\mathrm{i},\sigma} \varepsilon_{\mathrm{i}} n_{\mathrm{i}\sigma} + \sum_{\mathrm{i}} U_{\mathrm{i}} n_{\mathrm{i}\uparrow} n_{\mathrm{i}\downarrow} + \sum_{\langle \mathrm{i},\mathrm{j}\rangle,\sigma} t_{\mathrm{ij}} a^{\dagger}_{\mathrm{i}\sigma} a_{\mathrm{j}\sigma} , \tag{1.1}$$

*Permanent address: Dipartimento di Fisica and Unitá CISM del Politecnico, Torino, Italy.

where $a^\dagger_{\mathbf{i}\sigma}$, $a_{\mathbf{i}\sigma}$ denote the fermionic creation and annihilation operators for spin $\sigma = \uparrow, \downarrow$ at site $\mathbf{i}$ of a d-dimensional lattice Λ, $n_{\mathbf{i}\sigma} = a^\dagger_{\mathbf{i}\sigma} a_{\mathbf{i}\sigma}$, and $t_{\mathbf{ij}} = t^*_{\mathbf{ji}}$ are nearest-neighbour (n.n.) hopping integrals. In (1.1) $\Sigma_{\langle \mathbf{i},\mathbf{j}\rangle}$ reminds us that the sum is restricted to n.n. pairs in Λ.

Some of the extensions of the model, proposed to enhance the possibility of a superconducting phase, include next-nearest-neighbour (n.n.n.) hopping terms as well as coulomb interactions between n.n. sites of the form $\Sigma_{\langle \mathbf{i},\mathbf{j}\rangle,\sigma,\sigma'} U'_{\mathbf{ij}} n_{\mathbf{i}\sigma} n_{\mathbf{j}\sigma'}$. We shall discuss terms of this and related forms in the second part of the paper. Initially we disregard them, in that our first approach is of a conventional mean-field-theory type, in which the former would simply imply a redefinition of the hopping parameters $t_{\mathbf{ij}}$ whereas the latter would require the inclusion in $t_{\mathbf{ij}}$ of a further contribution from the $U'_{\mathbf{ij}}$ as well as a redefinition of the site self-energies $\varepsilon_{\mathbf{i}}$. (This was shown[6] to be the case also in the framework of an effective tight-binding scheme.)

In this paper we intend to show how (1.1), when Fermi-linearized according to the scheme discussed in Ref. 7, exhibits a dynamical superalgebra, as does an extension of the model to a superlattice consisting of the sum of two sublattices, when dealt with in a generalized mean field scheme where a cluster of two sites (one from each sublattice) is treated exactly instead of a single site. Based on these results we conjecture that a more general model than (1.1) should be considered, whose hamiltonian be obtained from a conserved fermionic charge as a supersymmetric hamiltonian. It turns out that this can indeed be done, and that several possible candidates on one hand do correspond to hamiltonians considered in the literature, whose extra terms with respect to (1.1) belong to the categories of couplings mentioned above, and have interesting physical meaning; on the other hand they indicate that the superconductive phase transition might be ascribed to spontaneous breaking of supersymmetry.

In Sec. 2, after briefly reviewing the technique and results of Ref. 7, we apply them first to the model (1.1), then to the model extended to the superlattice. Self-consistency equations for the order parameters, both of *magnetic* and of *superconductive* type, are analyzed, leading to the determination of the critical temperature. Also the structure of the resulting superalgebra is briefly discussed. In Sec. 3 the generators of this superalgebra are used to construct supersymmetry charges whereby the most general supersymmetric hamiltonians are obtained. The role of vanishing order parameters at critical temperature in breaking supersymmetry is discussed. Section 4 gives a few concluding remarks and comments.

2. Dynamical Superalgebras of the Fermi-Linearized Models

2.1. *The Fermi linearization scheme*

We review here the linearization procedure first introduced in Ref. 7, which we refer to as the Fermi linearization scheme. Given a generic many-fermion hamiltonian

$$H = \sum_i \varepsilon_i a_i^\dagger a_i + \frac{1}{2} \sum_{i,j,k,l} \langle ij|V|\kappa l\rangle \, a_i^\dagger a_j^\dagger a_l a_k \,, \tag{2.1}$$

where the label i is a multi-index, [typically one has $i \equiv (\mathbf{i}, \sigma)$ with $\mathbf{i} \in \Lambda^\#$ and $\sigma \in \mathbf{Z}_2 \equiv \{\uparrow,\downarrow\}$, $\Lambda^\#$ denoting the wave-vector dual space associated with Λ] and $\{a_i, a_j\} = 0; \{a_i, a_j^\dagger\} = \delta_{i,j}$, the *conventional* linearization process goes as follows:

(a) One assumes that in the identity for operators A and B,

$$AB \equiv (A - \langle A\rangle)(B - \langle B\rangle) + \langle A\rangle\, B + A\,\langle B\rangle - \langle A\rangle\,\langle B\rangle\,;$$

where $\langle A\rangle$, $\langle B\rangle$ are c-numbers, $\langle\bullet\rangle$ denoting either expectation value in some suitable equilibrium quantum state or statistical mechanical average (e.g. in the Gibbs canonical ensemble), the terms $(A - \langle A\rangle)$ and $(B - \langle B\rangle)$ are *small* in some norm sense, so that their product can be neglected. Thus the identity above is replaced by the approximate form

$$AB \approx \langle A\rangle\, B + A\,\langle B\rangle - \langle A\rangle\,\langle B\rangle \qquad (*)\,.$$

(b) A and B are identified successively with all different pairs of commuting operators entering the four-fermion interaction terms of (2.1), and use of the approximate identity $(*)$ transforms H into a bilinear form, which can be recognized[8] as an element of a Lie algebra $\mathscr{L}$.

For instance the BCS theory is readily recovered retaining in H only terms of the form (i) (notice: we use here the following notation: if $\kappa \equiv (\mathbf{k}, \uparrow)$ then $-\kappa \equiv (-\mathbf{k}, \downarrow)$)

(i) $$\frac{1}{2} \sum_{i,j} \langle i, -i|V|j, -j\rangle \, a_i^\dagger a_{-i}^\dagger a_{-j} a_j;$$

while with little extra effort additional interactions of umklapp type may be included by retaining terms of the form

(ii) $$\frac{1}{2}\sum_{i,j}\langle i,j|V|-j,-i\rangle\, a_i^\dagger a_j^\dagger a_{-i}a_{-j}\,; \qquad 2(\mathbf{i}+\mathbf{j})\in L^{\#}\,;$$

where $L^{\#}$ is the reciprocal lattice of $\Lambda^{\#}$.

Upon selecting successively $A(=B^\dagger)$ equal to $a_i^\dagger a_{-i}^\dagger$ and $a_i^\dagger a_{-i}$ in (∗), and introducing the two order parameters

$$\Delta_\kappa := \frac{1}{2}\sum_j \langle \kappa,-\kappa|V|j,-j\rangle\langle a_j a_{-j}\rangle\,,$$

and

$$v_\kappa := \frac{1}{2}\sum_j \langle \kappa,j|V|-j,-\kappa\rangle\langle a_j^\dagger a_{-j}\rangle$$

the hamiltonian H is replaced by the new *reduced* one

$$H^{(\text{red})} = \sum_\kappa H_\kappa; \qquad H_\kappa = (\varepsilon_\kappa n_\kappa + \varepsilon_{-\kappa}n_{-\kappa}) + \{\Delta_\kappa a_\kappa^\dagger a_{-\kappa}^\dagger + v_\kappa a_\kappa^\dagger a_{-\kappa} + \text{h.c.}\}\,. \tag{2.2}$$

H_κ is readily seen to be an element of su(2) ⊕ su(2), so that it can be diagonalized in a straightforward way by a *Bogolubov* rotation in the associated group space, giving the familiar form

$$H_\kappa \mapsto \sqrt{\varepsilon_\kappa\varepsilon_{-\kappa}+|\Delta_\kappa|^2}\,(n_\kappa+n_{-\kappa}-1)+|v_\kappa|^2(n_\kappa-n_{-\kappa})\,, \tag{2.3}$$

where the two remaining operators in brackets are the diagonal elements of the two su(2)'s.

In Ref. 7 we extended the approximate identity (∗) to the case in which the operators A and B anticommute. The consistency of (∗) implies that $\langle A\rangle$ and $\langle B\rangle$ should anticommute both with Fermi operators and among themselves.

This may be achieved by considering them to belong to a Banach-Grassmann algebra $\mathscr{G}$. $\mathscr{G}$ has a natural gradation : $\mathscr{G}=\mathscr{G}_0\oplus\mathscr{G}_e$ in two non-intersecting subsets of elements which are products of odd or even numbers of anticommuting objects, respectively. (It should be noted that when the linearization procedure is applied to the hamiltonian H in (2.1) it can never happen that A equals B or $B^\dagger$. It is therefore *not* inconsistent to restrict the definition of $\mathscr{G}$ by requiring that $\langle A\rangle\langle A\rangle^*$ $(=-\langle A\rangle^*\langle A\rangle)$ and $\langle B\rangle\langle B\rangle^*$ $(=-\langle B\rangle^*\langle B\rangle)$ be *non-nilpotent* c-numbers.)

multiplied by elements of $\mathscr{G}_0$; this permits us to recognize the linearized Hamiltonian as an element of a Lie superalgebra $\mathscr{S}$.

Thus, for example, retaining other terms from the general hamiltonian (2.1), namely momentum non-conserving terms of the CDW-*umklapp* form:

$$\text{(iii)} \qquad \frac{1}{2}\sum_{i,\kappa} \langle i, -i|V|\kappa, i\rangle\, a_i^\dagger a_{-i}^\dagger a_i a_\kappa; \qquad (\mathbf{i}+\boldsymbol{\kappa}) \in L^\#;$$

$$\text{(iv)} \qquad \frac{1}{2}\sum_{i,\kappa} \langle i, -i|V|\kappa, -i\rangle\, a_i^\dagger a_{-i}^\dagger a_{-i} a_\kappa; \qquad (\mathbf{i}-\boldsymbol{\kappa}) \in L^\#,$$

the new resulting linearized hamiltonian was recognized in Ref. 7 to be an element of the superalgebra su(2|2). Moreover, it was realized that one could construct in the fermionic sector $\mathscr{F}\{\mathrm{su}(2|2)\}$ of this superalgebra an element Q, together with it conjugate $Q^\dagger$, whose square is zero and such that

$$H_\kappa = \{Q, Q^\dagger\}, \tag{2.4}$$

and

$$[H_\kappa, Q] = 0; \tag{2.5}$$

indicating that H_κ describes a spontaneously broken supersymmetric model (because Q does not annihilate the ground state). H_κ describes, however, a system with unbroken supersymmetry if the pairing order parameter Δ_κ vanishes. In other words, in the framework described, the superconducting phase transition may be considered as due to a spontaneous breaking of supersymmetry.

2.2. *The Fermi-linearized Hubbard model*

We apply now the Fermi-linearization scheme discussed in Sec. 2.1 to the Hubbard hamiltonian (1.1). The linearization is performed only on the hopping terms $a_{\mathrm{i},\sigma}^\dagger a_{\mathrm{j},\sigma}$ which are replaced by $\langle a_{\mathrm{i},\sigma}^\dagger\rangle a_{\mathrm{j},\sigma} + a_{\mathrm{i},\sigma}^\dagger \langle a_{\mathrm{j},\sigma}\rangle - \langle a_{\mathrm{i},\sigma}^\dagger\rangle\langle a_{\mathrm{j},\sigma}\rangle$. Upon defining

$$\vartheta \equiv \vartheta_{\mathrm{i},\sigma} = \sum_{\mathrm{j=n.n.(i)}} t_{\mathrm{ji}} \langle a_{\mathrm{j},\sigma}^\dagger\rangle \in \mathscr{G}_0; \tag{2.6}$$

and assuming further that $\vartheta_{\mathrm{i}\uparrow} = \vartheta_{\mathrm{i}\downarrow}$ (time reversal invariance), and that $\varepsilon_\mathrm{i} \equiv \varepsilon_{\mathrm{i},\sigma}$ is independent of σ, the linearized hamiltonian takes the form $H_{\mathrm{Hub}}^{(\mathrm{lin})} = \sum_\mathrm{i} H^{(\mathrm{i})}$ with

$$H^{(i)} = \varepsilon(n_+ + n_-) + Un_+n_- + \vartheta(a_+ + a_-) - \vartheta^*(a_+^\dagger + a_-^\dagger)\,; \tag{2.7}$$

where we dropped the index i and wrote ± for ↑, ↓.

$H^{(i)}$ is an element of a superalgebra $\mathcal{S}_i$ – hence $H_{\rm Hub}^{(\rm lin)} \in \mathcal{S} = \oplus_i \mathcal{S}_i$. $\mathcal{S}_i$ is the 16-dimensional superalgebra u(2|2) which can be thought of as the extension, by the Cartan element n_+n_-, of su(2|2). To be more precise, $H^{(i)}$ sits in a smaller superalgebra $\mathcal{A}$ generated by $\{n_+ + n_-, n_+n_-, a_+^\dagger a_- + a_-^\dagger a_+, \mathbf{1}; a_+ + a_-, a_+^\dagger + a_-^\dagger, n_+a_- + n_-a_+, n_+a_-^\dagger + n_-a_+^\dagger\}$, which is an 8-dimensional subalgebra of $\mathcal{S}_i$. $\mathcal{A}$ is isomorphic to the centralizer $C_D(\mathrm{u}(2|2)) = \mathrm{u}(1|1) \oplus \mathrm{u}(1|1)$, where $D = (n_+ + n_- + a_+a_-^\dagger + a_-a_+^\dagger)$. However, $\mathcal{A}$, unlike $\mathcal{S}_i$, does not contain in its bosonic sector the Lie subalgebra su(2) necessary to rotate $H^{(i)}$ to diagonal form in Fock space.

It is worth pointing out that $H^{(i)}$ is also an element of the Lie algebra su(4) under *commutation*, however since now the order parameters of the theory are anticommuting numbers in $\mathcal{G}$, it is more natural to consider $\mathcal{S}_i$ as the dynamical (super) algebra of the system.

Once more diagonalization is obtained by implementation of a generalized Bogolubov automorphism $\Phi_B : \mathcal{S}_i \to \mathcal{S}_i$. The procedure is more transparent if one performs it in two steps. First consider the adjoint action of the skew hermitian fermionic element $Z^{(f)} \in \mathcal{F}(\mathcal{S}_i)$

$$Z^{(f)} := \{\lambda(a_+ + a_-) + \mu(n_+a_- + n_-a_+)\} - \mathrm{h.c.}\,; \tag{2.8}$$

where λ and μ are indeterminates $\in \mathcal{G}_0$, on $H^{(i)}$. This rotates $H^{(i)}$ into a form which — selecting λ and μ appropriately — no longer contains fermionic terms. With $\lambda = -\vartheta\varepsilon$ and $\mu = c\vartheta$, where $c = U/\varepsilon(U+\varepsilon)$,
one obtains:

$$\exp(\mathrm{ad}\, Z^{(f)})(H^{(i)}) := H_b$$

$$= \varepsilon(n_+ + n_-) + Un_+n_- + c\vartheta\vartheta^*(n_+ + n_- + a_+a_-^\dagger + a_-a_+^\dagger) - \frac{2}{3}\vartheta\vartheta^*\mathbf{1}\,. \tag{2.9}$$

Notice that $\vartheta\vartheta^* \in \mathcal{G}_e$.

H_b is an element of the 3-dimensional *abelian* subalgebra generated by $\{n_+ + n_-, n_+n_-, a_+a_-^\dagger + a_-a_-^\dagger\}$, and its spectrum is easily obtained by conventional methods. It gives however a deeper insight in the structure of the system to rotate H_b into a form H_0 which is manifestly diagonal in the Fock basis $\{|n_+\rangle \oplus |n_-\rangle; n_\pm \in \mathbb{Z}_2\}$ by the adjoint action of the bosonic operator

$$Z^{(b)} := \varphi(a_+ a_-^\dagger - a_- a_+^\dagger)\,. \tag{2.10}$$

By selecting $\varphi = (2m + 1)\dfrac{\pi}{4}$; $m \in \mathbb{Z}$, we obtain:

$$\begin{aligned}\exp(\mathrm{ad}\, Z^{(b)})(H_b) &:= H_0 \\ &= \varepsilon(n_+ + n_-) + U n_+ n_- + 2c\vartheta\vartheta^* n_{(-1)^{m+1}} - \frac{2}{\varepsilon}\vartheta\vartheta^* \mathbf{1}\,.\end{aligned} \tag{2.11}$$

The spectrum is obviously independent of the choice of the integer m.

2.3. *The self-consistency equations*

We compute statistical-mechanical averages in the canonical ensemble:

$$\begin{aligned}\langle\bullet\rangle &:= \mathrm{tr}\{\bullet \exp[-\beta H^{(\mathrm{l})}]\}/\mathrm{tr}\{\exp[-\beta H^{(\mathrm{l})}]\} \\ &= \frac{\mathrm{tr}\{\exp(-\beta H_0)\exp(\mathrm{ad}\, Z^{(b)})(\exp(\mathrm{ad}\, Z^{(f)})(\bullet))\}}{\mathrm{tr}\{\exp(-\beta H_0)\}}\,,\end{aligned} \tag{2.12}$$

where $\beta = (\kappa_B T)^{-1}$, T denoting the temperature and κ_B the Boltzmann's constant. The model defined by (2.7) has no *superconductive* order parameter, in that $\langle a_+ a_- \rangle$ is identically zero. However it does have a *magnetic* order parameter which is interesting to consider. We evaluate first

$$\begin{aligned}&\exp(\mathrm{ad}\, Z^{(b)})(\exp(\mathrm{ad}\, Z^{(f)})[\vartheta(a_+ + a_-)]) \\ &= 2\vartheta\vartheta^*\{-\frac{1}{\varepsilon}\mathbf{1} + c n_{(-1)^{m+1}}\} + \sqrt{2}\,\vartheta(-1)^{\left[\!\left[\frac{m+1}{2}\right]\!\right]} a_{(-1)^m}\end{aligned} \tag{2.13}$$

whence, recalling that a_+, a_- have vanishing trace in the basis in which H_0 is diagonal,

$$\langle\vartheta(a_+ + a_-)\rangle = -\langle\vartheta^*(a_+^\dagger + a_-^\dagger)\rangle^* = -\frac{2\vartheta\vartheta^*}{\varepsilon(U+\varepsilon)}\left\{\varepsilon + \frac{U}{\mathcal{Z}_r}[1 + \exp(-\beta\varepsilon)]\right\}, \tag{2.14}$$

where

$$\mathcal{Z}_r = 1 + \exp(-\beta\varepsilon) + \exp(-2\beta c\vartheta\vartheta^*)[\exp(-\beta\varepsilon) + \exp(-\beta(2\varepsilon + U))], \tag{2.15}$$

is the *reduced* partition function.

Self-consistency is imposed by requiring that (2.6) is satisfied. For example in the case in which complete n.n. symmetry is assumed (i.e. $t_{ij} = t/q$, $\forall \mathbf{i}$, and $\forall \mathbf{j}$ = n.n. (i) in Λ, q coordination number and $\langle a_{i\sigma}\rangle$ independent of $\mathbf{i}$ and σ), it reads $\langle a_+ + a_-\rangle = 2\vartheta^*/t$. Notice that in this case the result no longer depends on the dimensionality. Upon setting

$$\kappa := \frac{(U+\varepsilon)(t+\varepsilon)}{tU}, \tag{2.16}$$

one gets

$$\vartheta\vartheta^* = -\frac{1}{2\beta c}\ln\left\{\frac{\kappa[1+\exp(-\beta\varepsilon)]}{(1-\kappa)\exp(-\beta\varepsilon)[1+\exp(-\beta(U+\varepsilon))]}\right\}, \tag{2.17}$$

for $\beta > \beta c$ and $\vartheta\vartheta^* = 0$ for $\beta < \beta_c$, where β_c is the solution of the equation

$$\frac{1+\exp(-\beta_c(U+\varepsilon))}{1+\exp(\beta_c\varepsilon)} = \frac{\kappa}{1-\kappa}. \tag{2.18}$$

Figure 1 shows a plot of $\vartheta\vartheta^*$ vs. $T\,(T_c := (\kappa_B\beta_c)^{-1})$. One can see that $\vartheta\vartheta^*$ has indeed the features of a magnetic order parameter.

The behaviour of the corresponding specific heat is given in Fig. 2, which is straightforwardly obtained by derivation with respect to T of the internal energy

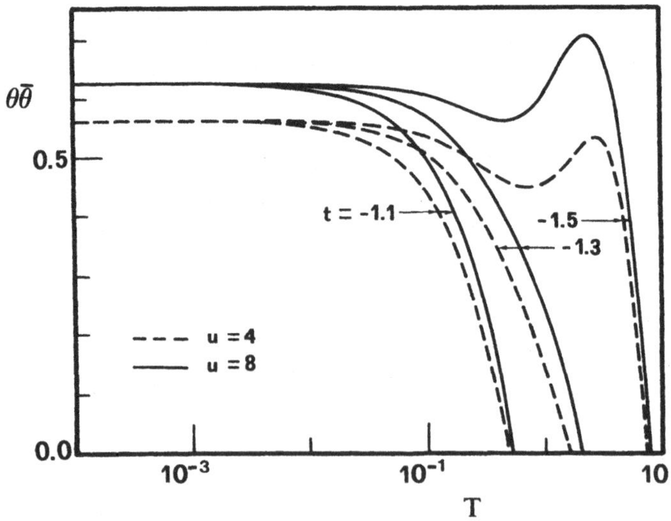

Fig. 1. $\vartheta\vartheta^*$ vs. T (semilogarithmic scale) for $U = 4$, 8, and $t = -1.1$, -1.3, -1.5. Notice that T_c depends strongly on t and weakly on U (units $\varepsilon = 1$).

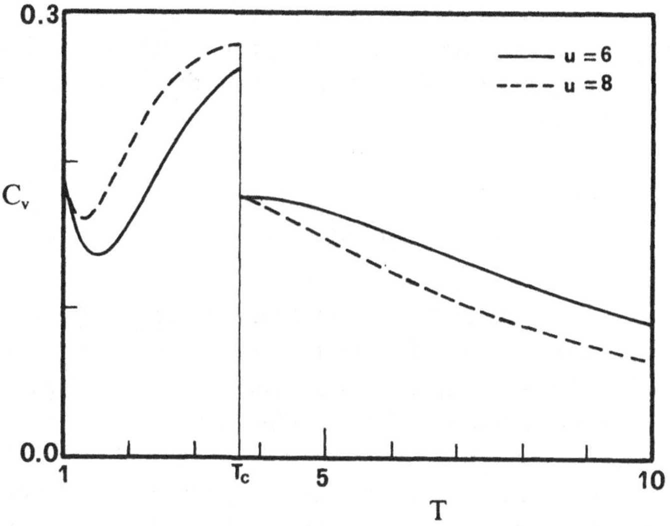

Fig. 2. C_v vs. T for $t = -1.4$, and $U = 6, 8$.

$\mathscr{U} = \langle H^{(i)} \rangle \equiv \exp(-\beta\varepsilon)N_u/D_u$, where — upon defining $\mathscr{L}_\beta$ by writing (2.17) as $\vartheta\vartheta^* := -(2\beta c)^{-1} \ln \mathscr{L}_\beta$ —

$$N_u = \begin{cases} \varepsilon + \mathscr{L}_\beta[\varepsilon + (U+2\varepsilon)\exp(-\beta(U+\varepsilon)) - \kappa_B T \ln \mathscr{L}_\beta(1+\exp(-\beta(U+\varepsilon))], \\ \qquad\qquad\qquad\qquad\qquad\qquad \text{if } \beta > \beta_c; \\ 2\varepsilon + (U+2\varepsilon)\exp(-\beta(U+\varepsilon)), \quad \text{if } \beta < \beta_c; \end{cases} \tag{2.19}$$

and

$$D_u = \begin{cases} 1 + \exp(-\beta\varepsilon)[1 + \mathscr{L}_\beta(1 + \exp(-\beta(U+\varepsilon)))], & \text{if } \beta > \beta_c; \\ 1 + 2\exp(-\beta\varepsilon) + \exp(-\beta(U+2\varepsilon)), & \text{if } \beta < \beta_c. \end{cases} \tag{2.20}$$

For completeness we recall that

$$\exp(\operatorname{ad} Z^{(b)})\,(\exp(\operatorname{ad} Z^{(f)})\,(a_+ a_-)) = a_+ a_-\,, \tag{2.21}$$

implying that $\langle a_+ a_- \rangle = 0$, as already stated.

2.4. *The linearized Hubbard model on a superlattice*

We think now of Λ as a superlattice, a disjoint sum of two sublattices Λ_E and Λ_O each of whose sites has as its n.n.'s sites belonging to the other sublattice. E and O stand for *even* and *odd* respectively, and we shall often use the same symbols to denote the whole sets of indices labeling sites in the corresponding sublattice: this will permit us to use a notation similar to a one-dimensional case in the theory that we shall develop in this section, even though the theory itself holds in any number of dimensions. With

$$\Lambda = \bigcup_{\alpha \in \{E, O\}} \Lambda_\alpha , \qquad (2.22)$$

the model we consider here is defined by a hamiltonian which is the sum of:

(i) a Hamiltonian $H_\alpha^{(\mathrm{Hub})}$ (see (1.1)) for each sublattice (α = E, O);

(ii) a coupling between sublattices which consists of a hopping term with equal amplitudes for both parallel-spin and spin-flip processes.

The approach we use is once more Fermi linearization, restricted, however, to the two sublattice terms, whereas the coupling term is treated *exactly*. In other words, we adopt a *cluster* mean field approach, namely one in which the set (cluster) of two n.n.'s is dealt with exactly, while the coupling of the two sites in the cluster with the rest of the lattice sites is handled in the approximate (i.e. linearized) way discussed in Sec. 2.2. Thus we are left with a cluster hamiltonian of the form:

$$H_c = H_\mathrm{E}^{(l)} + H_\mathrm{O}^{(l)} + \tau A_\mathrm{O}^\dagger A_\mathrm{E} + \tau^* A_\mathrm{E}^\dagger A_\mathrm{O} ; \qquad (2.23)$$

where

$$A_\alpha := a_{\alpha\uparrow} + a_{\alpha\downarrow} ; \qquad (2.24)$$

and

$$H_\alpha^{(l)} = \varepsilon_\alpha N_\alpha + U_\alpha P_\alpha + \vartheta_\alpha A_\alpha - \vartheta_\alpha^* A_\alpha^\dagger ; \qquad (2.25)$$

with

$$N_\alpha := n_{\alpha\uparrow} + n_{\alpha\downarrow}; \qquad P_\alpha := n_{\alpha\uparrow} n_{\alpha\downarrow} . \qquad (2.26)$$

The hamiltonians $H_\alpha^{(l)}$ are the linearized Hubbard hamiltonians for each sublattice, τ is a complex c-number describing the hopping amplitude between the two sublattices, and $\vartheta_\alpha \in \mathscr{G}_0$. The index α assumes values E and O.

It is worth pointing out that the possible difference in hopping amplitude between parallel-spin and spin-flip processes can be thought of as taken care of by ϑ_α in the frame of the Fermi linearization scheme.

Once more the hamiltonian H_c is an element of a superalgebra, $\tilde{\mathscr{S}}$, generated by

$\{\mathcal{A}_E, \mathcal{A}_O, A_O^\dagger A_E, A_E^\dagger A_O\}$. $\tilde{\mathcal{S}} \sim u(2|1) \oplus u(2|1) \oplus u(2|1) \oplus u(2|1)$ is a subalgebra of $u(8|8)$. The latter contains the graded algebra necessary to perform the generalized Bogolubov transformation which diagonalizes H_c in Fock space.

Upon defining, besides the operators given in (2.24) and (2.26), the new operators

$$C_\alpha := a_{\alpha\uparrow} a_{\alpha\downarrow}^\dagger + a_{\alpha\downarrow} a_{\alpha\uparrow}^\dagger, \qquad B_\alpha := n_{\alpha\uparrow} a_{\alpha\downarrow} + n_{\alpha\downarrow} a_{\alpha\uparrow}, \tag{2.27}$$

and

$$D_\alpha := N_\alpha + C_\alpha \tag{2.28}$$

one can check the following commutation and anticommutation relations (we write them without indices, it being implied that fermionic operators with different indices anticommute among themselves, and fermionic operators commute with bosonic operators of different index):

$$\begin{array}{lll} \{A,A^\dagger\} = 2I; & \{B,B^\dagger\} = D; & \{A,B^\dagger\} = D; \\ \{A,A\} = 0; & \{B,B\} = 0; & \{A,B\} = 0; \\ [A,C] = -A; & [A,N] = A; & [A,P] = B; \\ [B,C] = -B; & [B,N] = B; & [B,P] = B; \\ [D,\bullet] = 0; & [P,N] = 0. & \end{array} \tag{2.29}$$

The diagonalization of H_c proceeds as for the case discussed in Sec. 2.2, with one minor yet extremely relevant difference. All the factors containing $\langle A_E \rangle \langle A_O \rangle$, $\langle A_E \rangle \langle A_O^\dagger \rangle$ and their conjugates are no longer eliminated by the anticommutation relations, and terms of the form $\vartheta_E \vartheta_O$, $\vartheta_E \vartheta_O^\dagger$ and their conjugates therefore remain in play. We do not present here the calculations — which are much longer and more cumbersome than in the case of the Hubbard model on a single lattice — in detail, but simply report the results, which we believe are very instructive from the physical point of view.

The hamiltonian *rotated* to a Cartan subalgebra is

$$H_0 = \sum_{\substack{\{E,O\} \\ (\mathrm{mod}\ 2)}} \left\{ (\varepsilon_\alpha N_\alpha + \tilde{U}_\alpha P_\alpha + d_\alpha D_\alpha) + \frac{1}{2} \delta D_\alpha D_{\alpha+1} \right.$$

$$\left. + (-)^\alpha \frac{1}{2} \gamma P_\alpha D_{\alpha+1} + \frac{1}{2} \eta N_\alpha D_{\alpha+1} \right\} + eI; \tag{2.30}$$

where d_α, δ, $\tilde{U}_\alpha$, γ, η and e are simple functions of ε_α, U_α, τ, τ^* and $\vartheta_\alpha \vartheta_\beta^*$, $\forall \alpha$ and β.

What makes this particular model interesting is that even though there are still only two *primitive* order parameters ϑ_α (α = E, O) available to describe the theory, and two self-consistency equations which determine them as functions of T, the model now exhibits a *pairing phase* in addition to the previous magnetic one. More precisely, in a phase for which $\vartheta_\alpha \vartheta_\alpha^*$ is not identically zero for both α = E and α = O, we would find that

(a) $\langle a_{E\sigma} a_{O\sigma'}^\dagger \rangle = \langle a_{O\sigma'} a_{E\sigma}^{\dagger\dagger} \rangle^* \neq 0 \; \forall \sigma, \sigma'$, since it is proportional to $\vartheta_E \vartheta_O^*$ $(= [\vartheta_O \vartheta_E^*]^*)$;

(b) $|\langle a_{E\sigma} a_{O\sigma'} \rangle| \neq 0$ and $|\langle a_{\alpha\sigma} a_{\alpha\sigma'}^\dagger \rangle| \neq 0 \; \forall \sigma, \sigma'$ and $\forall \alpha$, in that they are proportional to $(|\vartheta_E \vartheta_E^* \vartheta_O \vartheta_O^*|)^{1/2}$, and $[(\vartheta_\alpha \vartheta_\alpha^*)^2]^{1/2}$ respectively.

In other words, the model appears to show a simultaneous antiferromagnetic and superconductive phase transition. Detailed analysis of the critical temperature as well as of the behaviour of the specific heat for this case will be given elsewhere.[9]

3. Supersymmetric Extensions of the Hubbard Model

In this section we shall deal with $d = 2$ only, showing how, in this case, there is a particular class of superlattices of the sort introduced in Sec. 2.4 naturally lending themselves to the construction of new hamiltonians which have the following features:

(i) the hamiltonians are supersymmetric in the sense of Eqs. (2.4) and (2.5) (whether or not the supersymmetry is exact or spontaneously broken depends on the ground state);

(ii) the conserved fermionic supercharges whereby the hamiltonians are constructed are elements of the fermionic sector of the dynamical superalgebras introduced in Secs. (2.2) and (2.4), or obvious generalizations thereof.

It is interesting to point out that the structure of a lattice realizing the above scheme is indeed that of most of the known compounds (essentially copper oxides with different rare earth additions, all appearing as a stack of weakly coupled two-dimensional planes of Cu and O atoms) exhibiting high-T_c superconductive phase transitions.

Let Λ_2 be a 2-d square lattice, whose sites are labelled by pairs of integers $(i|j)$ denoting their cartesian coordinates in units of lattice spacings in the x and y directions respectively, with the property that:

(a) (*odd*|*odd*) sites $[i = 1, j = 1 \pmod 2]$ are *empty*;

(b) (*odd*|*even*) and (*even*|*odd*) sites $[i + j = 1 \pmod 2]$ are occupied by one type of fermion (e.g. oxygen electrons, in the states described by the orbitals $O(2p_y)$ and $O(2p_x)$ respectively);

(c) (*even*|*even*) sites $[i = 0, j = 0 \pmod 2]$ are occupied by a different type of fermion (e.g. copper electrons, in the state represented by the orbital $Cu(3d_{x^2-y^2})$).

Such a lattice is shown schematically in Fig. 3. One can see that it belongs to the class of superlattices defined in Sec. 2.4, with the additional feature that one of the sublattices has only half of its sites occupied.

We now construct the fermionic charge $Q \in \mathscr{F}(\langle \cup_{c \in \Lambda_2} \mathscr{F}[\mathrm{u}(8|8)_c] \rangle)$

$$Q = \sum_{\substack{i=0,\, j=0 \pmod 2 \\ \sigma}} \Big\{ n_{(i|j),\sigma} \sum_{\sigma'} [\alpha_{\sigma'}(a_{(i|j\pm 1),\sigma'} + a_{(i\pm 1|j),\sigma'}) + \text{h.c.}]$$

$$+ [\beta_\sigma(n_{(i|j+1),\sigma} a_{(i|j+1),-\sigma} + n_{(i+1|j),\sigma} a_{(i+1|j),-\sigma}) + \text{h.c.}] \qquad (3.1)$$

$$+ [\lambda_\sigma a_{(i|j),\sigma} + \mu_\sigma(a_{(i+1|j),\sigma} + a_{(i|j+1),\sigma}) + \text{h.c.}] \Big\} ;$$

where $\alpha_\sigma, \beta_\sigma, \lambda_\sigma, \mu_\sigma$ and their conjugates are complex numbers yet to be determined in such a way as to get the desired hamiltonian.

In doing explicitly the calculations, it turns out to be slightly simpler to adopt the convention of assuming Q hermitian (i.e. $Q^\dagger = Q$) and $H = Q^2$ — different from (2.4), (2.5), but obviously equivalent. Two cases appear to be interesting in the sense described in the introduction, namely of obtaining hamiltonians which differ from the Hubbard hamiltonian or its extension to the superlattice discussed in Sec. 2.4 as little as possible, and at most by some physically meaningful factors. They correspond to the following choices:

(1) $\beta_\sigma = i\kappa\alpha_{-\sigma}$ with κ a real constant;

(2) the phase of β_σ equals the phase of α_σ and $\alpha_\sigma = \alpha_\sigma(l|j)$ is site-dependent ((*even*|*even*) sites only) : $\alpha_\sigma(l+2|j) = i\alpha_\sigma(l|j)$, from which it follows that $\alpha_\sigma(l|j) = \exp(i\,\frac{\pi}{4}(l+j))\alpha_\sigma$, where by α_σ we shall henceforth mean only $\alpha_\sigma(0|0)$ (and analogously for β_σ).

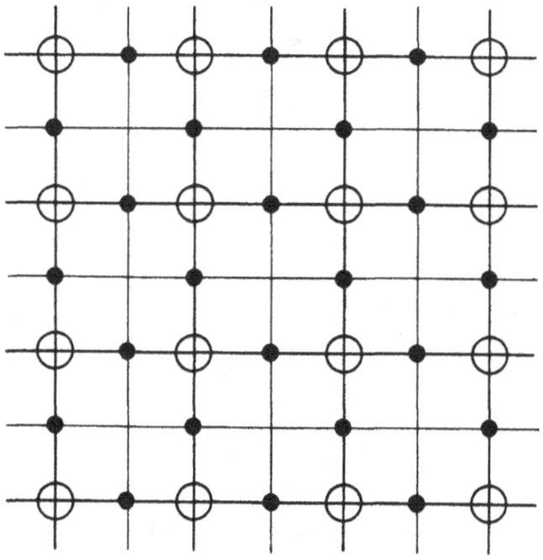

Fig. 3. The lattice Λ_2.

In case (1) the resulting hamiltonian is the sum of eight contributions of different types:

$$H_{(1)} = \sum_{\substack{i=0,\, j=0 (\mathrm{mod}\, 2) \\ \sigma}} \sum_{l=1}^{8} \mathcal{H}^{(l)}_{(i|j),\sigma} \tag{3.2}$$

where:

$$\mathcal{H}^{(1)}_{(i|j),\sigma} = \varepsilon_{ee} n_{(i|j),\sigma} \tag{3.3}$$

with

$$\varepsilon_{ee} = \sum_{\sigma'} (2|\alpha_{\sigma'}|^2 + \alpha_{\sigma'}\mu^*_{\sigma'} + \alpha^*_{\sigma'}\mu_{\sigma'}) \tag{3.4}$$

representing the self-energy of (*even*|*even*) sites;

$$\mathcal{H}^{(2)}_{(i|j),\sigma} = \varepsilon_{eo}(n_{(i|j\pm 1),\sigma} + n_{(i\pm 1|j),\sigma}) \tag{3.5}$$

with

$$\varepsilon_{eo} = \beta_\sigma \mu^*_{-\sigma} + \beta^*_\sigma \mu_{-\sigma} + |\beta_\sigma|^2 \tag{3.6}$$

representing the self-energy of (*even*|*odd*) and (*odd*|*even*) sites;

$$\mathcal{H}^{(3)}_{(i|j),\sigma} = U_{ee} n_{(i|j)\uparrow} n_{(i|j)\downarrow} \tag{3.7}$$

with

$$U_{ee} = \sum_{\sigma'} 4|\alpha_{\sigma'}|^2 \tag{3.8}$$

representing the on-site Coulomb repulsion at (*even*|*even*) sites;

$$\mathcal{H}^{(4)}_{(i|j),\sigma} = \sum_{\sigma'} V^{(\sigma'\sigma')}_{ee} n_{(i|j),\sigma}(n_{(i|j\pm 2),\sigma'} + n_{(i\pm 2|j),\sigma'}) \tag{3.9}$$

with

$$V^{(\sigma\sigma')}_{ee} = \alpha^*_\sigma \alpha_{\sigma'} + \alpha_\sigma \alpha^*_{\sigma'} \tag{3.10}$$

representing the Coulomb repulsion between n.n. (*even*|*even*) sites;

$$\mathscr{H}^{(5)}_{(i|j),\sigma} = \sum_{\sigma'} [t^{(\sigma\sigma')}_{\text{eeo}} a_{(i|j),\sigma}(a^{\dagger}_{(i\pm 1|j),\sigma'} + a^{\dagger}_{(i|j\pm 1),\sigma'}) + \text{h.c.}] \tag{3.11}$$

with

$$t^{(\sigma\sigma')}_{\text{eeo}} = \lambda_{\sigma}\alpha_{\sigma'} \tag{3.12}$$

representing the hopping amplitude between n.n. (*even*|*even*) and (*even*|*odd*) (or (*odd*|*even*)) sites;

$$\mathscr{H}^{(6)}_{(i|j),\sigma} = \sum_{\sigma'} [p^{(\sigma\sigma')}_{\text{eeo}} a_{(i|j),\sigma}(a_{(i\pm 1|j),\sigma'} + a_{(i|j\pm 1),\sigma'}) + \text{h.c.}] \tag{3.13}$$

with

$$p^{(\sigma\sigma')}_{\text{eeo}} = -\lambda_{\sigma}\alpha^{*}_{\sigma'} \tag{3.14}$$

representing the *pairing* between n.n. (*even*|*even*) and (*even*|*odd*) (or, again, (*odd*|*even*)) sites;

$$\mathscr{H}^{(7)}_{(i|j),\sigma} = \pi_{\text{eo}}(a_{(i\pm 1|j)\uparrow}a_{(i\pm 1|j)\downarrow} + a_{(i|j\pm 1)\uparrow}a_{(i|j\pm 1)\downarrow}) + \text{h.c.} \tag{3.15}$$

with

$$\pi_{\text{eo}} = \mu_{\uparrow}\beta_{\uparrow} - \mu_{\downarrow}\beta_{\downarrow} \tag{3.16}$$

representing the amplitude for on-site pair formation on (*even*|*odd*) and (*odd*|*even*) sites; and

$$\mathscr{H}^{(8)}_{(i|j),\sigma} = s_{\text{eo}}(a_{(i\pm 1|j)\uparrow}a^{\dagger}_{(i\pm 1|j)\downarrow} + a_{(i|j\pm 1)\uparrow}a^{\dagger}_{(i|j\pm 1)\downarrow}) + \text{h.c.} \tag{3.17}$$

with

$$s_{\text{eo}} = \mu_{\uparrow}\beta^{*}_{\uparrow} + \mu^{*}_{\downarrow}\beta_{\downarrow} + \beta_{\downarrow}\beta^{*}_{\uparrow} \tag{3.18}$$

representing the on-site spin-flip hopping amplitude on (*even*|*odd*) and (*odd*|*even*) sites. One can easily check that the terms $\mathscr{H}^{(l)}_{(i|j),\sigma}$ with l up to 5 do indeed add up to a Hubbard hamiltonian. It may also be of some interest to point out a few features of the coefficients : ε_{ee} and ε_{eo} are different from each other; $V^{(\sigma\sigma')}_{\text{ee}}$

couples both parallel- and antiparallel-spin particles. Moreover, among the extra terms generated, $p_{\rm eeo}^{(\sigma\sigma')}$ could be eliminated by selecting $\lambda_\sigma = 0$ (however this would make also $t_{\rm eeo}^{(\sigma\sigma')}$ vanish), whereas $\pi_{\rm eo}$ is zero if both β_σ and μ_σ are chosen to be independent of σ. It is worth noticing that $\pi_{\rm eo}$ and $s_{\rm eo}$ would be simultaneously eliminated if one allowed $\varepsilon_{\rm eo}$ to be zero (i.e. if one could consider the self-energy of (*even*|*odd*) sites as negligible).

In case (2) the resulting hamiltonian has, with respect to (3.2), two additional contributions and one missing. The last is the term representing Coulomb repulsion between n.n. (*even*|*even*) sites. Thus we can write:

$$H_{(2)} = H_{(1)}|_{V=0} + \sum_{\substack{i=0,\, j=0 \,(\mathrm{mod}\ 2)\\ \sigma}} \sum_{l=9}^{10} \mathscr{H}^{(l)}_{(i|j),\sigma}\,; \tag{3.19}$$

where

$$\mathscr{H}^{(9)}_{(i|j),\sigma} = \sum_{\sigma'} U_{\rm eeo}\, n_{(i|j),\sigma}\,(n_{(i\pm 1|j),\sigma'} + n_{(i|j\pm 1),\sigma'}) \tag{3.20}$$

with

$$U_{\rm eeo} = \alpha_{\sigma'}\beta^{*}_{-\sigma'} + \alpha^{*}_{\sigma'}\beta_{-\sigma'} \tag{3.21}$$

representing the Coulomb repulsion between n.n. (*even*|*even*) and (*even*|*odd*) (or (*odd*|*even*)) sites, and

$$\mathscr{H}^{(10)}_{(i|j),\sigma} = \sum_{\sigma'} \tau_{\rm eo}\, n_{(i|j),\sigma'}\,(a_{(i\pm 1|j),\sigma} a^{\dagger}_{(i\pm 1|j),-\sigma} + a_{(i|j\pm 1),\sigma} a_{(i|j\pm 1),-\sigma} + \mathrm{h.c.}) \tag{3.22}$$

with

$$\tau_{\rm eo} = \alpha_{\sigma'}\beta^{*}_{\sigma'} + \alpha^{*}_{\sigma'}\beta_{\sigma'} \tag{3.23}$$

representing the amplitude for processes of on-site hopping (at sites (*even*|*odd*) and (*odd*|*even*)) with spin-flip, controlled by the occupancy of the n.n. (*even*|*even*) sites.

$H_{(2)}$ includes all the terms of the conventional extended Peierls-Hubbard model[5,10] as well as of the expanded version of the latter due to Kivelson, Su, Schrieffer and Heeger,[11] and most of the terms of the further generalized version of these proposed by Campbell, Tinka Gammel and Loh.[12] This is discussed in much greater detail in Ref. 9, in particular in view of how it bears on the possible diagonalization of models including off-diagonal Coulomb terms, whose importance in understanding many-body systems characterized by strong electron-electron interactions has long been recognized.

Here we simply want to emphasize the following considerations. If linearization of $H_{(1)}$ is performed (fermionic for the factors $\mathcal{H}^{(l)}_{(i|j),\sigma}$ $l = 6,7,8$, bosonic for $\mathcal{H}^{(4)}_{(i|j),\sigma}$), one gets a cluster mean field-model — with hamiltonian $\in u(8|8)$ — as in Sec. 2.4. On the other hand, if the cluster mean-field approach is applied directly to the model defined by $H_{(1)}$, namely if the mutual interaction of a pair of (*even*|*odd*) — (*even*|*even*) sites is taken into account exactly, whereas all other interactions along the bonds connecting the cluster with the rest of Λ are treated in linearized form (fermionic or bosonic, depending on their form), then one gets the same model *plus* two extra-terms, bosonic, one proportional to C_α, the other a pairing term proportional to $A_O A_E$ + h.c..

Thus, not only can diagonalization proceed in the way discussed above, but one can expect a thermodynamic phase with two-site order parameters $\langle a_{\alpha\uparrow} a_{\beta\downarrow}\rangle$ and $\langle a_{\alpha\uparrow} a^\dagger_{\beta\downarrow}\rangle$ and their conjugates both different from zero. Moreover, one can infer from the form of $H_{(1)}$ that when such order parameters vanish, then Q could annihilate the vacuum. In other words, a spontaneous breaking of supersymmetry would be associated with the appearance of a phase which shows both pairing and magnetic order.

4. Conclusions

We have shown that the Fermi linearized form of the Hubbard model and of its extension to a superlattice, as well as of its generalizations obtained by constructing supersymmetric models out of charges living in the fermionic sector of the dynamical superalgebras of these, exhibit the existence of a superconductive-magnetic phase whose appearance is associated with spontaneous supersymmetry breaking. Also, these same models can be easily diagonalized by means of algebraic techniques, made natural by the simple structure of the dynamical superalgebra. We suggest that spontaneous supersymmetry breaking could be a possible mechanism to account for high T_c superconductive phase transition in systems, such as the ceramic rare earth doped copper oxides, well described by Hubbard-like hamiltonians. Work is in progress along these lines.

Acknowledgements

One of the authors (M.R.) wishes to thank the Director and the Faculty of the Institute for Advanced Study, Princeton, NJ for the kind hospitality extended to him during the completion of this work.

References

1. J. G. Bednorz and K. A. Müller, *Z. Phys.* **B64** (1986), 189; C. W. Chu, P. H. Hor, R. L. Meng, L. Gao, Z. J. Huang, and Y. Q. Wang, *Phys. Rev. Lett.* **58** (1987) 405.
2. V. J. Emery, *Phys. Rev. Lett.* **58** (1987) 2794.
3. P. W. Anderson, *Science* **235** (1987) 1196; P. W. Anderson, G. Baskaran, Z. Zou, and T. Hsu, *Phys. Rev. Lett.* **58** (1987) 2790.
4. N. F. Mott, *Proc. Phys. Soc.* **A62** (1949) 416; *Canad. J. Phys.* **34** (1956) 1356 and *Phil. Mag.* **6** (1961) 287.
5. J. Hubbard, *Proc. Roy. Soc. London* **A276** (1963) 238.
6. L. F. Mattheiss, *Phys. Rev. Lett.* **58** (1987) 1028.
7. A. Montorsi, M. Rasetti and A. I. Solomon, *Phys. Rev. Lett.* **59** (1987) 2243.
8. A. I. Solomon, *J. Phys.* **A14** (1981) 2177.
9. A. Montorsi, M. Rasetti and A. I. Solomon, submitted to *Phys. Rev. Lett.* (Dec. 1988).
10. R. G. Parr, *J. Chem. Phys.* **20** (1952) 1499; R. Pariser and R. G. Parr, *J. Chem. Phys.* **21** (1953) 767; J. A. People, *Proc. Roy. Soc. London* **A68** (1955) 81.
11. S. Kivelson, W. P. Su, J. R. Schrieffer, and A. J. Heeger, *Phys. Rev. Lett.* **58** (1987) 1899.
12. D. K. Campbell, J. Tinka Gammel, and E. Y. Loh, Jr., "The extended Peierls Hubbard model: Off-diagonal terms", in *Proc. of the Int. Conference on Synthetic Metals*, 1988, to be published in *Synthetic Metals*.

Progress of Theoretical Physics, Vol. 30, No. 3, September 1963

Electron Correlation and Ferromagnetism of Transition Metals

Junjiro KANAMORI

Department of Physics
Osaka University, Osaka

(Received May 14, 1963)

The electron correlation in a narrow energy band is discussed taking into account the multiple scattering between two electrons. The discussion is an adaptation of Brueckner's theory of nuclear matter. It is assumed that electrons interact with each other only when they are at the same atom. The effect of the electron correlation depends in an intricate way on the energy spectrum of a given band. An approximate expression of the effective magnitude of the interaction is derived. The condition for the occurrence of ferromagnetism is investigated for various types of bands. The ferromagnetism of Ni and the paramagnetism of Pd can be understood reasonably through the present approach The degeneracy of the *d* bands is taken into account in the discussion of these metals.

§ 1. Introduction

The purpose of this paper is to present a semi-quantitative discussion of the electron correlation and its bearing on the ferromagnetism of transition metals. Although the importance of the role of the electron correlation in metallic ferromagnetism is emphasized by many authors,[1),2)] all explicit calculations have been confined to the free electron model. This model, however, is hardly justified for the *d* electrons. The present approach is essentially based on the contrasting *tight binding* approximation, which may be justified for metals near the end of the transition metal series. The discussion is intended to clarify the role of the intra-atomic exchange interaction of the *d* electrons. Ni and Pd will be the main subjects of this discussion.

In 1936 Slater[3)] discussed the ferromagnetism of Ni by use of the Hartree-Fock approximation in calculating the interaction energy of the *d* electrons, concluding that the origin of the ferromagnetism is in the intra-atomic exchange interaction. His theory has been criticized since then by several authors particularly on the fact that the electron correlation is not taken into account in the calculation. It was argued, for example by Wohlfarth[2)] that the electron correlation reduced the intra-atomic interaction to the extent that it was less important than the inter-atomic interaction. On the other hand, Slater showed that the necessary amount of the exchange interaction to produce the ferromagnetism is of the order of the intra-atomic exchange interaction between two *d* electrons in *different* atomic *d* orbitals, which is characterized by the exchange integral of the order of 0.6 ev. If, however, the Hartree-Fock approximation is honestly applied, the exchange interaction between two d band electrons involves the Coulomb self-energy of an atomic d orbital, which is of the order of 10 ev. This arises from the fact that two band electrons having antiparallel spins can enter into the same atomic orbital, while those with parallel spins cannot. The importance of the intra-atomic interaction, therefore, depends critically on how and to what extent this Coulomb self-energy effect is reduced by the electron correlation. We calculate this reduction by taking into account the multiple scattering between two electrons. The calculation is essentially an adaptation of Brueckner's theory of nuclear matter and He^3.[4] When the number of electrons is small, the main feature of the electron correlation is the modification of electron trajectories by the two-body collision. The present approach takes into account this effect fully, and it will serve to semi-quantitative understanding of the electron correlation effect even when the electron number is moderately high.

The s electrons, which are free electrons to good approximation, will shield the interaction between two d electrons to a considerable extent, when the d electrons are at different atoms. It is assumed in the present paper that the interaction between two d electrons vanishes unless they are at the same atom. Within the same atom the interaction will not be shielded appreciably. When the Bohm-Pines theory is applied to Ni which has 0.6 $4s$ electrons per atom, the shielded interaction changes its sign at about the distance between the nearest-neighboring lattice sites. Though this result may not be taken for a quantitative verification, the above-mentioned assumption can be regarded as a semi-quantitatively correct simplification to make the calculation feasible. In calculating the matrix elements of the interaction, we shall neglect the overlap between the atomic orbitals belonging to the neighboring atoms. In other words, the atomic orbitals are replaced by the corresponding Wannier functions. This last assumption may be justified for a nondegenerate band in the limit of tightly bound electrons. It will be re-examined later (§ 4), when the degeneracy of the d bands is taken into account.

With the above-mentioned assumptions we can show that the correlation effect reduces the intra-atomic interaction (the Coulomb self-energy) to the magnitude of the order of the band-width when the self-energy is much larger than the band width. This result may be understood physically as follows. When the intra-atomic interaction is large, electrons will avoid entering into the same atom by the sacrifice of the one electron energy of the order of the band width. Thus this increase of the one electron energy corresponds to the effective magnitude of the interaction. Since the reduction is more effective for smaller number of electrons, the paramagnetic state is stable for a sufficiently small number of electrons regardless of the energy spectrum of the band. Also the condition for the occurrence of ferromagnetism is more stringent than in the Hartree-Fock approximation. It will be shown that the occurrence of ferromagnetism for such a small number of holes as in Ni requires a special characteristic of the state-density vs energy relation.

When the electron correlation is taken into account, the definition of the one electron energy is not a simple problem. It is assumed in the present paper that one can choose an approximate one electron energy spectrum which suffices for practical purposes. This energy spectrum is assumed to be independent of spin states and to include self-consistently the effect of the interaction among electrons that does not depend on the spin states of the interacting electrons. This problem will be discussed in more detail in the following sections.

In §§ 2 and 3 we discuss the case of nondegenerate band, calculating the multiple scattering in § 2 and discussing the stability of the paramagnetic state in § 3. The possibility of other types of spin ordering than ferromagnetism is neglected in the latter discussion. Section 4 deals with the discussion of Ni and Pd. In § 5 Ni-Cu alloys and other cases will be discussed briefly.

§ 2. Multiple scattering between two electrons in a nondegenerate band

We discuss first the multiple scattering between two electrons in the absence of other electrons. The Hamiltonian governing the motion of two electrons is represented by

$$H = H_0(1) + H_0(2) + V(1, 2), \tag{1}$$

where 1 and 2 denote the coordinates of electrons 1 and 2, respectively, H_0 is the one electron energy, $V(1, 2)$ the interaction between two electrons. The eigenvalue of H_0 associated with a given wave vector $\boldsymbol{k}$ is denoted by $\varepsilon(\boldsymbol{k})$. The corresponding wave function $\varphi(x, \boldsymbol{k})$ with $x=1$ or 2 is connected to the Wannier function associated with a given lattice site $\boldsymbol{R}$, $W(x, \boldsymbol{R})$, by the well-known relations,

$$\varphi(x, \boldsymbol{k}) = (1/N)^{1/2}\textstyle\sum_{\boldsymbol{R}} W(x, \boldsymbol{R})\exp(i\boldsymbol{k}\cdot\boldsymbol{R}) \tag{2}$$

and

$$W(x, \boldsymbol{R}) = (1/N)^{1/2}\textstyle\sum_{\boldsymbol{k}} \varphi(x, \boldsymbol{k})\exp(-i\boldsymbol{k}\cdot\boldsymbol{R}), \tag{3}$$

where N is the number of the lattice sites in a given volume.

As was mentioned in § 1, the matrix elements of $V(1, 2)$ referred to the Wannier functions are assumed to be

$$\int W^*(1, \boldsymbol{R}_1)\, W^*(2, \boldsymbol{R}_2)\, V(1, 2)\, W(1, \boldsymbol{R}_3)\, W(2, \boldsymbol{R}_4)\, dv_1\, dv_2$$

$$= U \text{ if } \boldsymbol{R}_1 = \boldsymbol{R}_2 = \boldsymbol{R}_3 = \boldsymbol{R}_4, \quad = 0 \text{ otherwise.} \tag{4}$$

U defined by (4) will be identified with the Coulomb self-energy of the atomic orbital. The matrix elements of $V(1, 2)$ referred to φ's are easily calculated by the use of (2) and (4) to be

$$\int \varphi^*(1, \boldsymbol{k}_1)\varphi^*(2, \boldsymbol{k}_2)V(1, 2)\varphi(1, \boldsymbol{k}_3)\varphi(2, \boldsymbol{k}_4)dv_1dv_2$$
$$= (U/N)\delta(\boldsymbol{k}_1, \boldsymbol{k}_2; \boldsymbol{k}_3, \boldsymbol{k}_4) \tag{5}$$

with

$$\delta(\boldsymbol{k}_1, \boldsymbol{k}_2; \boldsymbol{k}_3, \boldsymbol{k}_4) = 1 \text{ if } \boldsymbol{k}_1 + \boldsymbol{k}_2 = \boldsymbol{k}_3 + \boldsymbol{k}_4 + \boldsymbol{K}, \tag{6}$$
$$= 0 \text{ otherwise,}$$

where $\boldsymbol{K}$ represents a reciprocal lattice vector.

Let $|\boldsymbol{k}_1\sigma_1, \boldsymbol{k}_2\sigma_2\rangle$ be the antisymmetrized wave function of the state where the one electron states specified by $\boldsymbol{k}_1$ with the spin coordinate σ_1 and $\boldsymbol{k}_2$ with σ_2 are occupied. In the Hartree-Fock approximation, the interaction energy of two electrons in this state is given by

$$\Delta E_{HF}(\boldsymbol{k}_1\sigma_1, \boldsymbol{k}_2\sigma_2) = (U/N)(1-\delta(\sigma_1\sigma_2)), \tag{7}$$

where $\delta(\sigma_1\sigma_2) = 1$ for parallel spins $(\sigma_1 = \sigma_2)$, $\delta(\sigma_1\sigma_2) = 0$ for antiparallel spins $(\sigma_1 \neq \sigma_2)$. In the case of parallel spins, the wave function $|\boldsymbol{k}_1\sigma_1, \boldsymbol{k}_2\sigma_2\rangle$ is an eigenfunction of the total Hamiltonian H. In the case of antiparallel spins, however, the eigenfunction of H is of the form given by

$$\Psi(\boldsymbol{k}_1, \boldsymbol{k}_2) = |\boldsymbol{k}_1, \boldsymbol{k}_2\rangle + (1/N)\textstyle\sum_{k_3k_4}\Gamma_{k_3k_4}\delta(\boldsymbol{k}_1\,\boldsymbol{k}_2;\ \boldsymbol{k}_3\,\boldsymbol{k}_4)\,|\boldsymbol{k}_3\,\boldsymbol{k}_4\rangle, \tag{8}$$

where the spin indices are omitted for simplicity. Inserting this expression into the Schroedinger equation, we can easily determine Γ and the energy eigenvalue to be

$$\Gamma_{k_3k_4} = -\{U/(1+UG(\boldsymbol{k}_1, \boldsymbol{k}_2)\}\{1/(\varepsilon(\boldsymbol{k}_3)+\varepsilon(\boldsymbol{k}_4)-\varepsilon(\boldsymbol{k}_1)-\varepsilon(\boldsymbol{k}_2))\} \tag{9}$$

and

$$E(\boldsymbol{k}_1, \boldsymbol{k}_2) = \varepsilon(\boldsymbol{k}_1) + \varepsilon(\boldsymbol{k}_2) + \Delta E(\boldsymbol{k}_1, \boldsymbol{k}_2)$$

with

$$\Delta E(\boldsymbol{k}_1, \boldsymbol{k}_2) = (U/N)\{1/(1+UG(\boldsymbol{k}_1, \boldsymbol{k}_2))\}, \tag{10}$$

where $G(\boldsymbol{k}_1, \boldsymbol{k}_2)$ is defined by

$$G(\boldsymbol{k}_1, \boldsymbol{k}_2) = (1/N)\textstyle\sum_{k_3,k_4}\delta(\boldsymbol{k}_1\,\boldsymbol{k}_2;\ \boldsymbol{k}_3\,\boldsymbol{k}_4)/(\varepsilon(\boldsymbol{k}_3)+\varepsilon(\boldsymbol{k}_4)-\varepsilon(\boldsymbol{k}_1)-\varepsilon(\boldsymbol{k}_2)). \tag{11}$$

In (9) and (11), ΔE in the factor $1/[\varepsilon(\boldsymbol{k}_3)+\varepsilon(\boldsymbol{k}_4)-\varepsilon(\boldsymbol{k}_1)-\varepsilon(\boldsymbol{k}_2)-\Delta E]$ is neglected because it is of the order of $1/N$.*) Comparing ΔE given by (10) with ΔE_{HF} of (7), we can see that the reduction of the interaction energy due to the

---

*) When the energy of the initial state, $\varepsilon(\boldsymbol{k}_1)+\varepsilon(\boldsymbol{k}_2)$, is high enough to yield $1+UG(\boldsymbol{k}_1, \boldsymbol{k}_2) \lesssim 0$ (G becomes negative for sufficiently high initial energy), ΔE cannot be neglected, being not of the order of $1/N$. In such a case the exact solution yields a quasi-bound state of two electrons. In the present discussion we are interested only in the low-lying initial states for which ΔE is of the order of $1/N$.

electron correlation is indicated by the factor $1/(1+UG(\boldsymbol{k}_1, \boldsymbol{k}_2))$.

In order to take into account the presence of other electrons than the interacting pair, we make the following modification of the above calculation. In the first place, we restrict the states $\boldsymbol{k}_3$ and $\boldsymbol{k}_4$ in the sum of (8) and (11) to the unoccupied states above the Fermi level, since the occupied states below the Fermi level are not available for the scattering. In the second place, we assume that other electrons affect the motion of the interacting pair only through a self-consistently defined potential energy. This potential energy is understood to be included already in the one electron energy H_0 of (1). We may define the modified one electron energy $\varepsilon(\boldsymbol{k})$ by

$$\varepsilon(\boldsymbol{k}) = \varepsilon_0(\boldsymbol{k}) + \sum_{k'} \Delta E(\boldsymbol{k}, \boldsymbol{k}'), \tag{12}$$

where $\varepsilon_0(\boldsymbol{k})$ is the one electron energy in the absence of other electrons, and the sum of the second term is taken over the occupied states below the Fermi level. It is $\varepsilon(\boldsymbol{k})$, not $\varepsilon_0(\boldsymbol{k})$, that should enter into the expressions of ΔE and $G(\boldsymbol{k}_1, \boldsymbol{k}_2)$, (10) and (11). One might regard (12) as the equation of self-consistency to define $\varepsilon(\boldsymbol{k})$. As was disussed by Brueckner, however, the energy of excited levels may not be determined uniquely, since it may depend on the states below the Fermi level from which the electrons are excited. To make the whole calculation self-consistently, one may follow Brueckner and Gammel's procedure faithfully, reformulating the scattering problem in terms of the reaction matrices and solving an infinite ladder of self-consistency equations. We assume in the present paper, however, that (12) defines the self-consistent one electron energy spectrum for either the occupied or the unoccupied levels.

With the above-mentioned assumptions, the effective value of U in the paramagnetic state can be defined as

$$U_{\text{eff}}(\boldsymbol{k}_1, \boldsymbol{k}_2) = U \cdot 1/(1+UG(\boldsymbol{k}_1, \boldsymbol{k}_2)), \tag{13}$$

where the sum in (11) which defines $G(\boldsymbol{k}_1, \boldsymbol{k}_2)$ is understood to be taken over the unoccupied states. Since $G(\boldsymbol{k}_1, \boldsymbol{k}_2)$ is generally of the order of $1/W$, where W is the band width, (13) yields

$$U_{\text{eff}} \approx W \text{ if } U \gg W, \tag{14}$$

whose physical meaning was discussed in § 1. If the bottom of the band corresponds to a point of high symmetry such as $\boldsymbol{k}=0$ or $\boldsymbol{k}=\boldsymbol{K}/2$, G of the pair of electrons, both of which occupy the state of lowest energy, can be written as

$$G(0,\ 0) = (1/2)\int_{\varepsilon_F}^{W} \{\eta(\varepsilon)/\varepsilon\}\, d\varepsilon, \tag{15}$$

where $\eta(\varepsilon)$ is the state density per unit energy per atom per spin, and ε_F is the Fermi energy. Except for a negligibly small number of pairs, $G(\boldsymbol{k}_1, \boldsymbol{k}_2)$ does not differ much from $G(0, 0)$. This is because the scattering process is the s wave scattering that does not depend on the relative momentum of the interacting pair, and also because the energy of low-lying excited levels involved in $G(\boldsymbol{k}_1, \boldsymbol{k}_2)$ is generally separated from the unperturbed energy by an energy of the order of ε_F by the condition of the crystal momentum conservation (6). Thus we obtain an approximate estimate of U_{eff},

$$U^0_{\text{eff}} = U/(1+UG(0,0)). \tag{16}$$

The difference between (16) and the average of (13) over $\boldsymbol{k}_1$ and $\boldsymbol{k}_2$ below the Fermi level is estimated by numerical calculation to be within 5 percent for the case of $\eta=$constant. The details of the band having a constant state density will be discussed in § 4.

§ 3. The stability of the paramagnetic state

In the Hartree-Fock approximation, the paramagnetic state becomes unstable if the condition,

$$U\eta(\varepsilon_F) > 1, \tag{17}$$

is satisfied. This condition is obtained by comparing the energy of the paramagnetic state with that of the ferromagnetic state having an infinitesimal magnetization. The corresponding condition in the present approximation is given to good approximation by

$$U^0_{\text{eff}}\,\eta(\varepsilon_F) > 1. \tag{18}$$

In order to derive this condition, we compare the energy of the paramagnetic state first with the state in which the + spin levels are occupied up to the energy ε_F^+ and the − spin levels up to the energy ε_F^-. ε_F^+ and ε_F^- satisfy the relation, $\varepsilon_F^+ - \varepsilon_F = \varepsilon_F - \varepsilon_F^- = \Delta\varepsilon$ to the first order of $\Delta\varepsilon$. The energy difference between the paramagnetic state and the ferromagnetic state consists of two terms,*)

$$\Delta E_1 = \text{the difference of } \varepsilon_0 \text{ in (12)} \tag{19}$$

and

$$\Delta E_2 = \int_0^{\varepsilon_F^+}\int_0^{\varepsilon_F^-} \tilde{U}_{\text{eff}}(\varepsilon_1, \varepsilon_2; \varepsilon_F^+, \varepsilon_F^-)\,\eta(\varepsilon_1)\,\eta(\varepsilon_2)\,d\varepsilon_1\,d\varepsilon_2$$

*) In the state having infinitesimal magnetization, the one electron energy defined by (12) is shifted compared with $\varepsilon(\boldsymbol{k})$ of the paramagnetic state through the second term of (12). Since the shift may not be a constant for all states, there will be a change in the functional form of $\eta(\varepsilon)$. This change of η is neglected in deriving (19) and (20), because the shift is approximately constant for the states having the same spin (it is of opposite sign for opposite spin state). ε in (20) represents the energy measured from the bottom of the band which is different for different spin in the ferromagnetic state.

$$-\int_0^{\varepsilon_F}\int_0^{\varepsilon_F} \tilde{U}_{\rm eff}(\varepsilon_1, \varepsilon_2\,;\, \varepsilon_F, \varepsilon_F)\,\eta(\varepsilon_1)\,\eta(\varepsilon_2)\,d\varepsilon_1\,d\varepsilon_2, \tag{20}$$

where $\tilde{U}_{\rm eff}$ is the average of $U_{\rm eff}$ over the states having the energies ε_1 and ε_2; the dependence of $\tilde{U}_{\rm eff}$ on ε_F's, which arises from the dependence of G on the Fermi energies, is explicitly indicated.

ΔE_2 can be rewritten up to the second order of $\Delta\varepsilon$

as

$$\Delta E_2 = \int_0^{\varepsilon_F^+}\int_0^{\varepsilon_F^-} \{\tilde{U}_{\rm eff}(\varepsilon_1, \varepsilon_2\,;\, \varepsilon_F^+, \varepsilon_F^-) - \tilde{U}_{\rm eff}(\varepsilon_1, \varepsilon_2\,;\, \varepsilon_F, \varepsilon_F)\}\,\eta(\varepsilon_1)\,\eta(\varepsilon_2)\,d\varepsilon_1\,d\varepsilon_2$$

$$- (\Delta\varepsilon)^2\eta(\varepsilon_F)^2\,\tilde{U}_{\rm eff}(\varepsilon_F, \varepsilon_F\,;\, \varepsilon_F, \varepsilon_F)$$

$$+ (\Delta\varepsilon)^2\eta(\varepsilon_F)\int_0^{\varepsilon_F} \{\partial\tilde{U}_{\rm eff}(\varepsilon_1 = \varepsilon_F, \varepsilon_2\,;\, \varepsilon_F, \varepsilon_F)/\partial\,\varepsilon_1\}\,\eta(\varepsilon_2)\,d\varepsilon_2. \tag{21}$$

The last term of (21) may be interpreted as a part of the one electron energy; adding it to ΔE_1 given by (19), we obtain $(\Delta\varepsilon)^2\eta(\varepsilon_F)$. The first term of (21) corresponds to the energy change arising from a modification of the electron correlation. Since the modification arises in the present approximation from a change in the availability of the states near the Fermi surface for the scattering, and since these states do not contribute much to the electron correlation because of the condition of the crystal momentum conservation, we may expect generally that the first term is small. It is shown in the Appendix that the first term is actually small in the case of η = constant. Neglecting thus the first term, we obtain for the condition of the occurrence of the ferromagnetism

$$\tilde{U}_{\rm eff}(\varepsilon_F, \varepsilon_F)\,\eta(\varepsilon_F) > 1. \tag{22}$$

The condition (22) is too stringent, because we assumed a special type of ferromagnetism. Let $\varepsilon(\boldsymbol{k})$ be dependent on the magnitude of the wave vector only. $G(\boldsymbol{k}_1, \boldsymbol{k}_2)$ with both $\boldsymbol{k}_1$ and $\boldsymbol{k}_2$ lying at the Fermi surface diverges for $\boldsymbol{k}_1 = -\boldsymbol{k}_2$ and is smallest for $\boldsymbol{k}_1 = \boldsymbol{k}_2$ according to (11). Correspondingly, $U_{\rm eff}$ for $\boldsymbol{k}_1 = -\boldsymbol{k}_2$ vanishes, and $U_{\rm eff}$ for $\boldsymbol{k}_1 = \boldsymbol{k}_2$ will be the largest for the states at the Fermi surface. If an electron is removed from the $-$spin level with $\boldsymbol{k}$ at the Fermi surface and put in the $+$spin level with a wave vector close to $\boldsymbol{k}$, the gain of the interaction energy is specified by $U_{\rm eff}(\boldsymbol{k}, \boldsymbol{k})$ which is larger than the average, $\tilde{U}_{\rm eff}(\varepsilon_F, \varepsilon_F)$. It is shown in the Appendix that $U_{\rm eff}(\boldsymbol{k}, \boldsymbol{k})$ with $\boldsymbol{k}$ at the Fermi surface is equal to $U^0_{\rm eff}$ given by (16) to good approximation in the case of the band having a constant state density. In this case, $\tilde{U}_{\rm eff}(\varepsilon_F, \varepsilon_F)$ is smaller by about 10 percent than $U_{\rm eff}(\boldsymbol{k}, \boldsymbol{k})$. Thus $U_{\rm eff}$ in the condition (18) may be understood to be given by (16).

In the following we assume the condition (18) with U^0_{eff} defined by (16). From (15) and (16) we can see that G increases and U^0_{eff} decreases with decreasing number of electrons. Thus there is a lower limit for the number of electrons to satisfy the condition (18) even when $U=\infty$. Of various state density functions, this critical number of electrons is calculated for the cases of $U=\infty$ and $U=2W$. The result is summarized in Table I, where n is defined to be the total number of electrons of both spins per atom and $\eta(\varepsilon)$ is normalized to satisfy $\int_0^W \eta(\varepsilon)d\varepsilon=1$. Also the minimum value of U required to satisfy the condition (18) for the optimum number of electrons ($n=1$) is listed in the Table. Though the present approximation is not valid for $n\sim1$, the result may be useful for semi-quantitative discussions.

Table I.

$\eta(\varepsilon)$	$WG(0, 0)$, $x=\varepsilon_F/W$	$n_{cr}(U=\infty)$	$n_{cr}(U=2W)$	U_{cr}/W
$1/W$	$(1/2)\log(1/x)$	0.271	0.446 (0.000)	1.53
$2\varepsilon/W^2$	$1-x$	0.222	0.500 (0.125)	0.89
$2(W-\varepsilon)/W^2$	$x-\log x-1$	0.233	0.440 (0.000)	1.12
$4\varepsilon/W^2 (\varepsilon<W/2)$, $4(\varepsilon-W/2)/W^2 (\varepsilon>W/2)$	$2\log 2-2x$	0.214	0.395 (0.063)	0.62
$6\varepsilon(W-\varepsilon)/W^3$	$(3/2)(1-x)^2$	0.208	0.367 (0.047)	0.89

η is the state density per atom per unit energy per spin, W the band width, ε_F the Fermi energy. n_{cr} is the minimum number of electrons per atom required for obtaining ferromagnetism, U_{cr} is the minimum of U required to produce ferromagnetism for the optimum number of electrons ($n=1$), the figures in parenthesis are the corresponding values obtained in the Hartree-Fock approximation.

§ 4. Ferromagnetism of Ni and Pd

In Ni and Pd, the d bands are occupied by about 0.6 holes per atom. The calculations of the $3d$ bands made by Fletcher,[5] Segall,[6] and Yamashita et al.[7] indicate that these holes are distributed mostly among the three d bands whose atomic orbitals correspond to the $d\varepsilon$ orbitals, i.e. $d(xy)$, $d(yz)$, and $d(zx)$ orbitals, where xy, etc. denote the angular part of the d orbitals. These three bands are degenerate by cubic symmetry. The states of highest energy (the bottom of the band for holes) of the $d(xy)$ band are specified by the $\boldsymbol{k}$ vectors, $(2\pi/a)(\pm1, 0, 0)$ and $(2\pi/a)(0, \pm1, 0)$, where a is the lattice parameter of the cubic unit cell, those of the $d(yz)$ band by $(2\pi/a)(0, \pm1, 0)$ and $(2\pi/a)(0, 0, \pm1)$, and those of the $d(zx)$ band by $(2\pi/a)(0, 0, \pm1)$ and $(2\pi/a)(\pm1, 0, 0)$, two of the three bands being degenerate at each of these points. At a general point in the $\boldsymbol{k}$ space, these degenerate bands are mixed with each other by the periodic potential. We shall neglect this band mixing in the following discussion, primarily because the mixing is small in the vicinity of the top of the bands and will not be essential for the intended semi-quantitative discussion. It might be worth while, however, to discuss briefly the definition of the Wannier function in degenerate bands.

If the band mixing is taken into account, the eigenfunction is generally a linear combination of the Bloch orbitals, each of which is constructed with a given atomic d orbital. Correspondingly, the degenerate energy eigenvalues are split by the mixing. When an energy band is defined by the requirement that the energy is a continuous function within a band, the level of the highest energy at each point in the $\boldsymbol{k}$ space belongs always to the first band, the second highest to the second band, and so on. With the energy band thus defined, it is impossible to assign a unique atomic orbital to each band and accordingly the Wannier function does not correspond to any atomic orbital. The energy band is invariant with respect to the transformation of the coordinate axes belonging to the cubic group in the sense that the operation transforms a given wave function into another wave function of the same band. Thus the Wannier function, which is a sum of all wave functions belonging to the band, is invariant under the operation, while the atomic d orbitals are not invariant. The Wannier function associated with a given lattice point consists of the d orbitals of surrounding atoms; the envelope of their amplitudes is zero at the central atom, perhaps having a maximum at the nearest neighbors, then decreasing with increasing distance from the central atom. On the other hand, if the band mixing is neglected, the $d(xy)$ band, etc. are not invariant with respect to the coordinate transformation, and the Wannier functions correspond to the atomic orbitals.

Treating the three d bands separately, we assign 0.2 holes to each band. The intra-band interaction, which is defined as the interaction between two holes belonging to the same band, is characterized by the Coulomb self-energy U as in the case of a single band. On the other hand, the inter-band interaction may be divided into three scattering processes; the process in which the interacting electrons (holes) remain in the original bands after scattering is characterized by the Coulomb integral defined by

$$U' = \int W^*(1)\, W'^*(2)\, V(1,2)\, W(1)\, W'(2)\, dv_1 dv_2\,; \tag{23}$$

the process in which the electrons are exchanged between two bands is characterized by the exchange integral defined by

$$J = \int W^*(1)\, W'^*(2)\, V(1,2)\, W'(1)\, W(2)\, dv_1\, dv_2\,; \tag{24}$$

the process which corresponds to the transfer of a pair of electrons from one band to another is characterized by the integral,

$$J' = \int W^*(1)\, W^*(2)\, V(1,2)\, W'(1)\, W'(2)\, dv_1\, dv_2. \tag{25}$$

In (23), (24), and (25), W and W' represent the Wannier functions of two different bands associated with the same atom. If W and W' are identified with the corresponding atomic d orbitals, integrals other than those mentioned above vanish by symmetry. U' is of the same order of magnitude as U. J and J', which are equal to each other within the present approximation, are much smaller than U or U', since they correspond to the exchange integral between different atomic d orbitals of the order of 0.6 ev.

The inter-band interactions will be reduced by the electron correlation to roughly the same extent as the intra-band interaction. This reduction can be discussed by adapting the theory developed in § 2. If only the scattering process characterized by (23) is considered, the effective value of U' is obtained by replacing in (13) U by U' and G by G' defined by

$$G'(\boldsymbol{k}_1, \boldsymbol{k}_2) = (1/N)\sum\delta(\boldsymbol{k}_1\boldsymbol{k}_2; \boldsymbol{k}_3\boldsymbol{k}_4)/(\varepsilon(\boldsymbol{k}_3) + \varepsilon'(\boldsymbol{k}_4) - \varepsilon(\boldsymbol{k}_1) - \varepsilon'(\boldsymbol{k}_2)), \quad (26)$$

where $\varepsilon(\boldsymbol{k})$ and $\varepsilon'(\boldsymbol{k})$ represent the one electron energies of a given pair of the bands. Since the interaction is independent of the spin states of interacting electrons, we may suppose that its effect is self-consistently incorporated in the one electron energy.

The effect of other scattering processes is negligibly small. If the exchange scattering is taken into account, the interaction energy of a pair of electrons with parallel spins, which belong to different bands, is modified to be given by

$$\begin{aligned}\Delta E &= (1/N)\{(U'-J)/[1+(U'-J)G']\}\\ &\cong (1/N)\{U'/(1+U'G') - J/(1+U'G')^2\},\end{aligned} \quad (27)$$

and ΔE of antiparallel spins by

$$\begin{aligned}\Delta E &= (1/2N)\{(U'+J)/[1+(U'+J)G']\\ &\qquad + (U'-J)/[1+(U'-J)G']\},\\ &\cong (1/N)\{U'/(1+U'G') - J^2G'/(1+U'G')^3\}.\end{aligned} \quad (28)$$

Comparing (27) and (28) with corresponding expressions in the Hartree-Fock approximation, we can see that the difference of the inter-band interaction between the case of parallel spins and that of antiparallel spins is reduced approximately by the factor $1/(1+U'G')^2$ which is estimated to be less than 0.1. Also the transfer process given by (25), which is effective only for antiparallel spins, modifies the intra-band interaction only slightly.*) The smallness of the above-mentioned effects may be understood qualitatively from the argument that the effective magnitude of the inter-band exchange interaction is reduced by the electron correlation to the same extent as the intra-band interaction is reduced, since the reduction arises in both cases from the decrease of the probability of finding two holes at the same atom.

---

*) The effect of the transfer process can be taken into account approximately by replacing U by $U(1-2J^2/U)$ in (10).

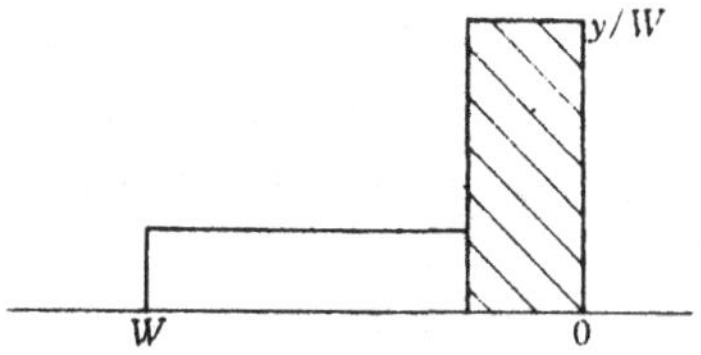

Fig. 1. A state density vs energy curve which favors the ferromagnetic state. The hatched region (the high density region) contains x states per atom per spin in total. The state density of the high density region is defined to be y/W; the state density of the low density region can be expressed in terms of x and y.

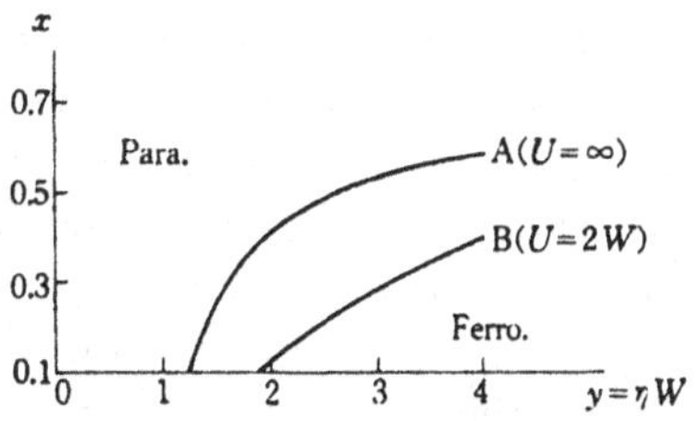

Fig. 2. The region in the x-y plane (x and y are defined in Fig. 1.) in which the ferromagnetic state is favored energetically for $n=0.2$. The condition $U_{\rm eff}\eta=1$ is satisfied on curve A for $U=\infty$ and on curve B for $U=2W$.

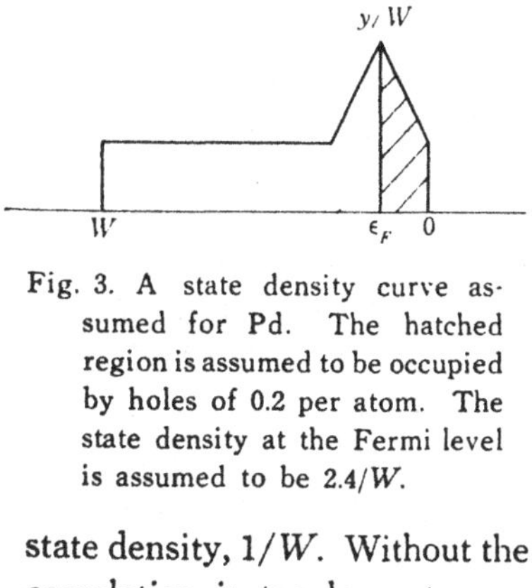

Fig. 3. A state density curve assumed for Pd. The hatched region is assumed to be occupied by holes of 0.2 per atom. The state density at the Fermi level is assumed to be $2.4/W$.

With the considerations so far discussed, the problem is now reduced to that of a single band. According to the available calculations of the d bands in Ni, the band width W is 3 to 4 ev. U of Ni is estimated by Van Vleck[1] to be about 7.6 ev. Thus we may assume $U \cong 2W$ for Ni. Table I shows that simple state density functions such as those listed in the Table do not yield ferromagnetism for $n=0.2$. To satisfy the condition (18), we must have a high state density at the Fermi level, and furthermore, a relatively low average state density, $1/W$. Without the latter condition the reduction of $U_{\rm eff}$ by the electron correlation is too large to produce ferromagnetism. A simplified state density vs energy curve which satisfies the requirements is shown in Fig. 1.

Although the state density function $\eta(\varepsilon)$ defined implicitly by (12) is not necessarily equivalent to that obtained in the usual band calculation in which the charge neutrality of an atom is assumed, the difference will be small when the number of electrons is small. The state density of the d bands calculated by Fletcher is actually characterized by the region of high state density at the top of the bands. A simplified model of the d bands was proposed by Gautier.[8] If the $d(xy)$ band is taken as the representative band, the energy of this band is nearly independent of $\boldsymbol{k}_z$ in the vicinity of the top specified by $(2\pi/a)$ $(\pm 1, 0, 0)$ and $(2\pi/a)$ $(0, \pm 1, 0)$, since the overlap of the atomic $d(xy)$ orbitals of neighboring atoms in the z direction is small there.[8] Thus the energy is approximately represented by the kinetic energy of a two-dimensional gas. The constancy of the state density near the top of the band can be easily verified with this simplified model. Since the correlation reduction of U does not depend much on the details of a given band, the state density curve shown in Fig. 1 can be regarded as an approximation which has the essential feature of the $d\varepsilon$ bands in Ni.

According to the specific heat measurement at low temperatures, the state density of ferromagnetic Ni at the Fermi level is about 3.1 states per atom per ev. Subtracting the contribution of the s band of about 0.3 states per atom per ev and dividing by 3, we obtain an approximate estimate of $\eta(\varepsilon)$ to be about 0.9 per atom per ev per spin. Since there will be a small contribution by other $(d\gamma)$ bands and also since the state density seems to be at maximum at the Fermi level of ferromagnetic Ni, the averaged η to be used for the state density curve of Fig. 1 will be around 0.75 states per atom per ev per spin. This corresponds to $y=\eta W=3$ if W is taken to be 4 ev.

With the state density curve of Fig. 1 and by the use of (16), the condition for the occurrence of ferromagnetism, (18), is investigated for various pairs of x and y which are defined in Fig. 1. The curve in the x-y plane on which the condition $U^0_{\text{eff}}\eta=1$ is satisfied is shown in Fig. 2. We can see from the figure that the ferromagnetic state is favored energetically for $y=3$ and $U=2W$ if x is smaller than 0.28. According to Fletcher's calculation x seems to be smaller than 0.25. Thus we may conclude that the ferromagnetic state has lower energy than the paramagnetic state in Ni. This conclusion does not depend on our choice of the band width. The magnitude of U^0_{eff} is mainly determined by U and y/W besides x, but only slightly dependent on W.*)

The specific heat measurement[9] indicates that the state density at the Fermi level of paramagnetic Pd is about 4 states per atom per ev (including both spin states). Subtracting again 0.3 states per atom of the s band, and dividing by 6, we obtain $\eta(\varepsilon_F)\cong 0.6$ per atom per ev per spin. U of Pd will be about the same as that of Ni. If we assume $W=4$ ev and the state density curve of Fig. 1, we obtain $y=2.4$. The corresponding x which favors the ferromagnetic state is less than 0.19 states per atom per spin. This seems to indicate that the paramagnetic state is very likely lower in energy than the ferromagnetic state within the present approximation. This conclusion does not depend on the assumed shape of the state density curve. If the state density curve shown in Fig. 3 is taken, the paramagnetic state is still found to be stable. This state density curve resembles more closely the one which is suggested by the specific heat measurement of Pd-Rh and Pd-Ag alloys.[9]

---

*) If a larger W is assumed, y becomes larger since $\eta=y/W$ is fixed. Then, however, U/W should be smaller, which cancels the effect of larger y.

§ 5. Supplementary discussions

The ferromagnetism of Ni-Cu alloys with Cu concentration less than 60 percent was regarded as a support for the simple theory of band electron ferromagnetism. Since the present theory denies the occurrence of ferromagnetism for such a small number of *d* holes as in the case of 50 percent Ni-50 percent Cu, the observed ferromagnetism might look a contradiction with the theory. However, the repulsive interaction between the *d* holes and the Cu atoms reduces the effective volume available for the *d* hole correlation and thereby the effectiveness of the hole correlation. Thus U_{eff} in Ni-Cu alloys will remain about the same as in Ni in spite of the smallness of the number of *d* holes.

In Co and Fe, where more than one *d* holes per atom are present, the exchange interaction between different atomic *d* orbitals will make some contribution to the energy difference between the ferromagnetic state and the paramagnetic state. The discussion of these metals, however, requires both detailed knowledge of the *d* band and a more refined treatment of the electron correlation. The discussion of light transition metals such as V and Ti will be also not feasible, since a simple tight binding approximation such as that employed in the present paper will not be justified in these metals, and the inter-atomic interaction neglected in the present paper might play an important role.

The auther would like to express his sincere thanks to Professor T. Nagamiya and Dr. M. Tachiki for valuable discussions. The present research was supported in part by the Scientific Research Fund of the Education Ministry.

Appendix

As was mentioned in the text, the energy spectrum of a two-dimensional gas given by

$$\varepsilon = \alpha(k_x^2 + k_y^2) \tag{A1}$$

yields the constant state density per unit energy. We define vectors $\boldsymbol{u}$ and $\boldsymbol{v}$ (two-dimensional) by

$$\begin{aligned} \boldsymbol{k}_1 &= \boldsymbol{u}_1 + \boldsymbol{v}_1, & \boldsymbol{k}_3 &= \boldsymbol{u}_2 + \boldsymbol{v}_2, \\ \boldsymbol{k}_2 &= \boldsymbol{u}_1 - \boldsymbol{v}_1, & \boldsymbol{k}_4 &= \boldsymbol{u}_2 - \boldsymbol{v}_2. \end{aligned} \tag{A2}$$

The condition of the momentum conservation requires $\boldsymbol{u}_1 = \boldsymbol{u}_2$. In terms of $\boldsymbol{u}$ and $\boldsymbol{v}$, $G(\boldsymbol{k}_1, \boldsymbol{k}_2)$ is expressed as

$$G(\boldsymbol{k}_1, \boldsymbol{k}_2) = (\eta_0/2\pi)\int dv_2 \{v_2/(v_2^2 - v_1^2)\} \cdot f(v_2\,;\, u, k_F), \tag{A3}$$

where η_0 is the state density per unit energy per atom per spin, and $f(v_2\,;\, u, k_F)$ is the weight function that is obtained by the integration of the angular part of v_2. From the conditions, $k_3^2 > k_F^2$ and $k_4^2 > k_F^2$, we obtain

$$f(v_2; u, k_F) = 0 \quad \text{for} \quad v_2 \le (k_F^2 - u^2)^{1/2},$$
$$= 4 \sin^{-1}\{(u^2 + v_2^2 - k_F^2)/2v_2 u\}$$
$$\text{for} \quad (k_F^2 - u^2)^{1/2} \le v_2 \le k_F + u,$$
$$= 2\pi \quad \text{for} \quad k_F + u \le v_2. \tag{A4}$$

Putting $u=0$ and $v_1=0$, we obtain from (A3)

$$G(0, 0) = \eta_0 \int_{k_F} dv_2/v_2. \tag{A5}$$

The difference between (A3) and (A5) is given by

$$\Delta G(\boldsymbol{k}_1, \boldsymbol{k}_2) = \eta_0 \int_{k_F+u} v_1^2/v_2(v_2^2 - v_1^2) \cdot dv_2$$
$$+ (2\eta_0/\pi) \int_{(k_F^2-u^2)^{1/2}}^{k_F+u} \{v_2 dv_2/(v_2^2 - v_1^2)\} \sin^{-1}\{(u^2 + v_2^2 - k_F^2)/2uv_2\}$$
$$- \eta_0 \int_{k_F}^{k_F+u} dv_2/v_2. \tag{A6}$$

The upper limit of the integral of the first term in (A6) will be assumed to be independent of u. The resulting error is negligible for small k_F. Putting $u = k_F$ and $v_1 = 0$, we obtain $G(\boldsymbol{k}, \boldsymbol{k})$ with $\boldsymbol{k}$ lying at the Fermi surface to be

$$\Delta G(\boldsymbol{k}, \boldsymbol{k}) = (2\eta_0/\pi) \int_0^{2k_F} (dv_2/v_2) \sin^{-1}(v_2/2k_F) - \eta_0 \int_{k_F}^{2k_F} (dv_2/v_2). \tag{A7}$$

It can be shown easily that (A7) vanishes.

In order to estimate the first term of (21) of the text, we expand U_{eff} in powers of ΔG to obtain

$$U_{\text{eff}}(\boldsymbol{k}_1, \boldsymbol{k}_2) = U/(1 + UG(0, 0)) - \{U^2/(1 + UG(0, 0))^2\} \Delta G(\boldsymbol{k}_1, \boldsymbol{k}_2). \tag{A8}$$

The first term of (21) is now expressed to the first order of ΔG by

$$\Delta E_2' = -U_{\text{eff}}^{0\,2} \Delta I_{\text{av}} \tag{A9}$$

with U_{eff}^0 defined by (16) and ΔI_{av} defined by

$$\Delta I_{\text{av}} = \left\langle \int_{(k_F^2-u^2)^{1/2}}^{k_F^+ + u} (\eta_0/2\pi)(v_2 dv_2/(v_2^2 - v_1^2))(f(v_2; u, k_F^+, k_F^-) - f(v_2; u, k_F)) \right\rangle_{\text{av}}, \tag{A10}$$

where the average is taken over v_1 and u of occupied states; k_F^+ and k_F^- satisfy the relation $k_F^2 - k_F^{-2} = k_F^{+2} - k_F^2 \equiv \Delta$; $f(v_2; u, k_F^+, k_F^-)$ is given by

$$f(v_2; u, k_F^+, k_F^-) = 2[\sin^{-1}\{(u^2+v_2^2-k_F^2-\Delta)/2uv_2\} + \sin^{-1}\{(u^2+v_2^2-k_F^2+\Delta)/2uv_2\}]$$
$$\text{for } (k_F^2-u^2)^{1/2} \leq v_2 \leq k_F^- + u,$$
$$= \pi + 2\sin^{-1}\{(u^2+v_2^2-k_F^2-\Delta)/2uv_2\}$$
$$\text{for } k_F^- + u \leq v_2 \leq k_F^+ + u. \qquad \text{(A11)}$$

The contribution to the integral (A10) is made mainly by the region $v_2 \sim k_F + u$. Defining $g(x)$ by

$$g(x) = \sin^{-1}x \quad \text{for} \quad x \leq 1,$$
$$= \pi/2 \quad \text{for} \quad x \geq 1,$$

we can easily show that

$$\int_0^{1+\delta} dx\{(1/2)(g(x+\delta)+g(x-\delta)) - g(x)\} = -\delta^2/2 + 0(\delta^4). \qquad \text{(A12)}$$

In (A12) the region $x<1$ makes a positive contribution which is overcome by the contribution from $x \sim 1$. Thus taking the contribution from $v_2 \sim k_F + u$ in (A10), we obtain a rough estimate of (A9) given by

$$\Delta E_2' = (1/8\pi) U_{\text{eff}}^{0\,2} \eta_0^2 \cdot \eta_0 (\alpha\Delta)^2. \qquad \text{(A13)}$$

Comparing with the corresponding increase of the one electron energy given by $\eta_0(\alpha\Delta)^2$, we can see that (A13) amounts to only 4 percent of the one electron energy increase for $U_{\text{eff}}^0 \eta_0 = 1$. A more exact estimate shows that $\Delta E_2'$ is even smaller than the above estimate.

References

1) J. H. Van Vleck, Rev. Mod. Phys. **25** (1953), 220.
2) E. P. Wohlfarth, Rev. Mod. Phys. **25** (1953), 211.
3) J. C. Slater, Phys. Rev. **49** (1936), 537.
4) K. A. Brueckner and C. A. Levinson, Phys. Rev. **97** (1955), 1344.
K. A. Brueckner and J. L. Gammel, Phys. Rev. **109** (1958), 1023, 1040.
5) G. C. Fletcher, Proc. Phys. Soc. **65** (1952), 192.
6) B. Segall, Phys. Rev. **125** (1962), 102.
7) J. Yamashita, M. Fukuchi and S. Wakoh, J. Phys. Soc. Japan **18** (1963), 999.
8) F. Gautier, J. phys. radium **23** (1962), 738.
9) F. E. Hoare, J. C. Matthews and J. C. Walling, Proc. Roy. Soc. **A216** (1953), 502.
F. E. Hoare and J. C. Matthews, Proc. Roy Soc. **212A** (1952), 137.
D. W. Budworth, F. E. Hoare, and J. Preston, Proc. Roy. Soc. **257A** (1960), 250.

PHYSICAL REVIEW B VOLUME 22, NUMBER 11 1 DECEMBER 1980

Renormalization-group study of the Hubbard model

Jorge E. Hirsch*
The James Franck Institute and Department of Physics, University of Chicago, Chicago, Illinois 60637
(Received 7 September 1979)

The ground state of the half-filled Hubbard model is studied using a real-space renormalization-group technique. We first study in detail the one-dimensional case. We find the ground state to be insulating for all finite coupling, in agreement with the exact solution. We compute the ground-state energy, localization length, energy gap, magnitude of the local moment, and spin-density autocorrelation function. For those quantities that are exactly known we find good agreement with the exact results. Using a simple extension of our one-dimensional calculation, we are able to study approximately two- and three-dimensional lattices. We find a Mott transition at finite interaction for these cases. The critical exponents for these transitions are found to satisfy an approximate interdimensional scaling law.

I. INTRODUCTION

The Hubbard model[1] is the simplest model one can study to examine the effects of correlations between electrons in narrow energy bands. The Hamiltonian consists of a nearest-neighbor hopping term and an electron-electron repulsion U which acts when two electrons are sitting at the same site. In this paper we study the half-filled Hubbard model using a zero-temperature renormalization-group (RG) technique. The method consists in constructing iteratively a variational ground state by dividing the system into cells and keeping at each step only the lowest-lying energy states in each cell. This method has been extensively used for studying spin systems at zero temperature.[2-7] One of the purposes of this work is to show that it is useful also to study systems with fermions defined on a lattice with short-ranged interactions. We concentrate mainly on the one-dimensional (1D) chain. For this case, a closed expression has been found for the ground state.[8] For the half-filled band, the results show that there is a Mott transition at $U = 0$; the system is insulating for any nonzero U. Our approximate analysis reproduces correctly this feature of the exact solution. We compute the ground-state energy and the magnitude of the local moment, and find good agreement with the exact results.[8,9] We also study the behavior of the localization length and the energy gap near the Mott transition. The method correctly predicts an essential singularity at zero coupling, although it fails to reproduce the detailed behavior near the singularity. We can also compute arbitrary ground-state correlation functions, which cannot be simply obtained from the exact solution. As an illustration, we show the static q-dependent spin-spin correlation function.

It should be mentioned that there exists a related calculation for the 1D Hubbard model,[10] where the authors keep many energy states at each iteration, and are thus able to study approximately temperature-dependent properties. In this paper, however, we concentrate on ground-state properties and we emphasize the fact that we want the renormalized Hamiltonian to be of the same form as the original one, which makes the physics of the problem more transparent.

By a simple extension of our one-dimensional calculation, we are able to treat approximately two- and three-dimensional (2D and 3D) hypercubic lattices. It is generally thought that because of the special form of the free-electron Fermi surface for these lattices when only nearest-neighbor hopping is considered in the Hamiltonian, the ground state is insulating for any nonzero U, as in one dimension.[11,12] However, any small distortion of the Fermi surface, introducing further-than-nearest-neighbor hopping, will change this result. Our approximate method is not sensitive to that special feature of the nearest-neighbor-only case, and it predicts a Mott transition for finite U, whether we include longer-range hopping or not. In any case, any realistic model of a metal should include longer-range hopping,[13] so that our results are relevant for those cases. We obtain the critical exponents for the transition and verify that an approximate interdimensional scaling law, recently proposed for disordered electronic systems[14] and critical phenomena[15] is well satisfied.

II. ONE DIMENSION

In one dimension, the Hubbard Hamiltonian is given by

$$H=-t\sum_{i,\sigma}[c_{i\sigma}^{\dagger}c_{i+1,\sigma}+c_{i+1,\sigma}^{\dagger}c_{i,\sigma}]+U\sum_{i}n_{i\uparrow}n_{i\downarrow}-\mu\sum_{i,\sigma}n_{i\sigma}+\tfrac{1}{2}UN\ , \tag{1}$$

where $c_{i\sigma}^{\dagger}$ and $c_{i\sigma}$ are the creation and annihilation operators for an electron of spin σ at site i, and $n_{i\sigma}=c_{i\sigma}^{\dagger}c_{i\sigma}$. t gives the kinetic energy in the band and U is the Coulomb repulsion between electrons on the same site. The chemical potential μ is $\frac{1}{2}U$ for the half-filled band case.[16]

We divide the chain into (nonoverlapping) cells of three sites (we want an odd number of sites per cell if we want the renormalized Hamiltonian to describe fermions) and the Hamiltonian into an intracell part, H_0, and an intercell coupling V. V is the hopping part of the Hamiltonian that transfers an electron from one cell to the next. The intracell Hamiltonian, H_0, can be written as a sum over cells (labeled by p) of cell Hamiltonians

$$H=\sum_{p}H_0^p\ . \tag{2}$$

Next, we diagonalize exactly the intracell Hamiltonian H_0^p. Our Hilbert space has four states per site: $|0\rangle$; $c_{i\uparrow}^{\dagger}|0\rangle\equiv|+\rangle$; $c_{i\downarrow}^{\dagger}|0\rangle\equiv|-\rangle$; and $c_{i\uparrow}^{\dagger}c_{i\downarrow}^{\dagger}|0\rangle\equiv|+-\rangle$ so that we have 64 states per cell. There are, however, several conserved quantities: number of particles, z component of spin (S_z), total spin (S), and parity. We are interested in the half-filled band case, so that we consider the states with $n=2$, 3, and 4 particles; we find one nondegenerate ground state in each of the subspaces $n=2$ and 4, with $S=0$ and $S_z=0$, and two degenerate ground states in the subspace $n=3$, with $S=\frac{1}{2}$, and $S_z=\pm\frac{1}{2}$. These states are

$$\begin{aligned}
|0'\rangle &= a_1|1,0\rangle+a_2|2,0\rangle+a_3|3,0\rangle, & E_{0'}&=-2\mu+\lambda_a\ ,\\
|+'\rangle &= b_1|1,+\rangle+b_2|2,+\rangle+b_3|3,+\rangle, & E_{+'}&=-3\mu+\lambda_b\ ,\\
|-'\rangle &= b_1|1,-\rangle+b_2|2,-\rangle+b_3|3,-\rangle, & E_{-'}&=-3\mu+\lambda_b\ ,\\
|+-'\rangle &= a_1|1,+-\rangle+a_2|2,+-\rangle+a_3|3,+-\rangle, & E_{+-'}&=-4\mu+\lambda_a+U\ .
\end{aligned} \tag{3}$$

TABLE I. States that form the lowest-lying eigenstates for three-site cells in the 1D Hubbard model.

n	S_z	S	States
2	0	0	$\|1,0\rangle=\frac{1}{\sqrt{2}}(\|+\rangle\|0\rangle\|-\rangle-\|-\rangle\|0\rangle\|+\rangle)$ $\|2,0\rangle=\frac{1}{2}(\|+\rangle\|-\rangle\|0\rangle-\|-\rangle\|+\rangle\|0\rangle+\|0\rangle\|+\rangle\|-\rangle-\|0\rangle\|-\rangle\|+\rangle)$ $\|3,0\rangle=\frac{1}{\sqrt{6}}(\|+-\rangle\|0\rangle\|0\rangle+2\|0\rangle\|+-\rangle\|0\rangle+\|0\rangle\|0\rangle+-\rangle)$
3	$\frac{1}{2}$	$\frac{1}{2}$	$\|1,+\rangle=\frac{1}{\sqrt{2}}(-\|+-\rangle\|+\rangle\|0\rangle+\|0\rangle\|+\rangle\|+-\rangle)$ $\|2,+\rangle=\frac{1}{2}(-\|+-\rangle\|0\rangle\|+\rangle-\|0\rangle\|+-\rangle\|+\rangle+\|+\rangle\|0\rangle\|+-\rangle+\|+\rangle\|+-\rangle\|0\rangle)$ $\|3,+\rangle=\frac{1}{\sqrt{6}}(\|+\rangle\|+\rangle\|-\rangle-2\|+\rangle\|-\rangle\|+\rangle+\|-\rangle\|+\rangle\|+\rangle)$
3	$-\frac{1}{2}$	$\frac{1}{2}$	$\|1,-\rangle=\frac{1}{\sqrt{2}}(-\|+-\rangle\|+\rangle\|0\rangle+\|0\rangle\|-\rangle\|+-\rangle)$ $\|2,-\rangle=\frac{1}{2}(-\|+-\rangle\|0\rangle\|-\rangle-\|0\rangle\|+-\rangle\|-\rangle+\|-\rangle\|0\rangle\|+-\rangle+\|-\rangle\|+-\rangle\|0\rangle)$ $\|3,-\rangle=\frac{1}{\sqrt{6}}(-\|-\rangle\|-\rangle\|+\rangle+2\|-\rangle\|+\rangle\|-\rangle-\|+\rangle\|-\rangle\|-\rangle)$
4	0	0	$\|1,+-\rangle=\frac{1}{\sqrt{2}}(\|-\rangle\|+-\rangle\|+\rangle-\|+\rangle\|+-\rangle\|-\rangle)$ $\|2,+-\rangle=\frac{1}{2}(\|+\rangle\|-\rangle\|+-\rangle+\|+-\rangle\|+\rangle\|-\rangle-\|-\rangle\|+\rangle\|+-\rangle-\|+-\rangle\|+\rangle\|-\rangle)$ $\|3,+-\rangle=\frac{1}{\sqrt{6}}(\|0\rangle\|+-\rangle\|+-\rangle+2\|+-\rangle\|0\rangle\|+-\rangle+\|+-\rangle\|+-\rangle\|0\rangle)$

The states $|1,0\rangle$, $|2,0\rangle$, etc., are listed in Table I. The quantities λ_a and (a_1,a_2,a_3) are the lowest eigenvalue and corresponding eigenvector of

$$A=\begin{pmatrix} 0 & -\sqrt{2}t & 0 \\ -\sqrt{2}t & 0 & -\sqrt{6}t \\ 0 & -\sqrt{6}t & U \end{pmatrix} , \quad (4)$$

while λ_b and (b_1,b_2,b_3) are the lowest eigenvalue and corresponding eigenvector of

$$B=\begin{pmatrix} U & -\sqrt{2}t & 0 \\ -\sqrt{2}t & U & -\sqrt{6}t \\ 0 & -\sqrt{6}t & 0 \end{pmatrix} . \quad (5)$$

We will keep only these states per cell in defining our renormalized Hamiltonian. Note that these 4 states very much resemble the original site states $|0\rangle$, $|+\rangle$, $|-\rangle$, $|+-\rangle$ in that the spin quantum numbers are the same and their occupation number also if we subtract two from n in all the new states. We define new cell-fermion operators by the relations

$$\begin{aligned} c_\uparrow^{\dagger\prime}|0'\rangle &= |+'\rangle , \\ c_\downarrow^{\dagger\prime}|0'\rangle &= |-'\rangle , \\ c_\uparrow^{\dagger\prime}c_\downarrow^{\dagger\prime}|0'\rangle &= -c_\downarrow^{\dagger\prime}c_\uparrow^{\dagger\prime}|0'\rangle = |+-'\rangle . \end{aligned} \quad (6)$$

The intracell Hamiltonian restricted to the subspace of these four states only can be written in terms of the new cell-fermion operators as

$$\begin{aligned} H_0' &= E_{0'} + (E_{+'} - E_{0'})(n_\uparrow' + n_\downarrow') \\ &\quad + (E_{+-'} + E_{0'} - 2E_{+'})n_\uparrow' n_\downarrow' \end{aligned} \quad (7)$$

with $n_\sigma' = c_\sigma^{\dagger\prime} c_\sigma'$, so that it has the same form as the original Hamiltonian for one site. For obtaining the intercell coupling we compute the matrix elements of the old fermion operators on the boundary site of the cell, c_σ^b, with the states we are keeping, and we find

$$\begin{aligned} \langle 0'|c_\uparrow^b|+'\rangle &= \langle -'|c_\uparrow^b|+-'\rangle = \langle 0'|c_\downarrow^b|-'\rangle \\ &= -\langle +'|c_\downarrow^b|+-'\rangle = \lambda \end{aligned} \quad (8)$$

with

$$\lambda = \frac{1}{2\sqrt{2}}(a_1b_2 + a_2b_1) + \frac{3}{2\sqrt{6}}(a_2b_3 + a_3b_2) \quad (9)$$

so that we can identify

$$c_\sigma^b = \lambda c_\sigma' \quad (10)$$

and our renormalized Hamiltonian has exactly the same form as the original one (except for an additive constant). We can then iterate this procedure, and obtain after the nth iteration a Hamiltonian of the form

$$\begin{aligned} H^{(n)} &= -t_n \sum_{i,\sigma} (c_{i\sigma}^\dagger c_{i+1,\sigma} + c_{i+1,\sigma}^\dagger c_{i,\sigma}) \\ &\quad + U_n \sum_i n_{i\uparrow} n_{i\downarrow} - \mu_n \sum_{i,\sigma} n_{i\sigma} + \sum_i d_n , \end{aligned} \quad (11)$$

with the coefficients determined by the recursion relations

$$\begin{aligned} U_{n+1} &= U_n + 2(\lambda_a^{(n)} - \lambda_b^{(n)}) , \quad t_{n+1} = \lambda_n^2 t_n , \\ \mu_{n+1} &= \mu_n + \lambda_a^{(n)} - \lambda_b^{(n)} , \quad d_{n+1} = 3d_n + \lambda_a^{(n)} - 2\mu_n , \end{aligned} \quad (12)$$

with λ_n, $\lambda_a^{(n)}$, $\lambda_b^{(n)}$ given by Eq. (8) and the diagonalization of (4) and (5), with t_n, and U_n replacing t and U. The initial conditions are

$$U_0 = U , \quad t_0 = t , \quad \mu_0 = \tfrac{1}{2}U , \quad d_0 = \tfrac{1}{2}U . \quad (13)$$

Note that the relation $\mu_n = \frac{1}{2}U_n$ is preserved at all steps of our iteration if we start with $\mu_0 = \frac{1}{2}U_0$.

We analyze the recursion relations for $y = U/t$ and find only two fixed points, $y=0$ and ∞. That is, the ground state is analytic as a function of y, except at the origin, as found in the exact solution. Starting with any nonzero y, the recursion relations lead to the $y=\infty$ fixed point; i.e., at each iteration the intercell hopping term becomes weaker with respect to the Coulomb interaction. This shows that the ground state is insulating for any nonzero interaction, so that the Mott transition occurs for $U=0$.

From the constant term in the Hamiltonian we obtain the ground-state energy

$$E_G = \lim_{n\to\infty} \frac{d_n}{3^n} = \sum_{n=1}^{\infty} \frac{\lambda_a^{(n)} - 2\mu_n}{3^n} + \tfrac{1}{2}U . \quad (14)$$

This is shown in Fig. 1, compared with the exact results of Lieb and Wu.[8] This method always yields an upper bound to the exact ground-state energy. The reason is that the Hamiltonian after n steps, $H^{(n)}$, equals the original Hamiltonian H truncated to some subspace of the original Hilbert space. Thus, the ground-state energy of $H^{(n)}$ equals the expectation value of the ground state of $H^{(n)}$ with the original Hamiltonian H and is, by the variational principle, an upper bound to the true ground-state energy

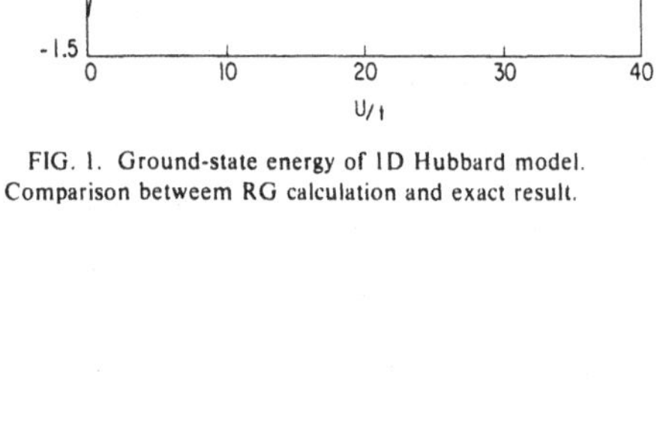

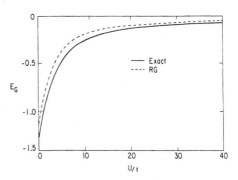

FIG. 1. Ground-state energy of 1D Hubbard model. Comparison betweem RG calculation and exact result.

of H. When n increases, the constant term in $H^{(n)}$ grows as 3^n while the rest of $H^{(n)}$ remains of the same order, so that the ground-state energy will be given by the constant term in $H^{(n)}$ as $n \to \infty$.

We have also computed the magnitude of the local moment, defined by

$$L_0 = \langle S^2 \rangle = \tfrac{3}{4} \langle n_\uparrow + n_\downarrow - 2n_\uparrow n_\downarrow \rangle \ . \tag{15}$$

This quantity is $\frac{3}{4}$ in the $U=\infty$ limit, when the electrons are completely localized, and $\frac{3}{8}$ in the free-electron limit $U=0$ and it gives therefore an idea of the degree of localization of the electrons for all U. To compute the average of an operator, we calculate the part of it contained in the subspace spanned by the states we are keeping by calculating the matrix elements of the operator with those states. This gives the renormalized operator, and its ground-state average will be approximately the same as that of the original operator. For a site operator like the local moment, we take it always at the center of the cell to minimize end effects. The renormalized operator for the local moment has the same form as the original one, except for an additive and a multiplicative constant. We obtain the recursion relation

$$L_0 = \tfrac{3}{4} a_2^2 + (b_1^2 + b_3^2 - a_2^2) L_0' \ , \tag{16}$$

where L_0' is the local moment for the system described by the renormalized Hamiltonian. Iterating this relation we obtain L_0. Our results are plotted in Fig. 2 and compared with the exact results, which can be obtained from the exact ground-state energy by[9]

$$L_0 = \tfrac{3}{4} - \tfrac{3}{2} \frac{dE_G}{dU} \ . \tag{17}$$

We obtain the limits $U=0$ and ∞ correctly, and the agreement for all U is very good.

We want to define a length that measures the distance over which the electrons are localized in the ground state. Consider the correlation function

$$F(R) = \frac{1}{N} \sum_i \langle c_{i+R,\sigma}^\dagger c_{i,\sigma} \rangle \ . \tag{18}$$

This function will presumably decay exponentially in the large U region, as can be seen from the following perturbation-theory argument: start from the ground state for $U=\infty$, with all sites singly occupied (the antiferromagnetic Heisenberg chain ground state in the 1D case) and construct the ground state for large but finite U in perturbation theory in t/U. For obtaining a nonvanishing result for $F(R=na)$ (a the lattice spacing) we have to go to nth order in perturbation theory, so that

$$F(r) \sim \left[\frac{t}{U}\right]^n \sim e^{-R|\ln t/U|} \ . \tag{19}$$

On the other hand, in the free-electron limit

$$F(R) = \frac{1}{N} \sum_k e^{ikR} \langle n_k \rangle = \frac{1}{N} \sum_{k<kF} e^{ikR} \ . \tag{20}$$

It is easy to convince oneself that $F(R)$ will decay algebraically (with oscillations) for any reasonable Fermi surface. In particular, one obtains $F(R) \sim 1/R^d$ for a d-dimensional hypercubic lattice with nearest-neighbor hopping only and R in the direction of an axis, and $F(R) \sim 1/R^{(d+1)/2}$ for a spherical Fermi surface in d dimensions. This suggests that quite generally $F(R)$ will have an algebraic decay

$$F(R) \sim 1/R^\eta \tag{21}$$

on the metallic side of the Mott transition (with η possibly depending on the coupling strength) and an exponential decay

$$F(R) \sim e^{-R/\xi} \tag{22}$$

on the insulating side. ξ gives then a measure of the localization of the electrons in the insulator and it diverges at the Mott transition.

Let us compute this correlation function for the one-dimensional case with our RG approach: by the method previously discussed we obtain

$$F(R) = \tfrac{2}{3} \lambda^2 F'(R/3) \tag{23}$$

so that for $R = 3^n$ we obtain after n iterations

$$F(3^n) = (\tfrac{2}{3})^n \left[\prod_{j=1}^{n} \lambda_j\right] \langle c_{1\sigma}^\dagger c_{2\sigma} \rangle H^{(n)} \ . \tag{24}$$

In the free-electron limit $\lambda = 1/\sqrt{2}$, so that we obtain for the exponent

$$\eta = \frac{-\ln \frac{2}{3} \lambda^2}{\ln 3} = 1 \tag{25}$$

in agreement with the exact result. In the case of

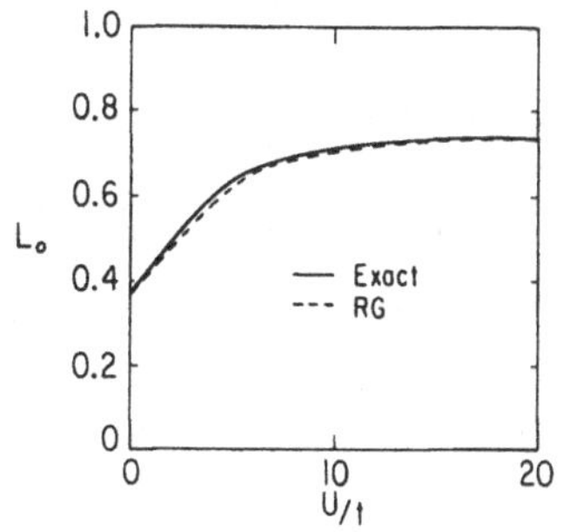

FIG. 2. Magnitude of the local moment in 1D Hubbard model. Comparison between RG calculation and exact result.

nonzero U, we scale to the large-U regime; taking into account that the nearest-neighbor average goes to zero as

$$\langle c_{1\sigma}^{\dagger}c_{2\sigma}\rangle_{H(n)} \sim t_n/U_n \tag{26}$$

and that the large-U recursion relations are

$$\frac{U_{n+1}}{t_{n+1}} = \frac{1}{\lambda_n^2}\frac{U_n}{t_n} \tag{27}$$

with $\lambda_n = \sqrt{3/4}$, we obtain

$$F(3^n) \sim (\tfrac{2}{3}\lambda^4)^n \ . \tag{28}$$

We see that we again obtain algebraic decay of the correlation function, this time with an exponent

$$\eta = \frac{-\ln\frac{2}{3}\lambda^4}{\ln 3} = 3.42 \ . \tag{29}$$

This failure of the RG method to display exponential decay of the correlation functions is well known to occur in other models also and has been discussed in detail by Fradkin and Raby.[17] Nevertheless, we obtain a much faster decay of the correlation function in the large-U region.

Even within this limitation of our method we can obtain an estimate of the localization length near the Mott transition as the distance over which a crossover occurs from the free-electron rate of decay of $F(R)$ to a faster rate of decay: start with a system with small y_0 and assume after n_0 iterations we have scaled to the large coupling region $y \sim 1$; i.e., $U \sim t$. The localization length will be given approximately by

$$\xi \sim 3^{n_0} \ . \tag{30}$$

For obtaining n_0, we consider the recursion relation for small y, which is

$$y_{n+1} = y_n + a y_n^3 \tag{31}$$

with $a = 0.047$. We can then write

$$dy = a y^3 d_n \tag{32}$$

and integrating we find

$$n_0 \approx \frac{1}{2 a y_0^2} \tag{33}$$

so that the localization length goes as

$$\xi \sim e^{bt^2/U^2} \tag{34}$$

with $b = 11.7$. We see that the localization length diverges extremely rapidly for small U/t (faster than any power) due to the very slow growth of y under iterations for small y.

A similar behavior is obtained for the energy gap of the system. The electrical conductivity at zero temperature is determined by the energy gap between the ground state and the lowest excited state that is connected to the ground state by a nonzero matrix element of the current operator. Within our RG treatment, this energy gap is given by the limiting value of U after infinite iterations: $E_g = U_\infty$. (The recursion relations for t and U separately give $t_n \to 0$, $U_n \to U_\infty > 0$ as $n \to \infty$.) For small initial U/t we obtain the behavior

$$E_g \sim e^{-ct^2/U^2} \tag{35}$$

with $c = 7.4$. The behavior of the localization length and the energy gap show that the system has an essential singularity at $U = 0$ in one dimension.

The energy gap for the 1D Hubbard model can be calculated exactly from the Lieb and Wu solution, and one finds for small U[18]

$$E_g \sim e^{-2\pi t/U} \ .$$

That is, although the RG method succeeds in predicting an essential singularity at $U = 0$, it predicts the wrong power in the exponent (note, however, that the coefficient c in the exponent is close to the exact result). The behavior (35) is related to the fact that the quadratic term in y_n is missing in the recursion relation (31). This cannot be corrected by taking larger cells in the RG calculation, since it has been shown by Pfeuty (private communication) that the same behavior would be obtained by taking odd-site cells of any size, if only the four states around the half-filled band occupation are kept. Clearly, further work is needed in this direction.

Finally, we have computed the q-dependent spin-spin correlation function, defined by

$$G(q) = \frac{1}{N}\sum_{i,j} e^{iq(R_i - R_j)}\langle \sigma_z^i \sigma_z^j \rangle \ . \tag{36}$$

By the method previously discussed we obtain a recursion relation that involves only the same correlation function for wave vector three times as large and the local moment, both in the renormalized system

$$G(q) = 1 + \tfrac{4}{3}e_3\cos(qa) + \tfrac{2}{3}e_5\cos(2qa) + \tfrac{4}{9}[4e_4\cos(qa) + 2e_6\cos(2qa)]L_0' + \tfrac{1}{3}[2e_1^2 + e_2^2 + 4e_1e_2\cos(qa) + 2e_1^2\cos(2qa)][G'(3q) - 1] \ , \tag{37}$$

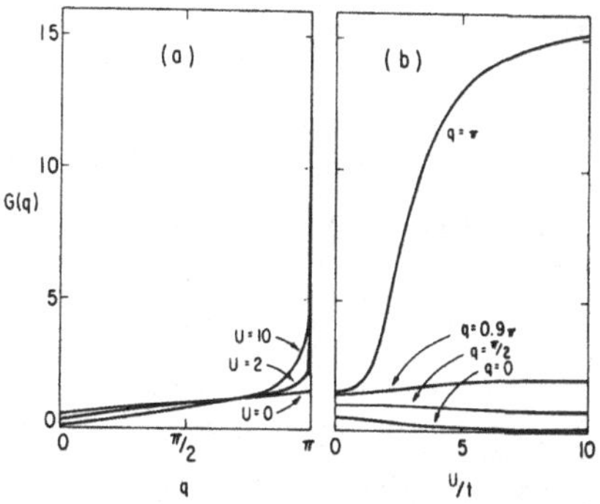

FIG. 3. Spin-spin correlation function $G(q)$ in 1D Hubbard model from the RG calculation. (a) For fixed U as a function of q. (b) For fixed q as a function of U.

with

$$
\begin{aligned}
e_1 &= \tfrac{1}{2}b_2^2 + \tfrac{2}{3}b_3^2 \ , \\
e_2 &= b_1^2 - \tfrac{1}{3}b_3^2 \ , \\
e_3 &= -\tfrac{1}{2}a_2^2 \ , \\
e_4 &= -\tfrac{2}{3}b_3^2 + \tfrac{1}{2}a_2^2 \ , \\
e_5 &= -a_1^2 \ , \\
e_6 &= \tfrac{1}{3}b_3^2 + a_1^2 \ ,
\end{aligned}
\qquad (38)
$$

so that iterating Eq. (37) together with Eq. (16) we obtain the correlation function. In Fig. 3 we plot $G(q)$: (a) as a function of q for several values of U; and (b) as a function of U for several values of q. Note the sharp increase in the correlation function for large U and $q \sim \pi$. In the large U limit, the system is equivalent to an antiferromagnetic Heisenberg model with coupling t^2/U and the increase in $G(q)$ at $q \sim \pi$ signals the tendency to antiferromagnetic ordering. In the antiferromagnetic Heisenberg chain, the ground-state spin-spin correlation function is believed to decay with distance as $\langle \sigma_z^0 \sigma_z^{R=na} \rangle \sim (-1)^n/R$ (Ref. 19), and hence the $q=\pi$ Fourier transform is divergent. We obtain in the large U limit $\langle \sigma_z^0 \sigma_z^R \rangle \sim (-1)^n/R^{1.07}$, which is close to the above result. Unfortunately, this small error in the exponent causes our calculated $G(q)$ to go to a large finite value instead of diverging for $q \to \pi$, and $U \to \infty$.

III. TWO AND THREE DIMENSIONS

We next apply our method to a two-dimensional square lattice. The smallest sensible choice for a cell in this case is a 3×3 square (we want an odd number of sites per cell). This is, however, not simple, since we have to deal with $4^9 = 262\,144$ states (although there are simplifications due to symmetries). For that reason, we implement the simplifying scheme shown in Fig. 4. We perform our transformation in two steps; in the first step we take three-site linear cells in the x direction and treat all the couplings in the y direction approximately, and in the second step we do the same procedure with a 90° rotation. Remarkably, we recover after these two steps an isotropic system.

As mentioned earlier, this system with only nearest-neighbor hopping is believed to exhibit the same behavior as the 1D model, a Mott transition at $U=0$. Any small distortion of the free-electron Fermi surface, however, by introducing longer range hopping, will change this very special feature. We have performed calculations with nearest-neighbor hopping only and including small second- and third-nearest-neighbor hopping terms and find a Mott transition for finite U in all cases. That is, our method is insensitive to that subtle property of the model with nearest-neighbor hopping only. Nevertheless, a Mott transition for a finite U is what one would expect in any higher-dimensional real material. Since our results do not change significantly by introducing second- and third-nearest-neighbor hopping terms, we will discuss for simplicity the nearest-neighbor hopping model only. Our recursion relations are now

$$
\begin{aligned}
U_{n+1} &= \bar{U}_{n+1} + 2(\bar{\lambda}_a^{(n+1)} - \bar{\lambda}_b^{(n+1)}) \ , \\
t_{n+1} &= 2\bar{\lambda}_{n+1}^2 \bar{t}_{n+1} \ ,
\end{aligned}
\qquad (39)
$$

where $\bar{U}_{n+1}$ and $\bar{t}_{n+1}$ are obtained from U_n and t_n by the 1D recursion relations (II) and $\bar{\lambda}_a^{(n+1)}$, $\bar{\lambda}_b^{(n+1)}$, and $\bar{\lambda}_{n+1}$ are given by the same functions of $\bar{U}_{n+1}$ and $\bar{t}_{n+1}$ as in the 1D case. Analyzing these recursion relations, we now find a nontrivial fixed point at

$$
\left(\frac{U}{t}\right)_c = 3.72 \qquad (40)
$$

which describes a Mott transition; below this point, we scale to the free-electron fixed point $U/t=0$, above this to the strong interaction limit $U/t=\infty$. The value one obtains from the Hubbard III calcula-

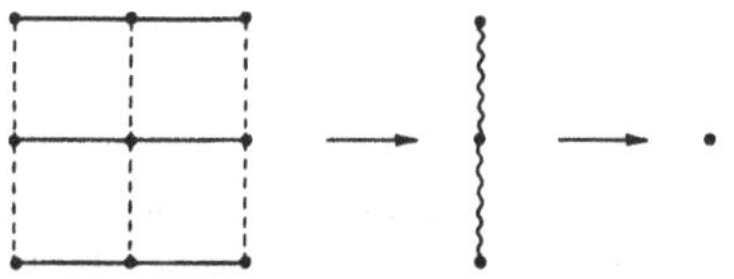

FIG. 4. Approximate RG transformation in 2D. First we take three-site cells in the x direction, then in the y direction. After the first step the couplings in the x and y direction are different, but they are again equal after the second step.

tion[1] for a two-dimensional lattice with a parabolic density of states is $(U/t)_c = 6.93$. This discrepancy is not surprising, if we take into account that the densities of states for both calculations are quite different. Also, our method probably underestimates y_c since we are neglecting transitions to higher states that could enhance delocalization.

We study the behavior of the localization length near the fixed point. In this case, the localization length has a power-law behavior

$$\xi \sim |y - y_c|^{-\nu} . \tag{41}$$

We can obtain the exponent ν by the same arguments discussed in the 1D case or by the standard method of linearizing the recursion relations around the fixed point; we find

$$\nu = 1.40 . \tag{42}$$

For the energy gap above the transition, we also obtain an algebraic behavior

$$E_g = U_\infty \sim |U - U_c|^s . \tag{43}$$

Plotting $\log E_g$ vs $\log |U - U_c|$ we obtain

$$s = 1.34 . \tag{44}$$

This can also be obtained via the scaling relation $s = \nu z$, where z is the "dynamic exponent" which is related to the factor by which the energy scales at the fixed point under the RG transformation, $c(l):c(l) = 1/l^z$ (l being the length scale change). In this case, $z = 0.963$.

It is interesting to compare our result for the gap exponent with the exponent obtained from Hubbard's calculation. For that case, one obtains $s = 1.5$, which is close to our result. In another calculation, however,[20] it is predicted that the gap exponent at the Mott transition should be $s = 0.5$. This result is obtained by taking into account the localization of the pseudoparticle states near the band edges of the Hubbard band. Our result, however, lends support to the original value of s, at least in two dimensions.

Finally, we can also extend this calculation to three dimensions by the same method, although one might expect it to be much less accurate. Again we find a Mott transition, now at

$$\left(\frac{U}{t}\right)_c = 4.76 \tag{45}$$

and the critical exponents are $\nu = 0.826$ and $s = 1.08$.

It is interesting to note that our exponents satisfy an approximate interdimensional scaling law recently proposed for electronic problems in disordered media[14] and for critical phenomena[15]

$$\nu_d = \nu_{d-1} \frac{5-d}{4-d-2\nu_{d-1}} . \tag{46}$$

In one dimension, we have $\nu_1 = \infty$; substituting in Eq. (44), we find $\nu_2 = 1.5$, which is satisfied by our result [Eq. (40)] within 7%. Putting $\nu_2 = 1.40$, we find $\nu_3 = 0.74$, which differs from our result by about 10%. Note that even for the Ising model this scaling relation is only approximately satisfied. These arguments lend further support to our higher-dimensional calculation.

We have also computed the exponent η in the metallic region for the two- and three-dimensional cases. We find $\eta = 2$ and 3, respectively, independently of the coupling, which are the exact results for $U = 0$. This happens because we scale always to the $U = 0$ fixed point on the metallic side of the transition. However, one cannot rule out the possibility that there may be a line of fixed points in the metallic region with continuously varying critical exponents which is missed in this simple RG analysis.

IV. SUMMARY

In summary, we have studied the half-filled Hubbard model at zero temperature using a real-space RG technique. Although the method is approximate, we believe it gives a good qualitative description of the behavior of these models, as it has for other cases. In one dimension, we obtained correctly the fact that the ground state is insulating for any nonzero U. We calculated the ground-state energy and the magnitude of the local moment and found good agreement with the exact results. We defined and calculated the localization length for the electrons in the ground state, and found that it diverges exponentially fast as $U \to 0$. A similar behavior is found for the energy gap, which goes to zero exponentially fast, in agreement with the exact solution. However, the power of the exponent is not given correctly by the RG method. We also calculated the spin-spin static correlation function and found a strong tendency towards antiferromagnetic ordering at large U, although we failed to obtain the divergency for $U = \infty$ that occurs in the antiferromagnetic Heisenberg chain. We extended the calculation to two and three dimensions and found a Mott transition at a finite U for these cases. The critical exponents for the transition were found to satisfy an approximate interdimensional scaling law recently proposed. We believe that the method will be useful for studying other fermionic Hamiltonians in one and higher dimensions.

ACKNOWLEDGMENTS

I acknowledge helpful discussions with Professor G. Mazenko and his constant encouragement, as well as his careful reading of the manuscript. A useful discussion with Professor J. Hertz is also acknowledged. This work was supported by the Materials Research Laboratory program of the National Science Foundation at the University of Chicago. In addition, the author was supported by a Victor J. Andrew Memorial Fellowship.

---

*Present address: Institute for Theoretical Physics, University of California, Santa Barbara, Calif. 93106.

[1]J. Hubbard, Proc. R. Soc. London Ser. A 276, 238 (1963); 281, 401 (1964).

[2]S. D. Drell, M. Weinstein, and S. Yankielowicz, Phys. Rev. D 14, 1769 (1977).

[3]R. Jullien, P. Pfeuty, J. N. Fields, and S. Doniach, Phys. Rev. B 18, 3568 (1978); R. Jullien, J. N. Fields, and S. Doniach, *ibid.* 16, 4889 (1977).

[4]J. E. Hirsch and G. F. Mazenko, Phys. Rev. B 19, 2656 (1979); J. E. Hirsch, Phys. Rev. B 20, 3907 (1979).

[5]J. N. Fields, Phys. Rev. B 19, 2637 (1979); J. N. Fields, H. W. J. Blöte, and J. C. Bonner, J. Appl. Phys. 50, 1807 (1979).

[6]R. Jullien and P. Pfeuty, Phys. Rev. B 19, 4646 (1979); W. A. Penson, R. Jullien, and P. Pfeuty, *ibid.* 19, 4653(1979).

[7]D. C. Mattis and J. Gallardo (unpublished).

[8]E. Lieb and F. Wu, Phys. Rev. Lett. 20, 1445 (1968).

[9]H. Shiba and P. A. Pincus, Phys. Rev. B 5, 1966 (1972).

[10]S. T. Chui and J. W. Bray, Phys. Rev. B 18, 2426 (1978).

[11]P. Richmond, Solid State Commun. 7, 997 (1969).

[12]P. M. Chaikin, P. Pincus, and G. Beni, J. Phys. C 8, L65 (1975).

[13]B. H. Brandow, J. Phys. C 8, L357 (1975).

[14]K. F. Freed, J. Phys. C 12, L17 (1979).

[15]Y. Imry, G. Deutscher, and D. J. Bergmann, Phys. Rev. A 7, 744 (1973).

[16]S. Methfessel and D. Mattis, in *Handbuch der Physik*, edited by S. Flügge (Springer Verlag, Berlin, 1968), Vol. 18, Pt. 1.

[17]E. Fradkin and S. Raby, Phys. Rev. D 20, 2566 (1979).

[18]A. A. Ovchinnikov, Sov. Phys. JETP 30, 1160 (1970).

[19]A. Luther and I. Peschel, Phys. Rev. B 12, 3908 (1975).

[20]F. Yonezawa and M. Watabe, Phys. Rev. B 8, 4540 (1973).

Chapter 5

NUMERICAL MONTE CARLO AND EXACT DIAGONALIZATION RESULTS

The lack of exact results in more than one dimension has also stimulated the growth of numerical investigations on finite systems, both by exact diagonalization and by quantum Monte Carlo (QMC) simulations. The latter are able to manage relatively large systems, with respect to the former, which, on the other hand, provide much more precise results and can be used in turn for testing the efficiency of the QMC simulations. Monte Carlo simulations have been used at different stages in the solution of the Hubbard model, both in *ab initio* numerical approaches to system of interacting electrons, or as an essential tool within the framework of the variational approach[r5.1],[5.6], to calculate correctly variational expectation values (variational Monte Carlo, VMC).

While the major drawback of exact diagonalization studies is the exponential growth of dimension of configuration space with the size of the system, the two central problems one has to deal with, using QMC methods, are the so-called *fermion sign problem*, and the achievment of numerical stability in calculations at low temperatures.

The *sign problem* consists in the fact that, particularly for large values of the interaction, some configurations turn out to be ascribed a probability which is negative. The particle-hole symmetry of the Hubbard model consents to prove[5.1], by means of a Hubbard-Stratonovich formulation, that this is never the case for a half-filled band. Within this framework, the numerical QMC study of the 2-dimensional Hubbard model at half-filling shows in particular that the system, at $T = 0$, is an antiferromagnetically ordered insulator for any value of the coupling constant.

The numerical instability at low temperature is mainly due to the fact that the eigenvalues of the fermion matrix — whose determinant has to be evaluated in order to compute the grand-canonical partition function and related quantities — grow exponentially with the inverse temperature[5.3]. The various algorithms developed in order to avoid such problems[5.2],[5.3] have a common key feature in the separation of the different energy scales appearing in the fermion matrix by means of an orthogonalization procedure. Applied to the study of the 2-dimensional Hubbard model[r5.2],[5.3], such algorithms allow to obtain some of its ground state and low temperature properties. In particular, at half-filling evidence is found of a long-range antiferromagnetic order even at $T \neq 0$, whereas at a quarter-filling the system seems to be non-magnetic. The d-wave pairing is found to be attractive near half-filling, although no signals appear of a phase transition to a superconducting state.

The problem of the existence of a superconducting phase has been also investigated within the VMC approach. In [5.4] the Gutzwiller wavefunction was extended to include antiferromagnetic long-range order, and it was found[r5.1] that the paramagnetic wavefunction is stable against antiferromagnetism only for a finite number of holes in a half-filled band. Then in [5.5] the stability of both the paramagnetic Gutzwiller wavefunction and the antiferromagnetic wavefuction against formation of s- and d-pairs has been studied, in a nearly half-filled band and for a square lattice. It is found that the s-wave pairing is always unfavoured, whereas d-wave pairing is favoured only with respect to paramagnetic wavefunction.

Noticeably, the pairing for a two-dimensional Hubbard model was also explored by means of exact diagonalization on small clusters[5.7]. Indeed, it has been found on an eight-site cluster that, among the various pairings which can take place, the most favourable involves a mixture of s- and d-wave states, which, however, is suppressed by the Hubbard interaction. Thus it is once more suggested that superconductivity cannot be described by a one-band Hubbard model.

[r5.1] H. Yokoyama, and H. Shiba, *J. Phys. Soc. Japan* **56**, 1490 (1987); 3570 (1987).

[r5.2] S. Sorella, A. Parola, M. Parrinello, and E. Tosatti, *Int. J. Mod. Phys.* **B3**, 1875 (1989).

PHYSICAL REVIEW B VOLUME 31, NUMBER 7 1 APRIL 1985

Two-dimensional Hubbard model: Numerical simulation study

J. E. Hirsch
Department of Physics, University of California, San Diego, La Jolla, California 92093
(Received 1 October 1984)

We have studied the two-dimensional Hubbard model on a square lattice with nearest-neighbor hopping. We first discuss the properties of the model within the mean-field approximation: Because of the form of the band structure, some peculiar features are found. We then discuss the simulation algorithm used and compare simulation results with exact results for 6-site chains to test the reliability of the approach. We present results for thermodynamic properties and correlation functions for lattices up to 8×8 in spatial size. The system is found to be an antiferromagnetic insulator for all values of the coupling constant at zero temperature in the half-filled-band case, but the long-range order is much smaller than predicted by mean-field theory. We perform a finite-size-scaling analysis to determine the character of the transition at zero coupling. For non-half-filled-band cases, our results suggest that the system is always paramagnetic, in contradiction with Hartree-Fock predictions. The system does not show tendency to ferromagnetism nor triplet superconductivity in the parameter range studied. We also discuss some properties of the attractive Hubbard model in the half-filled-band case.

I. INTRODUCTION

The Hubbard model[1] is defined by the lattice Hamiltonian:

$$H=\sum_{\substack{i,j\\ \sigma}} t_{ij}(c_{i\sigma}^{\dagger}c_{j\sigma}+\text{H.c.})+U\sum_{i} n_{i\uparrow}n_{i\downarrow}-\mu\sum_{i}(n_{i\uparrow}+n_{i\downarrow}) \;. \tag{1.1}$$

It describes a single s band in a tight-binding basis, with a local electron-electron repulsion U for electrons of opposite spin at the same atomic orbital. The model is thought to be appropriate to describe the main features of electron correlations in narrow energy bands, leading to collective effects such as itinerant magnetism and metal-insulator transition, and has been often used to describe real materials exhibiting these phenomena.[2] A detailed justification for Eq. (1) as a model for narrow-band systems has been given by Hubbard.

Although simple in appearance, the model cannot be solved exactly except in one dimension.[3] Even there, the exact solution provides only partial information about the system. In more than one dimension, the model is not exactly solvable and a variety of approximate techniques have been used to study it, among others mean-field theory,[4] Green's-function decoupling schemes,[1] functional integral formulations,[5] and variational approaches.[6] These techniques are uncontrolled (except perhaps for weak coupling) and often give conflicting results, so that it is fair to say that no general agreement exists on what the properties of the model are.

An important approach to the problem is the study of (small) finite systems. In one dimension, this was first done by Shiba and Pincus for chains of up to 6 sites.[7] More recently, exact diagonalization of chains of up to 12 sites have been performed.[8] Because in exact diagonalizations the computer time needed for a calculation increases exponentially with the size of the system, the method becomes of limited use in more than one dimension. Results for higher-dimensional lattices have been recently reported by Kawabata[9] (for systems of up to 8 sites) and by Takahashi[10] for the somewhat simpler case where $U=\infty$ (up to 12-site systems). However, these systems are too small to allow one to draw conclusions about the properties of the model in the thermodynamic limit.

An alternative method to study finite lattices is Monte Carlo simulations. In one dimension, an algorithm exists where the computer time increases *linearly* with the size of the system,[11] and chains of up to 40 sites have been studied with modest amounts of computer time.[11–13] In more than one dimension, no algorithm where the computer time increases linearly with the size is known. Here, we have used a discrete Hubbard-Stratonovich transformation to convert the problem into one of free electrons interacting with a time-dependent Ising field,[14] together with an exact updating algorithm for the fermion Green's function[15] to compute the relative weights of the Ising configurations. The computer time in this algorithm increases with the *cube* of the size of the system, and we have used it to study lattices of up to 64 sites (8×8 two-dimensional lattices). This is well beyond the reach of exact diagonalization techniques at present, and large enough to allow us to draw some conclusions about the infinite system. We have recently reported results of simulations on this model in a short communication.[16] Here, we discuss the approach in more detail and present additional results.

We discuss results of our study for the two-dimensional square lattice with nearest-neighbor hopping only. The properties of the model are sensitive to the band structure, and two distinct features of our model in the half-filled-band case, namely nesting of the Fermi surface and a singularity in the density of states at the Fermi energy, determine much of its properties. A different band structure (including next-nearest-neighbor hopping, for example) should have rather different properties, and will be discussed in a future publication. This feature of fermion models restricts the "universality classes" more than, for example, in classical spin models, where the properties are usually independent of the details of the short-range couplings. Nevertheless, we believe our results for the half-filled band should apply at least qualitatively for models with nested Fermi surface. For the non-half-filled band, we believe our conclusions are quite general.

We compare our results with predictions of Hartree-Fock theory, thus illustrating the effect of fluctuations in changing the mean-field solution. A comparison with more sophisticated approximate theories provides information about the reliability of these schemes and will be discussed elsewhere.

The main question addressed in this paper concerns magnetism due to itinerant electrons. Can the Hubbard model provide a sensible description of it? Because our model is two dimensional, we can only have magnetic long-range order in the ground state; at $T>0$, a continuous symmetry cannot be broken in two dimensions. We find that the model does exhibit long-range antiferromagnetic order in the ground state for the half-filled-band case, even in the presence of a substantial delocalization of the electrons. The system, however, does not show "metallic magnetism:" only for the insulating half-filled case there appears to be magnetic long-range order. In addition, it does not display any tendency to *ferromagnetic* correlations, let alone ferromagnetism, in the range of interaction and band filling studied. We believe that with a modified band-structure tendency to ferromagnetism could be enhanced, but it is unlikely that ferromagnetic long-range order will appear. We also do not find any tendency towards triplet or singlet pairing in our model.

Concerning the metal-insulator transition, the model discussed here is an insulator for any $U>0$ (in the half-filled case) due to the nesting of the Fermi surface. In that respect, the model is similar to the one-dimensional Hubbard model where nesting always occurs.

The paper is organized as follows. In the next section we discuss some features of the model in the noninteracting case, the Hartree-Fock solution, and the strong-coupling limit. In Sec. III we discuss the simulation method used and compare simulation results with exact results for a 6-site system with the exact diagonalization results of Shiba, to test the reliability of our approach. In Sec. IV we present results of simulations for the half-filled case, and in Sec. V for some non-half-filled band cases. We discuss briefly the implications of our results for the two-dimensional attractive Hubbard model in Sec. VI, and summarize our conclusions in Sec. VII.

II. THE MODEL

We consider the model defined by the Hamiltonian:

$$H=-t\sum_{\substack{\langle i,j\rangle\\ \sigma}}(c_{i\sigma}^{\dagger}c_{j\sigma}+\text{H.c.})+U\sum_i n_{i\uparrow}n_{i\downarrow}-\mu\sum_i(n_{i\uparrow}+n_{i\downarrow})\ , \tag{2.1}$$

where $\langle i,j\rangle$ denotes nearest neighbors and the sum runs over sites of a two-dimensional square lattice. The chemical potential μ is $U/2$ for a half-filled band. We consider here the case of repulsive interactions ($U\geq 0$).

The single-particle eigenstates for the noninteracting case ($U=0$) have energies

$$\epsilon_k=-2t(\cos k_x+\cos k_y)-\mu\ ,$$
$$k_{x,y}=\frac{2\pi}{N_{x,y}}n_{x,y},\quad \frac{-N_{x,y}}{2}\leq n_{x,y}<\frac{N_{x,y}}{2}\ , \tag{2.2}$$

so that the bandwidth is $W=8t$. The ground state for the noninteracting system is obtained by filling the negative energy states. Figure 1 shows the Fermi surface for various band fillings for an infinite two-dimensional lattice. The Fermi surface in the half-filled case is "nested," i.e., a reciprocal-lattice vector $[(\pi/a,\pi/a)$ or $(-\pi/a,\pi/a)]$ maps an entire section of the Fermi surface onto another. This is due to the fact that our lattice is bipartite, and the kinetic energy connects only one sublattice to the other. As is well known, this has important consequences for the properties of the model.[17]

Another important feature of the noninteracting system appears in the density of states, defined by

$$g(\epsilon)=\sum_k \delta[\epsilon-\epsilon(k)]\ , \tag{2.3}$$

which is shown in Fig. 2. The density of states displays a logarithmic singularity $g(\epsilon)\sim\ln(\epsilon/4t)$ for small ϵ. From topological arguments, one can show that such a singularity will always occur in a two-dimensional system. However, the fact that it occurs at the Fermi energy for the half-filled case, for the same energy where nesting occurs, is special to the nearest-neighbor-hopping model considered here.

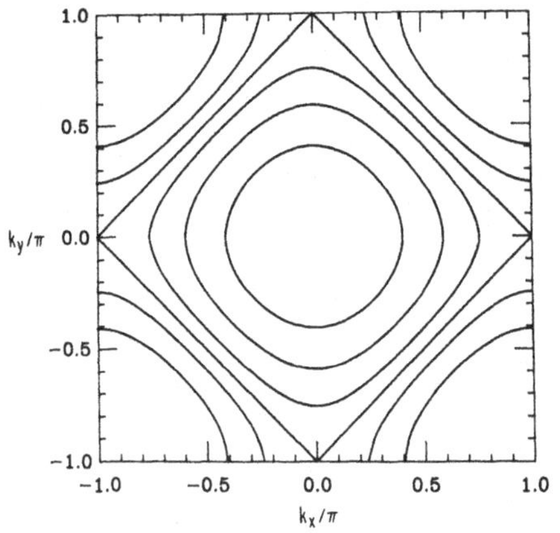

FIG. 1. Fermi surfaces for electrons on a two-dimensional square lattice with nearest-neighbor hopping only. Band fillings are $\rho=0.25, 0.5, \ldots, 1.5$ starting from the inner surface. Note that the Fermi surface for the half-filled case is nested.

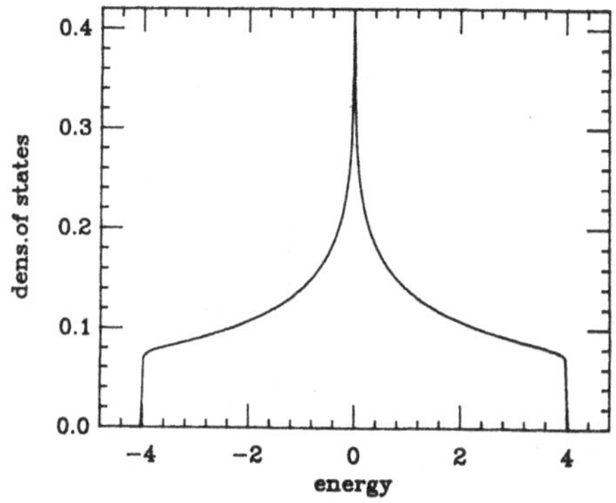

FIG. 2. Density of states for noninteracting electrons on a two-dimensional square lattice with nearest-neighbor hopping. The singularity at the origin is logarithmic.

The q-dependent zero-frequency susceptibility (per spin) for the noninteracting case is given by

$$\chi_0(q)=-\frac{1}{N}\sum_k \frac{f(\epsilon_{k+q})-f(\epsilon_k)}{\epsilon_{k+q}-\epsilon_k} \tag{2.4}$$

with $f(\epsilon)$ the Fermi function:

$$f(\epsilon)=\frac{1}{e^{\beta\epsilon}+1} . \tag{2.5}$$

For $q=0$, we have

$$\chi_0(q=0)=\int d\epsilon g(\epsilon)\frac{\partial f}{\partial \epsilon} \underset{T\to 0}{\to} g(\epsilon_F) , \tag{2.6}$$

i.e., the usual Pauli result. In the half-filled case, the susceptibility diverges as $T\to 0$ due to the singularity in the density of states, as

$$\chi_0(q=0)\sim -\ln\left|\frac{T}{t}\right| . \tag{2.7}$$

The staggered susceptibility is given by

$$\chi_0(q=\pi)=\frac{1}{N}\sum_k \frac{f(-\epsilon_k)-f(\epsilon_k)}{2\epsilon_k}$$

$$=\int d\epsilon g(\epsilon)\frac{f(-\epsilon)-f(\epsilon)}{2\epsilon} . \tag{2.8}$$

In the usual case where one has a nested Fermi surface this gives a logarithmic divergence of the susceptibility at low temperatures. Here, however, we have in addition the singularity in the density of states, and the low-temperature behavior in the half-filled case is

$$\chi_0(q=\pi)\sim \left|\ln\frac{T}{t}\right|^2 , \tag{2.9}$$

i.e., a stronger divergence than for the $q=0$ susceptibility. For the non-half-filled case, the q-dependent susceptibility is finite as $T\to 0$.

For nonzero U, the magnetic susceptibility within the random-phase approximation (RPA) is given by the sum of particle-hole ladder diagrams[18] as

$$\chi(q)=\frac{2\chi_0(q)}{1-U\chi_0(q)} . \tag{2.10}$$

For the half-filled case, the divergences in χ_0 indicate instabilities for arbitrarily small values of U at both $q=0$ and $q=(\pi,\pi)$. Because the divergence in χ_0 is stronger for $q=(\pi,\pi)$, RPA predicts a transition to an antiferromagnetic phase at a higher temperature than for the ferromagnetic phase in the half-filled case. For the non-half-filled case, RPA predicts a transition at a *finite* value of U, since the susceptibility is nondivergent. The transition will be to states defined by the wave vector q for which χ_0 is maximum, which is a decreasing function of the band filling. However, because we are dealing with a two-dimensional model with a continuous symmetry, it is clear that no transition to a state with magnetic long-range order can occur expect possibly for $T=0$.

We now discuss the Hartree-Fock (HF) solution for this model. This has been discussed in detail by Penn for the three-dimensional case.[4] Within the Hartree-Fock approximation, the Hamiltonian is

$$H_{\rm HF}=-t\sum_{\substack{\langle i,j\rangle \\ \sigma}} c_{i\sigma}^{\dagger}c_{j\sigma}$$

$$+U\sum_i (\langle n_{i\uparrow}\rangle n_{i\downarrow}+n_{i\uparrow}\langle n_{i\downarrow}\rangle-\langle n_{i\uparrow}\rangle\langle n_{i\downarrow}\rangle)$$

$$-\mu\sum_i (n_{i\uparrow}+n_{i\downarrow}) . \tag{2.11}$$

For the half-filled case, the appropriate solution is the antiferromagnetic one:

$$\langle n_{i\uparrow}\rangle = n+(-1)^i m ,$$
$$\langle n_{i\downarrow}\rangle = n-(-1)^i m , \tag{2.12}$$

which yields the gap equation:

$$1=\frac{U}{4\pi^2}\int d^2k\frac{1}{(\epsilon_k^2+\Delta^2)^{1/2}} , \tag{2.13a}$$

$$\Delta=Um , \tag{2.13b}$$

which has a solution with $m\neq 0$ for arbitrarily small U due to the nesting of the Fermi surface. The features of the singularity are, however, different from the usual case. Rewriting Eq. (2.13a) as

$$1=U\int_0^{4t} d\epsilon\frac{\rho(\epsilon)}{(\epsilon^2+\Delta^2)^{1/2}} \tag{2.14}$$

and due to the singularity in the density of states, we find for the gap

$$\Delta\sim te^{-2\pi\sqrt{t/U}} \quad (d=2) . \tag{2.15}$$

In contrast, both in one and three dimensions there is no singularity at the Fermi energy for the half-filled case, and the gap within HF behaves as

$$\Delta \sim te^{-2\pi t/U} \quad (d=1,3)\ . \tag{2.16}$$

This particular feature of the model under consideration gives a stronger tendency to antiferromagnetic ordering than in the usual case.[19] It should be remarked that the behavior, Eq. (2.15), is a consequence of the fact that both the nesting of the Fermi surface and the Van Hove singularity in the density of states occur at the same energy, the Fermi energy for the half-filled case. One can easily construct other band structures in $d=2$ where both features occur at different energies, or where the nesting is absent altogether, which will then have rather different properties. The logarithmic singularity, however, is required by topology in $d=2$ so that it always occurs. In the absence of nesting, it may give a tendency for ferromagnetic correlations.

For the non-half-filled band case we consider, in addition to (2.12), the possibility of ferromagnetic solutions:

$$\langle n_{i\uparrow}\rangle = n+m\ ,\qquad \langle n_{i\downarrow}\rangle = n-m\ , \tag{2.17}$$

and paramagnetic ones, with $m=0$. The Hartree-Fock phase diagram, obtained by choosing the solution that gives the lowest energy (if more than one solution exists) is shown in Fig. 3, and is rather similar to the one obtained in $d=3$ by Penn. Note that for large U and fillings close to 1 ferromagnetism is predicted to occur. This is also in agreement with the exact results by Nagaoka,[20] who found the ferromagnetic state to be the ground state for $U=\infty$ and one hole in a half-filled-band system. However, we will see that our simulations do not give any indication that there is a tendency towards ferromagnetic ordering.

Finally, we review the strong-coupling limit of the model Eq. (2.1) for the half-filled-band case. To second order in the hopping the model is equivalent to an antiferromagnetic Heisenberg model, defined by the Hamiltonian

$$H_{\text{eff}} = +\frac{4t^2}{U}\sum_{\langle i,j\rangle} \mathbf{S}_i\cdot\mathbf{S}_j \tag{2.18}$$

with **S** the Pauli matrices. According to the finite lattice calculations of Oitmaa and Betts,[21] this model has antiferromagnetic long-range order in the ground state, approximately 50% reduced from the classical Néeel state due to quantum fluctuations. The existence of long-range antiferromagnetic order for the case $0<U<\infty$, where both spin and charge fluctuations occur, was an open question; in this work we believe we have established that the model indeed has antiferromagnetic order for $0<U<\infty$ in the half-filled-band case.

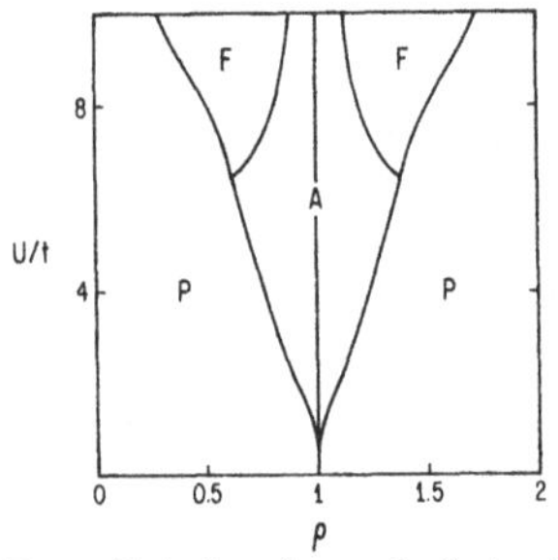

FIG. 3. Hartree-Fock phase diagram for the two-dimensional Hubbard model. A, F, and P denote antiferromagnetic, ferromagnetic, and paramagnetic ground states, respectively.

The behavior of the magnetic susceptibility at low temperatures in the limit where Eq. (2.18) is valid follows from a spin-wave analysis. Assuming a linear dispersion relation for the spin-wave energy, $\epsilon(k)\sim ck$, we obtain

$$\chi \sim \frac{1}{T}\int kdk\frac{e^{\beta k}}{(e^{\beta k}-1)^2} \sim T\ln T \tag{2.19}$$

at low temperatures. The staggered susceptibility should be the same as the susceptibility for a ferromagnet, with $\epsilon(k)\sim ck^2$, and yields

$$\chi(q=2k_F)\sim\frac{1}{T}\ , \tag{2.20}$$

which happens to be the same as the Curie law because we are in two dimensions. Note that the susceptibility in the large-U limit vanishes as $T\to 0$ [Eq. (2.19)], in contrast to the $U=0$ limit where it diverges [Eq. (2.7)].

III. THE SIMULATION

The simulation was constructed using an Ising functional integral formulation recently introduced. The partition function is written as

$$Z=\text{Tr}e^{-\beta H}=\text{Tr}\prod_{l=1}^{L}e^{-\Delta\tau H}\cong\text{Tr}\prod_{l=1}^{L}e^{-\Delta\tau H_0}\exp\left[-\Delta\tau\left[U\sum n_{i\uparrow}n_{i\downarrow}-\mu\sum(n_{i\uparrow}+n_{i\downarrow})\right]\right] \tag{3.1}$$

with H_0 the kinetic energy and $\beta=L\Delta\tau$. The error in the breakup in Eq. (3.1) is of order $O(\Delta\tau^2 tU)$. The electron-electron interaction was eliminated using the identity[14]

$$e^{-\Delta\tau U n_{i\uparrow}n_{i\downarrow}}=\text{Tr}_\sigma\exp\left[\lambda\sigma(n_{i\uparrow}-n_{i\downarrow})-\frac{\Delta\tau U}{2}(n_{i\uparrow}+n_{i\downarrow})\right] \tag{3.2}$$

with

$$\lambda=2\arctan\sqrt{\tanh(\Delta\tau U/4)} \tag{3.3}$$

and $\sigma=\pm 1$. The partition function is then

$$Z=\mathrm{Tr}_\sigma \mathrm{Tr}\prod_{l=1}^{L} e^{-\Delta\tau H_0}\exp\left[\lambda\sigma(n_{i\uparrow}-n_{i\downarrow})-\Delta\tau(\mu-U/2)(n_{i\uparrow}+n_{i\downarrow})\right]$$

$$=\mathrm{Tr}_\sigma \mathrm{Tr}\left[\prod_{l=1}^{L} e^{-\Delta\tau H_{0\uparrow}}\exp\{[\lambda\sigma-\Delta\tau(\mu-U/2)]n_{i\uparrow}\}\right]\left[\prod_{l=1}^{L} e^{-\Delta\tau H_{0\downarrow}}\exp\{[-\lambda\sigma-\Delta\tau(\mu-U/2)]n_{i\downarrow}\}\right]. \quad (3.4)$$

Denote

$$B_l(\alpha)=e^{-\Delta\tau K}e^{V^{\alpha}(l)}, \quad (3.5)$$

$$(K)_{ij}=\begin{cases}-t & \text{for } i,j \text{ nearest neighbors,}\\ 0 & \text{otherwise,}\end{cases} \quad (3.6a)$$

$$V^{\alpha}_{ij}(l)=\delta_{ij}[\lambda\alpha\sigma_i(l)-\Delta\tau(\mu-U/2)], \quad (3.6b)$$

and define the operators

$$D_l(\alpha)=e^{-\Delta\tau c_i^\dagger K_{ij}c_j}e^{c_i^\dagger V_i^{\alpha}(l)c_i}, \quad (3.7)$$

so that the partition function is

$$Z=\mathrm{Tr}_\sigma \mathrm{Tr}\prod_{\alpha=\pm 1}\prod_{l=1}^{L} D_l(\alpha). \quad (3.8)$$

We can take the trace over fermions explicitly, since there are only bilinear forms in fermion operators, and obtain

$$Z=\mathrm{Tr}_\sigma\prod_{\alpha}\det[1+B_L(\alpha)B_{L-1}(\alpha)\cdots B_1(\alpha)]$$

$$\equiv \mathrm{Tr}_\sigma \det O_\uparrow \det O_\downarrow. \quad (3.9)$$

This identity was proved by Blankenbecler *et al.*,[15] using Grassmann variables. For the reader unfamiliar with Grassmann algebras, we give an elementary derivation of this result in the Appendix.

The remaining sum over Ising spins in Eq. (3.9) is performed using a Monte Carlo technique, taking as Boltzmann weight the product of determinants in Eq. (3.9). For the case of a half-filled band, it is easy to show that this product is positive for arbitrary σ configurations. Consider the particle-hole transformation

$$d_{i\sigma}=(-1)^i c_{i\sigma}^\dagger,$$
$$c_{i\sigma}^\dagger c_{i\sigma}=1-d_{i\sigma}^\dagger d_{i\sigma}. \quad (3.10)$$

For the half-filled band, $\mu=U/2$ and we have from Eq. (3.4)

$$\det O_\uparrow=\mathrm{Tr}_c\prod_{i=1}^{L} e^{-\Delta\tau H_{0\uparrow}}e^{\lambda\sigma c_{i\uparrow}^\dagger c_{i\downarrow}}$$

$$=\mathrm{Tr}_d\prod_{i=1}^{L} e^{-\Delta\tau H_{0\uparrow}}e^{-\lambda\sigma d_{i\uparrow}^\dagger d_{i\downarrow}}e^{\lambda\sigma_i(l)}, \quad (3.11)$$

so that

$$\det O_\uparrow=e^{\lambda\sum_{i,l}\sigma_k(l)}\det O_\downarrow \quad (3.12)$$

and the product of determinants in Eq. (3.9) is positive definite, so that it can be used as a Boltzmann weight. For the non-half-filled band, we define the Boltzmann weight as

$$P(\sigma)=|\det O_\uparrow \det O_\downarrow| \quad (3.13)$$

and have to compute the average sign of the product of determinants. We find that the product of determinants does become negative for certain field configurations, but the average sign is always well behaved and does not go to zero rapidly as β or the lattice size increases, so that it does not represent a problem for doing Monte Carlo simulations.

We use the heat-bath algorithm to perform the sum over Ising spins. If R_α is the ratio of the new to the old determinant for fermion spin α on flipping a given Ising spin, it is flipped with probability

$$P=\frac{R_\uparrow R_\downarrow}{1+R_\uparrow R_\downarrow}. \quad (3.14)$$

To compute R_α, we use the procedure introduced by Blankenbecler, Scalapino, and Sugar[15] which involves updating the full Green's function exactly when a move is accepted. This takes the bulk of the computer time in the calculation, N^2 operations per update, with N the number of spatial sites. After several updatings, the Green's function degrades due to rounding errors and has to be recomputed from scratch. This makes a non-negligible difference in terms of computer time only if it has to be done every time slice or two. In practice, we started our simulations by recomputing G from scratch every 10 time slices and checking whether it had degraded by more than 1%. If so, we recomputed it more often, which we had to do for large values of the interaction.

We now consider the evaluation of average quantities. First, it is easy to show that[14]

$$\langle (n_{i\uparrow}(\tau)-n_{i\downarrow}(\tau))(n_{j\uparrow}(0)-n_{j\downarrow}(0))\rangle$$
$$=(1-e^{-\Delta\tau U})^{-1}\langle\sigma_i(\tau)\sigma_j(0)\rangle, \quad (3.15)$$

so that we obtain fermion spin-spin correlation functions simply from correlation functions of the Ising spins. For other correlation functions, we do not have enough information in the Ising variables but have to average over appropriate fermion matrices. For an equal time correlation of the operators P_i and Q_j we have

$$\langle\langle P_iQ_j\rangle\rangle=\frac{\mathrm{Tr}_\sigma \mathrm{Tr}P_iQ_j\prod_\alpha\prod_l D_l(\alpha)}{Z}$$

$$=\frac{\mathrm{Tr}_\sigma\langle P_iQ_j\rangle\det O_\uparrow\det O_\downarrow}{Z}, \quad (3.16)$$

$$\langle P_iQ_j\rangle=\frac{\mathrm{Tr}P_iQ_j\prod_{l,\alpha}D_l(\alpha)}{\det O^\uparrow\det O_\downarrow}. \quad (3.17)$$

It is easy to obtain the appropriate formulas by using the

transformation to normal modes, Eq. (A6), of the entire product of factors in (3.17). For example, consider the single-particle Green's functions (we omit spin indices for simplicity):

$$\langle c_i c_j^\dagger \rangle = \frac{\mathrm{Tr} c_i c_j^\dagger \prod_\nu e^{-c_\nu^\dagger l_\nu c_\nu}}{\prod_\nu (1+e^{-l_\nu})}$$

$$= \sum_{\nu'} \langle \nu' | i \rangle \langle j | \nu' \rangle \frac{\mathrm{Tr} c_{\nu'} c_{\nu'}^\dagger \prod_\nu e^{-c_\nu^\dagger l_\nu c_\nu}}{\prod_\nu (1+e^{-l_\nu})}$$

$$= \sum_{\nu'} \langle \nu' | i \rangle \langle j | \nu' \rangle \frac{1}{1+e^{-l_{\nu'}}}$$

$$= \left[\frac{1}{1+B_L B_{L-1} \cdots B_1} \right]_{ij} . \tag{3.18}$$

Similarly,

$$\langle c_i^\dagger c_j \rangle = \left[B_L \cdots B_1 \frac{1}{1+B_L \cdots B_1} \right]_{ji} . \tag{3.19}$$

For two-particle Green's functions, it is straightforward to show, by expanding in eigenstates, that Wick's theorem applies, i.e.,

$$\langle c_{i_1}^\dagger c_{i_2} c_{i_3}^\dagger c_{i_4} \rangle = \langle c_{i_1}^\dagger c_{i_2} \rangle \langle c_{i_3}^\dagger c_{i_4} \rangle + \langle c_{i_1}^\dagger c_{i_4} \rangle \langle c_{i_2} c_{i_3}^\dagger \rangle . \tag{3.20}$$

Note that this decoupling applies only to the "single-bracket" average (trace over fermions) and not to the full average, denoted by double angular brackets [Eq. (3.16)], which involves the additional trace over spins. For averages involving fermion operators of both spins we can simply factorize, since everything is diagonal in spins, for example,

$$\langle n_{i\uparrow} n_{j\downarrow} \rangle = \langle n_{i\uparrow} \rangle \langle n_{j\downarrow} \rangle . \tag{3.21}$$

Finally, we can obtain in a similar fashion time-dependent correlation functions, by inserting the operators at different points in the product over time slices Eq. (3.8). For example, a time-dependent Green's function is

$$\langle\langle c_i(l_1) c_j^\dagger(l_2) \rangle\rangle = \frac{\mathrm{Tr} D_L D_{L-1} \cdots D_{l_1+1} c_i D_{l_1} \cdots D_{l_2+1} c_j^\dagger D_{l_2} \cdots D_1}{\mathrm{Tr} D_L \cdots D_1} \cdot$$

$$= \frac{\mathrm{Tr} D_{l_2} \cdots D_1 D_L \cdots D_{l_2+1} [(D_{l_1} D_{l_1-1} \cdots D_{l_2+1})^{-1} c_i D_{l_1} \cdots D_{l_2+1}] c_j^\dagger}{\mathrm{Tr} D_{l_2} \cdots D_1 D_l \cdots D_{l_2+1}} . \tag{3.22}$$

By expanding in eigenstates of $D_{l_1} D_{l_1-1} \cdots D_{l_2+1}$, we find

$$(D_{l_1} D_{l_1-1} \cdots D_{l_2+1})^{-1} c_i D_{l_1} D_{l_1-1} \cdots D_{l_2+1} = \sum_k (B_{l_1} B_{l_1-1} \cdots B_{l_2+1})_{ik} c_k , \tag{3.23}$$

and replacing in (3.22)

$$\langle\langle c_i(l_1) c_j^\dagger(l_2) \rangle\rangle = \sum_k (B_{l_1} B_{l_1-1} \cdots B_{l_2+1})_{ik} \frac{\mathrm{Tr} D_{l_2} \cdots D_1 D_L \cdots D_{l_2+1} c_k c_j^\dagger}{\mathrm{Tr} D_{l_2} \cdots D_1 D_l \cdots D_{l_2+1}} \tag{3.24}$$

and using (3.18), we finally obtain

$$\langle\langle c_i(l_1) c_j^\dagger(l_2) \rangle\rangle = \left[B_{l_1} B_{l_1-1} \cdots B_{l_2+1} \frac{1}{1+B_{l_2} \cdots B_1 B_L \cdots B_{l_2+1}} \right]_{ij} . \tag{3.25}$$

Similarly,

$$\langle\langle c_i^\dagger(l_1) c_j(l_2) \rangle\rangle = \left[\frac{1}{1+B_{l_2} \cdots B_{l_2+1}} B_{l_2} \cdots B_{l_1+1} \right]_{ij} , \tag{3.26}$$

and for higher-order correlation functions Wick's theorem applies. From these formulas we can obtain arbitrary correlation functions of interest in the Hubbard model, such as charge and spin correlation functions and susceptibilities.

We have done a variety of checks on our simulation program to make sure it was running properly. For the noninteracting case our results should be exact for a finite lattice, and we have verified this by comparing the results from our simulation program with those obtained from a direct calculation for lattices up to 8×8 in spatial size and various temperatures. For the interacting case, we have earlier reported comparison with exact results for two sites,[14] where we found that the choice $\Delta\tau U=0.5$ gives reasonable accuracy (within a few percent). In Figs. 4 and 5 we show comparison of our simulation results for the local moment [Eq. (4.1)] and the magnetic susceptibility [Eq. (4.2)] for 6-site rings with the exact results of Shiba.[7] It can be seen that the agreement is excellent. These tests lead us to believe that the results to be presented in the next section for the two-dimensional Hubbard model are reliable.

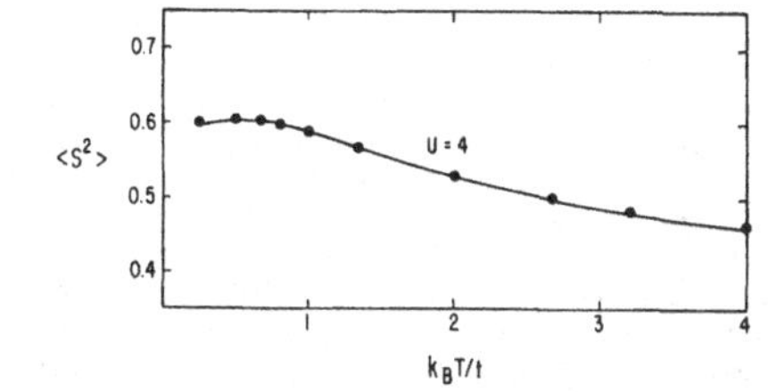

FIG. 4. Local moment versus temperature for a 6-site ring, $U=4$, with $\Delta\tau=0.125$. The solid line is the exact result of Shiba.

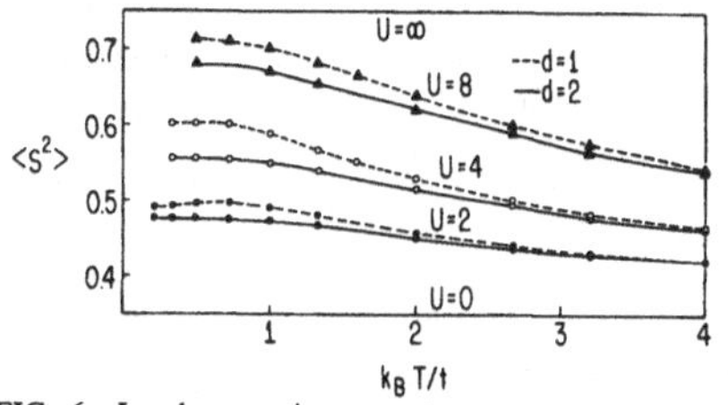

FIG. 6. Local magnetic moment versus temperature on a 6×6 two-dimensional lattice and a 6-site one-dimensional lattice.

IV. RESULTS FOR THE HALF-FILLED-BAND CASE

We have performed simulations in the half-filled-band sector for lattices up to 8×8 in spatial size (with periodic boundary conditions) and interaction strengths $U=2$, 4, and 8. The time slice size was taken to be $\Delta\tau=0.25$, 0.125, and 0.0625 for $U=2$, 4, and 8, respectively. Simulations were performed on a Vax 750 computer and a Cray 1S supercomputer. A sweep through a 6×6 lattice with 32 time slices took 70 sec on the Vax and 0.60 sec on the Cray, plus some fraction of this number depending on the measurements performed. Typically, 200 warm-up and 1000 measurements separated by two sweeps were performed for a given set of parameters. For low temperatures, the algorithm becomes unstable, and the lowest temperature we could study without running into accuracy problems was $\beta=4$ on the Cray (using single precision) and $\beta=4.5$ on the Vax (using double precision). Our program on the Cray with double precision was a factor of 15 to 20 slower so that it was impractical. Because more time slices are needed the larger the interaction and the algorithm becomes unstable more rapidly, we could not reach as low temperatures with $U=8$ as with $U=2$. Also, the statistical error becomes larger the larger the interaction.

Figure 6 shows the local magnetic moment, defined by

$$\langle S^2\rangle=\tfrac{1}{4}\langle\sigma_x^2+\sigma_y^2+\sigma_z^2\rangle=\tfrac{3}{4}\langle\sigma_z^2\rangle\ , \tag{4.1a}$$

$$\sigma_z^l=n_{i\uparrow}-n_{i\downarrow}\ . \tag{4.1b}$$

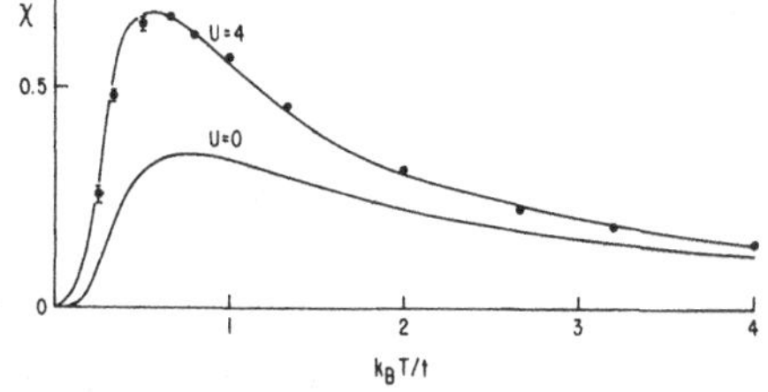

FIG. 5. Magnetic susceptibility versus temperature for a 6-site ring, $U=4$, with $\Delta\tau=0.125$. The solid line is the exact result of Shiba.

The local moment increases gradually as the temperature is lowered, indicating that the electrons are becoming more localized. There is no evidence of any abrupt change as a function of temperature. It also increases gradually as a function of U, and for $U=8$ (equals bandwidth) it is already quite close to the $U=\infty$ value at low temperatures. We also show results for the one-dimensional case for comparison. In units of the bandwidth, the Hubbard interaction is more effective in localizing the electrons as the dimensionality increases.

Figure 7 shows the magnetic susceptibility versus tem-

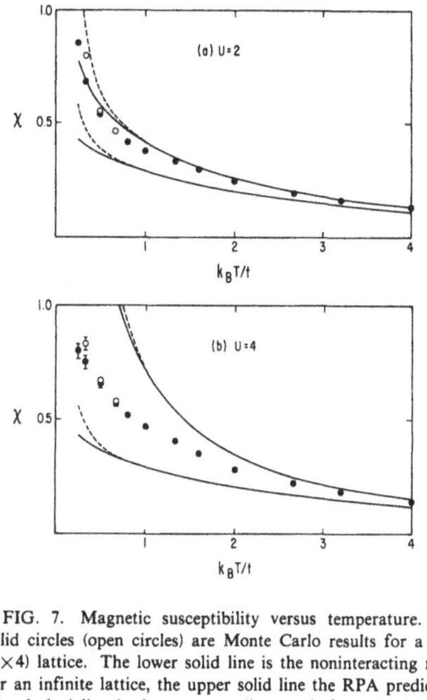

FIG. 7. Magnetic susceptibility versus temperature. The solid circles (open circles) are Monte Carlo results for a 6×6 (4×4) lattice. The lower solid line is the noninteracting result for an infinite lattice, the upper solid line the RPA prediction. The dashed line is the corresponding result for a 6×6 lattice, showing that finite-size effects start to appear around $T\sim0.75$. For the interacting case, finite-size effects appear at a somewhat lower temperature.

perature for $U=2$ and 4. Some results for $U=8$ were given in Ref. 16. The q-dependent (zero-frequency) susceptibility is defined by

$$\chi(q)=\frac{1}{N}\sum_{i,j}e^{iq(R_i-R_j)}\int_0^\beta d\tau\langle[n_{i\uparrow}(\tau)-n_{i\downarrow}(\tau)]\times[n_{j\uparrow}(0)-n_{j\downarrow}(0)]\rangle \quad (4.2)$$

and for $q=0$ it satisfies

$$\chi(q=0)=\beta S(q=0) \quad (4.3)$$

with $S(q)$ the magnetic structure factor,

$$S(q)=\frac{1}{N}\sum_{i,j}e^{iq(R_i-R_j)}\langle(n_{i\uparrow}-n_{i\downarrow})(n_{j\uparrow}-n_{j\downarrow})\rangle , \quad (4.4)$$

since the total magnetization commutes with the Hamiltonian. We evaluated the susceptibility using both sides of Eq. (4.3) and found agreement within statistical error. The susceptibility increases smoothly as the temperature is lowered, and there is a slight bulge (particularly for $U=4$) for T between 1 and 2, which is where the local moment is increasing more rapidly. As $T\to 0$, χ diverges logarithmically if $U=0$ [Eq. (2.7)] because of the logarithmic singularity in the density of states, and we expect it to go to zero for large U [Eq. (2.19)]. Although we do not see this behavior up to the lowest temperature studied, χ does increase less rapidly for $U=4$ than for $U=2$ at low temperatures. Except at the lowest temperatures, where the singularity in $\rho(\epsilon)$ plays a role, χ is enhanced more the larger U is, as one would expect. We also show in Fig. 7 the RPA results, Eq. (2.10). From comparison with the Monte Carlo results, we conclude that RPA is fairly accurate for $U=2$, but becomes rapidly inaccurate as U is increased, and it always predicts too large an enhancement of the susceptibility.

Figure 8 shows the staggered magnetic susceptibility versus temperature. Here, for the free case we have the combined effect of the singularity in the density of states and the nested Fermi surface, yielding a low-temperature behavior $\chi_{st}\sim\ln^2(T/t)$. For large U, we expect $\chi_{st}\sim 1/T$ at low temperatures, as discussed in Sec. II. Our results for finite U appear to follow the stronger divergence $\chi_{st}\sim 1/T$. Similarly, as for the susceptibility, RPA overestimates the effect of the interaction, but it is quite accurate for $U=2$ in the temperature range studied. However, RPA predicts a transition to an antiferromagnetic state at $T=0.33$ and 0.75 for $U=2$ and 4, respectively, and we find no evidence for it, as one would expect.

Figure 9 shows the internal energy versus temperature. It is a smoothly varying function of temperature and increases as U is increased. We have extrapolated the ground-state energy assuming a T^2 dependence at low temperature and using our Monte Carlo data for the lowest temperatures. The extrapolated results are $E=-1.17(2)$ for $U=2$, $E=-0.88(3)$ for $U=4$, and $E=-0.48(5)$ for $U=8$. The errors are estimates on the error due to the extrapolation, since the statistical error is very small. The extrapolated data are shown in the inset, where they are compared with results from the Hartree-Fock approximation and an exact lower bound obtained

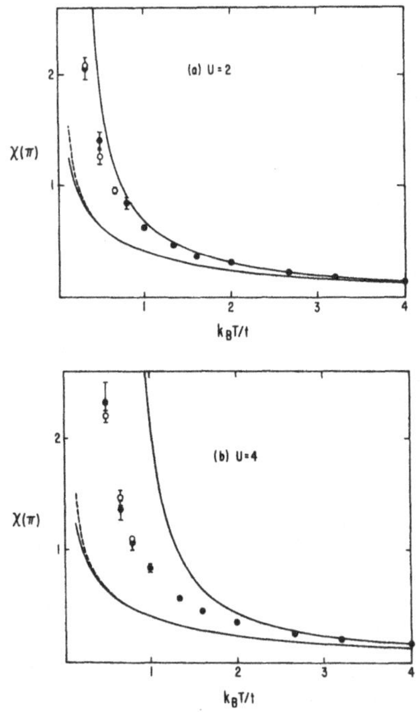

FIG. 8. Same as Fig. 7 for the staggered magnetic susceptibility.

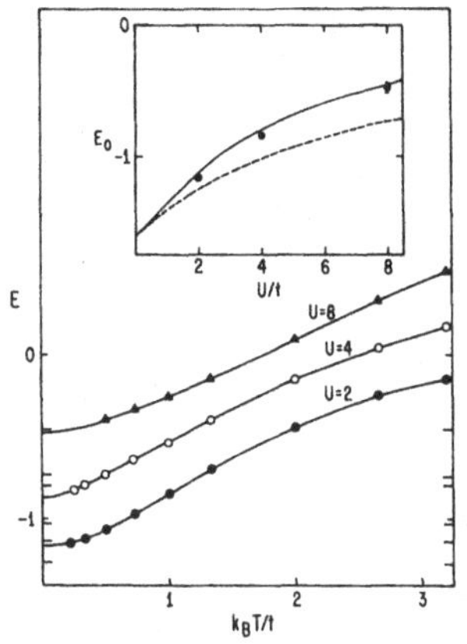

FIG. 9. Energy versus temperature for $U=2$, 4, and 8. The inset shows the extrapolated ground-state energy, compared with Hartree-Fock predictions (solid line) and the Langer-Mattis lower bound (dashed line).

by Langer and Mattis.[22]

At finite temperature, the model considered here cannot undergo a transition to a magnetically ordered state since a continuous symmetry (rotation is spin space) would be broken. At zero temperature, however, the existence of long-range magnetic order is an open question. Although our simulation method cannot deal with $T=0$ directly, we can go to sufficiently low temperatures so that correlations build up over the whole extent of our finite spatial lattice. When the thermal correlation length is larger than the spatial lattice size, the system behaves effectively as if at zero temperature. We can see how the long-range order builds up in the spin-spin correlation function. A picture of the spin-spin correlations $\langle \sigma_z^i \sigma_z^j \rangle$ in real space for an intermediate coupling case ($U=4$) at low temperatures ($\beta=4$) is shown in Fig. 10, on a 8×8 lattice. Note that there are definitely antiferromagnetic correlations extending over the whole lattice. For this low temperature, fluctuations are predominantly quantum rather than thermal (note that the local moment is essentially independent of T for $\beta > 1.5$ in Fig. 5). The reduction in the spin-spin correlation function from the perfect Néel state ($\langle S_z^i S_z^j \rangle = \pm 1$) is due to both charge fluctuations (since $U < \infty$) and spin fluctuations (even for $U \to \infty$, the antiferromagnetic Heisenberg model does not have perfect long-range order). The magnitude of the on-site charge fluctuation can be measured from

$$\langle (n_\uparrow + n_\downarrow)^2 \rangle - \langle n_\uparrow + n_\downarrow \rangle^2 , \qquad (4.5)$$

which is zero for $U=\infty$, and 0.5 for $U=0$, and is related to the local moment, Eq. (4.1). For the case shown in Fig. 10, it is 0.26. Another indication that the spins shown in Fig. 10 are not "localized" but "itinerant" is given by the average value of the kinetic energy of the electrons. For the case of Fig. 10, we find

$$\frac{\langle c_{i\sigma}^\dagger c_{j\sigma} \rangle_{U=4}}{\langle c_{i\sigma}^\dagger c_{j\sigma} \rangle_{U=0}} = 0.86 , \qquad (4.6)$$

which implies that the electrons are quite delocalized, at least on a short-range basis.

Figure 11 shows the Fourier transform of the spin-spin correlation function

$$S(q) = \frac{1}{N} \sum_{i,j} e^{i\mathbf{q}\cdot(\mathbf{R}_i - \mathbf{R}_j)} \langle (n_{i\uparrow} - n_{i\downarrow})(n_{j\uparrow} - n_{j\downarrow}) \rangle \qquad (4.7)$$

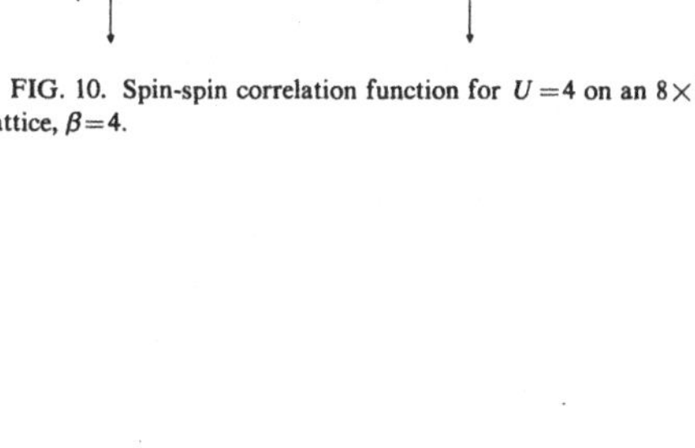

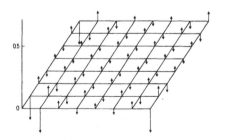

FIG. 10. Spin-spin correlation function for $U=4$ on an 8×8 lattice, $\beta=4$.

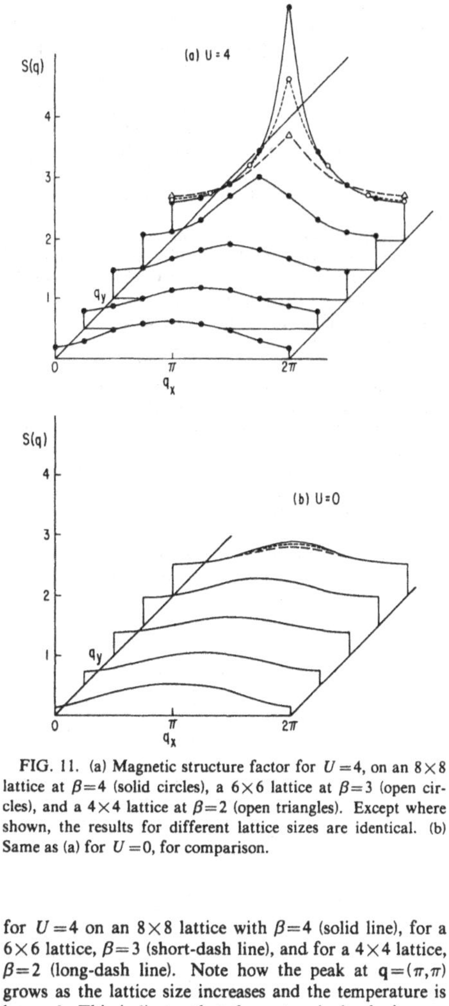

FIG. 11. (a) Magnetic structure factor for $U=4$, on an 8×8 lattice at $\beta=4$ (solid circles), a 6×6 lattice at $\beta=3$ (open circles), and a 4×4 lattice at $\beta=2$ (open triangles). Except where shown, the results for different lattice sizes are identical. (b) Same as (a) for $U=0$, for comparison.

for $U=4$ on an 8×8 lattice with $\beta=4$ (solid line), for a 6×6 lattice, $\beta=3$ (short-dash line), and for a 4×4 lattice, $\beta=2$ (long-dash line). Note how the peak at $\mathbf{q}=(\pi,\pi)$ grows as the lattice size increases and the temperature is lowered. This indicates that the system is developing antiferromagnetic order. The same correlation function for $U=0$ is shown for comparison in Fig. 11(b). For $T=0$, we will have for a sufficiently large lattice

$$S(\pi,\pi) = Nm^2 + S_c(\pi,\pi) , \qquad (4.8)$$

with m the staggered magnetization

$$m = \frac{1}{N} \sum_i (-1)^{R_i} \langle n_{i\uparrow} - n_{i\downarrow} \rangle , \qquad (4.9)$$

and S_c the connected structure factor. To extrapolate the long-range order, we plot $S(\pi,\pi)/N$ versus $1/N$, following Oitmaa and Betts.[21] According to (4.8), we expect $S(\pi,\pi)/N$ to follow a straight line if plotted versus $1/N$.

From the extrapolated value as $N \to \infty$, we obtain the square of the staggered magnetization. We expect this procedure to work if we are at sufficiently low temperatures such that the thermal coherence length, $v_F\beta$ (v_F denotes Fermi velocity), is much larger than the linear lattice size. We have taken $\beta = 0.75\sqrt{N}$, which was close to the lowest temperature we could study without having accuracy problems in the computation of the matrices, and values of $N = 8$, 10, 16, 26, and 36. For $N = 8$, 10, and 26, we used the "tilted" lattices discussed by Oitmaa and Betts.[21]

Figure 12 shows our results. For $U = 0$, the points extrapolate to 0 as they should. For $U = 2$ the extrapolation is already clearly finite. The slope of our lines is smaller than the Oitmaa-Betts line, indicating that $S_c(\pi,\pi)$ is smaller. Although our results are certainly a lower bound to the long-range order, there could be some effect of the finite temperature reducing the long-range order, and we believe our results could be underestimating the long-range order by up to 10%. Our results are substantially lower than the Hartree-Fock predictions, due to both charge and spin fluctuations. The fluctuations, however, do not destroy the order as they do in one dimension.

We have attempted to verify the scaling behavior at zero temperature predicted by Hartree-Fock theory for small U, Eq. (2.15), from the numerical simulation. Because the gap is difficult to extract, we focused instead on a spin-spin correlation function at large distances

$$w^2 = \langle \sigma_z(0)\sigma_z(N/2)\rangle \ . \tag{4.10}$$

Here, $N/2$ indicates the site that is furthest apart from site 0, and we expect $w \sim m$ for large N. According to HF theory, we would have

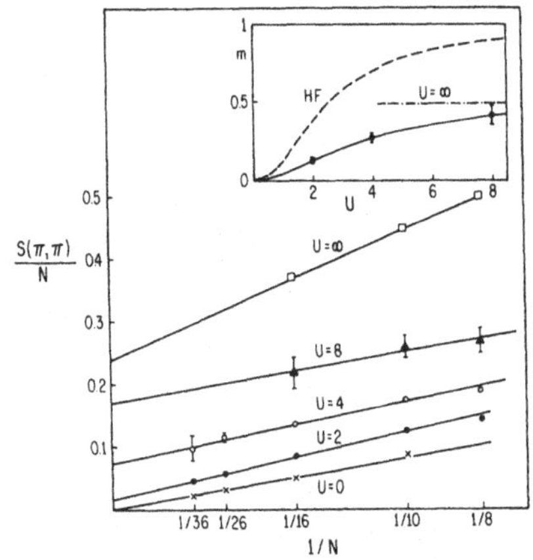

FIG. 12. Extrapolation of long-range antiferromagnetic order. The $U = \infty$ results are taken from Ref. 20. The inset shows the staggered magnetization m versus U, and the Hartree-Fock predictions (dashed line).

$$\Delta^{\rm HF} = Um \sim Uw \sim te^{-2\pi\sqrt{t/U}} \ . \tag{4.11}$$

In one dimension (1D), it is known that the HF gap is

$$\Delta_{\rm 1D}^{\rm HF} \sim te^{-2\pi/U} \tag{4.12}$$

and that fluctuations only modify the prefactor, since the exact gap goes as

$$\Delta_{\rm 1D} \sim \sqrt{Ut}\, e^{-2\pi/U} \ . \tag{4.13}$$

If the same happens in two dimensions (2D), we would have

$$\Delta_{\rm 2D} \sim Uw \sim \sqrt{tU}\, e^{-2\pi\sqrt{t/U}} \ . \tag{4.14}$$

We cannot verify Eqs. (4.11) or (4.14) directly on a small lattice because of finite-size effects. However, we can use a finite-size-scaling analysis. If Eq. (4.11) is valid for an infinite lattice, we expect for a finite lattice of *linear* dimension n:[23]

$$Uw_n = \frac{1}{n} f(ne^{-2\pi\sqrt{t/U}}) \ , \tag{4.15}$$

while if (4.14) is valid, we would have

$$\sqrt{U}\, w_n = \frac{1}{n} f(ne^{-2\pi\sqrt{t/U}}) \ . \tag{4.16}$$

To verify (4.16), we plot $\sqrt{U}\, w_{n_1}$ versus $1/\sqrt{U}$ and $\sqrt{U}\, w_{n_2}$ versus

$$\frac{1}{\sqrt{U}} - \frac{1}{2\pi}\ln\left[\frac{n_2}{n_1}\right] ,$$

for lattices of linear size n_1 and n_2; if (4.16) holds, the results should follow the same curve. If (4.15) holds, this should happen if we plot Uw_n instead.

Figure 13 shows such plots for a 4×4 lattice at $\beta = 3$, and a 6×6 lattice at $\beta = 4.5$. In Fig. 13(a) we show the results assuming (4.16), while in Fig. 13(c) we check the form (4.15). Our numerical results appear to support the former assumption. In Fig. 13(b) we show scaling for the HF gap on the same finite lattices, as a check on our procedure.

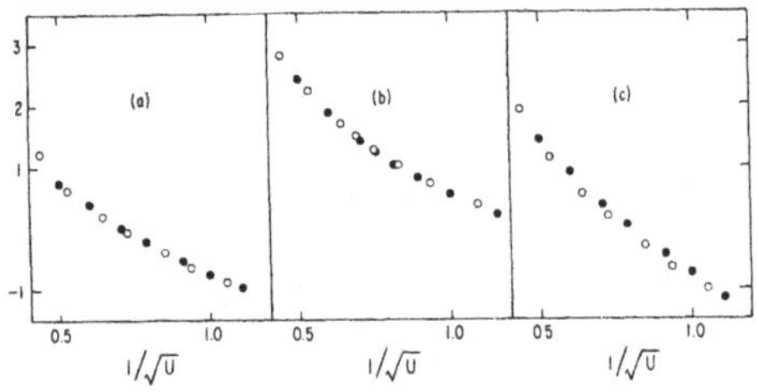

FIG. 13. Finite-size-scaling analysis for small U. (a) $\ln(n\sqrt{U}\, w_n)$, (b) $\ln(n\Delta_n^{\rm HF})$, and (c) $\ln(nUw_n)$. The solid circles are results for a 4×4 lattice plotted versus $1/\sqrt{U}$, the open circles are results for a 6×6 lattice plotted versus $1/\sqrt{U} - (1/2\pi)\ln\frac{6}{4}$. The points fall on the same curve on (a) but not on (c), indicating that Eq. (4.16) holds.

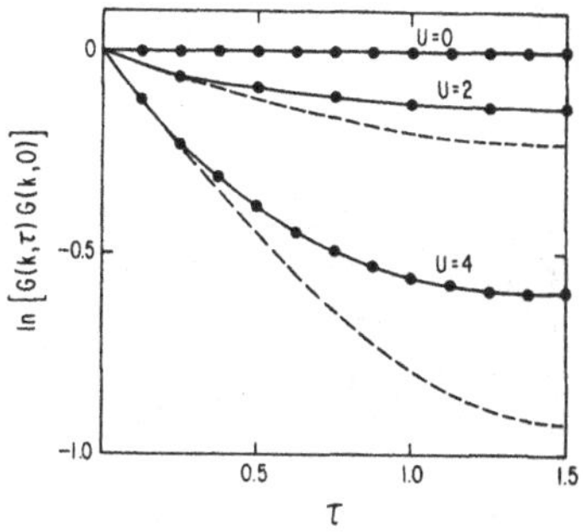

FIG. 14. Imaginary time dependence of Green's function for $k=(\pi/3,2\pi/3)$ 6×6 lattice, $\beta=3$. The dashed lines are the predictions of a rigid two-band picture fitted to the initial time decay.

To summarize, our numerical results suggest that the magnetization for small coupling behaves as

$$m \sim \sqrt{t/U}\, e^{-2\pi\sqrt{t/U}} , \tag{4.17a}$$

and assuming the relation $\Delta \propto Um$ between gap and magnetization holds, as in HF theory, we have for the gap in two dimensions,

$$\Delta_{2D} \sim \sqrt{tU}\, e^{-2\pi\sqrt{t/U}} , \tag{4.17b}$$

that is, it differs from the HF gap [Eq. (4.11)] by the prefactor only. We cannot completely rule out other possibilities; in particular, our numerical results are not inconsistent with the form $\Delta_{2D} \sim e^{-2\pi t/U}$ (with no U-dependent prefactor). However, we believe that our numerical results together with the HF prediction and the analogy with the one-dimensional case strongly suggest that (4.17) is valid. Further numerical and analytic work should be able to fully resolve this question.

It should be pointed out that the relation (4.17b) follows from (4.17a) only if we make the reasonable assumption that the gap and the magnetization scale in the same way at zero temperature, but it has not been checked directly. In addition, at finite temperature this relation breaks down, since the magnetization is destroyed by thermal fluctuations while the gap for charge excitations remains. We can see directly that the system develops a gap when U is nonzero by studying the imaginary time decay of various correlation functions. Figure 14 shows the Fourier transform of the time-dependent Green's function

$$G(k,\tau) = \frac{1}{N} \sum e^{ik(R_i-R_j)} \langle c_i(\tau) c_i^\dagger(0) \rangle$$
$$= \langle c_k(\tau) c_k^\dagger(0) \rangle \tag{4.18}$$

at $k=(\pi/3,2\pi/3)$ on a 6×6 lattice. Since this k lies on the Fermi surface for the noninteracting case, the Green's function does not decay for $U=0$. When U is turned on, $G(k,\tau)$ decays, indicating that the system develops a gap for charge excitations. We have attempted to fit the time decay to a simple exponential corrected for finite-temperature effects:

$$G(k,\tau) \propto e^{-\Delta\tau} + e^{-(\beta-\Delta)\tau} , \tag{4.19}$$

and to the form predicted by HF theory for $G(k,\tau)$ on a finite lattice at finite temperatures. Both procedures give essentially the same answer and are shown in Fig. 14 as dashed lines, with Δ chosen to fit the initial time decay of $G(k,\tau)$, $\Delta=0.57$ for $U=2$, and $\Delta=1.05$ for $U=4$. The exact results differ markedly from these mean-field results for long times, indicating that a rigid two-band picture is not adequate and that collective charge excitations with a gap smaller than the one predicted by a rigid-band picture exist.

Some mean-field calculations predict that the Hubbard model should undergo a sharp insulator-metal transition as the temperature is increased, leading to a metallic state at high temperatures.[24] We find no evidence for such a transition. Figure 15 shows the imaginary time decay of the current-current correlation function

$$J(\tau) = \langle j_x(\tau) j_x(0) \rangle , \tag{4.20a}$$

$$j_x = -i \sum_{i,j,\sigma} (x_i - x_j) c_{i\sigma}^\dagger c_{j\sigma} , \tag{4.20b}$$

and x_i the x component of the position vector at site i. The time dependence does not change qualitatively when the temperature is increased, and indicates that there is a finite gap for conductivity at all temperatures. This will give rise to a thermally-activated-type conductivity similar to a semiconductor at all temperatures. Our results support the picture of Bari and Kaplan[25] that the half-filled Hubbard model does not undergo a transition to a metallic state as the temperature is increased.

Finally, it is also interesting to study the occupation number in k space, $\langle c_k^\dagger c_k \rangle = 1 - G(k)$, at low temperatures. Figure 16 shows results for $\beta=4$ on an 8×8 lattice. The temperature is sufficiently low that the results are very close to the ground-state values. For $U=0$, the occupation number is essentially 1 inside the Fermi surface, 0.5 on the Fermi surface, and 0 outside. Note how the interaction alters this behavior: for $U=4$, the occupation number even at $k=0$ is somewhat smaller than 1.

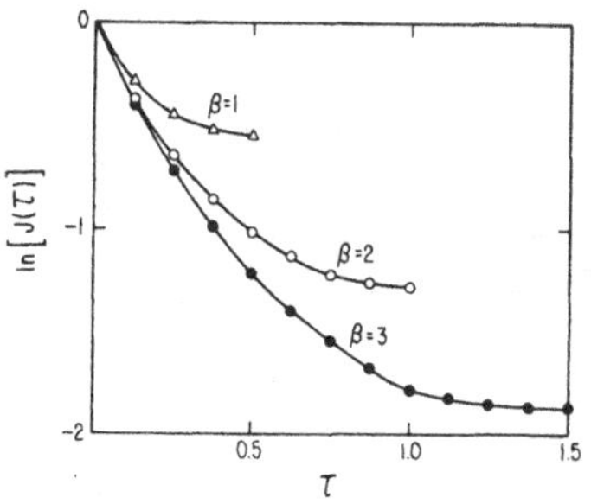

FIG. 15. Imaginary time dependence of current-current correlation function for $U=4$, $\beta=3$, 2, and 1 on a 6×6 lattice.

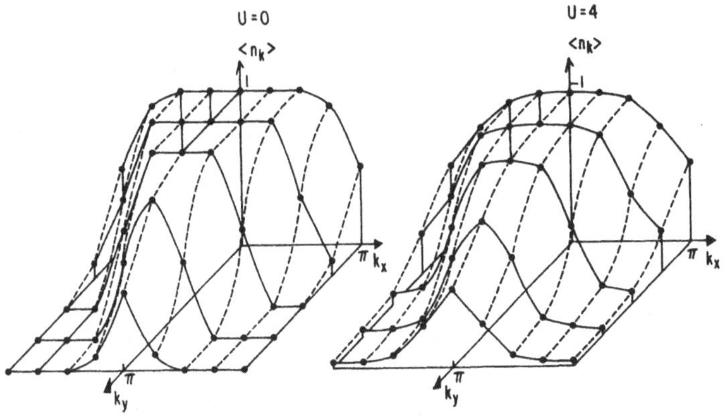

FIG. 16. Occupation number in k space, $n_k = \langle c_k^\dagger c_k \rangle$ on an 8×8 lattice at $\beta=4$. Note how the interaction causes a rounding, and even at $k=0$ the occupation number is less than unity due to the electron-electron interaction.

It is not possible to determine from such a small system whether a well-defined Fermi surface still exists. It is believed, however, from general arguments, that the Fermi surface is destroyed for $U\neq0$, since the system becomes an insulator.

V. NON-HALF-FILLED BAND

We have performed simulations for band fillings other than one-half for $U=4$ and $U=8$. As mentioned in Sec. III, the determinant does become negative for some cases if the band is not half-full, and it becomes necessary to compute the average sign. This happens particularly for large values of the interaction. Table I shows some results for the average sign for various cases. It can be seen that the average sign does become somewhat smaller as the temperature decreases, and does not seem to be very dependent on lattice size. As a function of band filling, it first decreases and then increases again as the band filling is further decreased. For the parameter range studied, the average sign did not become small enough to cause problems in the simulations. It is not clear whether as $\beta\to\infty$ the average sign vanishes, and if so how. It certainly does not appear to be vanishing exponentially, as in methods where the fermions are not integrated out, but could possibly be vanishing algebraically.

Figure 17 shows the behavior of the spin-spin correlation functions for $U=4$ and several band fillings, for 6×6 lattices at $\beta=4.5$. We also show the results for 4×4, $\beta=3$. Except for the case $\rho=1$, there is essentially no change in the spin-spin correlations in going to larger lattices and decreasing the temperature. This suggests that there is no magnetic order except for $\rho=1$. The peak value shifts from (π,π) as the filling is decreased. The results are very similar to the ones obtained with no correlations except for the $\rho=1$ case.

Figure 18 shows the case $\rho=0.9$ for 4×4 at $\beta=2$; 6×6, $\beta=3$; and 8×8, $\beta=4$. The peak is still at (π,π) here, since we are close to the half-filled band, and there is a small increment in going to larger lattices and increasing β. There is a large difference, however, with the cor-

TABLE I. Average value of the sign of the product of determinants for some cases where the band is non-half-filled. For the half-filled case, the sign is always positive.

U	Band filling	Lattice size	β	$\langle$Sign$\rangle$
4	0.9	4×4	2	1
			3	0.99
			4	0.94
		8×8	4	0.93
	0.84	4×4	3	0.99
		6×6	4.5	0.77
	0.65	4×4	3	0.99
		6×6	4.5	0.97
	0.48	4×4	3	1
		6×6	4.5	1
8	0.67	4×4	2	0.98
		6×6	3	0.61

TABLE II. On-site and nearest-neighbor spin-spin correlations for $U=8$, $\beta=3$ on a 4×4 lattice for various band fillings. $i+\delta$ denotes a nearest neighbor of site i. The number in parentheses is the statistical error in the last figure.

Band filling	$\langle\sigma_z^2\rangle$	$\langle\sigma_{z,i}\sigma_{z,i+\delta}\rangle$
1	0.899(3)	−0.31(4)
0.91	0.841(4)	−0.21(3)
0.81	0.750(2)	−0.12(4)
0.67	0.636(2)	−0.089(2)
0.60	0.565(2)	−0.079(5)
0.42	0.400(8)	−0.036(7)
0.25	0.247(6)	−0.017(2)

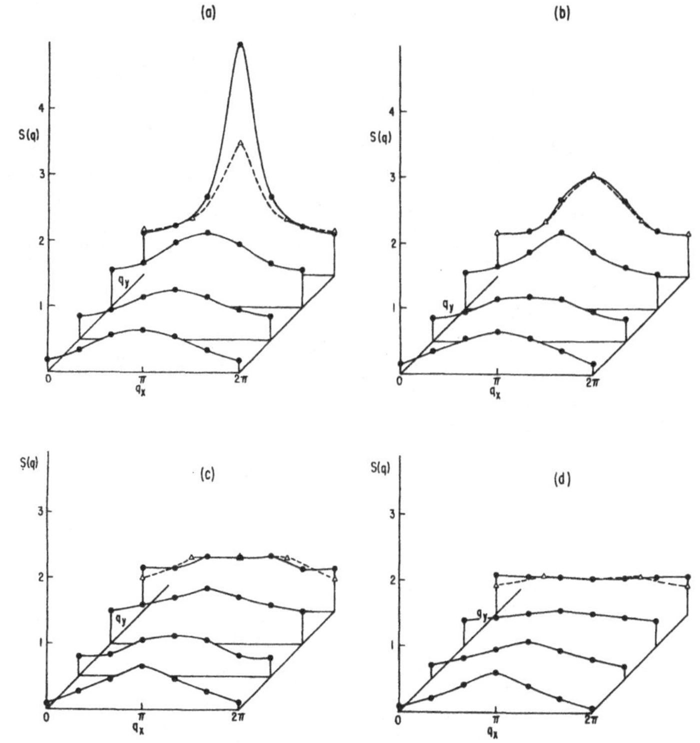

FIG. 17. Magnetic structure factor for $U=4$ on a 6×6 lattice, $\beta=4.5$ (solid lines) and a 4×4 lattice, $\beta=3$ (dashed lines) for several band fillings. (a) $\rho=1$, (b) $\rho=0.84$, (c) $\rho=0.65$, and (d) $\rho=0.48$.

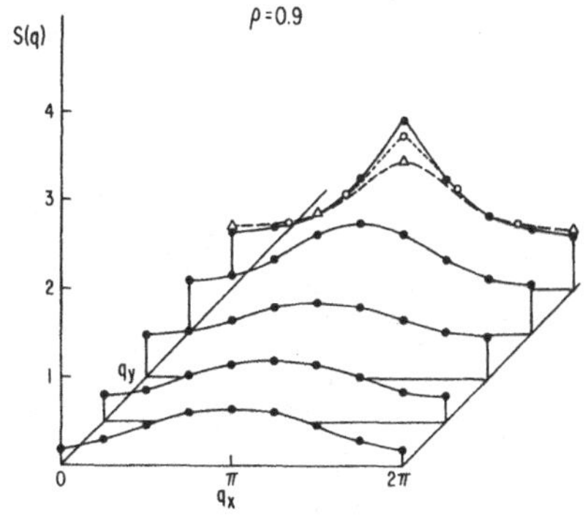

FIG. 18. Magnetic structure factor for $U=4$, $\rho=0.9$ on an 8×8 lattice, $\beta=4$ (solid circles); a 6×6 lattice, $\beta=3$ (open circles); and a 4×4 lattice, $\beta=2$ (open triangles).

responding half-filled case (Fig. 11), while Hartree-Fock theory predicts the antiferromagnetic order for $\rho=0.9$ to be still 75% of the value for $\rho=1$. Also, $S(\pi,\pi)$ appears to be saturating as N and β increase. In Fig. 19 we plot

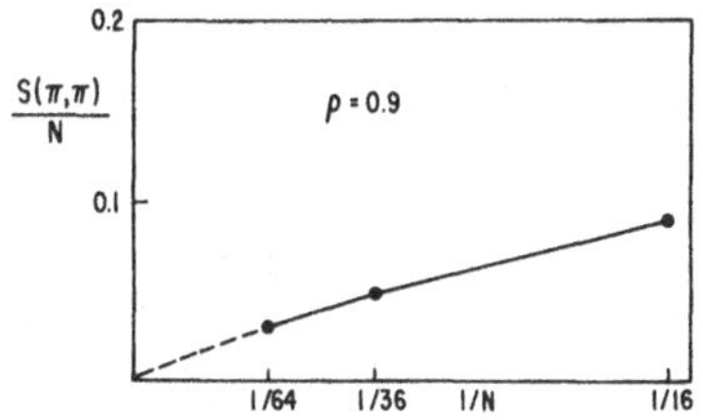

FIG. 19. Extrapolation of long-range antiferromagnetic order for $\rho=0.9$, $U=4$. The results suggest that no long-range order exists.

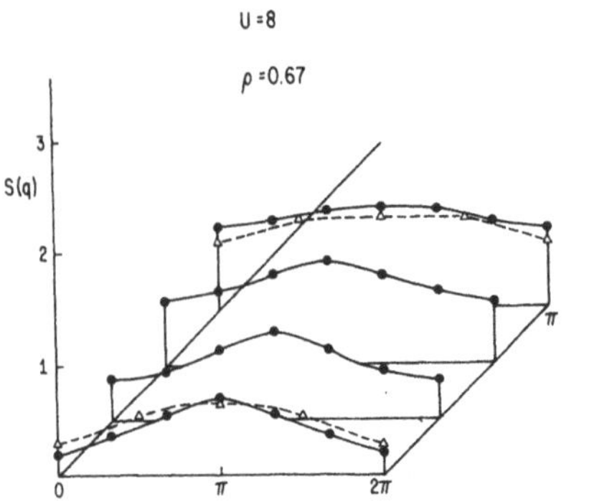

FIG. 20. Magnetic structure factor for $U=8$, $\rho=0.67$ on a 6×6 lattice, $\beta=3$, and a 4×4 lattice, $\beta=2$. Note that there is no indication of ferromagnetism, in contradiction with HF theory.

$S(\pi,\pi)/N$ versus $1/N$ for this case. Although not quite unambiguous, the results suggest that the system does not have antiferromagnetic order in the thermodynamic limit.

We now consider a case which according to mean-field theory should have a ferromagnetic ground state: $U=8$, $\rho=0.65$. The spin-spin correlations in real space show no evidence of even short-ranged ferromagnetic correlations, but rather weak antiferromagnetic correlations. The structure factor is shown in Fig. 20. It has a broad peak at (π,π) but rather weak dependence on lattice size and temperature, suggesting that there is no long-range order here either. In fact, the $q=0$ structure factor *decreases* as the temperature is lowered. We have also explored other band fillings for $U=8$, and found nowhere even short-ranged ferromagnetic correlations. Table II shows the on-site and nearest-neighbor correlations for $U=8$ and several band fillings. It can be seen that the nearest-neighbor correlations are always antiferromagnetic, becoming weaker as the band filling decreases. We conclude that the system shows no tendency for ferromagnetic correlations in the parameter range studied.

Finally, we have also studied pairing correlations in our simulation. We measured the singlet- and triplet-pairing susceptibility, given by

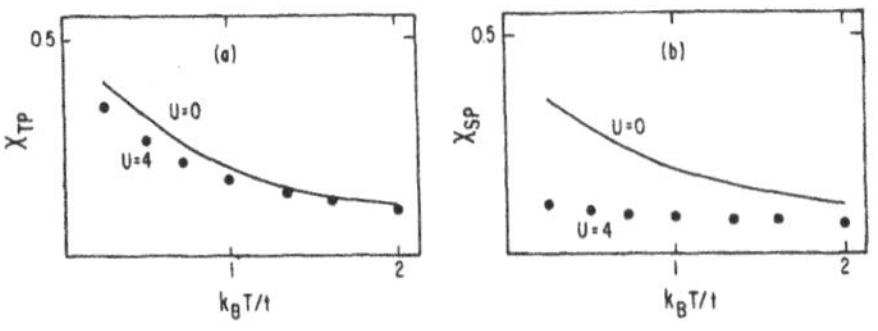

FIG. 21. (a) Triplet- and (b) singlet-pairing susceptibility at $q=0$ versus temperature for $U=0$ and $U=4$, $\rho\sim0.65$.

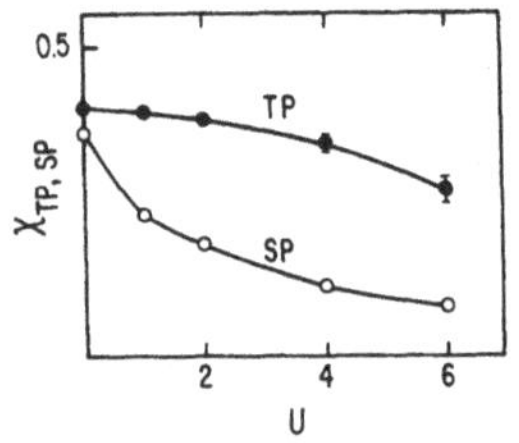

FIG. 22. Triplet- and singlet-pairing susceptibilities at $q=0$ versus U for $\beta=4$, $\rho\sim0.65$.

$$\chi_{SP}(q)=\frac{1}{N}\sum_{i,j}e^{iq(R_i-R_j)}\int_0^\beta d\tau\langle c_{i\uparrow}(\tau)c_{i\downarrow}(\tau)c_{j\downarrow}^\dagger(0)c_{j\uparrow}^\dagger(0)\rangle\ , \tag{5.1}$$

$$\chi_{TP}(q)=\frac{1}{N}\sum_{i,j}e^{iq(R_i-R_j)}\times\int_0^\beta d\tau\langle c_{i\uparrow}(\tau)c_{i+\hat{x},\uparrow}(\tau)c_{j+\hat{x},\uparrow}^\dagger(0)c_{j\uparrow}^\dagger(0)\rangle\ . \tag{5.2}$$

These susceptibilities peak at $q=0$, and diverge if the system undergoes a transition to a singlet or triplet superconducting state. Figure 21 shows the temperature dependence of $\chi_{TP}(q=0)$ and $\chi_{SP}(q=0)$ for $U=0$ and $U=4$, and band filling $\rho\sim0.65$. χ_{SP} is strongly suppressed by U, as one would expect. χ_{TP} for $U=4$ appears to grow in a similar way as for the noninteracting case (χ_{TP} at $U=0$ diverges logarithmically as $T\rightarrow0$), but it is smaller at all temperatures. Figure 22 shows the dependence of the triplet- and singlet-pairing susceptibilities on U for $\rho\sim0.65$ at a fixed low temperature; both susceptibilities are suppressed with U, although χ_{SP} much more rapidly. We find also that the suppression is larger for the band closer to half-full. We have also measured the triplet-pairing susceptibility for antiparallel spins, and obtained results very close to χ_{TP}.

Our results for χ_{TP} are surprising, since it has been argued that spin fluctuations should give rise to triplet pairing in a model with strong short-ranged repulsive interactions like the Hubbard model;[26] in particular, it is believed that in ^{3}He the short-ranged repulsion is the dominant mechanism causing superfluidity.[26] Our findings cast doubt on this picture, at least in two dimensions. For the ^{3}He case, we believe that the longer-range attractive tail in the potential is crucial, and simulations on a Hubbard model with nearest-neighbor attractive interactions are in progress.

VI. THE ATTRACTIVE HUBBARD MODEL

We now discuss briefly the properties of the attractive Hubbard model in the half-filled-band case. We can do this without extra work, since as is well known, a particle-hole transformation can map $U>0$ into $U<0$. The transformation is[27]

$$d_{i\uparrow}=c_{i\uparrow}\ ,$$
$$d_{i\downarrow}=c_{i\downarrow}^{\dagger}(-1)^{i}\ , \tag{6.1}$$

which takes $U\rightarrow -U$ and leaves the kinetic energy invariant. Under this transformation, S_z-S_z correlations are mapped onto charge-density correlations:

$$n_{i\uparrow}-n_{i\downarrow}\rightarrow n_{i\uparrow}+n_{i\downarrow}\ , \tag{6.2}$$

so that long-range antiferromagnetic order corresponds to a charge-density-wave (CDW) state. We have also, however, long-range antiferromagnetic order in the other directions of spin space; the spin operator in the x direction, for example, is

$$S_x^i=c_{i\uparrow}^{\dagger}c_{i\downarrow}+c_{i\downarrow}^{\dagger}c_{i\uparrow}\rightarrow d_{i\uparrow}^{\dagger}d_{i\downarrow}^{\dagger}+d_{i\downarrow}d_{i\uparrow}\ , \tag{6.3}$$

and long-range order in the x component of the spin corresponds to long-range singlet superconducting order in the attractive case. We conclude then that the ground state of the two-dimensional attractive Hubbard model (half-filled) is very peculiar in that it exhibits simultaneously CDW and superconducting long-range order. In the presence of any small perturbation, for example, a longer-range electron-electron interaction, one of the types of order will probably be destroyed and the other further stabilized. This is an interesting question to explore further.

VII. CONCLUSIONS

We have studied properties of the two-dimensional Hubbard model on a square lattice using Monte Carlo simulations. As mentioned in Sec. II, we expect some of the features found here to be very specific to the model studied and others to be more general. The purpose of this study was to provide answers to the following questions. (a) What are the properties of the simplest lattice model of interacting electrons in two dimensions? (b) How useful is the Hubbard model to describe electron correlations and its consequences for narrow-band systems? And, (c) how well does mean-field theory describe the model? We believe we have provided at least partial answers to these questions. Finally, another purpose of our work was to demonstrate that numerical simulations can be a useful tool to study interacting electron systems in more than one dimension.[28]

Our conclusions can be summarized by the phase diagram in Fig. 23. We have shown that for the half-filled band case the system exhibits antiferromagnetic long-range order for all values of the interactions. This conclusion is by no means obvious: In the $U=\infty$ limit, spin fluctuations reduce the long-range order to 50% of its classical value, as shown by Betts and Oitmaa; one might have expected that for any $U<\infty$, charge fluctuations destroy the order altogether, or that a critical value U_c exists below which no long-range order exists. Even though the susceptibility diverges for $U=0$, this does not prove that long-range order exists for any $U>0$; recall the case of one-dimensional spinless fermions with a nearest-neighbor repulsion V, where, even though the $V=0$ susceptibility is divergent, long-range order starts building up only for $V\geq 2t$. Our numerical results indicate that long-range order exists for any value of $U>0$, although substantially reduced from the mean-field-theory predictions. We have also studied the character of the transition at zero coupling using finite-size scaling, and concluded that the mean-field-theory predictions are essentially correct except for a prefactor, as occurs in one dimension.

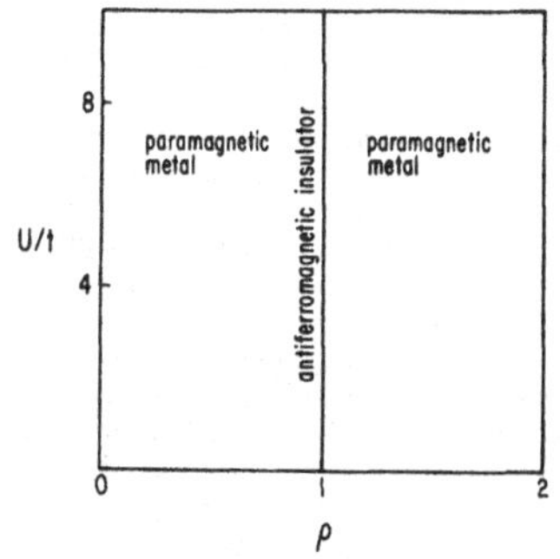

FIG. 23. Conjectured ground-state phase diagram for the two-dimensional Hubbard model.

For the non-half-filled band, although we have only explored a few points on the phase diagram, we believe our results strongly suggest that no long-range magnetic order exists. We have also found no indication of even short-ranged ferromagnetic correlations. Concerning this last point, our findings are in agreement with those of other authors,[10] who suggested that bipartite lattices are particularly unfavorable for ferromagnetic correlations.

Unfortunately, our phase diagram, Fig. 23, is much less interesting than the Hartree-Fock phase diagram, Fig. 3. It appears that an approximate (and qualitatively wrong) solution to a simple model does much better in describing features of real materials than the exact solution of the model. In fact, our model does not really describe itinerant magnetism: only for the insulating case do we find magnetic long-range order. Although this could be due to dimensionality (work on the three-dimensional Hubbard model is in progress), it is possible that a model to describe itinerant magnetism will have to necessarily include band degeneracy.

We have also explored pairing correlations and, in particular, triplet pairing. Recently, it has been suggested that triplet superconductivity could occur in strongly interacting fermion systems, driven by an electronic interaction mechanism.[29] Our results suggest that this is unlikely in a single-band model; it could, however, conceivably occur in models with more than one band.

Several questions about the two-dimensional Hubbard model remain open. The consequences of the singularity in the density of states coinciding with the nesting of the Fermi surface should be investigated theoretically beyond the RPA: a renormalization-group treatment appears possible. For strong coupling, the formulation we have used breaks down because of the finite time step $\Delta\tau$. In that regime, it should be more useful to construct first an effective Hamiltonian, valid to some order in t/U, to be studied directly by numerical simulations. Questions such as the range of validity of Nagaoka's theorem could then be addressed. Finally, changing the band structure by introducing, for example, a next-nearest-neighbor hopping term should lead to new and interesting physics. In particular, it will produce the singularity in the density of states to occur at an energy where nesting does not occur. This could possibly lead to ferromagnetism if the Fermi energy coincides with the singularity in the density of states. It could also lead to superconductivity if an effective attractive interaction between the electrons exists caused by coupling to boson degrees of freedom. In addition, since nesting will not occur in the half-filled case, it could lead to a Mott metal-insulator transition for a finite value of the interaction.

ACKNOWLEDGMENTS

I am grateful to the Condensed Matter Sciences Division at Los Alamos National Laboratory for providing funds for use of the Cray, to the Center of Nonlinear Studies at Los Alamos for its hospitality, and in particular to D. Campbell and J. Gubernatis for their assistance. I am also grateful to D. J. Scalapino for stimulating discussion. This work was supported by the National Science Foundation under Grant No. DMR-82-17881.

APPENDIX

We discuss here an elementary derivation of Eq. (3.9). It follows from the identity:

$$\mathrm{Tr}\, e^{-c_i^\dagger A_{ij} c_j} e^{-c_i^\dagger B_{ij} c_j} = \det(1+e^{-A}e^{-B}) , \tag{A1}$$

where A and B are arbitrary matrices, and the summation over indices is understood. To prove (A1), we first prove the identity:

$$e^{-c_i^\dagger A_{ij} c_j} e^{-c_i^\dagger B_{ij} c_j} = e^{-\sum_\nu c_\nu^\dagger l_\nu c_\nu} , \tag{A2}$$

where $\lambda_\nu = e^{-l_\nu}$ are the eigenvalues of the matrix $e^{-A}e^{-B}$. From (A2), Eq. (A1) follows immediately, since

$$\begin{aligned} \mathrm{Tr}\, e^{-\sum_\nu c_\nu^\dagger l_\nu c_\nu} &= \mathrm{Tr} \prod_\nu e^{-c_\nu^\dagger l_\nu c_\nu} \\ &= \prod_\nu (1+e^{-l_\nu}) \\ &= \det(1+e^{-A}e^{-B}) . \end{aligned} \tag{A3}$$

To prove (A2), we show that an arbitrary many-particle state propagates in the same way using the expression on either side. Consider first a single-particle state

$$|\phi\rangle = \sum_j a_j c_j^\dagger |0\rangle \tag{A4}$$

with a_j arbitrary numbers, and $|0\rangle$ the vacuum state. Let $|\mu\rangle$ be the basis where the matrix B is diagonal, i.e.,

$$B = \sum_\mu |\mu\rangle b_\mu \langle \mu| , \tag{A5}$$

and define new fermion coordinates

$$\begin{aligned} c_\mu &= \sum_j \langle \mu | j \rangle c_j , \\ c_\mu^\dagger & \sum_j \langle j | \mu \rangle c_j^\dagger , \end{aligned} \tag{A6a}$$

with inverse

$$\begin{aligned} c_j &= \sum_\mu \langle j | \mu \rangle c_\mu , \\ c_j^\dagger &= \sum_\mu \langle \mu | j \rangle c_\mu^\dagger . \end{aligned} \tag{A6b}$$

We can write the exponential of B, using the properties of fermion operators, as

$$e^{-c_i^\dagger B_{ij} c_j} = e^{-c_\mu^\dagger b_\mu c_\mu} = \prod_\mu [1+(e^{-b_\mu}-1)c_\mu^\dagger c_\mu] . \tag{A7}$$

On applying this to the state (A4), expanding $c_j^\dagger$ in terms of $c_\mu^\dagger$, and using fermion anticommutation relations, we find

$$\begin{aligned} e^{-c_i^\dagger B_{ij} c_j} |\phi\rangle &= \sum_j a_j' c_j^\dagger |0\rangle , \\ a_i' &= \sum_j (e^{-B})_{ij} a_j . \end{aligned} \tag{A8}$$

Similarly, in operating with both factors on the left-hand side of Eq. (A2), one finds after some algebra

$$\begin{aligned} e^{-c_i^\dagger A_{ij} c_j} e^{-c_i^\dagger B_{ij} c_j} |\phi\rangle &= \sum_j a_j'' c_j^\dagger |0\rangle , \\ a_i'' &= \sum_j (e^{-A}e^{-B})_{ij} a_j , \end{aligned} \tag{A9}$$

i.e., the amplitude of the propagated state is obtained by multiplying the original amplitude by the *product* of the matrices. Equation (A9) is valid in any basis, in particular in the one where $e^{-A}e^{-B}$ is diagonal. If we start then with a state that is an eigenstate of $e^{-A}e^{-B}$:

$$|\phi\rangle = c_\nu^\dagger |0\rangle ,$$

then

$$e^{-c_i^\dagger A_{ij} c_j} e^{-c_i^\dagger B_{ij} c_j} = (e^{-A}e^{-B})_{\nu\nu} c_\nu^\dagger |0\rangle = e^{-l_\nu} c_\nu^\dagger |0\rangle ,$$

which is the same of course as we obtain from the right-hand side of Eq. (A2), using the relation (3.16). Thus, we have proved Eq. (A2) when applied to single-particle states, and it only remains to be shown that if we have more than one particle they propagate independently. Consider first the propagation by one factor. If we take a two-particle state

$$|\phi\rangle = c_{\mu_1}^\dagger c_{\mu_2}^\dagger |0\rangle \tag{A10}$$

and propagate it with B, we have

$$e^{-c_i^\dagger B_{ij} c_j}\,|\phi\rangle = \prod_\mu [1+(e^{-B_\mu}-1)c_\mu^\dagger c_\mu] c_{\mu_1}^\dagger c_{\mu_2}^\dagger\,|0\rangle$$
$$= e^{-B_{\mu_1}} e^{-B_{\mu_2}} c_{\mu_1}^\dagger c_{\mu_2}^\dagger\,|0\rangle \,. \tag{A11}$$

Equation (A11) clearly holds if $\mu_1 \neq \mu_2$, since we pair μ_1 and μ_2 with its corresponding factor, and also if $\mu_1=\mu_2$, since then both sides are zero due to the Pauli principle. Clearly then, the propagation of an arbitrary two-particle state is a superposition of the propagation of each particle independently, and similarly for many-particle states. By using the argument repeatedly, it follows also for propagation through more than one factor, which completes the proof of Eq. (A1). Of course, this is then trivially extended to more than two factors.

[1]J. Hubbard, Proc. R. Soc. London, Ser. A **276**, 283 (1963); **281**, 401 (1964).

[2]See, for example, *Electron Correlation and Magnetism in Narrow-Band Systems*, edited by T. Moriya (Springer, New York, 1981), and references therein.

[3]E. Lieb and F. Wu, Phys. Rev. Lett. **20**, 1445 (1968).

[4]D. Penn, Phys. Rev. **142**, 350 (1966).

[5]M. Cyrot, J. Phys. (Paris) **33**, 125 (1972).

[6]M. C. Gutzwiller, Phys. Rev. **137**, A1726 (1965).

[7]H. Shiba and P. Pincus, Phys. Rev. B **5**, 1966 (1972); H. Shiba, Prog. Theor. Phys. **48**, 2171 (1972).

[8]Z. G. Soos and S. Ramasesha, Phys. Rev. B **29**, 5410 (1984).

[9]A. Kawabata, in Ref. 2, p. 172.

[10]M. Takahashi, J. Phys. Soc. Jpn. **51**, 3475 (1982).

[11]J. E. Hirsch, D. J. Scalapino, R. L. Sugar, and R. Blankenbecler, Phys. Rev. B **26**, 5033 (1982).

[12]J. E. Hirsch and D. J. Scalapino, Phys. Rev. B **27**, 7169 (1983); **29**, 5554 (1984).

[13]J. E. Hirsch, Phys. Rev. Lett. **53**, 2327 (1984), and unpublished.

[14]J. E. Hirsch, Phys. Rev. B **28**, 4059 (1983).

[15]R. Blankenbecler, D. J. Scalapino, and R. L. Sugar, Phys. Rev. D **24**, 2278 (1981); D. J. Scalapino and R. L. Sugar, Phys. Rev. B **24**, 4295 (1981).

[16]J. E. Hirsch, Phys. Rev. Lett. **51**, 1900 (1983).

[17]See, for example, P. M. Chaikin, P. Pincus, and G. Beni, J. Phys. C **8**, L65 (1975); P. Richmond, Solid State Commun. **7**, 997 (1969); D. Mattis and W. Langer, Phys. Rev. Lett. **25**, 376 (1970).

[18]S. Doniach and E. Sondheimer, *Green's Functions for Solid State Physicists* (Benjamin, Reading, Mass., 1982), p. 162.

[19]This feature was also found recently by J. Gubernatis, D. Scalapino, R. Sugar, and D. Toussaint (private communication) in a two-dimensional spin-polarized fermion model, where it causes an enhanced tendency to charge-density-wave formation.

[20]Y. Nagaoka, Phys. Rev. **147**, 392 (1966).

[21]J. Oitmaa and D. Betts, Can. J. Phys. **56**, 897 (1978).

[22]W. D. Langer and D. C. Mattis, Phys. Lett. **36A**, 139 (1971).

[23]H. H. Roomany and H. W. Wyld, Phys. Rev. D **21**, 3341 (1980).

[24]S. Doniach, Adv. Phys. **18**, 819 (1969), and references therein.

[25]R. Bari and T. Kaplan, Phys. Rev. B **6**, 4623 (1972).

[26]K. Levin and O. T. Valls, Phys. Rep. **98**, 1 (1983), and references therein.

[27]V. J. Emery, Phys. Rev. B **14**, 2989 (1976).

[28]Results of simulations of a two-dimensional spin-polarized fermion model have been recently reported by D. Scalapino, R. Sugar, and W. Toussaint, Phys. Rev. B **29**, 5253 (1984).

[29]P. W. Anderson, Phys. Rev. B **30**, 1549 (1984).

EUROPHYSICS LETTERS 1 April 1989

Europhys. Lett., 8 (7), pp. 663-668 (1989)

A Novel Technique for the Simulation of Interacting Fermion Systems.

S. SORELLA, S. BARONI, R. CAR and M. PARRINELLO

Scuola Internazionale Superiore di Studi Avanzati
Strada Costiera 11, I-34014 Trieste, Italy

(received 5 December 1988; accepted in final form 23 January 1989)

PACS. 71.20A – Developments in mathematical and computational techniques.
PACS. 05.30F – Quantum statistical mechanics.
PACS. 74.65 – Insulator-superconductor transition.

Abstract. – A new method for the simulation of ground-state properties of interacting fermions is introduced. A trial wave function, which is assumed to be a Slater determinant, is propagated to large imaginary times. The quantum many-body propagator is represented by a coherent superposition of single-particle propagators by means of a Hubbard-Stratonovich transformation. The resulting functional integral is performed by stochastic methods based on Langevin dynamics. Numerical stability is achieved by orthonormalizing the propagating single-particle orbitals entering the Slater determinant. The problem of the positiveness of the statistical weight is addressed and solved in most cases. Illustrative examples are given for the 1D and 2D Hubbard models.

Considerable attention is presently being paid to the development of numerical methods that allow the investigation of interacting fermion systems [1]. In one way or another, all these methods face severe difficulties which are related to the antisimmetry of the fermionic wave functions.

In Green's function Monte Carlo (MC) and related methods [1, 2], these difficulties are partially circumvented by assuming a given nodal structure for the ground-state wave function. This restriction can be lifted in principle. However, the procedure for relaxing the nodal surface is computationally demanding and prone to numerical instability. This has limited the application of this more exact procedure to only a few cases.

Another class of methods is based on the MC sampling of the finite-temperature partition function $\mathcal{Z} = \mathrm{Tr}(\exp[-\beta H])$ [3-5], where $\beta = T^{-1}$ is the inverse temperature and H the Hamiltonian of the system. The ground-state properties are obtained by taking the $T \to 0$ limit. No assumption is made on the wave functions, but the $T \to 0$ limit is difficult to perform, since numerical instabilities appear in the low-T regime. These instabilities reflect the ill-conditioned nature of the determinant which is obtained once the fermionic degrees of freedom are traced out. Another severe problem is that the distribution that one has to sample by MC methods is non–positive-definite.

In this letter, we shall show that a method recently proposed by Koonin *et al.* [6] and applied so far only to a simple boson system can be successfully extended to treat fermions. Our method does not assume any particular structure for the wave functions; the calculations are stable and give very accurate results. The problem of nonpositive definite distributions can be circumvented in most cases.

Following Koonin *et al.* [6] we introduce a projected partition function

$$Q = \langle \Psi_T | \exp[-\beta H] | \Psi_T \rangle , \tag{1}$$

where $|\Psi_T\rangle$ is a trial wave function nonorthogonal to the ground state $|\Psi_0\rangle$ and β can be thought of as an imaginary time. Since the imaginary time propagator $\exp[-\beta H]$ for $\beta \to \infty$ projects from $|\Psi_T\rangle$ its component along $|\Psi_0\rangle$ in terms of Q the ground-state energy is given by

$$E_0 \underset{\beta\to\infty}{=} -\frac{1}{\beta} \ln Q . \tag{2}$$

Expressions can be found for other ground-state properties, by differentiating eq. (2) with respect to appropriate external fields coupled to the quantities of interest.

In order to demonstrate the soundness of our procedure, we will apply our method to the 1D and 2D Hubbard model, of great current interest for its possible relevance to the physics of high-T_c superconductors [7]. The Hubbard model is described by the Hamiltonian $H = -\sum_{\langle ij \rangle, \alpha} c_{i\alpha}^\dagger c_{j\alpha} + U \sum_i c_{i\uparrow}^\dagger c_{i\uparrow} c_{i\downarrow}^\dagger c_{i\downarrow}$, where $\sum_{\langle ij \rangle}$ indicates nearest-neighbours sums, the indices run over the M lattice sites, $c_{i\alpha}^\dagger (c_{i\alpha})$ are the usual creation (annihilation) operators at site i with spin α and the two-body interaction is on site and repulsive ($U > 0$). Note, however, that our method can be extended to treat Hamiltonians with two-body interactions of fairly general range and sign.

In order to evaluate $\exp[-\beta H] |\Psi_T\rangle$ we split the imaginary-time propagator $\exp[-\beta H]$ into a product of P short-time propagators, and apply the Hubbard-Stratonovich (HS) transformation to each of them. Finally eq. (1) reads

$$Q \approx \int \mathrm{d}\sigma \exp\left[-\frac{1}{2}\sigma^2\right] \langle |\Psi_T| U_\sigma | \Psi_T \rangle , \tag{3}$$

where

$$\mathrm{d}\sigma = \prod_{j=1}^{P} \prod_{r=1}^{M} \frac{\mathrm{d}\sigma_r(j)}{\sqrt{2\pi}}, \quad \sigma^2 = \sum_{r,j} \sigma_r(j)^2 ,$$

and U_σ is the (discretized) imaginary time propagator in the time-dependent external magnetic field $\sigma_r(j)$:

$$U_\sigma \equiv U_\sigma(\beta, 0) = \exp\left[-\frac{U\beta N}{2}\right] \prod_{j=1}^{P} \exp\left[-H_0 \frac{\Delta r}{2}\right] \exp\left[-\lambda \sum_r \sigma_r(j)\, m_r\right] \exp\left[-H_0 \frac{\Delta r}{2}\right], \tag{4}$$

where H_0 is the one-body part in the Hamiltonian H, $\lambda = \sqrt{U(\beta/P)}$, $m_r = c_{r\uparrow}^\dagger c_{r\uparrow} - c_{r\downarrow}^\dagger c_{r\downarrow}$ is the magnetization operator at site r, and N is the total numer of electrons. Equation (3) becomes exact in the $P \to \infty$ limit. $U_\sigma(\beta, 0)$ can be written as a product of operators that act separately on the different spin components: $U_\sigma = U_\sigma^\uparrow U_\sigma^\downarrow$.

The most convenient form for $|\Psi_T\rangle$ is to assume that it is a Slater determinant of single-particle orbitals $|\phi_i^\alpha\rangle$ for each spin component. With this choice Q becomes $Q =$

$= \int d\sigma \exp[-\sigma^2/2] \det A_\sigma^\uparrow \det A_\sigma^\downarrow$, where A_σ^α are square matrices of components $(A_\sigma^\alpha)_{m,n} = \langle \phi_m^\alpha | U_\sigma^\alpha | \phi_n^\alpha \rangle$. These are evaluated by propagating the single-particle orbitals with the method of Fest *et al.* [8], *i.e.* by using FFT techniques and calculating potential energy propagators in real space and kinetic-energy propagators in reciprocal space, where they are diagonal in both cases. Other break-ups of the Hubbard Hamiltonians [9], or different forms of the HS transformation [4] may lead to more efficient algorithms, which are, however, difficult to extend to other Hamiltonians. Our approach is instead of more general applicability.

If we assume for the moment that $\langle \Psi_T | U_\sigma | \Psi_T \rangle$ is positive definite Q takes the form of a classical partition function: $Q = \int d\sigma \exp[-V_{\text{eff}}(\sigma)]$, where $V_{\text{eff}}(\sigma) = \frac{1}{2}\sigma^2 - \ln(\det A_\sigma^\uparrow \times \det A_\sigma^\downarrow)$. The sampling of this distribution can be done by standard means. We have preferred here to use a force biased method based on Langevin dynamics [10, 11], rather than the standard MC procedure given the highly nonlocal nature of the interaction which makes the MC update of a single degree of freedom rather costly.

In the large-β limit, a straightforward implementation of our approach undergoes numerical instabilities since the single-particle orbitals $|\phi_n^\alpha\rangle$ tend to become parallel under the action of U_σ, albeit remaining linearly independent. The numerical calculation of a determinant having all the columns nearly parallel is ill-conditioned in finite precision arithmetics. Therefore, for large enough β all the useful information is lost. This numerical instability can be successfully removed without any approximation if one takes the precaution of orthonormalizing during the propagation the single-particle orbitals. This operation is permissible because a Slater determinant of nonorthonormal orbitals can always be replaced by a determinant of orthogonalized orbitals, times a multiplicative constant. Since the orthogonalization procedure is rather costly, in practice one does not perform this operation at every imaginary time step, but at the largest interval which is compatible with the stability of the calculation.

Once the algorithm has been set up one has to find for each operator of interest, O, appropriate estimators, namely functions $E_O(\sigma)$, such that

$$\langle \Psi_0 | O | \Psi_0 \rangle \underset{\beta\to\infty}{=} Q^{-1} \int d\sigma \exp\left[-\frac{1}{2}\sigma^2\right] \det(A^\uparrow)\det(A_\downarrow) E_O(\sigma) . \tag{5}$$

For instance the indicator of the density-matrix operator $c_r^\dagger c_{r'}$ is given by

$$E_{c_r^\dagger c_{r'}} = \sum_{m,n,\alpha} \phi_m^{\alpha\rangle}(r, \beta/2) A_{m,n}^{-1} \phi_n^{\alpha\langle}(r', \beta/2) , \tag{6}$$

where $|\phi_m^{\alpha\rangle}(\beta/2)\rangle = U_\sigma(\beta/2, 0)|\phi_m^\alpha\rangle$, and $|\phi_m^{\alpha\langle}(\beta/2)\rangle = U_\sigma(\beta/2, \beta)|\phi_m^\alpha\rangle$.

So far we have assumed that $\langle \Psi_T | U_\sigma | \Psi_T \rangle$ is a positive definite quantity. This is in general not true. The standard procedure in this case would be to introduce the auxiliary partition function

$$Q_M = \int d\sigma \exp\left[-\frac{1}{2}\sigma^2\right] |\langle \Psi_T | U_\sigma | \Psi_T \rangle| \tag{7}$$

and to relate averages in the Q_M ensemble denoted by $\langle\ \rangle_M$ to the needed averages in the Q ensemble by

$$\langle A \rangle = \frac{\langle A \times s \rangle_M}{\langle s \rangle_M} , \tag{8}$$

where $s = \text{sign}\,\langle \Psi_T | U_\sigma | \Psi_T \rangle$. This method relies on the fact that $\langle s \rangle_M \neq 0$. However, very often $\langle s \rangle_M$ is very small and its variance very large. Thus estimates based on eq. (8) are subject to a large numerical uncertainty and require long simulation runs in order to converge to a reliable value. We show below that for the calculation of ground-state properties this is an unnecessarily complicated procedure.

To this end, we consider first the $\beta \to \infty$ limit of $Q/Q_M = \langle s \rangle_M$. It can be shown [12] that $\langle s \rangle_M$ is either bounded from below by $|\langle \Psi_0 | \Psi_T \rangle|^2$, or it vanishes exponentially with β for $\beta \to \infty$. In the latter case any calculation is hopelessly difficult as it would be a calculation based on eq. (8). However, if the following condition is satisfied for any β:

$$\langle s \rangle_M \geqslant |\langle \Psi_0 | \Psi_T \rangle|^2 , \tag{9}$$

from the asymptotic behaviour of Q, it follows that $Q_M = h_M(\beta) \exp[-\beta E_0]$, where $h_M(\beta)$ is bounded for $\beta \to \infty$. Thus the ground-state energy can be obtained as: $E_0 \underset{\beta\to\infty}{=} -\frac{1}{\beta} \ln Q_M$. The consequence of this is that Q_M can replace Q in the calculation of ground-state properties. The use of Q_M, however, presents some technical difficulties since the corresponding classical potential energy is singular along the nodal surface of $\langle \Psi_T | U_\sigma | \Psi_T \rangle$ and this breaks up the integration domain into disconnected regions separated by infinite potential barriers. This difficulty can be removed and the statistical quality of the results further improved by introducing a third partition function

$$Q_N = \int \mathrm{d}\sigma \exp\left[-\frac{1}{2}\sigma^2\right] \langle \phi_T | U_\sigma^\dagger U_\sigma | \phi_T \rangle^{1/2} . \tag{10}$$

It will be shown elsewhere [12] that Q_N satisfies the following inequalities:

$$Q_M(\beta) \leqslant Q_N(\beta) \leqslant D Q_M(2\beta)^{1/2} , \tag{11}$$

where D is the (finite) dimension of the Hilbert space. If $\langle s \rangle_M \geqslant |\langle \Psi_0 | \Psi_T \rangle|^2$, then $Q_N(\beta) = h_N(\beta) \exp[-E_0 \beta]$ and one must have again $E_0 \underset{\beta\to\infty}{=} -\frac{1}{\beta} \ln Q_N$. Therefore, Q_N can be used to calculate ground-state properties, if for $\beta \to \infty$ $\langle s \rangle_M \neq 0$.

In order to check the above theory, we have performed calculations on 1D and 2D Hubbard models. The 1D case has been solved analytically by Lieb and Wu (LW) [13]. From their solution, we can extract ground-state energies for finite-size systems. Thus we have the possibility of a direct check of our method. In fig. 1 we display the total and kinetic energy for filling $\nu = N/M = 3/4$ as a function of U, and $M = 8$.

The comparison with the LW results for the same size system is very favourable and can be further improved by increasing the statistics as shown for the particular case $U = 8$. In 1D, we have found in all the cases explored, that an appropriate choice of $|\Psi_T\rangle$ leads to a positive definite $\langle \Psi_T | U_\sigma | \Psi_T \rangle$. We have also verified that our choice of β, P, and of the Langevin integration time step lead to errors smaller than the statistical uncertainty.

In 2D, no exact solution is available except for the 2×2 case which can be mapped onto a 4 site one-dimensional ring [14]. In the 2D half-filled case it is always possible to find trial wave functions such that $\langle \Psi_T | U_\sigma | \Psi_T \rangle \geqslant 0$, thus the sampling presents no sign problem. Our findings are summarized in table I. They are in excellent agreement with the 2×2 data and in line with previous numerical estimates. Away from half-filling, we have not been able to find a trial wave function such that $\langle \Psi_T | U_\sigma | \Psi_T \rangle \geq 0$. We have therefore to check that we are in a favourable situation, namely that for $\beta \to \infty$, $\langle s \rangle_M$ does not vanish exponentially. This is

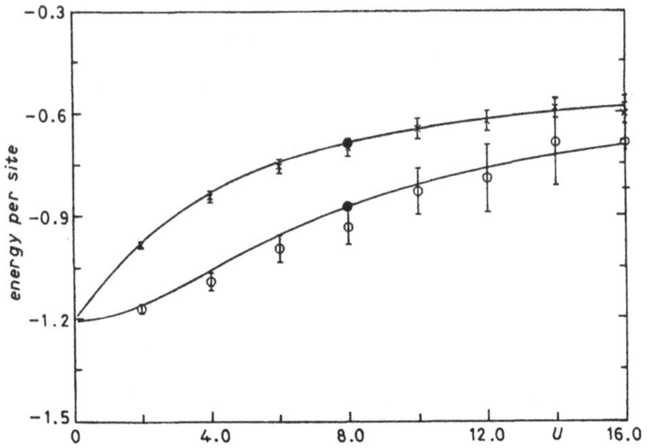

Fig. 1. – Total (upper curve) and kinetic (lower curve) energies for the 8-site 1D Hubbard model, for $\nu = 3/4$, as a function of U. The continuous line indicates the exact results by the method of Lieb and Wu. Open circles indicate simulations where averages were taken over ≈ 100 Langevin time steps. The full dots refer to a run of 2500 time steps. In this case, the error bar is smaller than the size of the dot.

TABLE I. – *Ground-state energy per site for a 2D square-lattice Hubbard model for different sizes and values of the interaction U. The electron density is always fixed to one electron per site. In typical simulations, β ranges from 10 to 20, $P \approx 4\beta U$, and the Langevin time step ranges from 0.1 to 0.3. Averages were taken over several thousands of configurations after suitable equilibration.*

Size	$U = 4$	$U = 8$	$U = 16$
2×2	$-1.40_5 \pm 0.06$	-1.04 ± 0.01	-0.65 ± 0.015
2×2 (*)	-1.414	-1.051	-0.6601
4×4	-0.85 ± 0.06	-0.53 ± 0.06	-0.26 ± 0.05
8×8	-0.86 ± 0.04	-0.53 ± 0.04	-0.28 ± 0.06

(*) Exact results from Lieb and Wu equations.

shown in fig. 2, where we plot $\langle s \rangle_M$ for a 4×4 system at $U = 8$, filling $\nu = 5/8$, and a Hartree-Fock trial wave function. We see that $\langle s \rangle_M$ tends to a finite limit, and we can, therefore, apply the theory developed above for obtaining estimates for the ground-state energy based on Q_N. After a linear extrapolation to $T = 0$ from values calculated for $\beta < 16$, we find: $E_0 = -1.09 \pm 0.01$. A consistent estimate can be obtained using eq. (8). However, in this case at lower temperatures the error bars are so large that we can extrapolate only from $\beta \leq 4$. Thus this second estimate is subject to much larger uncertainties. For larger sizes $\langle s \rangle_M$ becomes so small that an adequate estimation is impossible. One has in these cases to rely only on Q_N. Although we have not yet been able to prove that for a generic determinantal wave function eq. (9) is satisfied, we believe that this is indeed the case. Furthermore, the internal consistency of the theory can be checked numerically in many different ways. Based on these considerations, we have been able to perform calculations on systems of size much

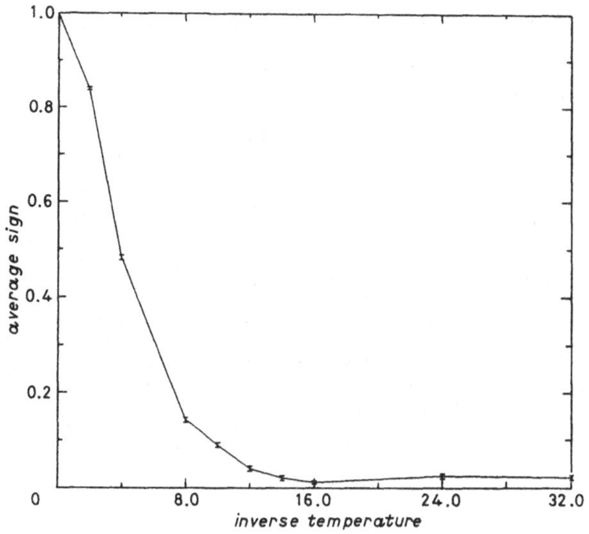

Fig. 2. – Values of $\langle s \rangle_M$ as a function of the inverse temperature β.

larger than previously thought possible [15]. A detailed study of the properties of 1D or 2D Hubbard model will be presented elsewhere.

* * *

It is a pleasure to thank E. TOSATTI for constant encouragement and useful advice. This work has been supported in part by the SISSA-CINECA collaboration project sponsored by the Italian Ministry for Public Education.

REFERENCES

[1] KALOS M. H., in *Monte Carlo Method in Quantum Physics*, edited by M. H. KALOS, NATO ASI Series (D. Riedel Publ. Co., Dordrecht) 1984, p. 19.
[2] CEPERLEY D. M., in *Recent Progress in Many-Body Theories*, edited by J. G. ZABOLITZKY, M. DE LIANO, M. FORTES and J. M. CLARK (Springer Verlag, Berlin) 1981, p. 262.
[3] BLANKENBECKER R., SCALAPINO D. J. and SUGAR R. L., *Phys. Rev. D*, 24 (1981) 2278.
[4] HIRSCH J. E., *Phys. Rev. B*, 31 (1985) 4403.
[5] MORGENSTERN I. and WORTZ D., *Z. Phys. B*, **61** (1985) 219.
[6] KOONIN S. E., SUGIYAMA G. and FRIEDRIECH H., *Proceedings of the International Symposium Bad Honnef*, edited by K. GOEKE and P. G. GREINHARD (Springer Verlag, Berlin) 1982, p. 214.
[7] ANDERSON P. W., *Science*, 235 (1987) 1196.
[8] FEST M. D., FLECK J. A. and STEIGER A., *J. Comp. Phys.*, **47** (1982) 412.
[9] DE RAEDT H., *Comp. Phys. Rep.*, **7** (1987) 1.
[10] PARISI G. and YOUGSHI W., *Scientia Sinica*, **24** (1981) 483.
[11] GUNSTEREN W. F. and BERENDSEN H. J., *Mol. Phys.*, **45** (1982) 637.
[12] SORELLA S., to be published.
[13] LIEB E. H. and WU F. Y., *Phys. Rev. Lett.*, **20** (1968) 1445.
[14] PAROLA A., private communication.
[15] SORELLA S., TOSATTI E., BARONI S., CAR R. and PARRINELLO M., *Int. J. Mod. Phys. B*, **1** (1988) 993.

PHYSICAL REVIEW B VOLUME 40, NUMBER 1 1 JULY 1989

Numerical study of the two-dimensional Hubbard model

S. R. White, D. J. Scalapino, and R. L. Sugar
Department of Physics, University of California, Santa Barbara, Santa Barbara, California 93106

E. Y. Loh and J. E. Gubernatis
Theoretical Division, Los Alamos National Laboratory, Los Alamos, New Mexico 87545

R. T. Scalettar
Department of Physics, University of Illinois, Urbana-Champaign, Urbana, Illinois 61801

(Received 19 December 1988)

We report on a numerical study of the two-dimensional Hubbard model and describe two new algorithms for the simulation of many-electron systems. These algorithms allow one to carry out simulations within the grand canonical ensemble at significantly lower temperatures than had previously been obtained and to calculate ground-state properties with fixed numbers of electrons. We present results for the two-dimensional Hubbard model with half- and quarter-filled bands. Our results support the existence of long-range antiferromagnetic order in the ground state at half-filling and its absence at quarter-filling. Results for the magnetic susceptibility and the momentum occupation along with an upper bound to the spin-wave spectrum are given. We find evidence for an attractive effective d-wave pairing interaction near half-filling but have not found evidence for a phase transition to a superconducting state.

I. INTRODUCTION

In this paper we report on a numerical study of the two-dimensional Hubbard model. We present and make use of two new algorithms for the simulation of many-electron systems which make use of matrix-decomposition methods to stabilize the calculation of fermion determinants and Green's functions. One of the algorithms allows simulations within the grand canonical ensemble to be performed at significantly lower temperatures than have previously been obtained. The other allows direct calculations of ground-state properties of systems with fixed numbers of electrons.[1]

A major problem in the numerical simulation of many-electron systems has been to carry out calculations at sufficiently low temperatures. The problems is essentially one of energy scales. The collective phenomena of interest, such as magnetism or superconductivity, have characteristic energy scales 10 to 1,000 times smaller than the bandwidth or Coulomb interaction strength. The difficulty in achieving temperatures low enough to probe these energy scales is the need to evaluate repeatedly the determinant or inverse of a fermion matrix that becomes increasingly ill-conditioned as the temperature is lowered. Indeed, the fermion matrix has eigenvalues which grow exponentially with the inverse temperature, β. These eigenvalues are associated with low-lying negative energy states which are rarely unfilled. On the other hand, for the grand canonical ensemble the fermion matrix has eigenvalues quite close to one, associated with states having large positive energy, which are rarely occupied. Of course, such a wide range of eigenvalues creates considerable difficulty in the numerical evaluation of the determinant or inverse of the fermion matrix.

Recently, algorithms have been developed which allow simulations to be carried out within the grand canonical ensemble at low temperatures.[2] In these approaches the original fermion matrix, which has dimension equal to the spatial volume of the lattice, V, is replaced by a larger, but generally better conditioned matrix. Since the computer time needed to calculate the determinant or inverse of a fermion matrix grows rapidly with its dimension, these algorithms tend to be slow at low temperatures. In this paper we describe a new algorithm for the grand canonical ensemble which allows us to work directly with the V-dimensional fermion matrix. Through the use of matrix-decomposition methods, this algorithm deals directly with the causes of the ill-conditioning, and remains stable at arbitrarily low temperatures. For temperatures 10^{-2} to 10^{-3} times the bandwidth, it is significantly faster than the algorithms of Ref. 2, and it has allowed us to perform simulations at significantly lower temperatures than have been previously reported. We also present a Monte Carlo algorithm for calculating ground-state properties which uses the same stabilization methods. This algorithm uses a fermion matrix of dimension equal to the number of electrons being simulated, and may have a significant advantage for low-density band fillings. However, for higher densities within the parameter regions we have studied, it is faster to obtain zero-temperature results using the grand-canonical ensemble at low temperatures, than it is to use the ground-state algorithm.

Our development of these algorithms was stimulated by the work of Sugiyama and Koonin[3] and of Sorella *et al.*,[4] who recently reported on simulations of ground-state properties of the two-dimensional Hubbard model. Their algorithm uses Langevin dynamics to project out the ground state from a trial state with a fixed number of electrons. A key elements in their algorithm is the use of an orthogonalization procedure to separate the diverse energy scales that are present in the fermion matrix, and thereby stabilize their calculation. We use a similar orthogonalization procedure to stabilize the calculation of the fermion matrices. Our zero-temperature approach differs from that of Sorella *et al.* in that it makes use of an exact updating, Monte Carlo method which avoids the finite step-size errors inherent in the Langevin equations. The grand-canonical algorithm furthers differs from the approach of Sorella *et al.* in that it allows calculations at finite temperature.

In Sec. II we introduce our algorithms by applying them to the two-dimensional Hubbard model. In Sec. III we give results for energies, momentum occupation, and various equal-time correlation functions for both half-filling, $\langle n \rangle = 1.0$, and quarter-filling, $\langle n \rangle = 0.5$. Low-temperature limits of the grand canonical ensemble are compared with zero-temperature results for fixed numbers of electrons. For the half-filled band our simulations are performed on large enough lattices to allow an extrapolation to the infinite system. We find evidence for long-range antiferromagnetic order in the ground state, in agreement with previous work of Hirsch and Tang.[5] Near half-filling we find evidence for an attractive effective *d*-wave pairing interaction, but not for a phase transition to a superconduction state. The problem associated with the sign of the fermion determinants for the non-half-filled band is also discussed. With the low-temperature stability of the new algorithm, this is the only major problem remaining in the simulation of many-electron systems.

II. THE ALGORITHMS

A. The grand-canonical ensemble

The basic approach for simulating the grand-canonical ensemble was formulated some time ago,[6] but we outline it here in order to make this paper reasonably self-contained. The expectation value of a physical observable, O, is given by

$$\langle O \rangle = \frac{\mathrm{Tr}(O e^{-\beta H})}{\mathrm{Tr}(e^{-\beta H})} , \tag{1}$$

where β is the inverse temperature, and H is the Hamiltonian. For the Hubbard model

$$H = -t \sum_{\langle ij \rangle,\sigma} (c_{i\sigma}^{\dagger} c_{j\sigma} + c_{j\sigma}^{\dagger} c_{i\sigma}) + U \sum_i (n_{i+} - \tfrac{1}{2})(n_{i-} - \tfrac{1}{2}) - \mu \sum_i (n_{i+} + n_{i-}) . \tag{2}$$

Here $c_{i\sigma}^{\dagger}$ and $c_{i\sigma}$ are the creation and annihilation operators for electrons with a z component of spin σ at lattice site i, and $n_{i\sigma} = c_{i\sigma}^{\dagger} c_{i\sigma}$. The sum $\langle ij \rangle$ is over all pairs of nearest neighbor lattice sites. t is the hopping parameter, μ the chemical potential, and U the Coulomb coupling constant, which we take to be positive.

In order to perform a numerical simulation, we must first carry out the traces over the fermion degrees of freedom. To this end we introduce a small imaginary-time step, $\Delta\tau$ by writing $\beta = \Delta\tau L$. The partition function can then be written in the form

$$Z = \mathrm{Tr}(e^{-\Delta\tau L H}) \simeq \mathrm{Tr}(e^{-\Delta\tau V} e^{-\Delta\tau K})^L , \tag{3}$$

with

$$K = -t \sum_{\langle ij \rangle,\sigma} (c_{i\sigma}^{\dagger} c_{j\sigma} + c_{j\sigma}^{\dagger} c_{i\sigma}) - \mu \sum_i (n_{i+} + n_{i-}) = \sum_{\langle ij \rangle,\sigma} c_{i\sigma}^{\dagger} k_{ij} c_{j\sigma} \tag{4}$$

and

$$V = U \sum_i (n_{i+} - \tfrac{1}{2})(n_{i-} - \tfrac{1}{2}) . \tag{5}$$

The Trotter formula[7] used in Eq. (3) introduces errors in measured quantities of order $\Delta\tau^2$.[8] This is the only source of systematic error in the calculations other than that caused by roundoff errors. The interaction terms, $\exp(-\Delta\tau V)$, can be made quadratic in the fermion creation and annihilation operators by introducing Hirsch's discrete Hubbard-Stratonovich transformation[9]

$$e^{-\Delta\tau U (n_{i+} - \frac{1}{2})(n_{i-} - \frac{1}{2})} = e^{-\Delta\tau U/4} \frac{1}{2} \sum_{s_{i,l} = \pm 1} e^{-\Delta\tau s_{i,l} \lambda (n_{i+} - n_{i-})} \tag{6}$$

at each lattice point, i, and each imaginary-time slice, l. λ is defined by the relation $\cosh(\Delta\tau\lambda) = \exp(\Delta\tau U/2)$. The transformation reduces the quartic self-interaction of the electrons to a quadratic interaction with the Hubbard-Stratonovich spin field, $s_{i,l}$. As a result, the trace over electron degrees of freedom can be performed yielding[6]

$$Z = \sum_{s_{i,l} = \pm 1} \det M^+ \det M^- , \tag{7}$$

with

$$M^{\sigma} = I + B_L^{\sigma} B_{L-1}^{\sigma}, \ldots, B_1^{\sigma} \tag{8}$$

and

$$B_l^{\pm} = e^{\mp \Delta\tau\lambda v(l)} e^{-\Delta\tau k} . \tag{9}$$

I is the $V \times V$ unit matrix and $v(l)_{ij} = \delta_{ij} s_{i,l}$. There is, of course, an analogous expression for the trace in the numerator of Eq. (1). The physical observable, O, ordinarily can be expressed in terms of Green's functions for the electrons propagating through the field $s_{i,l}$, that is in terms of the matrix of $1/M^{\sigma}$.[6]

Once the trace over the electron degrees of freedom has been performed, we can use standard Monte Carlo techniques to evaluate the right-hand side of Eq. (1). We wish to obtain a sequence of spin configurations, $\{s_{i,l}\}$, with a probability distribution $Z^{-1} \det M^+ \det M^-$. To this end, we sweep through the lattice many times updating one spin variable at a time. Notice that the determinant of M^{σ} is unchanged by a cyclic permutation of the B_l^{σ}. So in order to update the spins on the lth time slice we write

$$\det M^\sigma = \det(I + B_l^\sigma, \ldots, B_1^\sigma B_L^\sigma, \ldots, B_{l+1}^\sigma) = \det[I + A^\sigma(l)] . \quad (10)$$

Now in order to update the spin variable $s_{i,l}$ using any standard algorithm, such as heat bath or Metropolis, we must calculate the change in the fermion determinants when $s_{i,l} \to -s_{i,l}$. Under this change,

$$A^\sigma(l) \to A^\sigma(l)' = [I + \Delta^\sigma(i,l)] A^\sigma(l) , \quad (11)$$

where $\Delta^\sigma(i,l)$ is a matrix with only one nonzero element,

$$\Delta^\pm(i,l)_{jk} = \delta_{ji}\delta_{ki}(e^{\pm 2\Delta\tau\lambda s_{i,l}} - 1) . \quad (12)$$

The ratio of determinants after and before the spin flip is

$$R^\sigma = \frac{\det M^{\sigma'}}{\det M^\sigma} = \det[I + G^\sigma(l)\Delta^\sigma(i,l) A^\sigma(l)] = 1 + [1 - G^\sigma(l)_{ii}]\Delta^\sigma(i,l)_{ii} . \quad (13)$$

$G^\sigma(l)$ is the equal-time Green's function for an electron propagating through the field produced by the $s_{i,l}$.

$$G^\sigma(l)_{ij} = \langle T[c_{i\sigma}(l\Delta\tau) c_{j\sigma}^\dagger(l\Delta\tau)] \rangle = [I + A^\sigma(l)]_{ij}^{-1} . \quad (14)$$

Thus the calculation of the determinant ratios needed to carry out an individual Monte Carlo step is trivial provided one can evaluate the equal-time Green's functions.

If the equal-time Green's function $G^\sigma(l)$ is known, and the spin $s_{i,l}$ is flipped, then its new value, $G^\sigma(l)'$, can be evaluated through the relation

$$G^\sigma(l)' = G^\sigma(l) - G^\sigma(l)\Delta^\sigma(i,l) A^\sigma(l) G^\sigma(l)' = G^\sigma(l) - \frac{G^\sigma(l)\Delta^\sigma(i,l)[I - G^\sigma(l)]}{\{1 + [1 - G^\sigma(l)_{ii}]\Delta^\sigma(i,l)_{ii}\}} . \quad (15)$$

Since $\Delta^\sigma(i,l)$ has only one nonzero element, there is really no matrix multiplication in Eq. (15), and it takes V^2 operations to update the $V \times V$ matrix, $G^\sigma(l)$, after a spin is flipped.

After all the spins on a given time slice have been updated, we can obtain the equal-time Green's functions on the next time slice through the relations

$$G^\sigma(l+1) = B_{l+1}^\sigma G^\sigma(l) B_{l+1}^{\sigma -1} . \quad (16)$$

Since the B_l^σ are sparse matrices,[10] this operations takes of order V^2 numerical operations.

We have now assembled all of the ingredients necessary to carry out a simulation. Indeed, this algorithm has been used to study a wide variety of many-electron models. Since the computing time scales as V^3, its use has been restricted to one- and two-dimensional systems, and small three-dimensional ones. In past applications of the algorithm, numerical instabilities have prevented its use at low temperatures. For example, in the case of the two-dimensional Hubbard model, one is restricted to $\beta \le 4/t$ for moderate values of U. In order to remove these instabilities, we must first understand their cause. They are not associated with the updating of the equal-time Green's function given in Eq. (15). Indeed, we have found that by staying on the same time slice one can update tens of thousands of spins without accumulating numerical errors. On the other hand, the process of advancing the equal-time Green's function to a new time slice given in Eq. (16) does introduce numerical errors, so one must periodically recompute the Green's function from scratch. In order to do so, one must compute the matrices $A^\sigma(l)$ to sufficient accuracy to incorporate information concerning its small eigenvalues. This becomes increasingly difficult as the temperature is lowered. For example, for the two-dimensional Hubbard model with $U=0$, the $A^\sigma(l)$ have eigenvalues as large as $\exp(4t\beta)$ and as small as $\exp(-4t\beta)$. Since we are interested in performing simulations for β at least as large as $20/t$, it is clear that we must separate the contributions from the large and small eigenvalues, or else the latter will be completely swamped by round-off errors.

This problem can be dealt with in a relatively straightforward manner using matrix factorization methods.[1,11] Suppose that one can multiply m of the B_l^σ without losing numerical accuracy. We then use the Gram-Schmidt orthogonalization procedure to write this product in the form

$$a_1^\sigma(l) = B_{l+m}^\sigma B_{l+m-1}^\sigma, \ldots, B_{l+1}^\sigma = U_1^\sigma D_1^\sigma R_1^\sigma , \quad (17)$$

where U_1^σ is an orthogonal matrix, D_1^σ a diagonal matrix, and R_1^σ a right triangular matrix with diagonal elements equal to one. The orthogonal matrix U_1^σ is necessarily well conditioned; *a priori*, R_1^σ need not be well conditioned, but in practice it is. Only the diagonal matrix D_1^σ has large variations in the size of its elements.

We next form

$$a_2^\sigma(l) = B_{l+2m}^\sigma, \ldots, B_{l+1}^\sigma = B_{l+2m}^\sigma, \ldots, B_{l+m+1}^\sigma U_1^\sigma D_1^\sigma R_1^\sigma = U_2^\sigma D_2^\sigma R_2^\sigma . \quad (18)$$

The order of operations in Eq. (18) is important. We first multiply U_1^σ by $B_{l+2m}^\sigma, \ldots, B_{l+m+1}^\sigma$. By assumption, m is small enough so that this matrix can be computed accurately. We then multiply it on the right by D_1^σ. This only rescales the columns of the matrix, and thus does no harm to the numerical stability of the next step, a *UDR* decomposition of this partial product. We then multiply the resulting triangular matrix on the right by R_1^σ to obtain the last line of Eq. (18). This process is repeated L/m times to obtain

$$A^\sigma(l) = a_{L/m}^\sigma(l) = U_{L/m}^\sigma D_{L/m}^\sigma R_{L/m}^\sigma . \quad (19)$$

To form $G^\sigma(l)^{-1}$, we must add the unit matrix to $A^\sigma(l)$. Care must be taken to isolate the diagonal matrix $D_{L/m}^\sigma$, whose elements have large variations in size. We therefore write

$$G^\sigma(l)^{-1} = I + A^\sigma(l) = U_{L/m}^\sigma (U_{L/m}^{\sigma -1} R_{L/m}^{\sigma -1} + D_{L/m}^\sigma) R_{L/m}^\sigma = U^\sigma D^\sigma R^\sigma . \quad (20)$$

In order to go from the second to third line of Eq. (20),

we have made a final UDR decomposition of the quantity in square brackets. This decomposition is numerically stable because the large and small matrix elements are again separated. The formation of $G^{\sigma}(l)$ from the last line of Eq. (20) is, of course, trivial. We have tested the numerical stability of our method for evaluating $G^{\sigma}(l)$ for values of $t\beta$ as large as 100 without encountering problems.

We conclude this discussion by examining the scaling of the simulation time with respect to lattice size and temperature. The most time consuming step in the Monte Carlo process, aside from the occasional *ab initio* calculation of the equal-time Green's functions, is their updating via Eq. (15), which requires of order V^2 operations. Therefore, the computer time required to update each of the VL spin variables is proportional to V^3L. Just as it is possible to multiply m of the B_l^{σ} matrices without losing numerical accuracy, it is also possible to advance the Green's functions to a new time slice, using Eq. (16), m' times without introducing unacceptable errors. Thus, one must completely recalculate the Green's functions L/m' times per lattice sweep. Since the UDR decomposition requires V^3 operations, the *ab initio* calculation of the Green's functions takes of order

$$V^3(L/m')(L/m)\approx V^3(L/m)^2$$

operations per sweep. We find that m and m' are roughly equal and relatively independent of temperature; requiring $t\Delta\tau m \le 1.5$ provides a good margin of safety on a computer with 64-bit precision. As a result, if for $U=4$ we set $\Delta\tau=(8t)^{-1}$, which results in a systematic error from the Trotter approximation of a few percent, the times spent updating the spins and the time spent in the recomputation of the Green's function are comparable for β of order $20/t$. For inverse temperatures less than this, the V^3L term dominates and the computer times scales approximately linearly with β. Of course at very low temperatures it will scale like β^2, but as we will now discuss, we would then use a rather different approach.

B. Zero-temperature algorithm

It is possible to use the techniques just described to directly calculate the ground-state properties of systems with a fixed number of electrons. Let us denote by $|\Psi_0\rangle$ the ground-state wave function for N^+ spin up electrons and N^- spin down ones, and by $|\psi_0\rangle$, a trial state that has a nonzero overlap with $|\Psi_0\rangle$. Then the ground-state expectation value of a physical observable, O, can be written in the form

$$\frac{\langle\Psi_0|O|\Psi_0\rangle}{\langle\Psi_0|\Psi_0\rangle}=\lim_{\gamma,\gamma'\to\infty}\frac{\langle\psi_0|e^{-\gamma' H}Oe^{-\gamma H}|\psi_0\rangle}{\langle\psi_0|e^{-(\gamma+\gamma')H}|\psi_0\rangle}\ . \tag{21}$$

In order to evaluate the right-hand side of Eq. (21) stochastically, we write $\gamma+\gamma'=\Delta\tau L$, and use the Trotter approximation of Eq. (3) and the Hubbard-Stratonovich transformation of Eq. (6) to again integrate out the fermion degrees of freedom. We take the state $|\psi_0\rangle$ to be either one in which the electrons with z component of spin σ are localized on sites $i_1^{\sigma},\ldots,i_{N^{\sigma}}^{\sigma}$ or a filled Fermi sea, the correct ground state for $U=0$. Then we can write

$$\langle\psi_0|e^{-(\gamma+\gamma')H}|\psi_0\rangle=\sum_{s_{i,l}}\det A^+\det A^-\ , \tag{22}$$

where

$$A_{jk}^{\sigma}=\{B_L^{\sigma},\ldots,B_1^{\sigma}\}_{jk}\ , \tag{23}$$

and the indices j and k are restricted to run over $i_1^{\sigma},\ldots,i_{N^{\sigma}}^{\sigma}$, so the A^{σ} are $N^{\sigma}\times N^{\sigma}$ matrices.

In order to carry out the simulation we must generate a set of spin configurations distributed as $\det A^+\det A^-$. The procedure is basically the same as for the grand-canonical ensemble. Let us imagine updating spins on time slices $nm+1$ to $(n+1)m$. Again making use of successive Gram-Schmidt orthogonalizations, we can write

$$A^{\sigma}=L_L^{\sigma}D_L^{\sigma}U_L^{\sigma}B_{(n+1)m}^{\sigma},\ldots,B_{nm+1}^{\sigma}U_R^{\sigma}D_R^{\sigma}R_R^{\sigma}\ , \tag{24}$$

where U_R^{σ} is a $V\times N^{\sigma}$ matrix whose columns are mutually orthogonal, and U_L^{σ} and $N^{\sigma}\times V$ matrix whose rows are mutually orthogonal. $D_{R,L}^{\sigma}$ are $N^{\sigma}\times N^{\sigma}$ diagonal matrices, and R_R^{σ} and L_L^{σ} are $N^{\sigma}\times N^{\sigma}$ right and left triangular matrices with unit diagonal elements. Once more, the diagonal matrices are the only ones with large variations in the size of their matrix elements. Suppose we update $s_{i,l}$, where $(n+1)m\ge l\ge nm+1$. If this spin is flipped, then $A^{\sigma}\to A^{\sigma'}$ with

$$\begin{aligned}A^{\sigma'}&=A^{\sigma}+L_L^{\sigma}D_L^{\sigma}U_L^{\sigma}B_{(n+1)m}^{\sigma},\ldots,B_{l+1}^{\sigma}\Delta^{\sigma}(i,l)B_l^{\sigma},\ldots,B_{nm+1}^{\sigma}U_R^{\sigma}D_R^{\sigma}R_R^{\sigma}\\&=A^{\sigma}+L_L^{\sigma}D_L^{\sigma}W_L^{\sigma}\Delta^{\sigma}(i,l)W_R^{\sigma}D_R^{\sigma}R_R^{\sigma}\ .\end{aligned} \tag{25}$$

$\Delta^{\sigma}(i,l)$ is again given by Eq. (12). The determinant ratio needed to make the decision on flipping the spin is

$$\begin{aligned}R^{\sigma}&=\frac{\det A^{\sigma'}}{\det A^{\sigma}}=\det[I+\{W_L^{\sigma}W_R^{\sigma}\}^{-1}W_L^{\sigma}\Delta^{\sigma}(i,l)W_R^{\sigma}]\\&=1+[W_R^{\sigma}\{W_L^{\sigma}W_R^{\sigma}\}^{-1}W_L^{\sigma}]_{ii}\Delta^{\sigma}(i,l)_{ii}\ .\end{aligned} \tag{26}$$

Notice that R^{σ} is independent of the matrices $D_{L,R}^{\sigma}$, R_R^{σ}, and L_L^{σ}. In addition, note that the last line of Eq.(26) cannot be simplified further because the $W_{R,L}^{\sigma}$ are not square matrices.

With the use of Eq. (26) the updating steps can be performed trivially if one knows $\{W_L^{\sigma}W_R^{\sigma}\}^{-1}$, which plays a role analogous to that of the equal-time Green's function in the simulation of the grand-canonical ensemble. If $s_{i,l}$ is flipped, then W_L^{σ} remains unchanged and

$$W_R^\sigma \to [I+\Delta^\sigma(i,l)]W_R^\sigma = W_R^{\sigma'} \ . \tag{27}$$

We then see that

$$\{W_L^\sigma W_R^{\sigma'}\}^{-1} = \{W_L^\sigma W_R^\sigma\}^{-1} - \{W_L^\sigma W_R^\sigma\}^{-1} W_L^\sigma \Delta^\sigma(i,l) W_R^\sigma \{W_L^\sigma W_R^{\sigma'}\}^{-1} \ . \tag{28}$$

The updating of $\{W_L^\sigma W_R^\sigma\}$ can therefore be carried out in $(N^\sigma)^2$ steps. In moving from the l to the $l+1$ times slice, $W_R^\sigma \to B_{l+1}^\sigma W_R^\sigma$ and $W_L^\sigma \to W_L^\sigma B_{l+1}^{\sigma-1}$, while $W_L^\sigma W_R^\sigma$ and its inverse remain unchanged.

Suppose we begin on the first time slice and update spins in the direction of increasing l. After updating the first m slices we must make a UDR decomposition of $B_m^\sigma, \ldots, B_1^\sigma$. After updating the next m time slices we make a UDR decomposition of $B_{2m}^\sigma, \ldots, B_1^\sigma$, and so forth. We save each of these. After updating all time slices we work back in the direction of decreasing l. As we do so, we make LDU decompositions of $B_L^\sigma, \ldots, B_{L-m+1}^\sigma$, etc., and store these also. At any one time we need to store a total of L/m decomposed matrices. However, we need only make one decomposition for each block of m time slices that we update. As a result, we use of order

$$[(N^+)^3+(N^-)^3](L/m)$$

numerical operations per sweep of the lattice in performing LDU or UDR decompositions. We also use of order

$$[(N^+)^3+(N^-)^3]L$$

operations in the updating process including the updating of the inverse matrix via Eq. (28). Thus, the operation count for this algorithm truly scales linearly with β.

III. NUMERICAL RESULTS

The one-band Hubbard model represents an excellent testing ground for new Monte Carlo algorithms. Not only are the physical properties of this model of great current interest, but in addition there are a variety of results available for comparison. Here we apply the algorithms described above to the two-dimensional Hubbard model at half-filling, $\langle n \rangle = 1.0$, and at quarter-filling, $\langle n \rangle = 0.5$.

A. Half-filled band

As has been noted in previous studies,[12,13] for some configurations the determinants $\det M^+$ and $\det M^-$ in Eq. (7) can be negative. In the half-filled Hubbard model, however, particle-hole symmetry implies that the *product* $\det M^+ \det M^-$ is never negative.[12] Off half-filling the product is negative for some configurations, and one must use $|\det M^+ \det M^-|$ as the probability distribution for the Monte Carlo simulation. As we show in the next section, this can limit the temperatures that can be reached at some fillings. However, at half-filling with the algorithm described above, computer time appears to be the only limitation to the size of the lattice and the temperature that can be simulated.

At half-filling, in the strong coupling limit, to second order in t/U the Hubbard model is equivalent to an antiferromagnetic Heisenberg model with Hamiltonian

$$H = J \sum_{\langle ij \rangle} (\mathbf{S}_i \cdot \mathbf{S}_j - \tfrac{1}{4}) \ , \tag{29}$$

where $\langle ij \rangle$ denotes nearest neighbors and $J = 4t^2/U$. Even for moderate U, simulations have shown strong antiferromagnetic correlations. For the Heisenberg model, recent Monte Carlo simulations strongly support the view that there is long-range order in the ground state.[14] Simulations also suggest that the ground state of the Hubbard model with finite U has long-range order.[5,12] Our present results confirm this. For example, staggered spin correlations are clearly visible in Fig. 1(a), which

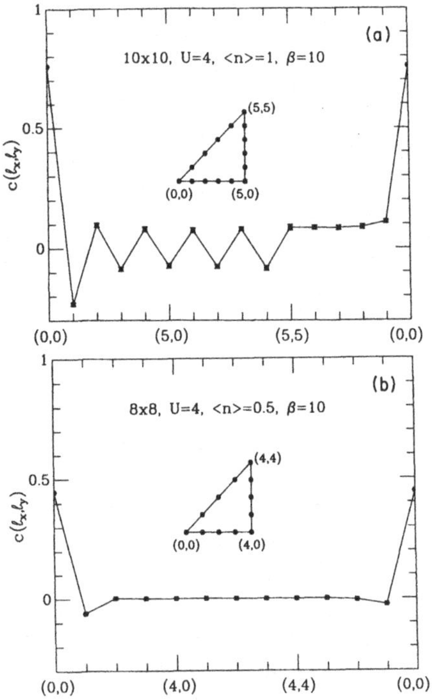

FIG. 1. Spin-spin correlation function $c(l_x, l_y)$. The horizontal axis traces out the triangular path showing in the center of the figure. Strong antiferromagnetic correlations are visible in (a), which is for a half-filled band, but are nearly absent in (b), which is at quarter-filling.

shows the spin-spin correlation function

$$c(l_x,l_y)=\frac{1}{N}\sum_i(-1)^{l_x+l_y}\langle(n_{i+l\uparrow}-n_{i+l\downarrow})(n_{i\uparrow}-n_{i\downarrow})\rangle \tag{30}$$

for a 10×10 lattice with $U=4$, $t=1$, and $\beta=16$. (In all our results $t=1$.) Except where otherwise noted, the calculations were done using the grand-canonical algorithm with $\Delta\tau=0.125$, which resulted in systematic errors of a few percent. At $\beta=16$, the correlation length is clearly much larger than the 10×10 lattice. In contrast, staggered spin-spin correlations are nearly absent in the quarter-filled results shown in Fig. 1(b). The Fourier transform of this correlation functions is the structure factor

$$S(\mathbf{q})=\sum_l e^{i\mathbf{q}\cdot\mathbf{l}}c(l_x,l_y)\,, \tag{31}$$

which is plotted in Fig. 2 for $\langle n\rangle=1.0$ with $q_x=q_y$. For a half-filled band, $S(\mathbf{q})$ peaks at $\mathbf{q}=(\pi,\pi)$, and $S(\pi,\pi)$ is plotted in Fig. 3 versus β for various sized lattices. The zero-temperature canonical ensemble results are shown as the symbols at $\beta=22$. From the figure we see that for the lattice sizes shown, by the time the inverse temperature β is 10 to 15, the system has very nearly reached the ground state. In the zero-temperature algorithm the effective temperature $\gamma+\gamma'$ in Eq. (21) needed for the system to be in the ground state depends on the trial state $|\psi_0\rangle$ used, but is typically ~30. The value of β roughly twice that of the grand-canonical method because the expectation value of the operator O in Eq. (21) is evaluated in the state $e^{-\gamma H}|\psi 0\rangle$, and projects out the ground state with an effective temperature γ. The strengths of the zero-temperature method appear to lie in doing low-density calculations and in problems where it is necessary to precisely set the number of particles. In practice, over the range of temperatures and band fillings we have studied, it is faster to obtain zero-temperature results by using the grand-canonical method at low temperature than the zero-temperature method. For the rest of the results present here the grand-canonical method was used.

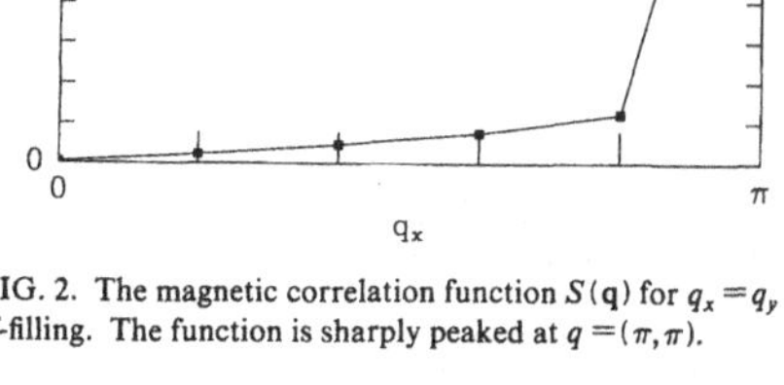

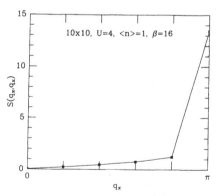

FIG. 2. The magnetic correlation function $S(\mathbf{q})$ for $q_x=q_y$ at half-filling. The function is sharply peaked at $q=(\pi,\pi)$.

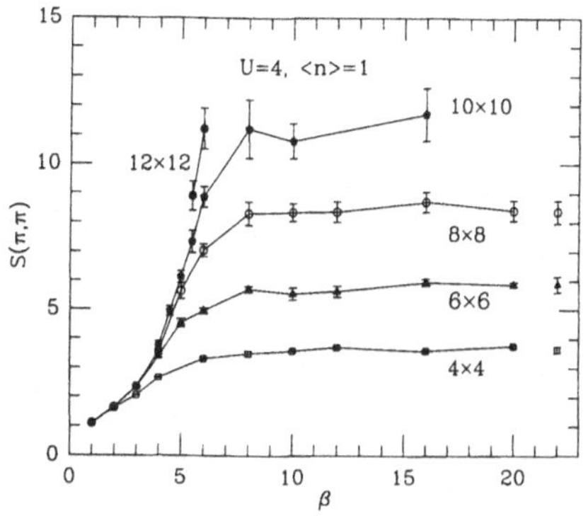

FIG. 3. The antiferromagnetic structure factor $S(\pi,\pi)$ as a function of inverse temperature for a variety of lattice sizes. The points at $\beta=22$ were done with the zero-temperature, fixed particle number algorithm, while the rest were obtained with the grand-canonical algorithm.

From Fig. 3 it is clear that the zero-temperature extrapolation of $S(\pi,\pi)$ increases with lattice size. Within spin-wave theory,

$$\frac{S(\pi,\pi)}{N}=\frac{m^2}{3}+O\left[\frac{1}{\sqrt{N}}\right], \tag{32}$$

with m the antiferromagnetic order parameter. This order parameter can also be obtained by extrapolating to infinite lattice size the zero-temperature limit of the spin-spin correlation function between the two most distant points on a lattice $c(N_x/2,N_x/2)$. Again, according to spin-wave theory,

$$c(N_x/2,N_x/2)=\frac{m^2}{3}+O\left[\frac{1}{\sqrt{N}}\right], \tag{33}$$

with $N=N_x^2$. Using these forms, Eqs. (32) and (33), results for $S(\pi,\pi)/N$ and $c(N_x/2,N_x/2)$ are plotted versus $N^{-1/2}=N_x^{-1}$ in Fig. 4. The values of m^2 which are obtained agree with each other (within statistical errors) as well as with the results reported in Ref. 13.

In Fig. 5 we show the total energy per site as a function of β for lattices ranging in size from 4×4 to 12×12. These lattices are of sufficient size to allow us to obtain the ground-state energy of the infinite system by extrapolation. The statistical errors for the energy are sufficiently small that systematic errors associated with finite $\Delta\tau$ become important. For that reason, we repeated some of the calculations shown in Fig. 5 for $\Delta\tau=0.167$ and 0.083. To a very good approximation, the error in the energy behaved as $\Delta\tau^2$, as expected. Furthermore,

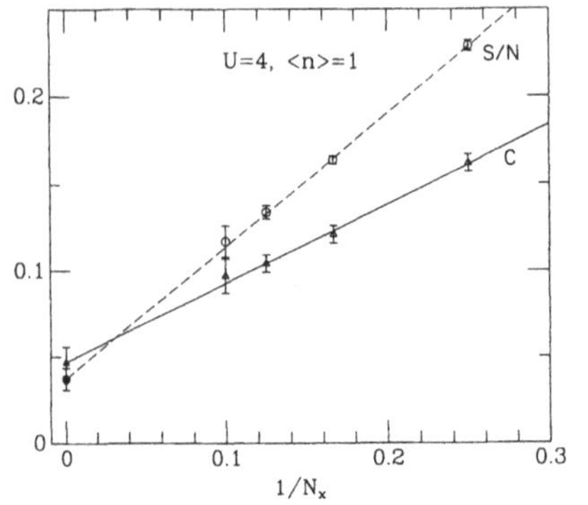

FIG. 4. Plots of $S(\pi,\pi)/N$ (circles) and $c(N_x/2,N_x/2)$ (triangles) vs $1/N_x$. The data was taken from grand-canonical runs at $\beta \geq 16$. The lines are least-squares fits to the data. The extrapolation of these fits to the infinite system are shown as the solid symbols.

the coefficient of the $\Delta\tau^2$ term did not appear to depend on lattice size.[8] From spin-wave theory[15] the finite-size correction to the energy from the infinite-system limit is proportional to N_x^{-3} at $T=0$. By performing a least squares fit to the form

$$E(\Delta\tau,N_x)=E(0,\infty)+a_1\Delta\tau^2+a_2N_x^{-3}\ , \tag{34}$$

we extracted the ground-state energy per site for the infinite system. The extrapolated energy was

$$E(0,\infty)=-0.864\pm0.001\ ,$$

with $a_1=-1.95\pm0.04$ and $a_2=1.59\pm0.13$. Assuming the $\Delta\tau$ errors vary slowly with temperature, using these results to correct for finite $\Delta\tau$ in Fig. 5 would shift the vertical axis by 0.03. The vertical axis on the right is labeled with this correction, and the extrapolated infinite-system result is indicated. A similar fit was performed for the kinetic energy alone. The extrapolated kinetic energy was $E_k(0,\infty)=-1.368\pm0.001$.

For moderate values of U, the staggered order parameter m is reduced by both the zero point spin fluctuations and charge fluctuations. The charge fluctuations also reduce the size of the local spin moment.[9] The squared local moment $\langle (n_{i+}-n_{i-})^2\rangle$ is plotted versus U in Fig. 6(a), and the related double occupancy $\langle n_{i+}n_{i-}\rangle$ is shown in Fig. 6(b). Clearly, as U increases, a local moment is formed and the double occupancy of a site decreases. At the same time, we expect the effective one-electron transfer to decrease. A measure of this reduction is given by the ratio of the ground-state expectation values of $\langle c_{i\sigma}^\dagger c_{j\sigma}+c_{j\sigma}^\dagger c_{i\sigma}\rangle$ in the presence of the interaction U to its noninteracting value for $U=0$. A plot of this ratio versus U for $\beta=16$ on a 4×4 lattice is shown in Fig. 7. This matrix element can also be obtained from

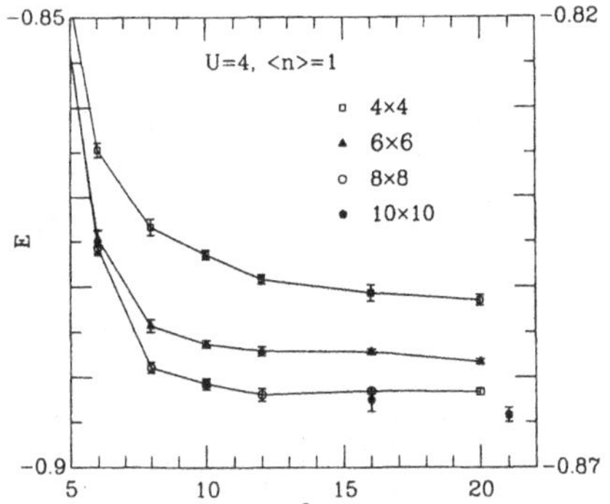

FIG. 5. Total energy $E=E_k+U\langle n^+n^-\rangle$ vs β for a variety of lattice sizes at half-filling. The runs shown have $\Delta\tau=0.125$, and these finite $\Delta\tau$ results correspond to the vertical scale on the left. Correcting the results to $\Delta\tau=0$ gives the scale on the right (see the text). The point at $\beta=21$ is the extrapolation to $T=0$ for an infinite system.

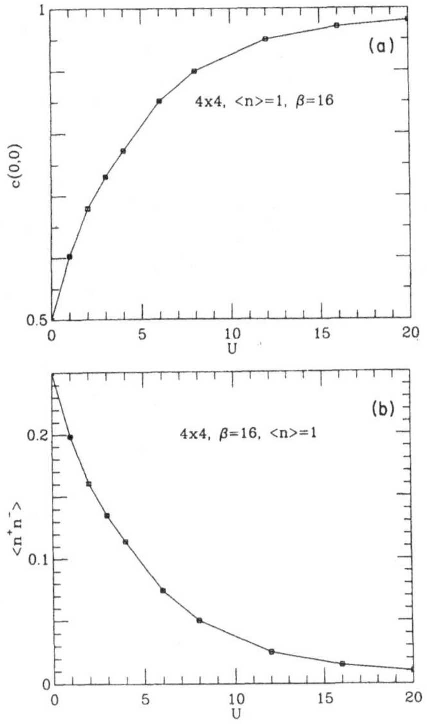

FIG. 6. (a) The squared local moment $c(0,0)=\langle (n_{i+}-n_{i-})^2\rangle$ and (b) the double occupancy $\langle n_{i+}n_{i-}\rangle$ as a function of U. As U increases, a local moment is formed and the double occupancy of a site decreases.

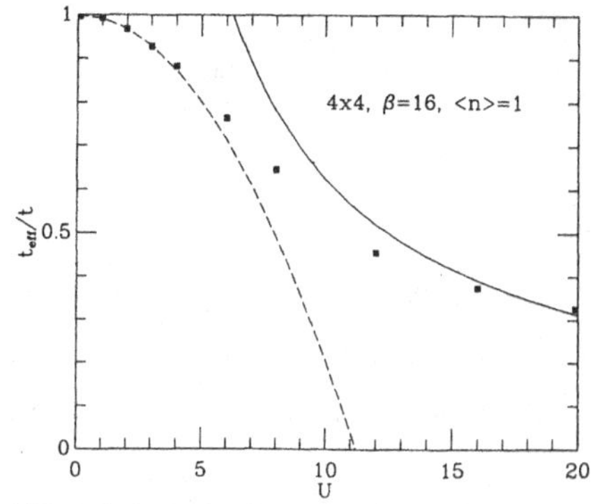

FIG. 7. Reduction in effective hopping as a function of U on a 4×4 lattice for $\beta=16$. The solid line is a strong coupling result. The dashed line is obtained from perturbation theory.

the ground-state energy E_0 by use of the Feynman-Hellman relation

$$\langle c^\dagger_{i\sigma}c_{j\sigma}+c^\dagger_{j\sigma}c_{i\sigma}\rangle=\frac{1}{2N}\frac{\partial E_0}{\partial t}\ . \tag{35}$$

Second-order perturbation theory gives

$$\frac{t_{\text{eff}}}{t}=\frac{\langle c^\dagger_{i\sigma}c_{j\sigma}+c^\dagger_{j\sigma}c_{i\sigma}\rangle_U}{\langle c_{i\sigma}c_{j\sigma}+c^\dagger_{j\sigma}c_{i\sigma}\rangle_0}=1-\left[\frac{U}{t}\right]^2\left[\frac{\Delta\varepsilon_2}{\Delta\varepsilon_0}\right], \tag{36}$$

with

$$\Delta\varepsilon_0=\frac{2}{N}\sum_k \varepsilon_k\ ,$$
$$\Delta\varepsilon_2=\frac{1}{N^3}\sum_{kk'q}\frac{f(\varepsilon_k)f(\varepsilon_{k'})[1-f(\varepsilon_{k+q})][1-f(\varepsilon_{k'-q})]}{\varepsilon_k+\varepsilon_{k'}-\varepsilon_{k+q}-\varepsilon_{k'-q}}\ . \tag{37}$$

Here,

$$\varepsilon_k=-2t(\cos k_x+\cos k_y)$$

and f is the Fermi factor. This result for t_{eff}/t is shown as the dashed curve in Fig. 7. In the strong coupling limit where (29) applies, the local moment $\langle m_z^2\rangle\rightarrow1$ and

$$\frac{t_{\text{eff}}}{t}=\frac{4\left[\frac{4t}{U}\right][|\langle \mathbf{S}_i\cdot\mathbf{S}_j\rangle|+\frac{1}{4}]}{|\Delta\varepsilon_0|}\ . \tag{38}$$

Using the value for $|\langle \mathbf{S}_i\cdot\mathbf{S}_j\rangle|=0.335$ obtained for the two-dimensional (2D) Heisenberg model by Reger and Young,[14] one finds the solid line plotted in Fig. 7. These results clearly show that a repulsive Coulomb interaction reduces the one-electron transfer and at the same time gives rise to local moments.

In the half-filled Hubbard model, the particle-hole charge-density fluctuations involve the creation of doubly occupied and empty sites. These excitations involve an energy gap set by U and at zero temperature are expected to lead to an insulating state with a divergent compressibility. The compressibility varies inversely with $\partial\langle n\rangle/\partial\mu$, which within our simulations appears to vanish at half-filling as $T\rightarrow0$. A second class of particle-hole excitations involves fluctuations of the spins. In this case, as T goes to zero, we expect these excitations to become the spin waves of the insulating antiferromagnetic state. A useful upper bound for the dispersion relation of these excitations can be obtained from $S(q)$ using a procedure similar to Feynman's treatment of the density fluctuations in ^{4}He.[16]

Consider the trial excited state

$$|\psi_q\rangle=S^z_q|\psi_0\rangle/\langle\psi_0|S^z_{-q}S^z_q|\psi_0\rangle\ , \tag{39}$$

with $|\psi_0\rangle$ the exact ground state and

$$S^z_q=\sum_l e^{iq\cdot l}(n_{l\uparrow}-n_{l\downarrow})\ . \tag{40}$$

Then the excitation energy of this state is

$$\bar\omega(q)=\langle\psi_q|H|\psi_q\rangle-\langle\psi_0|H|\psi_0\rangle$$
$$=\frac{2}{3}\frac{E_0}{N}\left[1-\frac{\varepsilon_q}{4}\right]/S(q)\ , \tag{41}$$

with (E_0/N) the ground-state energy per site and $S(q)$ the structure factor, Eq. (31). Figure 8 shows $\bar\omega(q)$ versus q for a 10×10 lattice with $U=4$. As the size increases, $S(\pi,\pi)$ increases, driving the excitation energy $\bar\omega(\pi,\pi)$ to zero and generating the expected spin-wave dispersion relation.

We have also calculated the temperature dependence of the $\mathbf{q}=0$ magnetic susceptibility

$$\chi(T)=\left\langle\left[\frac{1}{N}\sum_l(n_{l\uparrow}-n_{l\downarrow})\right]^2\right\rangle\Big/T\ . \tag{42}$$

Figure 9 shows χ versus T for 8×8 lattice with $U=4$. The behavior of $\chi(T)$ is similar to that expected for a two-dimensional Heisenberg antiferromagnet. However, for $U/t=4$, the moment is not fully developed [see Fig. 6(a)] and the itineracy and charge fluctuations still play a role.

B. Non-half-filled band

As discussed above, when the Hubbard model is doped off of half-filling, the finite-temperature simulations become more difficult because of sign problems. In cases where the (unnormalized) probability distribution $P(\{s\})$ is not positive-semidefinite, one uses $|P(\{s\})|$ as the probability distribution, and the expectation value of an observable O is calculated as

$$\langle O\rangle_P=\frac{\langle O\,\text{sgn}P\rangle_{|P|}}{\langle\text{sgn}P\rangle_{|P|}}\ , \tag{43}$$

where the subscript P and $|P|$ indicate averages taken in the distributions $P(\{s\})$ and $|P(\{s\})|$, respectively. However, if the average sign, $\langle\text{sgn}P\rangle_{|P|}$, is close to zero, this estimator for $\langle O\rangle_P$ is very noisy. Figure 10 shows the average sign versus $\langle n\rangle$ for a various lattice sizes and temperatures. The dramatic fall off of the average sign as

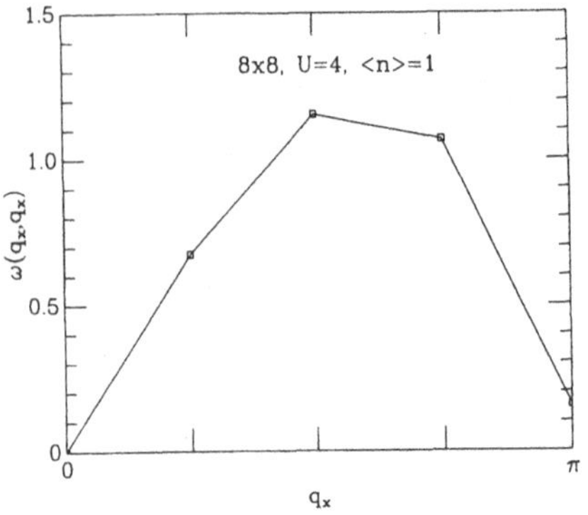

FIG. 8. Spin-wave energy $\omega(q_x,q_x)$ for a 10×10 lattice with $U=4$. This dispersion relation is generated using a procedure similar to Feynman's treatment of the density fluctuations in ^{4}He.

the system is doped away from half-filling makes it difficult to calculate in the interesting filling regime $0.7<\langle n\rangle<0.95$. As shown, in this region the average sign decreases as the spatial and temporal size of the lattice increases. The peak at a filling of 0.625 for the 4×4 lattice is striking. This filling corresponds to the presence of 10 particles, which for the noninteracting system just fills the five lowest k states, leaving a gap to the next empty k state.[17]

Clearly, from Fig. 10 one finds that there is a negligible sign problem for a quarter-filled band, $\langle n\rangle=0.5$. As shown in Fig. 1(b), the staggered spin-spin correlations are nearly absent for $\langle n\rangle=0.5$ and $U=4$. It is also interesting to compare the single-particle momentum occupation $\langle n_{ks}\rangle=\langle c_{ks}^{\dagger}c_{ks}\rangle$ for half-filling and quarter-filling. Figures 11(a) and 11(b) show $\langle n_{ks}\rangle$ versus $\mathbf{k}=(k,k)$ for an 8×8 lattice with $U=4$ and $\beta=10$ with fillings of $\langle n\rangle=1.0$ and 0.5, respectively. The dashed lines represent the noninteracting Fermi distributions at $\beta=10$. Qualitatively, the distribution for $\langle n\rangle=1.0$ is broadened more by the interaction than for $\langle n\rangle=0.5$

We have also studied the equal time d-wave pair-field correlation function

$$D_d=\langle\Delta_d\Delta_d^{\dagger}\rangle\ , \tag{44}$$

with

$$\Delta_d^{\dagger}=\frac{1}{2\sqrt{N}}\sum_{l,\delta}c_{l\uparrow}^{\dagger}c_{l+\delta\downarrow}^{\dagger}(-1)^{\delta}\ . \tag{45}$$

Here, δ sums over the four near neighbor sites of l and $(-1)^{\delta}$ gives the $+-+-$ sign alternation characteristic of a d-wave pairing amplitude. White et al.[18] found that by comparing susceptibilities with uncorrelated susceptibilities (susceptibilities with the interaction vertex moved), one can determine whether the effective pairing interaction is attractive or repulsive. The full and uncorrelated susceptibilities exhibit the same dressed single-quasiparticle effects: differences between them are due to the two-particle effective interaction. White et al. found that near half-filling the d-wave susceptibility was

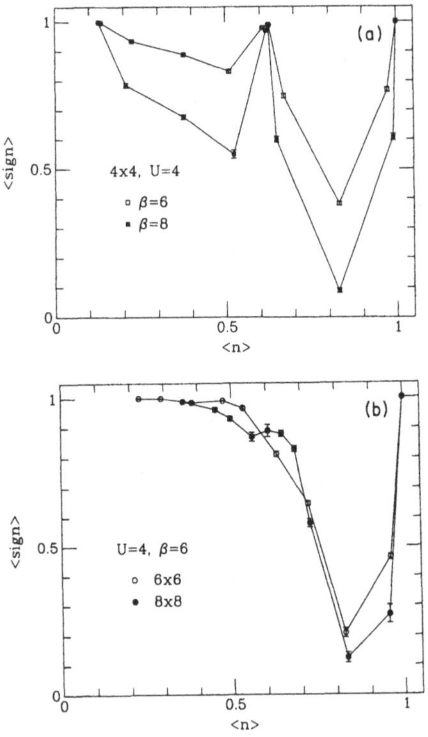

FIG. 10. (a) and (b) Average sign as a function of band-filling for various lattice sizes and temperatures.

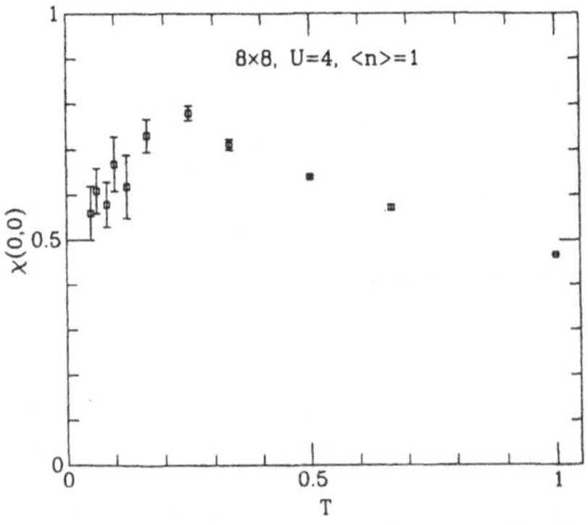

FIG. 9. The $\mathbf{q}=0$ magnetic susceptibility, χ, as a function of T on an 8×8 lattice with $U=4$.

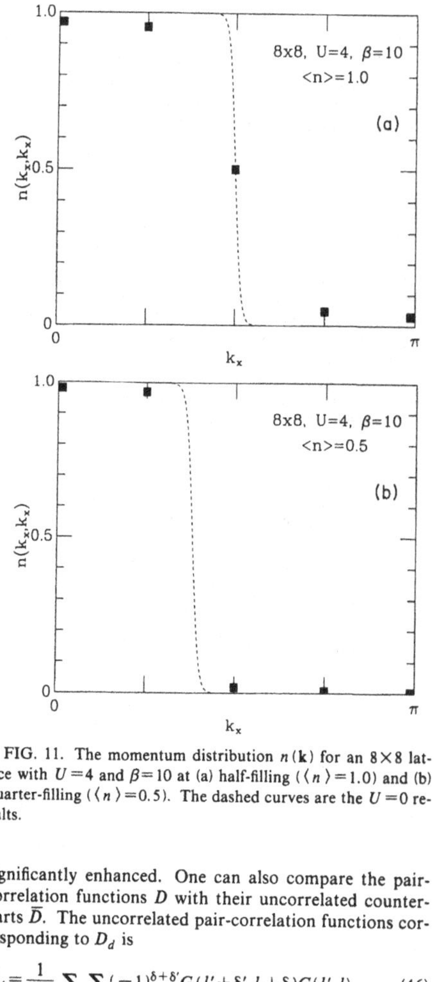

FIG. 11. The momentum distribution $n(\mathbf{k})$ for an 8×8 lattice with $U=4$ and $\beta=10$ at (a) half-filling ($\langle n\rangle=1.0$) and (b) quarter-filling ($\langle n\rangle=0.5$). The dashed curves are the $U=0$ results.

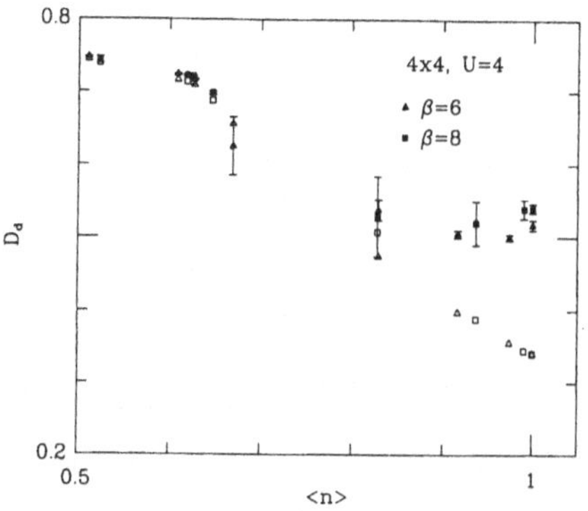

FIG. 12. d-wave pair correlation functions as a function of band-filling. The solid symbols show the full correlation functions, while the open symbols show the corresponding uncorrelated form for which the interaction vertex is removed.

significantly enhanced. One can also compare the pair-correlation functions D with their uncorrelated counterparts $\bar{D}$. The uncorrelated pair-correlation functions corresponding to D_d is

$$\bar{D}_d=\frac{1}{4N}\sum_{\delta,\delta'}\sum_{l,l'}(-1)^{\delta+\delta'}G(l'+\delta',l+\delta)G(l',l)\ , \quad (46)$$

where $G(l',l)\equiv\langle c_{l'\sigma}c^{\dagger}_{l\sigma}\rangle$. In Fig. 12 we show both D and $\bar{D}$ versus $\langle n\rangle$ for a 4×4 lattice. As in the case of the d-wave pair-field susceptibility, D_d is enhanced over $\bar{D}$ near half-filling. Unfortunately, the fillings where there is enhancement correspond to the region where the sign problem is worst for the simulation, so that very low temperatures cannot be reached (except exactly at half-filling). Hence, the question of whether the attractive interaction ever leads to superconductivity remains open.

IV. SUMMARY

In this paper we have introduced two algorithms which use matrix-decomposition methods to remove the numerical instabilities which have plagued simulations of many-electron systems at very low temperatures, and we have used these algorithms to study the properties of the two-dimensional Hubbard model. The first algorithm gives grand-canonical, finite-temperature results. The second gives ground-state results for a fixed number of electrons. Both algorithms can give ground-state results at half-filling. The grand-canonical algorithm is useful at all fillings, although sign problems, the biggest remaining obstacle to simulations of many-electron systems, can limit the temperatures which can be reached at some fillings. Away from half-filling the ground-state algorithm appears to have more limited usefulness, since sign problems may become overwhelming before the ground state is reached. The main benefit of the ground-state algorithm may lie in problems where the number of particles must be set precisely, such as in determining energy gaps. In addition, at very low fillings the ground-state algorithm may be faster than the grand canonical, because the dimension of the fermion matrix used is the number of particles rather than the number of sites.

Our numerical results on the single-band Hubbard model support previous conclusions concerning magnetic properties. In particular, our results support the existence of long-range order in the ground state at half-filling and the absence of long-range order at quarter-filling. The magnetic susceptibility for a half-filled band with $U/t=4$ shows a temperature dependence similar to that of a two-dimensional antiferromagnet. We have determined the ground-state energy, and obtained an upper bound to the spin wave excitation spectrum. At one-quarter filling, the system appears to be nonmagnetic. Near half-filling, fluctuations in the sign of the fermion determinant make it difficult to perform simulations on large lattices at low temperatures. We have given some results showing how the average sign of the determinant varies with filling, temperature, and lattice size. The behavior of the d-wave pair-field correlation function is consistent with an attractive d-wave pairing interaction near half-filling.

ACKNOWLEDGMENTS

We would like to thank A. Moreo for discussions and for providing most of the data on the 10×10 and 12×12 lattices shown in Fig. 3. This work was supported in part by National Science Foundation Grants Nos. PHY86-14185, DMR 86-15454, and DMR 86-12860, by Department of Energy Grant No. DE-FG03-88ER45197, and by the San Diego Supercomputer Center. S.R.W. gratefully acknowledges the support of IBM.

---

[1] An initial report on this work was given by E. Y. Loh, J. E. Gubernatis, R. T. Scalettar, S. R. White, and R. L. Sugar, in *Workshop on Interacting Electrons in Reduced Dimension*, edited by D. Baeriswyl and D. K. Campbell (Plenum, New York, in press).

[2] S. R. White, R. L. Sugar, and R. T. Scalettar, Phys. Rev. B **38**, 11695 (1988); J. E. Hirsch, *ibid.* **38**, 12023 (1988).

[3] G. Sugiyama and S. E. Koonin, Ann. Phys. **168**, 1 (1986).

[4] S. Sorella, S. Baroni, R. Car, and M. Parrinello, Europhys. Lett. **8**, 663 (1989); S. Sorella, E. Tosatti, S. Baroni, R. Car, and M. Parrinello Int. J. Mod. Phys. B **1**, 993 (1989).

[5] J. E. Hirsch and S. Tang (unpublished).

[6] R. Blankenbecler, D. J. Scalapino, and R. L. Sugar, Phys. Rev. D **24**, 2278 (1981).

[7] M. Suzuki, Prog. Theor. Phys. **56**, 1454 (1976).

[8] R. M. Fye, Phys. Rev. B **33**, 6271 (1986); R. M. Fye and R. T. Scalettar, *ibid.* **36**, 3833 (1987).

[9] J. E. Hirsch, Phys. Rev. B **31**, 4403 (1985).

[10] We use the Trotter approximation to write $\exp(-\Delta\tau K)$ as a product of four sparse matrices. In the "checkerboard" breakup we use, one of these matrices allows hopping between sites (i_x,i_y) and sites (i_x+1,i_y) with i_x even, a second with i_x odd. The third and fourth matrices allow for hopping between sites (i_x,i_y) and (i_x,i_y+1) with i_y even and odd.

[11] A detailed discussion of the use of matrix factorization methods to compute fermion Green's functons for both the grand-canonical ensemble and zero-temperature problems is planned to be given by E. Y. Loh, J. E. Gubernatis, R. T. Scalettar, S. R. White, and R. L. Sugar (unpublished).

[12] J. E. Hirsch, Phys. Rev. B **31**, 4403 (1985).

[13] S. R. White and J. W. Wilkins, Phys. Rev. B **37**, 5024 (1988).

[14] J. D. Reger and A. P. Young, Phys. Rev. B **37**, 5978 (1988).

[15] D. A. Huse, Phys. Rev. B **37**, 2380 (1988).

[16] R. P. Feynman, Phys. Rev. **94**, 262 (1954).

[17] S. Sorella has suggested that the sign problems appear to be reduced at fillings corresponding to nondegenerate ground states of the noninteracting systems (private communication).

[18] S. R. White, D. J. Scalapino, R. L. Sugar, N. E. Bickers, and R. T. Scalettar, Phys. Rev. B **39**, 839 (1989).

Journal of the Physical Society of Japan
Vol. 56, No. 10, October, 1987, pp. 3582-3592

Variational Monte-Carlo Studies of Hubbard Model. II

Hisatoshi YOKOYAMA and Hiroyuki SHIBA

Institute for Solid State Physics, University of Tokyo, Roppongi, Minato-ku, Tokyo 106

(Received June 3, 1987)

As a continuation of a previous paper [J. Phys. Soc. Jpn. **56** (1987) 1490], the variational Monte-Carlo method is extended to include the antiferromagnetic long-range order. The theory is based on the Gutzwiller-type correlation factor and its effect is exactly taken into account by the Monte-Carlo procedure. An application is made to the half-filled-band case of one-dimensional lattice (50 sites), two-dimensional square lattice (up to 20×20 sites) and three-dimensional simple cubic lattice (6×6×6 sites). The result is qualitatively different from previous studies relying on the random-phase-type "Gutzwiller approximation." The variational energy for two and three dimensions is favorably compared with Hirsch's quantum Monte-Carlo data.

§1. Introduction

The strong correlation effect between electrons has been for a long time an important theoretical problem in magnetism and metal-insulator transition. It also seems essential in some problems of current concern, such as heavy electron systems[1] and possibly recent high-T_c oxide superconductors.[2] A general solution of the correlation effect is not easy, however, in the intermediate and strong correlation regime. Even for the Hubbard model,[3-6] which is the simplest one, the ground state and other properties have not been established yet except for the one-dimensional (1D) case.[7-9]

It is our belief that among various approximate methods the variational approach[5] is very promising. Firstly, it is applicable, in principle, to the whole range of correlation strength in contrast to the perturbation theory. Secondly it is flexible enough to incorporate important features into the trial function, according to physical situations. One can bring these merits of this approach into full play only when the variational expectation values are calculated correctly.[10] This can be made possible actually by utilizing a computer for the Monte-Carlo calculation of expectation values. We call this approach a variational Monte-Carlo (VMC) method.

In previous papers, we demonstrated that the VMC method is in fact useful in the application to the periodic Anderson model[11] and the Hubbard model;[12] the latter paper will be referred to as I. In I the Hubbard model of 1D chain, square lattice (SQL) and simple cubic lattice (SCL) was studied with the VMC method by assuming a paramagnetic state and using the famous Gutzwiller wave function (GF).[5] The variational energy values calculated exactly by this method were compared with the conventional random-phase-type Gutzwiller approximation (GA). The VMC results have turned out to be different from the GA on various important points including the Brinkman-Rice transition.[6] Thereby a serious doubt has been cast on the validity of the GA. We have shown that the Monte-Carlo method is also useful to evaluate various physical quantities such as the momentum distribution function and the spin and charge correlation functions.

The purpose of the present paper is an extension of I to include antiferromagnetic long-range order for half-filled-band systems. The systems we take up are 1D chain, SQL and SCL; notice that the nesting condition is completely satisfied for the half-filled band in these systems. Therefore the antiferromagnetism tends to be realized easily here. In fact, as shown in I, the lowest energy within the paramagnetic GF was higher than the antiferromagnetic Hartree-Fock (HF) solution in strong coupling regime.

As a matter of fact, such a generalization of the GF is not new, but many authors[13-16] have tried it already. However they resorted to the GA or similar approximations; therefore in view of the results in I one should be very careful about the conclusions. Ogawa *et al.*,[13] for instance, concluded that the antiferromagnetic HF energy is not lowered at all by the Gutzwiller's correlation factor. We find with the VMC calculation that it is not the case.

Another motivation of this paper is that Hirsch has recently studied 2D[17] and 3D[18] Hubbard models with the half-filled band by using the quantum Monte-Carlo (QMC) method. The present VMC theory can offer an independent approach to the same problem, so that we think a comparison between these two results would be interesting.

This paper is arranged as follows. In §2 our model and trial wave function are introduced with a brief description of the VMC method. The VMC results are presented in §3 and compared with the GA, the QMC and other available theories. Summary and supplementary discussion are given in the last section. In Appendix the weak correlation limit is examined analytically.

§2. Model and Wave Function

The model in this paper is the Hubbard Hamiltonian with hopping restricted only to the nearest neighbors:

$$\mathcal{H}=-t\sum_{\langle ij\rangle}\sum_{\sigma}C_{i\sigma}^{\dagger}C_{j\sigma}+U\sum_{j}n_{j\uparrow}n_{j\downarrow},\qquad(2.1)$$

where t and U are positive, and t will be taken as the unit of energy henceforth. The electron density is fixed at the half-filled band sector; lattices we consider here are 1D chain, SQL and SCL, in each of which the nesting condition is satisfied. Taking these features into account, we incorporate the antiferromagnetic long-range order into the GF. In other words, as a variational trial function, we generalize the GF by replacing the plane wave state in the original GF with the antiferromagnetic one-particle state (abbreviated as AFGF).

$$|\Psi_{AF}\rangle=\prod_{j}\,[1-(1-g)n_{j\uparrow}n_{j\downarrow}]\,|\Phi_{AF}\rangle,\qquad(2.2)$$

where $|\Phi_{AF}\rangle$ is the Slater determinant of the ground state with the Hartree-Fock-type antiferromagnetic order

$$|\Phi_{AF}\rangle=\det\,[\varphi_{\uparrow}(\boldsymbol{k},\boldsymbol{r})]\,\det\,[\varphi_{\downarrow}(\boldsymbol{k},\boldsymbol{r})].\qquad(2.3)$$

The i, j-element of these determinants is written as

$$\varphi_{\sigma}(\boldsymbol{k}_i,\boldsymbol{r}_{j\sigma})=\exp\,(\mathrm{i}\boldsymbol{k}_i\boldsymbol{r}_{j\sigma})u_{ki}+\mathrm{sgn}\,(\sigma)\exp\,[\mathrm{i}(\boldsymbol{k}_i+\boldsymbol{K})\boldsymbol{r}_{j\sigma}]v_{ki},\qquad(2.4)$$

where

$$u_{ki}=[(1-\varepsilon_{ki}/\sqrt{\Delta^2+\varepsilon_{ki}^2})/2]^{1/2},\qquad(2.5a)$$

$$v_{ki}=[(1+\varepsilon_{ki}/\sqrt{\Delta^2+\varepsilon_{ki}^2})/2]^{1/2},\qquad(2.5b)$$

and sgn (σ) takes either $+1$ or -1 according as $\sigma=\uparrow$ or $\downarrow$; $\boldsymbol{k}_i$ and ε_{ki} denote the i-th wave vector and the corresponding tight-binding energy spectrum: $\varepsilon_k=-2\cos k_x$ (for 1D), $-2(\cos k_x+\cos k_y)$ (for SQL) or $-2(\cos k_x+\cos k_y+\cos k_z)$ (for SCL). $\boldsymbol{r}_{j\sigma}$ refers to the position of the j-th electron of spin σ, and $\boldsymbol{K}$ denotes the antiferromagnetic ordering wave vector, which is π, (π, π) or (π, π, π) for 1D chain, SQL or SCL respectively.

In this formulation the ordinary HF theory corresponds to $g=1$; there Δ is related to the antiferromagnetic long range-order parameter M_s by $\Delta=M_sU/2$. In the present theory, however, both Δ and g are variational parameters to be determined so as to minimize the total energy. The ordinary (paramagnetic) GF is also a special case of this function, i.e., $\Delta=0$, since eq. (2.3) is reduced then to the Slater determinant of free electrons.

Since the Monte-Carlo method used in evaluating variational expectation values is the same as in I, we explain only its main points. According to the spirit of the variation theory, the energy

$$E=\langle\Psi_{AF}|\mathcal{H}|\Psi_{AF}\rangle/\langle\Psi_{AF}|\Psi_{AF}\rangle,\qquad(2.6)$$

is evaluated by varying the parameters Δ and g; the lowest energy and corresponding values of Δ and g are then searched. The Monte-Carlo method is effective in the evaluation of expectation values: the summation over all the configurations is replaced by an average over a finite number of samples picked up by importance sampling in virtue of Metropolis algorithm. One has to collect sufficient samples to reduce the statistical error. The treatment of the Slater determinant is most time-consuming and limits the available system size. In practice we follow Ceperley *et al.*'s method[19] for updating the Slater determinant and evaluating the expectation values of physical quantities. The maximum size is practically 400~500 particles on a medium size computer (FACOM 360M) we use. The computational efficiency can be heightened by transforming the complex elements of the Slater determinant in eq. (2.3) into real numbers, which we actually used in the present work. For the 1D chain these system sizes are large enough; for 2D lattices we have checked the size dependence from the sites up to 400(20^2). It is found that the size dependence is quite insensitive in the intermediate and strong coupling region for the AFGF because of the presence of a staggered field; thus an extrapolation to the infinite lattice can be carried out easily. For 3D systems it is hard to change drastically the linear dimension of the system; however we have studied the lattices as large as 216 (6^3). In every case the results of the VMC calculation are exact within the variational wave function and controllable statistical fluctuations.

§3. Variational Monte-Carlo Results

Let us present our VMC results on 1D, 2D and 3D Hubbard models with the half-filled band. The result for 1D offers a chance to check the method in the light of the exact solution. The results for 2D and 3D are new; they can be compared with Hirsch's QMC data. As shown below, in all the systems the VMC result is sharply different from previous studies relying on the GA.

3.1 One-dimensional chain

To show how the variational procedure is carried out actually, we present first VMC calculations for the 1D chain (50 sites) with the periodic boundary condition. Figure 1 shows the results of the energy expectation value for various values of variational parameters. The value of U is taken as 4 in this case. The minimum of the energy is searched for in this way in the $\Delta - g$ plane. For $U=4$ the lowest energy is found to be -0.544 at $\Delta=0.3$ and $g=0.5$. Since the values of variational parameters are changed discretely in practice, an interpolation is performed to find the minimum. The value of energy thus determined hardly depends on the way of interpolation because the energy is stationary by the variation principle.

The AFGF includes the original paramagnetic GF as well as the HF solution as a special case. The HF result is identical with the $g=1$ curve. The VMC result for $g=1$ is shown with open circles. A good agreement between the two results gives support to reliability of the VMC calculation. The GF value corresponds to the point on the vertical axis in Fig. 1. The energy minimum within the paramagnetic GF, -0.518, is lowered further by about 0.026 with the AFGF. Clearly the AFGF energy is lower than the HF theory as expected.

The energy lowering from both paramagnetic GF and antiferromagnetic HF approximation can be confirmed definitely by the VMC method for $U>2$. However, for $U<2$ the energy of the paramagnetic GF and that of the AFGF are so close to each other that it is difficult to judge whether the energy minimum is really located at $\Delta \neq 0$. This is partly due to statistical error inevitable in the VMC calculation. For the small-U region one can examine the variational energy by an analytic approach of the linked cluster expansion method.[20] As described in the Appendix, the AFGF with a nonzero Δ gives the lowest energy as far as $U>0$.

In Fig. 2 the variational energy of the AFGF is presented as a function of U. The AFGF gives the lowest energy among various variational theories shown here, although it is still slightly higher than the exact solution.[7] Approximately speaking, for a small U the AFGF is very close to that of the GF as mentioned above; as U increases, it approaches the HF energy. The AFGF energy stays all the way below the GF and the HF. Notice that this feature is in sharp contrast with the result of the GA.[13] In fact in the latter theory its energy follows the GF in small-U region and switches suddenly to the HF above a critical value of U. This difference between the VMC result and

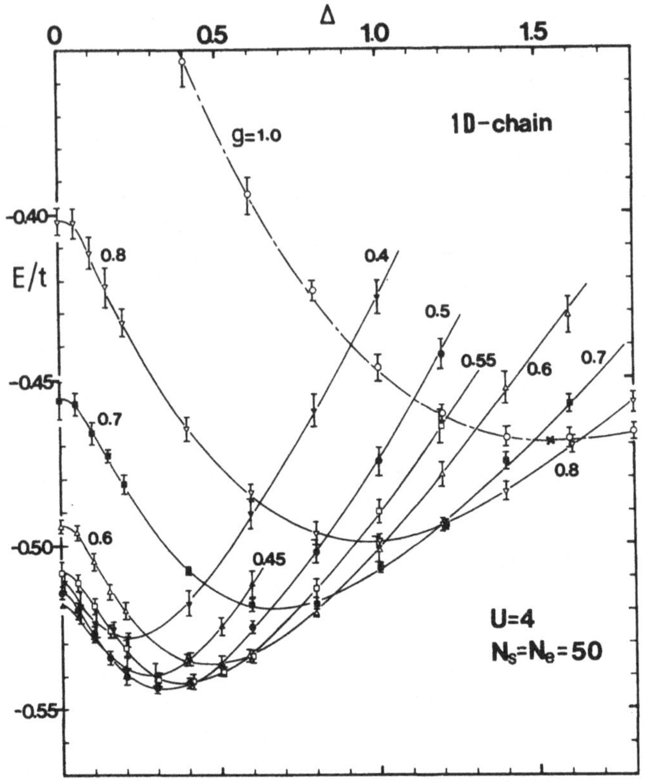

Fig. 1. The energy expectation value of one-dimensional lattice for AFGF as a function of variational parameters Δ and g. Each point represents an average over 10^4 MCS's on 50 sites. The standard deviation among 10 groups of 10^3 MCS's is shown with the error bar. The dot-dashed line for $g=1$ corresponds to the standard HF result for the infinite chain, on which the cross represents the minimum point where the self-consistency condition is satisfied.

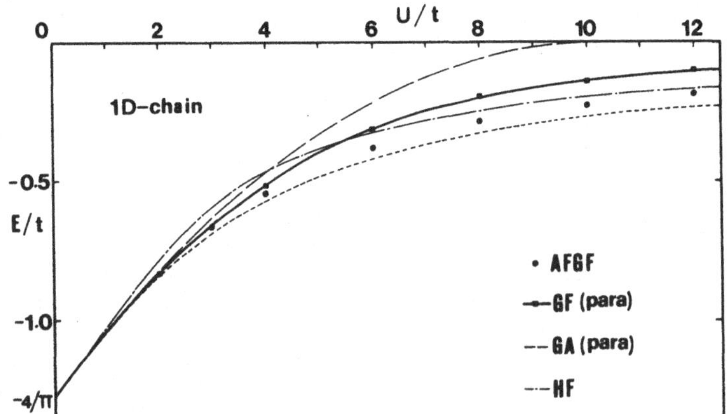

Fig. 2. The total energy as a function of U/t for 1D chain: GF (solid line) for 50 sites,[12] HF (dot-dashed line), AFGF (solid circle) for 50 sites, paramagnetic GA (dashed line) and the exact solution (broken line).[7]

GA remains also in 2D SQL and 3D SCL as we see shortly; it strongly suggests that in order to make the best use of the variational theory one should not rely on the GA but evaluate correctly the expectation values. In fact the Monte-Carlo technique serves well for this purpose.

The sublattice magnetization M_s has been calculated with the variational parameters corresponding to the energy minimum; M_s is defined as

$$M_s=\frac{1}{N_s}\,|\sum_j(-1)^j\langle n_{j\uparrow}-n_{j\downarrow}\rangle|, \qquad (3.1)$$

where N_s is the number of the sites. The resulting M_s vs U relation is given in Fig. 3. As mentioned before, it is not easy to estimate M_s reliably in a small-U region, for which the linked cluster expansion (LCE) method[20] is suitable instead. In the Appendix the latter method is described and the result is included in Fig. 3. Combining it with our VMC result, one can obtain M_s for the whole range of U/t.

According to the present calculation the ground state has an antiferromagnetic long-range order in 1D Hubbard model. This is contrary to what the exact solution[7] tells us. The main reason for this disagreement is that the AFGF does not take into account the spin fluctuation sufficiently. In fact a different variational approach, which is more appropriate for the strong correlation regime, leads us to conclude that the singlet liquid state (resonating-valence-bond state) is lower in energy than the antiferromagnetic state.[23]

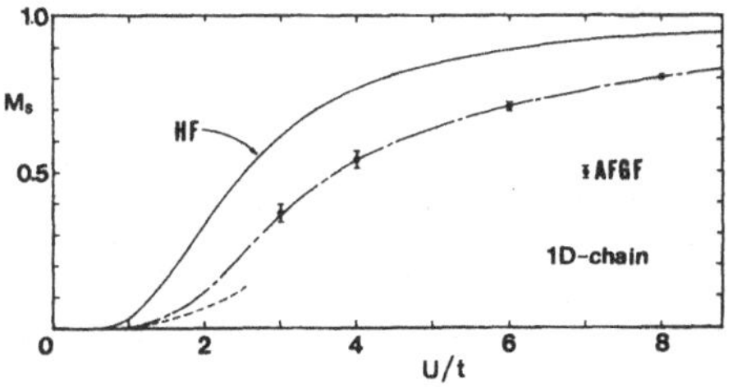

Fig. 3. Sublattice magnetization M_s as a function of U/t for 1D chain: HF (solid line) and AFGF for 50 sites (solid circle). The broken line shows the LCE for the weak correlation limit and the dot-dashed line is a smooth interpolation between the VMC and the LCE.

3.2 Two-dimensional square lattice

The exact solution is not available for two and three dimensions; therefore using the VMC approach with an appropriate wave function, we wish to offer a useful and approximate solution for these systems.

To carry out the VMC calculation for 2D SQL, we take up finite systems with size L^2 ($L=6\sim20$). Notice that the largest system $L=20$ contains $20^2=400$ sites. As for the boundary condition, we impose the periodic one in one direction (namely x-axis) and the antiperiodic one in the other direction (y-axis). This boundary condition is chosen so that the degeneracy of the lowest state in $|\Psi_{AF}\rangle$ is absent, while the case of periodic boundary condition in both directions does not meet this requirement.

Figure 4 shows the size dependence of the lowest energy and the corresponding sublattice magnetization for typical values of U. The lowest energy has been determined through a search in the $g-\Delta$ domain similar to Fig. 1. Both the ground state energy and the sublattice magnetization have turned out to depend on the system size in the weak coupling regime ($U<4$), since electrons move about relatively freely. As for the sublattice magnetization at $U=3$, the lattice size of 20^2 is still small to be judged as equivalent to the infinite lattice. However it is possible to determine the $L=\infty$ limit by an extrapolation. As regards the energy the system as large as 20^2 may be regarded sufficiently large even for $U=3$. On the other hand, as U increases, because of the increase of a staggered magnetization, the size dependence becomes less important. Thus for $U=8$, for instance, the results of both quantities for $L=8$ and $L=20$ coincide with each other within the statistical errors.

In Table I the values of the variational energy and the sublattice magnetization of the AFGF are collected together with the corresponding values of the variational parameters g_0 and Δ_0 for some values of U. The variational energy per site, E/t, in this table is compared with results of other methods in Fig. 5. For SQL the paramagnetic GF gives rather high energy in the intermediate and strong coupling regime. The energy is greatly lowered when the antiferromagnetic long-range order

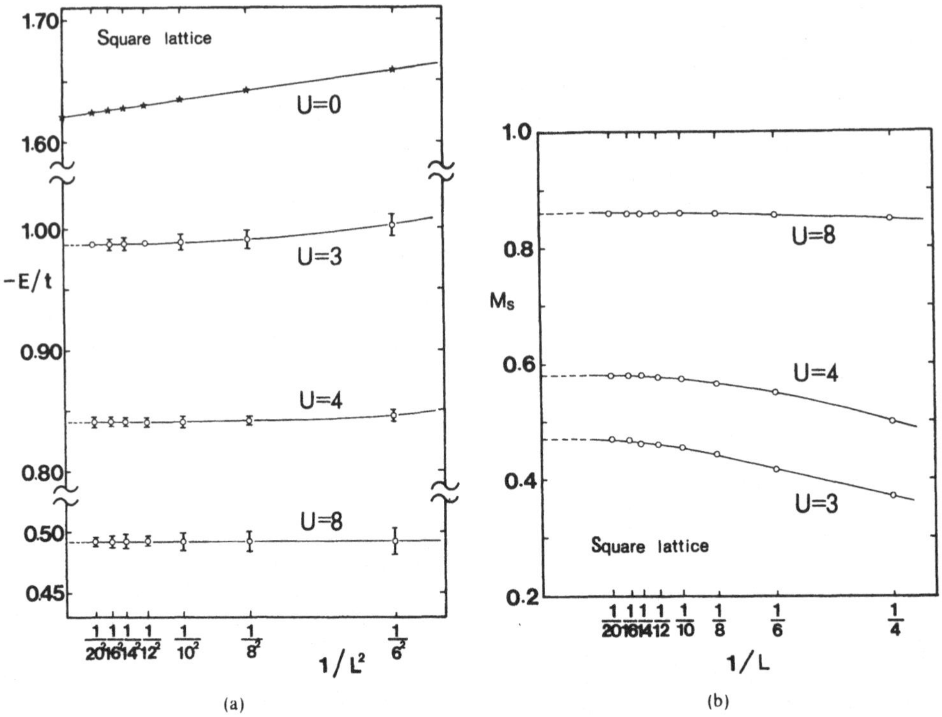

Fig. 4. The size dependence of (a) total energy and (b) sublattice magnetization for 2D square lattice.

Table I. Energy and sublattice magnetization for 2D square lattice ($L=\infty$).

U/t	3	4	6	8	10	12
g_0	0.70	0.65	0.55	0.50	0.45	0.40
Δ_0	0.35	0.45	0.90	1.3	1.5	1.6
$-E/t^*$	0.987(2)	0.841(3)	0.629(3)	0.493(3)	0.401(4)	0.336(5)
$M_s^\dagger$	0.47(4)	0.58(2)	0.77(1)	0.86(1)	0.90(1)	0.92(1)

* The figure in the bracket represents the statistical error in the last figure.

† The figure in the bracket represents the possible error in the last figure in determining g_0 and Δ_0 (and in the extrapolation for $U=3$).

is introduced in the HF approximation. The AFGF lowers the energy further both from the GF and the HF solution, as we have already observed for the 1D chain. The VMC investigation of the weak coupling regime ($U<2$) is difficult, but the linked cluster expansion method described in the Appendix confirms that the AF long-range order is present for $U>0$.

Our variational result of the AFGF must be useful in a comparison with the quantum Monte-Carlo (QMC) simulation by Hirsch,[17] which was done for SQL up to 8^2 and $U=2$, 4 and 8. Since our result should be an upper bound for the correct value, a good agreement between these two gives support of the accuracy of both the QMC simulation and our VMC theory.

Next M_s is plotted in Fig. 6 as a function of U and compared with the HF and the QMC methods. In the weak correlation region the VMC cannot give a reliable value for M_s

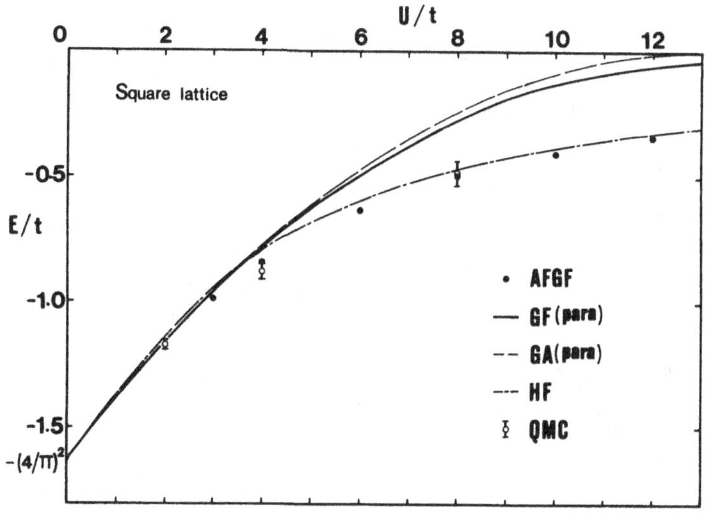

Fig. 5. The total energy as a function of U/t for SQL: paramagnetic GA (dashed line), GF (solid line),[12] HF (dot-dashed line), AFGF (solid circle) and Hirsch's QMC result (open circle).[17]

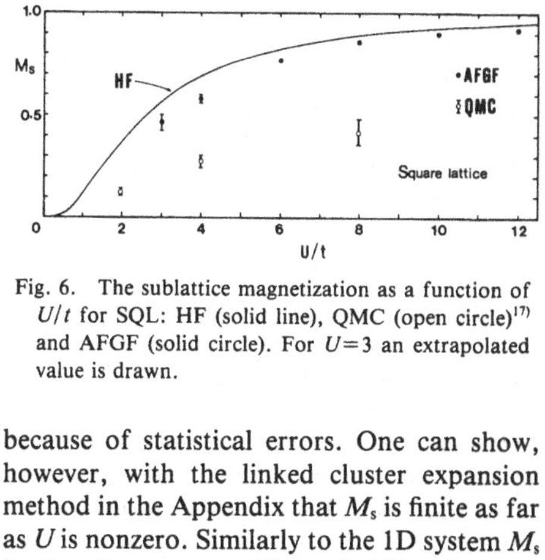

Fig. 6. The sublattice magnetization as a function of U/t for SQL: HF (solid line), QMC (open circle)[17] and AFGF (solid circle). For $U=3$ an extrapolated value is drawn.

because of statistical errors. One can show, however, with the linked cluster expansion method in the Appendix that M_s is finite as far as U is nonzero. Similarly to the 1D system M_s tends to be suppressed from the HF value over the whole region of U. As the correlation becomes stronger, however, the reduction of M_s decreases in correspondence with the energy approaching the HF result. On the other hand the QMC simulation[17] gives a much smaller value for M_s especially for a large U. Again it seems to suggest us that an introduction of the spin fluctuation into the wave function is needed to improve the variational description.

3.3 Three-dimensional simple cubic lattice

We have studied the AFGF for SCL on the 6^3 lattice, which contains 216 lattice points. As for the boundary condition we impose the periodic boundary condition for one direction and the antiperiodic one for the remaining two directions. The reason for this choice is that the state $|\Psi_{AF}\rangle$ has to be unique for the half-filled band, as in §3.2.

The expectation values of energy and sublattice magnetization are recapitulated in Table II for several values of U; corresponding variational parameters g_0 and Δ_0 are also included. In Fig. 7 the ground state energy is plotted as a function of U and compared with some other theories. All the theories on this figure is variational, so that they give an upper bound for the true value. We believe that the AFGF result must be close to the exact value, because

Table II. Energy and sublattice magnetization for 3D simple cubic lattice ($L=6$).

U/t	4	6	8	10	12
g_0	0.65	0.60	0.60	0.525	0.525
Δ_0	0.35	0.80	1.5	1.8	2.4
$-E/t^*$	1.161(4)	0.886(5)	0.704(7)	0.579(6)	0.491(7)
$M_s^{\dagger}$	0.38(4)	0.65(2)	0.82(1)	0.87(1)	0.91(1)

*,† Same as in Table I.

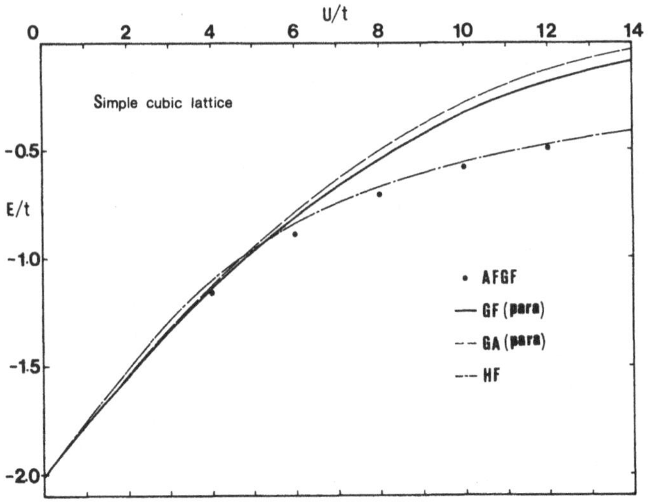

Fig. 7. The total energy as a function of U/t for SCL among GF (solid line) for 6^3 sites, HF (dot-dashed line) and AFGF (solid circle). The value of GF (paramagnetic) is normalized to the energy for $U=0$ (see also I).

the AFGF in 2D SQL was already in good agreement with Hirsch's QMC results. This belief is strengthened by considering that the antiferromagnetic long-range order should be certainly present in 3D, although it may be controversial in 2D.

Let us mention in this connection that the QMC results by Hirsch[18] are available for the energy in 3D (actually 4^3), but unfortunately they are limited to relatively high temperatures. An extrapolation, which is actually a little ambiguous, gives us an estimation of the zero-temperature limit; it looks consistent with our VMC results. A further development of the QMC study to lower temperatures and larger systems is highly expected.

The sublattice magnetization has also been determined as a function of U and is shown in Fig. 8; clearly the correlation effect suppresses the sublattice magnetization from the HF value. Comparing the results for the sublattice magnetization in 1D, 2D and 3D with each other (Figs. 3, 6 and 8), one may notice that the magnetization for a fixed value of U is not a monotonic function of the dimension. This is due to the effect of the density of states at the Fermi energy. In fact the 2D case is special in this respect: the density of states diverges logarithmically in 2D, while it is finite in 1D and 3D. Because of this divergence, the magnitude of the sublattice magnetization is larger in 2D than in 1D and 3D. This effect is clearly observed in the HF result.

§4. Summary and Supplementary Discussion

Following I we have studied with the variational Monte-Carlo (VMC) method one-dimensional (1D) chains, square lattices (SQL) and simple cubic lattices (SCL). As emphasized in I, the VMC method enables us to evaluate the expectation values without any approximation for a chosen wave function. In this paper we have extended the ordinary Gutzwiller function (GF) by introducing the antiferromagnetic long-range order into the wave function.

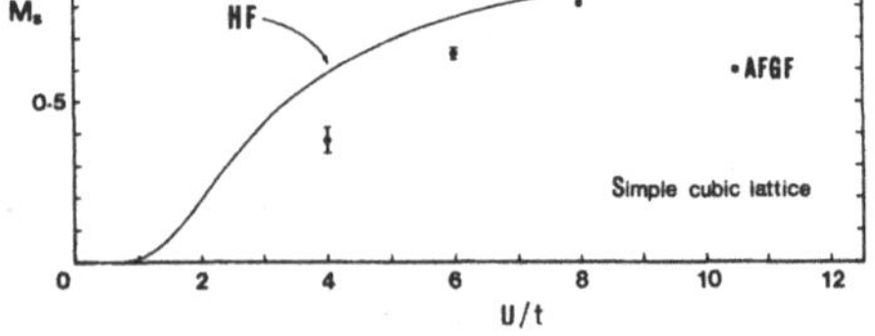

Fig. 8. Sublattice magnetization as a function of U/t for SCL: HF (solid line) and AFGF (solid circle).

Our VMC results have the following features:

(1) It has been demonstrated successfully that the VMC method is powerful to study many-body problems on a lattice.

(2) Common to all the lattices investigated, the antiferromagnetic state is always stabilized for $U>0$ and its energy stays lower than those of both the paramagnetic GF and the Hartree-Fock theory. Although it may sound trivial, it is actually in contrast to other studies resorting to the Gutzwiller approximation (GA) or similar ones.[13-16] Therefore the present study together with I demonstrates that one should go beyond the GA to establish a reliable variation theory. We emphasize in this connection that our theory is strictly variational, so that the energy is truly an upper bound for the exact energy.

(3) The variational energy for 2D SQL is very close to Hirsch's quantum Monte-Carlo data, thus giving partial support of the latter theory. The energy for 3D SCL is favorably compared with Hirsch's results, although the latter is limited to relatively high temperatures and its extrapolation to zero temperature is ambiguous to some extent.

(4) The theory based on eq. (2.2) predicts that in the large U regime the energy as well as the antiferromagnetic sublattice magnetization approaches the HF value asymptotically. Therefore the zero-point spin fluctuation effect, which should be present even in $U/t\gg 1$, has to be implemented for an improvement of the wave function.

In connection with the problem of (4) we note that there are two approaches in the VMC studies of the Hubbard model in intermediate and strong coupling regime. One way is to study directly the original Hamiltonian for a finite interaction strength,[12] which we have actually taken in this paper. The other is to start from the strong coupling limit. Along the latter line Kaplan, Horsch, and Flude[21] first studied the spin correlation for 1D half-filled chains; quite recently Gros, Joynt and Rice[22] have refined this approach on various important points. The two approaches are complementary with each other. However, as far as the strong correlation regime is concerned, the latter one is superior; because virtual processes creating a doubly occupied site next to an empty site are included in the effective Hamiltonian. Along this approach we have recently examined the stability of the singlet liquid state and antiferromagnetic state; it will be reported separately.[23]

Various interesting problems are left still for future studies.

(1) The ground state of the system with band-filling other than half-filled has to be studied. Especially for 2D SQL the systems with electron density slightly less than the half-filled band is interesting, where competition between spin order and singlet state is subtle and superconducting phase may appear.[2,24]

(2) The 2D triangular lattice and 2D square lattice with next nearest neighbor hopping would be similarly interesting from the viewpoint of magnetic frustration.[2,25]

(3) Various intersite correlations have to be taken into account for the variational wave function.

We hope to take up those problems in the next paper of this series, and to develop further the VMC studies on the Hubbard model.

Acknowledgments

One of the authors (H. Y.) is grateful to M. Ogata for useful advice on the calculations. This work is supported partly by a Grant-in-Aid from Ministry of Education, Science and Culture.

Appendix: Linked Cluster Expansion for Weak Correlation Regime

For the 1D chain, SQL and SCL with half-filled band the AFGF ($\Delta=0$ and $g\neq 1$) always gives the lowest state, no matter how small U may be. This can be proved most conveniently with the linked cluster expansion (LCE). For this end we first rewrite eq. (2.2) as

$$|\Psi_{AF}\rangle=e^{-S}|\Phi_{AF}\rangle, \qquad (A\cdot 1)$$

where $S=(\alpha/2)\sum_j n_{j\uparrow}n_{j\downarrow}$ and α is related to g by $g=e^{-\alpha/2}$. Using the linked cluster theorem, one obtains for the energy expectation value[20,26]

$$E\equiv\langle\Phi_{AF}|e^{-S}\mathcal{H}e^{-S}|\Phi_{AF}\rangle/\langle\Phi_{AF}|e^{-2S}|\Phi_{AF}\rangle$$
$$=\langle\Phi_{AF}|\mathcal{H}|\Phi_{AF}\rangle-\langle\Phi_{AF}|\{S,\mathcal{H}\}|\Phi_{AF}\rangle_c$$

$$+\frac{1}{2}\langle\Phi_{AF}|\{S,\{S,\mathcal{H}\}\}|\Phi_{AF}\rangle_c-\cdots$$
$$\equiv E_0(\Delta)+E_1(\Delta)\alpha+E_2(\Delta)\alpha^2+\cdots, \tag{A·2}$$

where $\langle\Phi_{AF}|\Phi_{AF}\rangle=1$ is assumed; $\{\ \ \}$ means the symmetric product. The subscript c denotes to take only the connected diagrams. The $\alpha=0$ limit of eq. (A·2), i.e. $E_0(\Delta)$, is simply the HF energy.

In the weak correlation limit, $U/t\ll 1$, the minimum of E must be located at a small Δ and a small α. Minimizing E first with respect to α, we obtain

$$E=E_0(\Delta)-\frac{1}{4}\frac{[E_1(\Delta)]^2}{E_2(\Delta)}, \tag{A·3}$$

and

$$\alpha=-\frac{1}{2}\frac{E_1(\Delta)}{E_2(\Delta)}. \tag{A·4}$$

The second term on the right-hand side is a correction to the HF theory. $E_n(\Delta)$ consists of the average of the hopping term, $t_n(\Delta)$, and that of the Coulomb term, $Ud_n(\Delta)$:

$$E_n(\Delta)=t_n(\Delta)+Ud_n(\Delta). \tag{A·5}$$

Since $t_2(0)\neq 0$ and $d_1(0)\neq 0$ hold, eq. (A·3) can be simplified further, for small U/t and Δ, as

$$E\simeq E_0(\Delta)-\frac{1}{4}\frac{[t_1(\Delta)+Ud_1(0)]^2}{t_2(0)}. \tag{A·6}$$

For the 1D case we can evaluate the right-hand side analytically as

$$\begin{cases} E_0(\Delta)=-\dfrac{2}{\pi}\left[\sqrt{4+\Delta^2}E(k)-\dfrac{\Delta^2}{\sqrt{4+\Delta^2}}K(k)\right]+\dfrac{1}{4}U(1-m_0^2), \\ t_1(\Delta)=\dfrac{1}{2}m_0\Delta\left(1-\dfrac{\Delta}{\sqrt{4+\Delta^2}}\right), \\ d_1(0)=-\dfrac{1}{12}, \\ t_2(0)=\dfrac{1}{4\pi}\left(1+\dfrac{4}{\pi^2}\right), \end{cases} \tag{A·7}$$

where

$$m_0=\frac{2}{\pi}\frac{\Delta}{\sqrt{4+\Delta^2}}K(k),$$

and k is the modulus given by $2/\sqrt{4+\Delta^2}$. $K(k)$ and $E(k)$ are, respectively, complete elliptic integrals of the first and second kind. Using the asymptotic expression for $K(k)$ and $E(k)$, we obtain the energy in the weak coupling limit:

$$E\simeq-\frac{4}{\pi}+\frac{1}{2\pi}\Delta^2\log\frac{8}{\Delta}+\frac{1}{4}U-\frac{1}{4\pi^2}U\Delta^2\left(\log\frac{8}{\Delta}\right)^2-\frac{\pi}{144}\frac{1}{1+\frac{4}{\pi^2}}U^2+\frac{1}{1+\frac{4}{\pi^2}}\frac{U}{12}\Delta^2\log\frac{8}{\Delta}, \tag{A·8}$$

where the last two terms are corrections to the HF approximation originating from the second term in eq. (A·6). We find from eq. (A·8) that the minimum of E is located at a finite Δ satisfying

$$1+\frac{1}{1+\frac{4}{\pi^2}}\frac{\pi}{6}U=\frac{U}{2\pi}\log\frac{8}{\Delta}. \tag{A·9}$$

Thus the energy of the antiferromagnetic state is lower than that of the paramagnetic ($\Delta=0$) state. The same conclusion holds for SQL and SCL as well.

The sublattice magnetization

$$M_s=\frac{1}{N_s}\left|\sum_j(-1)^j\langle n_{j\uparrow}-n_{j\downarrow}\rangle\right|,$$

can be studied similarly by expanding e^{-S} in terms of α as

$$\begin{aligned}\langle n_{j\sigma}\rangle &\equiv \langle \Phi_{AF}|\, e^{-S} n_{j\sigma}\, e^{-S}|\Phi_{AF}\rangle \\ &\quad /\langle \Phi_{AF}|\, e^{-2S}|\Phi_{AF}\rangle \\ &= \langle \Phi_{AF}|n_{j\sigma}|\Phi_{AF}\rangle \\ &\quad -\langle \Phi_{AF}|\{S, n_{j\sigma}\}|\Phi_{AF}\rangle_c + \cdots. \end{aligned} \tag{A·10}$$

Evaluating the right-hand side up to the first correction, we arrive at

$$M_s = m_0\left(1+\frac{\alpha}{2}+0(\alpha^2)\right), \tag{A·11}$$

in which α is given by eq. (A·4) and m_0 is the quantity defined after eq. (A·7). For the 1D lattice we have calculated M_s using eq. (A·11); it is plotted in Fig. 3.

References

1) See for instance, P. A. Lee, T. M. Rice, J. W. Serene, L. J. Sham and J. W. Wilkins: Comments on Cond. Matt. Phys. **12** (1986) 99; *Theory of Heavy Fermions and Valence Fluctuation*, ed. T. Kasuya and T. Saso (Springer, 1985).
2) P. W. Anderson: Science **235** (1987) 1196.
3) J. Hubbard: Proc. R. Soc. London **A276** (1963) 238.
4) J. Kanamori: Prog. Theor. Phys. **30** (1963) 275.
5) M. C. Gutzwiller: Phys. Rev. Lett. **10** (1963) 159; Phys. Rev. **134** (1965) A1726.
6) W. F. Brinkman and T. M. Rice: Phys. Rev. **B2** (1970) 4302; D. Vollhardt: Rev. Mod. Phys. **56** (1984) 99.
7) E. H. Lieb and F. Y. Wu: Phys. Rev. Lett. **20** (1968) 1445.
8) M. Takahashi: Prog. Theor. Phys. **42** (1969) 1098.
9) H. Shiba: Phys. Rev. **B6** (1972) 930.
10) Recently W. Metzner and D. Vollhardt Phys. Rev. Lett. **59** (1987) 121 and F. Gebhard and D. Vollhardt (preprint) have reported some exact analytic results for the original Gutzwiller function in the one-dimensional Hubbard model.
11) H. Shiba: J. Phys. Soc. Jpn. **55** (1986) 2765.
12) H. Yokoyama and H. Shiba: J. Phys. Soc. Jpn. **56** (1987) 1490. This paper is referred to as I.
13) T. Ogawa, K. Kanda and T. Matsubara: Prog. Theor. Phys. **53** (1975) 614.
14) F. Takano and M. Uchinami: Prog. Theor. Phys. **53** (1975) 1267.
15) J. Florencio Jr. and K. A. Chao: Phys. Rev. **B14** (1976) 3121; K. A. Chao: Solid State Commun. **22** (1977) 737.
16) H. Takano and A. Okiji: J. Phys. Soc. Jpn. **50** (1981) 2891.
17) J. E. Hirsch: Phys. Rev. Lett. **51** (1983) 1900; Phys. Rev. **B31** (1985) 4403.
18) J. E. Hirsch: Phys. Rev. **B35** (1987) 1851.
19) D. Ceperley, G. V. Chester and M. H. Kalos: Phys. Rev. **B16** (1977) 3081.
20) P. Horsch and P. Fulde: Z. Phys. **B36** (1979) 23.
21) T. A. Kaplan, P. Horsch and P. Fulder: Phys. Rev. Lett. **49** (1982) 889; P. Horsch and T. A. Kaplan; J. Phys. **C16** (1983) L1203.
22) C. Gros, R. Joynt and T. M. Rice: Phys. Rev. B **36** (1987) 381.
23) H. Yokoyama and H. Shiba: J. Phys. Soc. Jpn. **56** (1987) 3570.
24) J. E. Hirsch: Phys. Rev. Lett. **54** (1985) 1317.
25) See for instance, K. Hirakawa, H. Kadowaki and K. Ubukoshi: J. Phys. Soc. Jpn. **54** (1985) 3256.
26) D. Baeriswyl and K. Maki: Phys. Rev **B31** (1985) 6633.

Z. Phys. B – Condensed Matter 68, 425–432 (1987)

Condensed Zeitschrift für Physik B Matter

Superconducting Instability in the Large-U Limit of the Two-Dimensional Hubbard Model

C. Gros[1], R. Joynt[2], and T.M. Rice[1]
[1] Theoretische Physik, ETH-Hönggerberg, Zürich, Switzerland
[2] Department of Physics, University of Wisconsin-Madison, Madison, Wisconsin, USA

Received July 2, 1987

We have investigated numerically the pairing instabilities of Gutzwiller wavefunctions. These are equivalent to a certain form of the resonant valence bond wavefunction. The case considered is a nearly half-filled two dimensional band with interactions given by a Hubbard model with large on-site Coulomb interactions. We find that the paramagnetic normal state is unstable to d-wave pairing but stable against s-wave pairing. The antiferromagnetic state is marginally stable against both types of pairing. These results can be explained as an interference effect resulting in enhanced antiferromagnetic spin correlation in the paired state.

I. Introduction

The unexpected discovery of high T_c superconductivity in the Cu-oxides by Bednorz and Müller [1] has aroused unprecedented interest. Right at the outset Anderson [2] proposed that the superconductivity originated in hole-like carriers introduced by doping in a Mott insulating state and developed out of a resonant valence bond (RVB) spin state. There are numerous other proposals; for a discussion of these see Refs. 3, 4. The RVB state was proposed previously by Anderson [5] as a description of the groundstate of a frustrated 2-dimensional Heisenberg lattice (e.g. a triangular lattice). In La_2CuO_4 and related materials the lattice is essentially a 2-dim square lattice so that the questions of the competition between the paramagnetic (PM) RVB and antiferromagnetic (AF) states as well as the instability of both states to superconductivity when holes are introduced needs to be investigated. In this paper we address the latter question.

Several approaches have been taken by others. All are based on transforming the original Hubbard Hamiltonian to an effective Hamiltonian which acts in the restricted Hilbert space with no doubly occupied sites. In one type of approach this effective Hamiltonian is treated by a mean field method involving a Gor'kov factorization of the Heisenberg spin-spin coupling term [6–8]. In particular Baskaran, Zou and Anderson and Ruckenstein, Hirschfeld and Appel show that in the mean field theory a RVB state can be obtained at 1/2-filling and a superconducting state in the presence of holes. Zhang [9] has shown that this theory predicts that the "extended s-wave" [10] and the d-wave state have the same critical temperature at 1/2 filling but the d-wave has a higher T_c at lower electron densities. However, these calculations can be questioned since the condition that the system must stay within the restricted Hilbert space without double occupancy is only approximately obeyed. The physics of the constrained and unconstrained systems can be expected to be quite different.

Recent work by Kivelson, Rokhsar and Sethna [11] and by Zou and Anderson [12] proposes that a Bose condensation of holes in the RVB state is to be identified with the superconducting transition.

Another approach is based on looking for a bound state of two holes in an AF ordered lattice. Such binding has been investigated by Takahashi for a Heisenberg lattice [13] using a moment method. Hirsch [14] has considered the related problem of holes in the oxygen states but strongly coupled to the Cu-spins. This latter model has been put forward by Emery [15].

It is, however, possible to investigate a number of these questions by using the variational Monte-Carlo (MC) method. This method allows accurate numerical evaluation of expectation values and matrix elements in wavefunctions with strong local correlations. For example, it has been used to evaluate Gutzwiller wavefunctions for the Hubbard Hamiltonian in 1-dim by Horsch and Kaplan [16], by Yokoyama and Shiba (YS) [17] in 2-dim and for the Anderson Hamiltonian by Shiba [18]. Previously we pointed out that it was better to work with the effective Hamiltonian when the onsite Coulomb interaction was large and examined many properties of the Gutzwiller wave-function in 1-dim [19]. In this approach the conditions on the Hilbert subspace are fully obeyed. Very recently YS have used this method to investigate the stability of the PM Gutzwiller wavefunction against long range AF ordering and indeed find it is only in the presence of a finite density of holes that the PM state is stable. These authors and Anderson et al. [20] point out that the Gutzwiller state has the form of an RVB state – a point we return to below. In this paper we examine the stability of both PM and AF wavefunctions to Cooper pairing in the s- and d-wave channels. Our key result is that a d-wave pairing is favored in the PM state but not in the AF state. s-wave pairing is not favored. Note a number of authors, Lee and Read [27], Ohkawa [22], and Cyrot [8] have proposed d-wave pairing in high-T_c superconductors, the first-named authors approaching the problem from smaller values of U, the onsite Coulomb interaction parameter. A scaling theory starting from weak coupling also finds d-wave superconductivity to be favorable, as shown by Schulz [23]. Miyake et al. [24], Scalapino et al. [25], Cyrot [26], and Lavagna et al. [27] have all proposed d-wave pairing in heavy-fermion systems.

II. Method

We consider a nearly-half-filled band of electrons described by a Hubbard model with $t/U \ll 1$, where t is the overlap integral between nearest neighbor sites. We make a canonical transformation to a representation where doubly occupied sites are eliminated, to first order. Details of this procedure can be found in Ref. 19.

The effective Hamiltonian is

$$H_{\mathrm{eff}} = -t \sum_{\langle ij\rangle} (1-n_{i-\sigma}) C^+_{i\sigma} C_{j\sigma} (1-n_{j-\sigma}) + \text{h.c.} + 4(t^2/U) \sum_{\langle ij\rangle} (\mathbf{S}_i \cdot \mathbf{S}_j - n_i n_j/4) \tag{1}$$

in this representation, neglecting terms of higher order in the hole density and t/U, $n_i = \sum_\sigma n_{i\sigma}$. The kinetic energy comes from the hopping of the holes only and there is an AF coupling between the electron spins. The Gutzwiller wavefunction is defined as

$$|\psi_G\rangle = P_{D=0}|\psi_B\rangle = \prod_i (1-n_{i\uparrow} n_{i\downarrow})|\psi_B\rangle \tag{2}$$

where $|\psi_B\rangle = \prod_{|\mathbf{k}|<k_F,\sigma} C^+_{\mathbf{k}\sigma}|VAC\rangle$ is the band ground state and $P_{D=0}$ is a projection operator which eliminates all spatial configurations with doubly occupied sites. It has been shown [19] that this wavefunction has a very favourable spin correlation energy: for the half-filled case, e.g., $\langle\psi_G|\mathbf{S}_i \cdot \mathbf{S}_j|\psi_G\rangle = -0.442$ is within about 0.2% of the exact value in one-dimension, and in two dimensions the value of -0.275 is also lower than the value -0.25 in the Néel state. The kinetic energy in the less-than-half-filled case also compares very well with ground state estimates by other methods. We are therefore confident that the Gutzwiller wavefunction is an excellent variational candidate for the lowest energy state within the class of states lacking long-range magnetic order. It represents a Fermi liquid with a Fermi surface (at general values of the filling) which satisfies the Luttinger theorem. It has been pointed out by YS (also see Ref. 20) that it is also identical to a RVB state of Anderson [2]. We repeat the argument of YS here. A general unrenormalized RVB state may be written as

$$P_{D=0}\Big(\sum_{\mathbf{k}} a(\mathbf{k})\, C^+_{\mathbf{k}\uparrow} C^+_{-\mathbf{k}\downarrow}\Big)^{N/2}|VAC\rangle \tag{3}$$

with $\sum_k a(\mathbf{k}) = 0$. A particular choice [2] is $a(\mathbf{k}) = 1$, $k<k_F$ and $a(\mathbf{k}) = -1$, $k>k_F$. (We discuss a different RVB state, where $a(k)$ has a different form, in Sect. IV). The Gutzwiller state corresponds to $a(k)=1$, $k<k_F$; $a(k)=0$, $k>k_F$. However, adding a constant to $a(k)$ only changes the weight of a spatial configuration if it contains doubly occupied sites. Since these are projected out at the end anyway, the two forms for $a(k)$ lead to wavefunctions which differ only by a normalization factor. The same argument implies that the extended s-wave states, superconducting states in which $a(k)$ changes sign, are not distinct from ordinary s-wave states in projected wavefunction. YS have given a generalization of $|\psi_G\rangle$ which can have AF long range order. In Eq. (2), we substitute $c^+_{k\sigma}$ by $u_k c^+_{k\sigma} + \mathrm{sgn}(\sigma) v_k c^+_{k+Q\sigma}$ where $\mathbf{Q}=(\pi,\pi)$. $2u_k^2 = 1 - \varepsilon_k/(\varepsilon_k^2+\Delta^2)^{1/2}$ and $u_k^2+v_k^2=1$, where ε_k is the band energy, $\varepsilon_k = -2t(\cos k_x + \cos k_y)$, and Δ is the gap energy. Only the lower band states are occupied. For

$\Delta=0$, this reduces to the PM Gutzwiller state given above. For the half filled band, in one dimension, $\Delta=0$ is the stable state, whereas on the square lattice in two dimensions $\Delta>0$ and there is AF long-range order. The relative stability of the AF ordered state with $\Delta>0$ and the PM state with $\Delta=0$ may be influenced by the dimerization of phonons. To take this effect into account would require us to enlarge the Hilbert space and we will not consider that. Other effects which can influence the normal-state stability are next-nearest neighbor coupling and the presence of holes. We will investigate the stability of both normal PM and normal AF states against Cooper pairing.

This is done by the MC method, which yields the relevant operator expectation values. Technical details are given in Refs. 16, 18, and 19. The electrons are placed on a lattice of L sites with periodic boundary conditions. L is chosen in the form (j^2+1) where j is an odd integer. This assures that the half-filled ground state is not degenerate, and further that the Fermi surface consists of $4k$-values. This situation is illustrated in Fig. 1 for $L=26$. We now ask the question: Is this Fermi sea unstable to Cooper pair formation? To find out, we take the number of electrons to be $L-2$, leaving two holes free to move the through the lattice. The ground state in the manifold with zero total z-component of magnetization is then sixteen fold degenerate in the non-interacting case. If it is possible to lower the energy by constructing a coherent combination of Gutzwiller states in this manifold which lowers the total energy including interactions, then an instability is indicated. It should be stressed, however, that the present calculation does not yield a good estimate for the binding energy, since no real attempt to optimize the wavefunction is made. Instead, a specific form for the $a(\mathbf{k})$ is chosen in which $a(\mathbf{k})$ differs from the ground state configuration only for the uppermost set of $\{\mathbf{k}\}$ at the Fermi surface. Further optimization would involve us with more Slater determinants and is hindered by calculational limitations. Therefore, the results in this paper should not be interpreted as relating to the actual superconducting state. Rather, they are to be interpreted in the spirit of Cooper's calculations of pair binding, indicating only whether or not the normal state is stable with respect to the formation of superconducting correlations. A poor estimate of the actual superconducting condensation energy is thereby obtained. (The Cooper binding energy and the BCS condensation energy differ by an exponential factor in the case of conventional superconductivity.)

The wavefunctions we choose are parametrized by $c(\mathbf{k}_i)$, $i=1, 2, 3, 4$, where $|\mathbf{k}_i|=k_F$, and $\mathbf{k}_1=(k_x, k_y)$, $\mathbf{k}_2=(-k_y, k_x)$, $\mathbf{k}_3=(-k_x, -k_y)$, $\mathbf{k}_3=(k_y, -k_x)$. (See

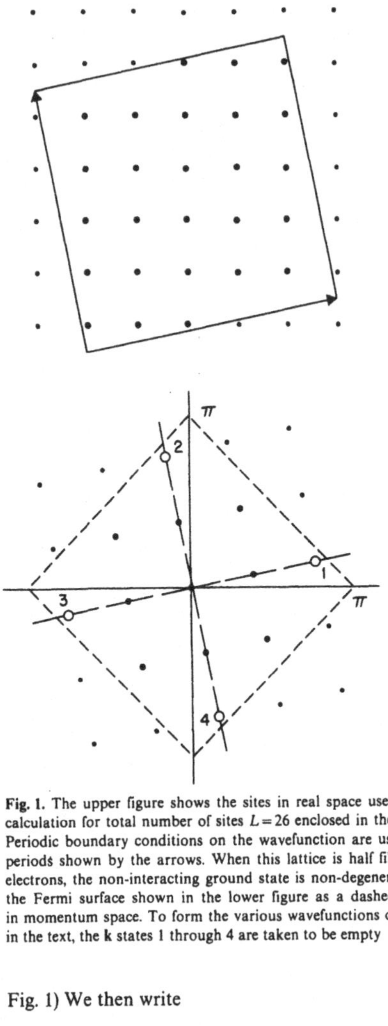

Fig. 1. The upper figure shows the sites in real space used in our calculation for total number of sites $L=26$ enclosed in the square. Periodic boundary conditions on the wavefunction are used, with periods shown by the arrows. When this lattice is half filled with electrons, the non-interacting ground state is non-degenerate with the Fermi surface shown in the lower figure as a dashed square in momentum space. To form the various wavefunctions described in the text, the **k** states 1 through 4 are taken to be empty

Fig. 1) We then write

$$|\Psi\rangle=c_1|\Psi_G^1\rangle+c_2|\Psi_G^2\rangle+c_3|\Psi_G^3\rangle+c_4|\Psi_G^4\rangle, \tag{4}$$

where $|\Psi_G^1\rangle$ is the Gutzwiller state (PM or AF) with the wavevectors $\mathbf{k}_1\uparrow$ and $-\mathbf{k}_1\downarrow$ vacant, etc. The s-state is given by $c_1=c_2=c_3=c_4=1$, and we will consider the d-state given by $c_1=c_3=-c_2=-c_4=1$. Their energies are to be compared to those of an incoherent combination of the same $|\Psi_G\rangle$, which represents the normal state wavefunction. The energy difference is the binding energy of the two holes in the variational wavefunction.

To obtain reliable results in two dimension it is necessary to do calculations for several different lattice sizes and extrapolate to the infinite lattice limit. All our results are plotted in this way. It is also very important to minimize statistical errors. We are calculating energy differences so that it would appear to be necessary to subtract two numbers which are much larger than their difference. The only feasible way to accomplish this with acceptable error is to calculate both energies in a *single* MC run. One can take advantage of the freedom within the MC method of partitioning the summand into two factors one of which determines transition rates, and the other is then measured. A full explanation is given in the context of calculating off-diagonal operator expectation values in Ref. 19. This enables us to compute the difference between normal and superconducting states directly, without the need to subtract quantities which have individually a large uncertainty.

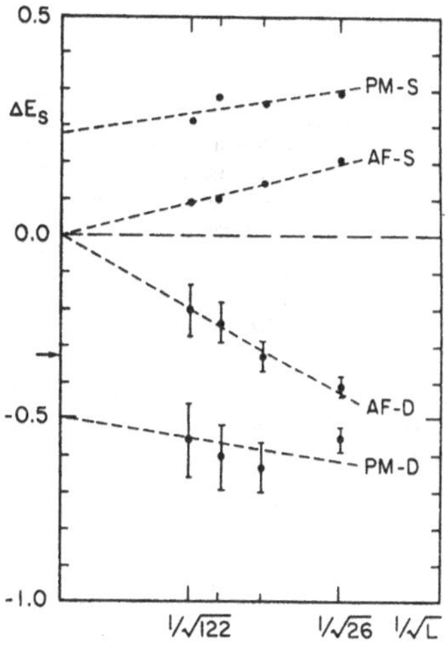

Fig. 2. Binding energies in units of $4t^2/U$ for various superconducting wavefunctions. PM$-S$ and AF$-S$ denote the s-wave states formed from paramagnetic and antiferromagnetic normal states, respectively. PM$-D$ and AF$-D$ are the corresponding d-wave states. The binding energy is the difference between the spin correlation energies in the superconducting and normal states. A negative result implies that the normal state is unstable. See Eq. (6) for a definition of ΔE_s

III. Results and Physical Interpretation

To set the stage for our results, we first review the underlying energetics of the PM and AF states. For the half-filled band, the Hamiltonian (1) reduces to a Heisenberg Hamiltonian

$$H_H = 4(t^2/U) \sum_{\langle ij \rangle} (\mathbf{S}_i \cdot \mathbf{S}_j - n_i n_j/4) \tag{5}$$

where $n_i n_j = 1$.

The energies calculated for the half-filled square lattice by YS are given in the table, and show that the AF state is the stable one. If we now inject a small number of holes into the system, the full Hamiltonian (1) must be used. If we keep in mind that the energy per bond is $4t^2/U$ times $(\mathbf{S}_i \cdot \mathbf{S}_j - 1/4) \approx 2t^2/U$, then we see from the table that the change in spin energy per hole is much larger than $8t^2/U$, which is the number one might naively expect from bond-breaking arguments. The kinetic energy, on the other hand, is not very different from that of a free hole. The holes are completely delocalized and therefore very efficient at disturbing the bonds between the spins. We aim to show that this tendency can be reduced by pairing the holes in d-wave configurations.

We compute first the spin correlation energy gain from the Cooper pairing:

$$\Delta E_s = \langle \psi_S | H_H | \psi_S \rangle - \langle \psi_N | H_H | \psi_N \rangle \tag{6}$$

for a series of lattices having up to 122 sites, with H_H given by Eq. (5). Note that in the thermodynamic limit we can ignore the change in the expectation value of the $n_i n_j$ term. The change in kinetic energy has also been calculated. We have found in every case that this change was very small: zero to within the accuracy of the calculation ($\sim 1\%$). This is partly due to the fact that the four $\mathbf{k}$'s we allow for the Cooper pair are all degenerate, but this choice does not preclude kinetic energy airising from the relative motion in the pair. That this energy is very small indicates stability of the d-wave state for a wide range of values of the parameters t and U. For the wavefunction used here, the change in spin energy may be taken as the total energy change.

Let us discuss first the paramagnetic normal state. Figure 2, together with the extrapolated numbers in the table, Rows 4 and 5, shows that there is a strong tendency toward binding in the d-wave channel. To understand the magnitude of the binding, again take a naive localized picture for the holes. The effect of putting two holes at random in a half-filled PM lattice (see Table 1, Row 2) is to break eight bonds, costing an energy $(32t^2/U)(\mathbf{S}_i \cdot \mathbf{S}_j - 1/4) \approx 17t^2/U$ relative to the half-filled state. The energy saved by putting the two holes in the d-wave paired state relative to the normal state thus corresponds to about 1.5 bonds. This is a large number. The pairing energy gained in this way is larger than one could get by localizing

Table 1. Summary of data and results. PM and AF denote paramagnetic and antiferromagnetic wavefunctions. Row 1 is the kinetic energy of a single hole introduced into a half-filled lattice. Row 2 is the spin correlation energy per bond in the half-filled lattice. Row 3 is the change in this quantity summed over all bonds when a hole is introduced into the half-filled state. Row 4, 5 are the change in spin energy when pairing correlations are introduced into the normal state wavefunctions, taking a Heisenberg interaction. Rows 6, 7 are the same change if the x and y components of the interaction are omitted. Rows 8, 9, and 10 are the nearest neighbor hole-hole correlation functions. The numbers in Rows 2 and 3 are taken from Ref. 17. In rows 2 through 7, to convert to energy units, multiply by $4t^2/U$, the energy per bond

	PM	AF
1. Kinetic energy/hole	$-2.72\,t$	$-2.52\,t$
2. $\sum_{\langle ij\rangle} \langle \mathbf{S}_i \cdot \mathbf{S}_j \rangle$/bond	-0.27	-0.32
3. $\sum_{\langle ij\rangle} \langle \mathbf{S}_i \cdot \mathbf{S}_j \rangle$/hole	$+19.2$	–
4. ΔE_s for d-wave	-0.5 ± 0.2	0.0 ± 0.05
5. ΔE_s for s-wave	$+0.3 \pm 0.1$	0.0 ± 0.05
6. ΔE_s^z for d-wave	-0.17 ± 0.07	-0.22 ± 0.1
7. ΔE_s^z for s-wave	-0.1 ± 0.03	$+0.08 \pm 0.02$
8. g in normal state	1.1 ± 0.1	1.6 ± 0.2
9. g in d-wave state	1.1 ± 0.1	6.9 ± 0.2
10. g in s-wave state	1.1 ± 0.1	0.0 ± 0.2

the holes on adjacent sites in the pair, ignoring phase relationships, and arguing that the number of broken bonds is thereby reduced. This only saves one bond and of course would cost considerable kinetic energy in the relative hole motion, far more than is present in our wavefunction.

The conclusion is inescapable that there is quantum-mechanical interference associated with the phases of the hole wavefunctions which gives rise to a surprisingly extensive rearrangement of the spins. Any picture, such as the fewer broken bonds picture, which ignores these phases, will not lead to the large energy actually involved. Further evidence for this point of view is offered by the data for the s-wave state, for which the effective interaction is positive. We discuss the interference question and the difference between s- and d-wave states in more detail at the end of this section.

The AF normal state, in contrast to the PM-state, is marginally stable to both s- and d-wave superconductivity. Figure 2 shows that there is no energy associated with the change due to Cooper pairing in the infinite-lattice limit. This is in agreement with the work of Takahashi [13] who studied the problem of two holes in an AF background by a variational method, and found no binding energy. In Fig. 3 we plot the correlation of only the z-components of spin ΔE_s^z and the extrapolations are given in the table, Rows 6 and 7. This would be proportional to the

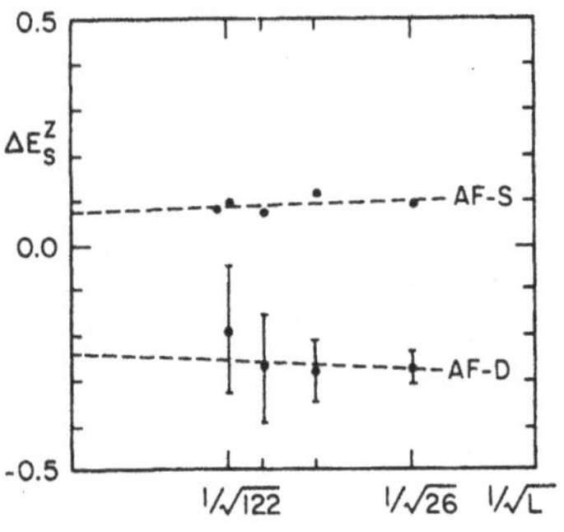

Fig. 3. Binding energies for the AF superconducting states for an Ising interaction in units of the Ising coupling constant. Error bars for the s-wave state have been omitted for clarity. For each lattice size they are the same as those for the d-wave state

total energy if the interactions were Ising-like. (This is only possible if there is very strong spin orbit coupling in the two-dimensional layer. For copper and oxygen that is very unlikely.) It is interesting that binding can take place for this case but only for the d-wave pairing. This indicate that arguments for binding [8, 14] based on the ability of the AF lattice to heal itself after two holes pass through, but not one, are valid for the Ising case. The Heisenberg interaction, on the other hand, gives the lattice without holes the ability to heal itself because of spin flip terms and apparently this is enough to unbind the pairs. This effect shows up in the nearest neighbor hole-hole correlation function:

$$g = L \sum_{\langle ij\rangle} \langle \psi | (1-n_i)(1-n_{j'}) | \psi \rangle. \tag{7}$$

This measures the probability that the two holes are adjacent and is normalized to unity for random positioning of the holes. For the three AF cases it is plotted in Fig. 4, and results given in the table, Rows 8–10. There is already an enhancement from randomness in the normal state, as was found previously in the one-dimensional normal PM state [19]. There is a very large enhancement in the d-wave case, which is the right superconducting wavefunction for Ising interaction. Some of the condensation energy must clearly come from the fewer broken bonds effect for this case, as well as from the absence of a healing effect. The s-wave wavefunction has *less* chance for adjacent holes even than the normal state; it seems likely that this accounts for at least part of the negative binding energy of this state.

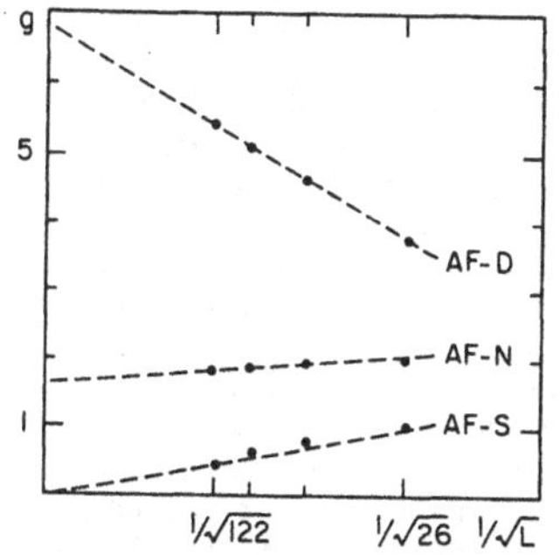

Fig. 4. The nearest-neighbor hole-hole correlation function is plotted as a function of the lattice size L for the AF states. A value of 1 corresponds to random placement of the holes

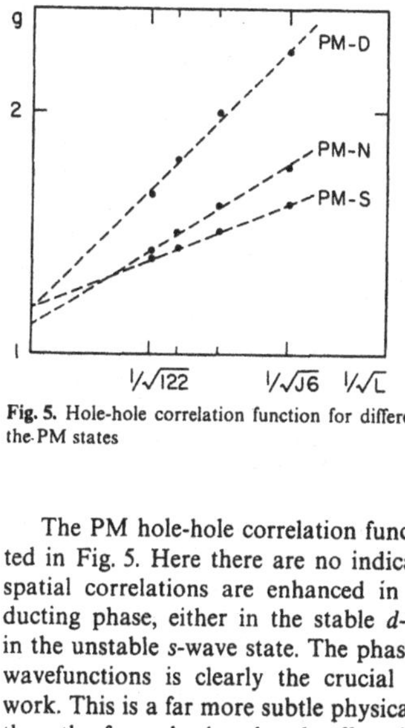

Fig. 5. Hole-hole correlation function for different lattice sizes for the PM states

The PM hole-hole correlation functions are plotted in Fig. 5. Here there are no indications that the spatial correlations are enhanced in the superconducting phase, either in the stable d-wave state, or in the unstable s-wave state. The phasing of the hole wavefunctions is clearly the crucial mechanism at work. This is a far more subtle physical phenomenon than the fewer broken bonds effect, but our results show clearly that this completely quantum-mechanical mechanism is the dominant factor at work in producing the d-wave instability for the PM system. Strong spatial correlations are not expected since we have paired holes in the outermost shell of $\mathbf{k}$ states, which corresponds to a long correlation length. However, we have demonstrated that such correlations are not crucial for having a large spin correlation energy. This fact also means that the pairing mechanisms is likely to remain effective even when the long-range Coulomb interaction is added to the problem.

This would change the Hamiltonian of Eq. (5), e.g., the coefficient of the final term would become positive. It is necessary to consider this effect, since the usual downward renormalization of the long-range Coulomb pseudopotential in electron-phonon superconductors due to mismatch of time scales will not take place in the present system. If the pairs are not tightly bound in space, as indicated by our results, then the long range repulsion may not greatly suppress the pairing instability.

The interference effect arises as follows. Consider a spin configuration which is AF ordered. Let the up (down) spins belong to the set $\mathbf{R}_i$ ($\mathbf{R}'_i$), where $R_{ix}+R_{iy}=\text{even}$, $R'_{ix}+R'_{iy}=\text{odd}$. The lattice constant is taken equal to one. The weight of such a configuration in the PM superconducting Gutzwiller wavefunction is proportional to

$$|c_1\Gamma_1^{\uparrow}\Gamma_1^{\downarrow}+c_2\Gamma_2^{\uparrow}\Gamma_2^{\downarrow}+c_3\Gamma_3^{\uparrow}\Gamma_3^{\downarrow}+c_4\Gamma_4^{\uparrow}\Gamma_4^{\downarrow}|^2 \qquad (8)$$

where $\Gamma_1^{\uparrow}$ is defined as the Slater determinant whose elements are of the form $\exp(\mathrm{i}\mathbf{k}_j\cdot\mathbf{R}_i)$ with $\mathbf{k}_j$ the set of all $\mathbf{k}$'s in the Fermi sea *except* $\mathbf{k}_1\cdot\Gamma_1^{\downarrow}$ is similar; it contains instead the R'_i and excludes $-\mathbf{k}_1$ instead of $\mathbf{k}_1$. Now, with the spin configuration fixed, compare $\Gamma_1^{\uparrow}$ and $\Gamma_2^{\uparrow}$. They differ only in one column, namely $\Gamma_1^{\uparrow}$ includes a column $\exp(\mathrm{i}\mathbf{k}_2\cdot\mathbf{R}_i)$, while $\Gamma_2^{\uparrow}$ includes a column $\exp(\mathrm{i}\mathbf{k}_1\cdot\mathbf{R}_i)$. The two matrices are otherwise the same. However

$$\mathbf{k}_2=\mathbf{k}_1+\mathbf{Q}, \qquad \mathbf{Q}\xrightarrow[L\to\infty]{}(\pi,\pi).$$

Hence

$$\exp(\mathrm{i}\mathbf{k}_2\cdot\mathbf{R}_i)=\exp(\mathrm{i}\mathbf{k}_1\cdot\mathbf{R}_i)\cdot\exp[\mathrm{i}\pi(R_{ix}+R_{iy})]$$
$$=\exp(i\mathbf{k}_1\cdot\mathbf{R}_i)$$

and $\Gamma_1^{\uparrow}=\Gamma_2^{\uparrow}$. Similarly, $\Gamma_1^{\downarrow}=-\Gamma_2^{\downarrow}$. Therefore, to maximize the amplitude (8), we need $c_1=-c_2$. It is clear that this argument can be extended to c_3 and c_4 and leads directly to the d-wave state. The s-wave $c_1=c_2=c_3=c_4$, on the other hand, gives destructive interference for the AF configuration and is therefore very unfavorable. The normal state wavefunction by definition has no interference; it therefore lies between the s- and d-wave states in energy. The relative amplitudes are in fact 4:2:0 for d-wave:normal:s-wave. In the PM case the configuration $R_{ix}+R_{iy}=\text{odd}$, $R'_{ix}+R'_{iy}=\text{even}$ has the same weight as the above configuration and then $\Gamma_1^{\uparrow}=-\Gamma_2^{\uparrow}$, $\Gamma_1^{\downarrow}=\Gamma_2^{\downarrow}$, etc., the d-wave interference is also constructive. The d-wave state is still a spin singlet and thus this argument, which in itself ensures only that there will be a large gain for the $\langle S_i^z S_j^z\rangle$ energy, applies also to the x and y parts of the spin-spin interaction. When we apply the argument to the AF wavefunctions there is a crucial difference. These are not spin singlets. Even though the d-wave gains energy in the z-component of the interaction, it is permitted to, and does, lose energy in the x- and y-(spin flip) components. The net gain turns out to be zero.

It is evident that a global change in amplitudes is involved in the comparison of the various states. This is not surprising since, in the wavefunctions considered, the holes are delocalized. This is probably a good approximation in spite of the fact that, in the AF lattice with a single hole, formation of a spin polaron is favored.

IV. Discussion

We would now like to compare our results in some detail to previous work. Hirsch [14] has proposed a binding mechanism within a model of Cu – O layers with holes on the 0 sites. His arguments apply also to a simple Hubbard model on a square lattice, and an AF ground state for the half-filled case. In the regime where $J_z > J_x$, J_y, the J's denoting the effective AF couplings he derives a linear attractive potential between holes for short separations based on the observation that the number of broken bonds in the AF lattice is proportional to the separation. This is consistent with our results: we also find binding in the $J_z > J_x$, J_y system, and a concomitant enhancement of the hole-hole correlation. Contrary to Hirsch, [14] however, we find pairing only for d-wave, not s-wave, states. Our results also indicate that the binding energy becomes zero rather than remaining finite for the isotropic $J_x = J_y = J_z$ AF coupling. This is also found by Takahashi [13].

Our results may also be compared to the RVB picture, because, as noted above, the Gutzwiller state is a form of RVB state. We should point out, however, that we found in previous work [19] that the insulating half-filled state did not have a linear dispersion relation in one dimension. If we make the identification of a "spinon" with a change in the **k**-distribution in (2) by a shift of one **k** from below to above the pseudo Fermi surface by an amount $\Delta \mathbf{k}$, then this excitation was found to have an energy proportional to $(\Delta \mathbf{k})^2$. This is in contradiction to the mean-field result of Zou and Anderson [22]. It is important also to keep in mind that excitations greated in this way are not orthogonal to one another or to the ground state.

In the less-than-half-filled case, the normal state considered in this paper is a Fermi liquid with a Fermi surface enclosing $N/2$ **k** values, where N is the number of electrons. It is this state which is unstable to d-wave superconductivity. If, on the other hand, one takes the view that the introduction of holes into the insulating state creates topological solitons but leaves the **k**-distribution fixed, then the number of **k**'s involved in the wavefunction exceeds $N/2$. This is perfectly possible within the expression (3) for the variational wavefunction. It has not yet proved possible to investigate such a function with our numerical technique. With regard to Bose condensation in general, we have concentrated on pairing instabilities, but nothing in our results rules out other (perhaps more exotic) instabilities of the Fermi liquid normal state.

It would be premature to compare our results in a serious way with experiment, since they relate to a model which may well be oversimplified. We only wish to point out that La_2CuO_4 has been found to be antiferromagnetic under some circumstances [28]. Antiferromagnetism and superconductivity never seem to occur in the same sample. This agrees with our finding that the PM, but not the AF, wavefunction has a superconducting instability in the presence of the physically realistic isotropic spin-spin interaction.

The calculations presented here can be extended in several directions. d-wave states of different symmetry should be investigated, as well as more general superconducting wavefunctions in which the momentum distribution is more strongly modified. Particularly interesting would be to discover what effect doping (N/L ratio, in our notation) has on the superconducting instability.

In summary, we have investigated the properties of variational Gutzwiller-type wavefunctions for the Hubbard model on a square lattice, concentrating on the non-degenerate nearly half-filled band. For wavefunctions which have AF long range order no energy is gained by condensation into a superconducting state. If an Ising-like AF coupling is added to the effective Hamiltonian, then d-wave superconductivity is obtained. Paramagnetic wavefunctions are unstable to d-wave superconductive pairing. s-wave superconductivity is unfavorable for all possible normal states. The main physical effect at work is constructive interference between different hole wavefunctions in the d-wave state.

We would like to thank P. Horsch for useful discussions and communication of unpublished results, F.C. Zhang and Y. Takahashi for discussions and H. Yogoyama and H. Shiba for sending us their results in advance of publication. R.J. thanks the Zentrum für Theoretische Studien, ETH Zurich for its hospitality during the course of this work.

References

1. Bednorz, J.G., Müller, K.A.: Z. Phys. B – Condensed Matter 64, 188 (1986)

2. Anderson, P.W.: Science **235**, 1196 (1987)
3. Rice, T.M.: Z. Phys. B – Condensed Matter **67**, 141 (1987)
4. Anderson, P.W., Abrahams, E.: Nature **327**, 363 (1987)
5. Anderson, P.W.: Mater. Res. Bull. **8**, 153 (1973)
6. Baskaran, G., Zou, Z., Anderson, P.W.: (to be published)
7. Ruckenstein, A., Hirschfeld, P., Appel, J.: (to be published)
8. Cyrot, M.: (to be published)
9. Zhang, F.C.: Private communication
10. Miyake, K., Matsuura, T., Jicha, H., Nagaoka, Y.: Prog. Theor. Phys. **72**, 1063 (1984);
. Hirsch, J.: Phys. Rev. Lett. **54**, 1317 (1985)
11. Kivelson, S., Rokhsar, D., Sethna, J.: Phys. Rev. B **35**, 8865 (1987)
12. Zou, S., Anderson, P.W.: (to be published)
13. Takahashi, Y.: (to be published)
14. Hirsch, J.: (to be published)
15. Emery, V.: Phys. Rev. Lett. **58**, 2794 (1987)
16. Horsch, P., Kaplan, T.A.: J. Phys. C **16**, L 1203 (1983)
17. Yokoyama, H., Shiba, H.: (to be published)
18. Shiba, H.: J. Phys. Soc. Jpn. **55**, 2765 (1986)
19. Gros, C., Joynt, R., Rice, T.M.: Phys. Rev. B **36**, 381 (1987)
20. Anderson, P.W., Baskaran, G., Zou, Z., Hsu, T.: (to be published)
21. Lee, P.A., Read, N.: Phys. Rev. Lett. **58**, 2691 (1987)
22. Ohkawa, F.: (to be published)
23. Schulz, H.J.: (to be published)
24. Miyake, K., Schmitt-Rink, S., Varma, C.M.: Phys. Rev. B **34**, 6554 (1986)
25. Scalapino, D.J., Loh, E., Hirsch, J.: Phys. Rev. B **34**, 8190 (1986)
26. Cyrot, M.: Solid State Commun. **60**, 253 (1986)
27. Lavagna, M., Millis, A.J., Lee, P.A.: Phys. Rev. Lett. **58**, 266 (1987)
28. Vaknin, D., Sinha, S.K., Moncton, D.E., Johnston, D.C., Newsam, J.M., Safinya, C.R., King, H.E.: (to be published)

C. Gros, T.M. Rice
Theoretische Physik
Eidgenössische Technische
Hochschule Zürich – ETH
Hönggerberg
CH-8093 Zürich
Switzerland

PHYSICAL REVIEW B VOLUME 42, NUMBER 10 1 OCTOBER 1990

Dynamics of quasiparticles in the two-dimensional Hubbard model

Daniel C. Mattis
Physics Department, University of Utah, Salt Lake City, Utah 84112

Michael Dzierzawa and Xenophon Zotos
Institut für der Kondensierten Materie, Universität Karlsruhe, Physiskhochhaus, Postfach 6980, 7500 Karlsruhe 1, Federal Republic of Germany

(Received 9 July 1990)

The Hubbard model at half-filling is a collective, antiferromagnetic insulator. We study added electrons or holes. The insulating energy gap and the dispersion of the added carriers are calculated variationally in two dimensions with use of a Monte Carlo evaluation of the electronic correlation functions in the insulating phase. Both E_g and m^* are found to be temperature dependent.

INTRODUCTION

For a variety of reasons, attention has recently focused on two-dimensional (2D) interacting electrons. At $N_{el}=N$, the number of sites, the Hubbard model is a collective, antiferromagnetic insulator with energy gap E_g. Away from half-filling there are a number of hypotheses but no consensus as to the nature, number, and dispersion of the carriers in the many-body system,[1] although it is widely conceded that solving this problem is prerequisite to any theory of high-T_c superconductivity. The evidence on the normal phase of CuO_2-based layered metals has been construed to favor a modified Fermi liquid[2] over ordinary band-structure theory.[3] Varma *et al.*[4] have characterized them as "marginal" Fermi liquids. Numerical analysis[5] of the strong-coupling-limit t-J model reveals quasiparticle peaks in the spectrum of added electrons or holes.

This paper analyzes the charge carriers in the Hubbard model by means of correlation functions that are accurately calculable *only* for the insulating half-filled band. When doping introduces a fraction x of charge carriers, we determine that they constitute a spin-$\frac{1}{2}$ Fermi liquid of xN, *not* $(1\pm x)N$, particles, with anisotropic dispersion and an effective mass m^* at the gap edge functionally related to E_g.

THE MODEL

The Hamiltonian for the Hubbard model $\mathcal{H}$ is

$$\mathcal{H}=-t\sum_{ij}(c_{i\sigma}^*c_{j\sigma}+\text{H.c.})+U\sum_i n_{i,+}n_{i,-}-(U/2)N_{el}\,, \quad (1)$$

with electrons hopping from sites i to nearest-neighbor sites j. With the chemical potential μ here set at $U/2$, on a bipartite lattice $\mathcal{H}$ is invariant under charge conjugation (followed by a phase transformation on the odd-numbered sublattice): $c_{r\sigma},c_{r\sigma}^*\to(-)^r c_{r,-\sigma}^*,(-)^r c_{r,-\sigma}$. This invariance creates extra symmetries.[6]

The exact ground state at half-filling is $|\Psi_0\rangle$,[7] its energy E_0. For a particle deleted or added, the energy is $E(\mathbf{k};t,U)\equiv E_0+e(\mathbf{k};t,U)\pm U/2$, its minimum lying on the star of a wave vector $\mathbf{k}_0$, independent of σ. It is known[1,2] that $\mathbf{k}_0$ is $(\pi/2,\pi/2)$ at all $t/U>0$,[5,8] both for deleted and for added particles. It follows that $E_g=U+2e_{del}(\mathbf{k}_0;t,U)$. The stability of the insulating phase is guaranteed by $E_g\geq 0$. However, if E_0 and $E(\mathbf{k};t,U)$, both $O(N)$, are obtained numerically, their differences, including E_g, which are $O(1)$, become difficult—if not impossible—to calculate reliably and other means must be sought.[9,10]

STATES OF ONE PARTICLE ADDED OR DELETED

We define

$$a^*(\mathbf{k},\sigma)\equiv N^{-1/2}\sum_i e^{i\mathbf{k}\cdot\mathbf{R}_i}\sum_j \Gamma_{ij}c_{i\sigma}^*\,,$$

with Γ_{ij} an operator incorporating the reaction of the medium at j to the extra carrier at i. For $N+1$ particles there are $2N$ orthogonal states $a^*(\mathbf{k},\sigma)|\Psi_0\rangle$ with which to estimate the $e(\mathbf{k};t,U)$ directly:

$$e\leq\frac{\langle\Psi_0|a(\mathbf{k},\sigma)[\mathcal{H},a^*(\mathbf{k},\sigma)]|\Psi_0\rangle}{\langle\Psi_0|a(\mathbf{k},\sigma)a^*(\mathbf{k},\sigma)|\Psi_0\rangle}+U/2\,. \quad (2)$$

The better the choice of Γ, the tighter is this variational bound. By successive applications of $(\mathcal{H}-E_0)$ to $c_{\mathbf{k}\sigma}^*|\Psi_0\rangle$, one determines the form of Γ at any given $\mathbf{k},\sigma$:

$$\sum_j\Gamma_{ij}(\mathbf{k},\sigma)=1+g_{0\mathbf{k}}n_{i,-\sigma}+\sum_{j\neq i}g_{1\mathbf{k}}(\mathbf{R}_{ij})(n_{j\uparrow}+n_{j\downarrow})+\cdots\,. \quad (3)$$

Because many-body perturbation theory in U fails about the singular point $U=0$, the g's have to be obtained variationally. But already the leading approximation $\Gamma=1$ yields sensible results, such as locating the minimum at $\mathbf{k}_0=(\pm\pi/2,\pm\pi/2)$.[5,11] The next term ($\propto g_{0\mathbf{k}}$) considerably improves the calculated energies without substantial complication. Higher-order terms which parametrize charge and spin polarization[12] of the environment assume increasing importance at large $U\gtrsim 8t$. But they also greatly complicate the programs, and we are forced to defer them for future investigations. In the present work, we opt for a simple operator designed for accuracy in the little-studied weak-coupling regime, $U<8t$; for *added* particles we choose

$$\Gamma_{ij}(\mathbf{k},\sigma)=[U_{\mathbf{k}}n_{r,-\sigma}+v_{\mathbf{k}}(1-n_{r,-\sigma})]\delta_{ij}\,.$$

By the aforementioned symmetry[6] ($\mathbf{k}\to-\mathbf{k}+\pi$, $\sigma\to-\sigma$, and $c^*\leftrightarrow c$), this also yields the operator for *deleted* particles.

At each $\mathbf{k}$ we determine $u_{\mathbf{k}}$ and $v_{\mathbf{k}}$ by optimizing $e(\mathbf{k})$. Using these, we also construct the Wannier operators of the composite particle. Their effective radius is found to be quite small.[13]

The numerator and denominator in (2) are each sums of correlation functions in the half-filled ground state of $\mathcal{H}$, and are calculated by quantum Monte Carlo.[14] Identities which keep our calculations to a manageable minimum include[7]

$$(\Psi_0|n_{i,-\sigma}c_{i\sigma}^*c_{j\sigma}|\Psi_0)\equiv\tfrac{1}{2}(\Psi_0|c_{i\sigma}^*c_{j\sigma}|\Psi_0)\neq0$$

for i,j on different sublattices, and $(\Psi_0|c_{i\sigma}^*c_{j\sigma}|\Psi_0)\equiv\frac{1}{2}\delta_{ij}$ if i,j are on the same sublattice. Our studies centered on the range of parameters $0<U/t\le8$, where neither the t-J model nor perturbation theory (the energies are not analytic in U at $U=0$) are applicable. To keep the systematic error as small as possible, we used different imaginary-time discretizations $\Delta\tau$, extrapolating the results to $\Delta\tau=0$. Simulations are on 8×8 lattices at $\beta=6/t$. At small U we verified that the β dependence was negligible by simulating also at $\beta=9/t$. Figure 1 (bottom curve) shows $e_{0,\mathrm{del}}(t,1)$, i.e., $e_{\mathrm{del}}(\mathbf{k}_0;t,U)$ at $\mathbf{k}_0=(\pm\pi/2,\pm\pi/2)$ and $U=1$, calculated at $t/U=0$ (where $e_{0,\mathrm{del}}\to0$), $\frac{1}{6}$, $\frac{1}{5}$, $\frac{1}{4}$, $\frac{1}{2}$, 1, and ∞ (where $e_{0,\mathrm{del}}/U\to-0.5$), and fitted to a smooth polynomial curve; accuracy is indicated by error bars. At $t/U<\frac{1}{8}$ our curve joins smoothly with the strong-coupling results of Dagotto *et al.*:[5] $e_{0,t\text{-}J}=-3.17t+2.83t(J/t)^{0.73}$ (dashed curve), with $J=4t^2/U$. While the $t-J$ model is closely related to the Hubbard model in strong coupling, it is not identical, and the upturn at $t/U>\frac{1}{8}$ in the dashed curve makes it clear that the two models differ substantially in weak coupling.

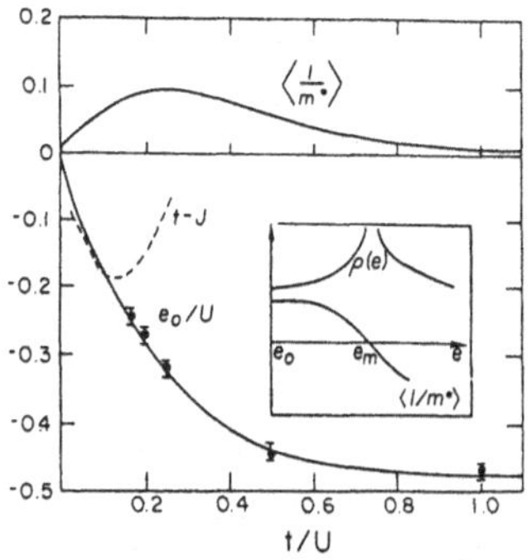

FIG. 1. Lower curve: $e_0(t,U)/U$ (e_0 plotted here is minimum energy to *delete* particle from half-filled band) vs t/U. Calculated points are indicated by error bars, curve is best polynomial fit. Dashed line marked t-J is from Dagotto *et al.* (Ref. 5). Upper curve: $\langle 1/m^*\rangle$ at $\mathbf{k}_0$, in units of $(1/m_0)(U/4\times13.6\ \mathrm{eV})\,(a/a_0)^2$ vs t/U as calculated from $e_0(t,U)$ using Eq. (5). (m_0 is the free-electron mass, U is in eV, and a/a_0 is the lattice parameter in units of the Bohr radius, 0.529 Å.) Inset: Schematic plots of $\langle 1/m^*\rangle$ and the density of states $\rho(e)$ (at fixed t,U) as functions of energy measured from e_0, demonstrating that $\langle 1/m^*\rangle$ changes sign at e_m where ρ has its maximum (Ref. 15).

MANY-PARTICLE STATES

Consider at first the *scattering states* of two added particles,

$$|\mathbf{k},\sigma;\mathbf{k}',\sigma')=a^*(\mathbf{k},\sigma)a^*(\mathbf{k}',\sigma')|\Psi_0)+N^{-1}\sum_{\mathbf{q}\sigma''\sigma'''}\mathcal{L}_{\sigma''\sigma'''}(\mathbf{q})a^*(\mathbf{k}+\mathbf{q},\sigma'')a^*(\mathbf{k}'-\mathbf{q},\sigma''')|\Psi_0)\,. \qquad (4)$$

As usual in scattering theory, the energy of the pair is additive: $e(\mathbf{k})+e(\mathbf{k}')+2(U/2)+O(1/N)$. But what is their statistics? If the Γ's were in the form of long strings, these could become entangled in 2D and lead to parastatistics (exotic representations of the braid group) under interchange of the two members of this pair. But in the present context, our variationally determined Γ's have a small radius.[13] Therefore, one readily forms wave packets such that the two particles are well separated in space and devoid of interaction with each other. From this it follows that the α^*'s anticommute. If the medium were spin polarized about each carrier, i.e., if the Γ's created a "Nagaoka bubble" of spin S about each carrier, the α^*'s would be creation operators for exotic fermions of spin $S\pm\frac{1}{2}$. Again, as our choice of Γ's carries only $S=0$ (as is appropriate in the weak-coupling regime), our fermions have the usual spin $\frac{1}{2}$.

Extending the arguments to a dilute fraction x of added or deleted particles, one concludes that they behave as an ordinary spin-$\frac{1}{2}$ Fermi liquid of xN charge carriers superposed onto a background of N non-current-carrying electrons. Their unusual dispersion is detailed below. Even if the residual interparticle forces were to favor bound states over the scattering states (4), as is claimed in some small-cluster calculations,[10] and this resulted in the many-body ground state of the presumed Fermi liquid becoming unstable against some condensed phase, the statistics of the particles under interchange and their dispersion would be required ingredients in constructing the new ground state.

DISPERSION OF QUASIPARTICLES

Figure 2 shows the calculated dispersion in $e(\mathbf{k};1,U)$ along two orthogonal axes centered at $\mathbf{k}_0$. The extreme anisotropy diminishes somewhat with increasing U. With increasing carrier concentration, the Fermi surface follows disjoint contours surrounding each $\mathbf{k}_0$, portions of which are show in Fig. 2 (inset). Such a topology was earlier conjectured by Lee.[2] The density of states $\rho(e)$ starts at a finite value ρ_0 ($\rho_0\propto m^*$) at e_0, and increases with in-

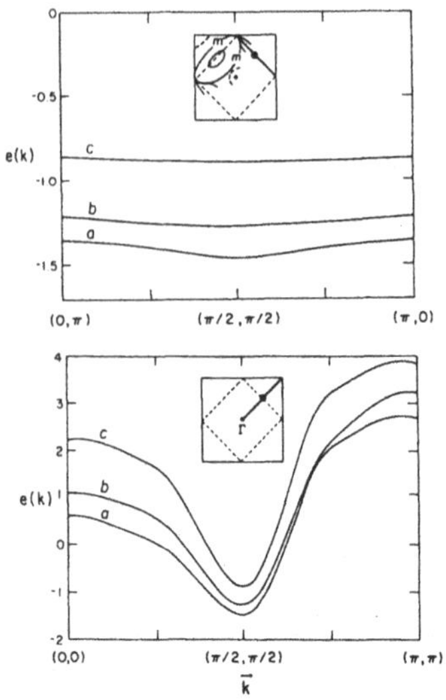

FIG. 2. Curves labeled a are for $U/t=6$, b are for $U/t=43$, and c are for $U/t=2$, respectively. Top: $e_{\rm del}(\mathbf{k})/t$ (energy to delete particle at $\mathbf{k}$ from half-filled band) plotted along perimeter of inscribed diamond. Inset: $\mathbf{k}_0=(\pi/2,\pi/2)$ shown as big dot. Portions of constant energy contours are shown; curves labeled m are the locus of e_m (cf. inset of Fig. 1) (Ref. 15). Bottom: $e_{\rm del}(\mathbf{k})/t$ along (1,1) direction. Note the anisotropy: m^* in this direction is smaller than along the perimeter by 1 order of magnitude.

creasing energy until, at e_m, the four contours merge. Their perimeter is then at a maximum, as is ρ.[15] For $e>e_m$, both $d\rho/de$ and $\langle 1/m^*\rangle$ become negative—shown schematically in Fig. 1 (inset). (By way of comparison, both in Hartree-Fock and in band-structure theory, where the many-body correlations are treated in an average way, e_m merges with e_0 and the low-energy dynamics are given incorrectly.[3])

THE EFFECTIVE MASS

We estimate the average inverse mass $\langle 1/m^*\rangle$ directly as a function of $\mathbf{k}$, t, U, and T by combining Galilean invariance with Feynman's theorem:[16]

$$\langle 1/m^*\rangle(\mathbf{k}) = -(t/2)\partial e(\mathbf{k};t,U)/\partial t\,. \quad (5)$$

$\langle 1/m^*\rangle(\mathbf{k}_0)$ as a function of t/U, plotted in the upper curve of Fig. 1, is found by differentiating the lower curve. Our results demonstrate that a correlated system can produce truly "heavy,"[17] yet delocalized, fermions at all U.

The addition or deletion of each particle affects not just the total energy of the medium, but its entropy as well. With $f(\mathbf{k})$ the excess free energy of an excess particle at $\mathbf{k}$, the $T\neq 0$ generalization of (5) is[16]

$$\langle 1/m^*\rangle(\mathbf{k}) = -(t/2)\partial f(\mathbf{k})/\partial t - (\beta/2)[\langle(\mathbf{v}-\langle\mathbf{v}\rangle)^2\rangle - \langle(\mathbf{v}-\langle\mathbf{v}\rangle_0)^2\rangle_0]\,, \quad (6)$$

where $\beta=1/kT$, $\mathbf{v}=(v_x,v_y)$ with $v_x=2t\sum_{k,\sigma}\sin k_x \times c^*(\mathbf{k},\sigma)c(\mathbf{k},\sigma)$, and $\langle\cdots\rangle$ indicates thermal average in the presence of the extra particle, while $\langle\cdots\rangle_0$ refers to such an average in the half-filled state.

CONCLUSION

By combining a variational ansatz with quantum Monte Carlo techniques, we have estimated the properties of low-lying eigenstates of a small number of extra holes or electrons introduced into the Mott insulating phase of the two-dimensional Hubbard model. In weak coupling, we conclude that they constitute a dilute fluid of spin-$\frac{1}{2}$ fermions with heavy, anisotropic, temperature- and energy-dependent masses.

ACKNOWLEDGMENTS

The authors thank the ISI (Torino) and Professor Mario Rasetti for his hospitality during the early stages of this work. Support was provided by U.S. Army Electronics Technology and Devices Laboratory, SLECT-E, Ft. Monmouth, New Jersey (D.C.M.), and by the Bundesministerium für Forschung und Technologie and Esprit Program No. 3041 (M.D. and X.Z.). Publication of this work was supported by ONR Grant No. N00014-86K-0710.

[1]For a perspective, see *High-Temperature Superconductivity*, Proceedings of the Los Alamos Symposium, 1989, edited by K. S. Bedell *et al.* (Addison-Wesley, Redwood, CA, 1990).

[2]G. A. Sawatzky, Nature (London) **342**, 480 (1989); P. E. Sulewski *et al.*, Phys. Rev. B **36**, 2357 (1987); also see review by P. A. Lee, in *High-Temperature Superconductivity* (Ref. 1), p. 96.

[3]In addition to some other inadequacies, band-structure theory yields the incorrect sign of the Hall constant; see review by N. P. Ong, in *Physical Properties of High-Temperature Superconductors II*, edited by D. M. Ginsberg (World Scientific, Singapore, 1990).

[4]C. M. Varma, P. B. Littlewood, S. Schmitt-Rink, E. Abrahams, and A. E. Ruckenstein, Phys. Rev. Lett. **63**, 1996 (1989).

[5]S. A. Trugman, Phys. Rev. B **41**, 892 (1990); E. Dagotto *et al.*, *ibid.* **41**, 2585 (1990); with $J\approx 4t^2/U$, the t-J model is believed to mimic the strong-coupling limit ($t\ll U$) of the Hubbard model.

[6]E. H. Lieb, Phys. Rev. Lett. **62**, 1201 (1989).

[7]$|\Psi_0\rangle$ is invariant under the point group of the lattice, translations, and charge conjugation. Its total momentum, angular momentum, and spin is zero for $N=$even. For details, see

Ref. 6.

[8]Although at $U=0$ the locus of $\mathbf{k}_0$ in the Brillouin zone is the inscribed diamond given by $(\cos k_x + \cos k_y)=0$, this degeneracy is lifted at all $U\neq 0$. See, for example, P. A. Lee, in *High-Temperature Superconductivity* (Ref. 1).

[9]The calculational error in each E is $O(1)$ for exact diagonalization but $O(\sqrt{N})$ if Monte Carlo is used. In the former case, the uncertainty in the energy differences $e(\mathbf{k})$ is serious and in the latter, fatal. Our procedure avoids this pitfall, as do some other recent schemes (see Ref. 10).

[10]E. Dagotto *et al.*, Phys. Rev. B **41**, 811 (1990), find a two-hole bound state with small binding energy on 4×4 clusters. (Note that such binding might not be sturdy against longer-range Coulomb forces.)

[11]According to Trugman (Ref. 5) at $t=U/8$ the fractional spectral intensity of the quasiparticle peak is only 0.358 (in the state which corresponds to $\Gamma=1$ in our notation.) Thus, quasiparticle eigenstates must include local-density and antiferromagnetic (Ref. 12) correlations, i.e., have $\Gamma\neq 1$.

[12]J. R. Schrieffer, X. G. Wen, and S. C. Zhang, Phys. Rev. Lett. **60**, 944 (1988); Z. Y. Weng, C. S. Ting, and T. K. Lee, Phys. Rev. B **41**, 1990 (1990).

[13]The Wannier operators are lattice Fourier transforms of the $a^*(\mathbf{k},\sigma)$. The Fourier transform $u(\mathbf{R}_{ij})$ of $u_{\mathbf{k}}/(\Psi_0|a(\mathbf{k},\sigma)\times a^*(\mathbf{k},\sigma)|\Psi_0)^{1/2}$, and the Fourier transform $v(\mathbf{R}_{ij})$ of $v_{\mathbf{k}}/(\Psi_0|a(\mathbf{k},\sigma)a^*(\mathbf{k},\sigma)|\Psi_0)^{1/2}$ have been computed by us. At $U=0$, the Wannier functions are trivially pointlike: $u(\mathbf{R})=v(\mathbf{R})=\delta(\mathbf{R})$. But even at $U/t=6$, where the correlations are consequential and $v(\mathbf{0})\approx 2u(\mathbf{0})$, the functions $v(\mathbf{R})$ and $u(\mathbf{R})$ decrease rapidly with distance with alternating signs, the effective radius being no more than a few interatomic spacings.

[14]Following R. Blankenbecler, D. J. Scalapino, and R. L. Sugar, Phys. Rev. D **24**, 2278 (1981), as extended to 2D by J. E. Hirsch, Phys. Rev. B **31**, 4403 (1985), and later by S. R. White *et al.*, *ibid.* **40**, 506 (1989).

[15]We compute $(e_m - e_0)/t$ to be 0 (at $U=0$), 0.03 (at $U=2t$, curves c in Figs. 2) 0.06 (at $U=4t$, curves b), and 0.12 (at $U=6t$, curves a).

[16]D. C. Mattis (unpublished).

[17]Z. Fisk *et al.*, Science **239**, 33 (1988).

PHYSICAL REVIEW B VOLUME 37, NUMBER 13 1 MAY 1988

Pairing in the two-dimensional Hubbard model: An exact diagonalization study

H. Q. Lin* and J. E. Hirsch
Department of Physics, University of California, San Diego, La Jolla, California 92093

D. J. Scalapino
Department of Physics, University of California, Santa Barbara, Santa Barbara, California 93106
(Received 24 August 1987)

We have studied the pair susceptibilities for all possible pair wave functions that fit on a two-dimensional (2D) eight-site Hubbard cluster by exact diagonalization of the Hamiltonian. Band fillings corresponding to four and six electrons were studied (two or four holes in the half-filled band) for a wide range of Hubbard interaction strengths and temperatures. Our results show that all pairing susceptibilities are suppressed by the Hubbard repulsion. We have also carried out perturbation-theory calculations which show that the leading-order U^2 contributions to the d-wave pair susceptibility suppresses d-wave pairing over a significant temperature range. These results are consistent with recent Monte Carlo results and provide further evidence suggesting that the 2D Hubbard model does not exhibit superconductivity.

I. INTRODUCTION

The question of whether the two-dimensional repulsive Hubbard model exhibits superconductivity is of great current interest, in view of the fact that it is one of the possible models to describe the recently discovered high-T_c oxide superconductors.[1] Although a variety of approximate calculations predict superconductivity in this model,[1-11] the need for calculations that do not rely on uncontrolled approximations clearly exists. In this paper we discuss results obtained from an exact calculation on eight-site clusters for a variety of interaction strengths and temperatures and two values of the band filling: $\rho=0.5$ and 0.75, corresponding to four and six electrons (i.e., a half-filled band with two and four holes, respectively). In another paper, results of Monte Carlo simulations of the two-dimensional Hubbard model are discussed.[12] The simulation approach does not reach as low a temperature or as large an interaction strength as the present study but applies to considerably larger clusters. Thus these calculations are complementary.

Most theoretical approaches have suggested extended s-wave or d-wave singlet pairing for the repulsive Hubbard model, involving pairs formed by nearest-neighbor electrons of opposite spins. In the present calculation we have considered all possible pairing of electrons that fit onto an eight-site lattice. We thus obtain an 8×8 pairing matrix, and the eigenvector associated with the largest eigenvalue of this matrix describes the most favorable pairing state. We find that usually the most favorable pairing involves a mixture of states with d- and s-wave symmetry[13] and that contributions from further than nearest-neighbor pairs can be significant. Most importantly, however, we find that the susceptibility associated with this and all other pairing is *suppressed* by the Hubbard interaction.

To shed further light on this question, we compute in perturbation theory the leading (U^2) contribution to the d-wave pair susceptibility, which was found to be the most favorable state near the half-filled band case in previous calculations.[3,11] We find the leading U^2 contribution to *suppress* d-wave pairing over a significant temperature range, consistent with our numerical results on small clusters. The results discussed here, together with Monte Carlo simulation results of Ref. 12, suggest that the two-dimensional repulsive Hubbard model does not exhibit superconductivity.

In Sec. II we describe the cluster formalism. We present our numerical results in Sec. III, perturbation theory results in Sec. IV, and conclude with a short discussion in Sec. V.

II. CLUSTER FORMALISM

The Hubbard model on a two-dimensional lattice is defined by the Hamiltonian

$$H=\sum_{i,j,\sigma} t_{ij}C_{i\sigma}^{\dagger}C_{j\sigma}+U\sum_{i} n_{i\uparrow}n_{i\downarrow}-\mu\sum_{i,\sigma} n_{i\sigma} \quad (1)$$

$$=\sum_{\mathbf{k},\sigma}\varepsilon_{\mathbf{k}}C_{\mathbf{k}\sigma}^{\dagger}C_{\mathbf{k}\sigma}+\frac{U}{N}\sum_{\mathbf{k}_1,\mathbf{k}_2,\mathbf{k}_3} C_{\mathbf{k}_1\uparrow}^{\dagger}C_{\mathbf{k}_2\uparrow}C_{\mathbf{k}_3\downarrow}^{\dagger}C_{\mathbf{k}_1-\mathbf{k}_3-\mathbf{k}_2\downarrow}\ , \quad (2)$$

where

$$\varepsilon_{\mathbf{k}}=-2t_x\cos k_x-2t_y\cos k_y+4t_2\cos k_x\cos k_y-\mu\ . \quad (3)$$

In Eq. (1), $C_{i\sigma}^{\dagger}$ ($C_{i\sigma}$) is the creation (annihilation) operator for spin $\sigma=\uparrow\downarrow$ at site i, and $C_{\mathbf{k}\sigma}^{\dagger}$ ($C_{\mathbf{k}\sigma}$) is its Fourier component. μ is the chemical potential, and t_{ij} is the hopping term. Here $t_{ij}=t_x$ for (i,j) nearest neighbors in the x direction, $t_{ij}=t_y$ for (i,j) nearest neighbors in the y direction, and $t_{ij}=t_2$ for (i,j) next-nearest neighbors.

We study this model on two different two-dimensional lattices of $N=8$ sites, as shown in Fig. 1, with periodic

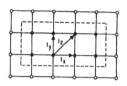

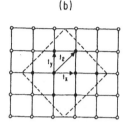

FIG. 1. Two two-dimensional eight-site lattices studied in this paper: (a) $L_x \times L_y = 4\times 2$, and (b) a tilted square. t_x and t_y are the nearest-neighbor hoppings in x and y direction, and t_2 is the next-nearest-neighbor hopping.

boundary conditions. What we are investigating is the possibility of the appearance of superconductivity due to the Coulomb repulsion energy U, as suggested by several authors.[1–11] We define a complete set of pairing operators

$$\Delta_{\mathbf{a}} = \frac{1}{\sqrt{N}} \sum_{\mathbf{r}} C_{\mathbf{r}+\mathbf{a}\uparrow} C_{\mathbf{r}\downarrow} = \frac{1}{\sqrt{N}} \sum_{\mathbf{k}} e^{i\mathbf{k}\cdot\mathbf{a}} C_{\mathbf{k}\uparrow} C_{-\mathbf{k}\downarrow} \ . \qquad (4)$$

where $\mathbf{a}$ is a displacement vector on the lattice. Using a linear combination of $\Delta_{\mathbf{a}}$'s, pair operators of various symmetries can be constructed. For example, operators with extended s-, d-, and p-wave symmetry involving only nearest-neighbor electrons have the form

$$\Delta_{s^*} = \Delta_{-\hat{x}} + \Delta_{\hat{x}} + \Delta_{-\hat{y}} + \Delta_{\hat{y}} \ , \qquad (5)$$

$$\Delta_{d} = \Delta_{-\hat{x}} + \Delta_{\hat{x}} - \Delta_{-\hat{y}} - \Delta_{\hat{y}} \ , \qquad (6)$$

and

$$\Delta_{p_x} = \Delta_{-\hat{x}} - \Delta_{\hat{x}} \ , \qquad (7a)$$

$$\Delta_{p_y} = \Delta_{-\hat{y}} - \Delta_{\hat{y}} \ , \qquad (7b)$$

respectively. We diagonalize an $N\times N$ pairing susceptibility matrix, defined by,

$$P_{\mathbf{ab}} = \int_0^{\beta} d\tau \langle \Delta_{\mathbf{a}}(\tau)\Delta_{\mathbf{b}}^{\dagger}(0)\rangle \qquad (8)$$

and study the behavior of its eigenvalues and eigenvectors as functions of the temperature and the Coulomb repulsion U. If superconductivity does exist in this model, the largest eigenvalue of $P_{\mathbf{ab}}$, corresponding to the susceptibility of the pairs described by its associated eigenvector, will diverge as the temperature goes to zero.

We calculate $P_{\mathbf{ab}}$ by diagonalizing the Hamiltonian and obtaining all the eigenvalues of H. For the eight-site Hubbard model the total Hilbert space is $4^8 = 65\,536$. Fortunately, we can use symmetries to reduce the dimensionality of matrices to be diagonalized. Three symmetries are used—the total particle number N_e, the total spin in the z direction S_z, and translational invariance. The biggest matrix in our calculation after using these symmetries is 628×628. Since we are interested in the low-temperature properties of $P_{\mathbf{ab}}$ we used a canonical ensemble. The pairing operator connects the (N_e-2)-particle space, as well as the (N_e+2)-particle space, to the N_e-particle space so the chemical potential μ which enters Eq. (1) was set by

$$\mu = \tfrac{1}{4}[E_0(N_e+2) - E_0(N_e-2)] \ . \qquad (9)$$

Here $E_0(N)$ is the ground-state energy of N electrons. When $\beta\Delta \gg 1$, where

$$\Delta = E_0(N_e+2) - E_0(N_e) - 2\mu$$
$$= E_0(N_e-2) - E_0(N_e) + 2\mu \ ,$$

our calculation is essentially equivalent to the usual grand canonical one. For the noninteracting case ($U=0$), the difference between the canonical ensemble and grand canonical ensemble is less than 2% when $\beta \geq 8$.

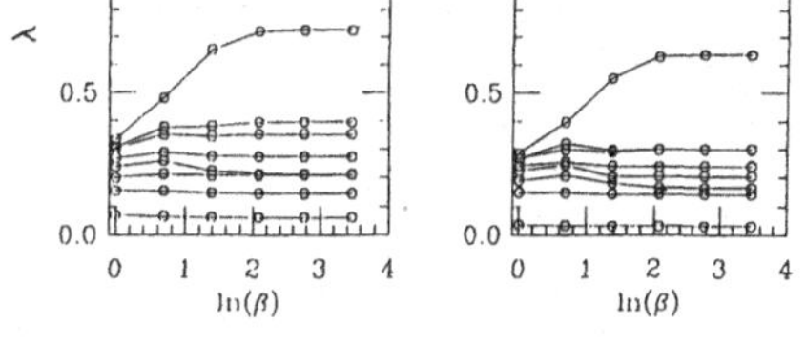

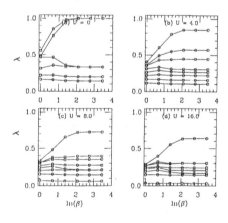

FIG. 2. All eigenvalues vs logarithm of the temperatures on a type-I (4×2) lattice with $N_e=4$, $t_x=1.0$, $t_y=0.25$, and $t_2=0$ for (a) $U=0$, (b) $U=4$, (c) $U=8$, and (d) $U=16$.

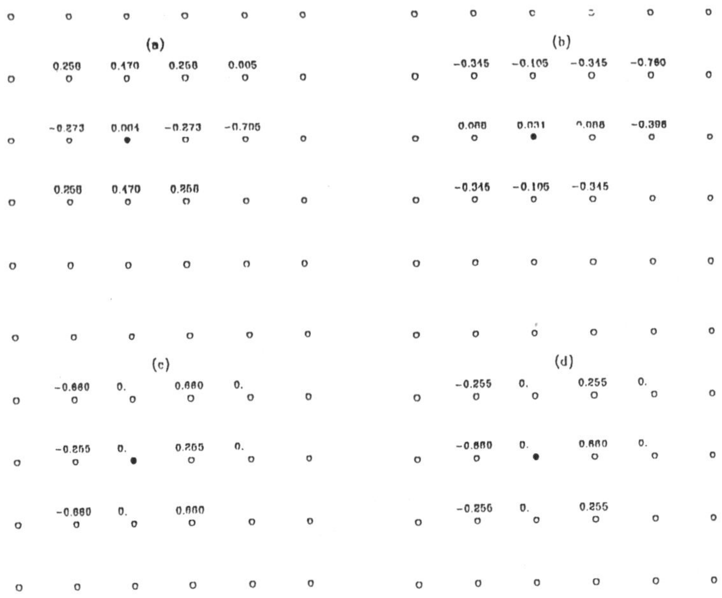

FIG. 3. Pairing structures of (a) the largest eigenvalue, (b) the second-largest eigenvalue, (c) the third-largest eigenvalue, and (d) the sixth-largest eigenvalue on a type-I (4×2) lattice with $N_e=4$, $t_x=1.0$, $t_y=0.25$, and $t_2=0$ for $U=16$, $\beta=32$.

III. NUMERICAL RESULTS

In this section we present numerical results for the eigenvalues of the pairing susceptibility matrix P_{ab} as functions of temperature T for various values of U. The pairing susceptibility matrix is positive definite, and its eigenvalues and eigenvectors depend on band filling and hoppings as well. We have concentrated on four cases, in the first three cases we use the type-I lattice (2×4), and in the last case we use the type-II lattice (tilted eight-site). Then all the eigenvalues λ are plotted versus $\ln(\beta)$ for given values of U (Figs. 2, 4, 6, and 8). The solid lines in the figures connect eigenvalues corresponding to the same eigenvector structure. Figures 3, 5, 7, and 9 show the structure of various eigenvectors. In the following we discuss each of the four cases.

Case (i)

$$N_e=4,\quad t_x=1.0,\quad t_y=0.25,\quad t_2=0 \ .$$

In case (i) the largest eigenvalue first increases and then saturates as the temperature decreases to zero. The largest eigenvalue is a decreasing function of increasing Coulomb repulsion U at low temperature, and so are the other eigenvalues, as shown in Fig. 2. Four eigenvectors are plotted in Fig. 3 for $U=16$, $\beta=32$. The eigenvectors

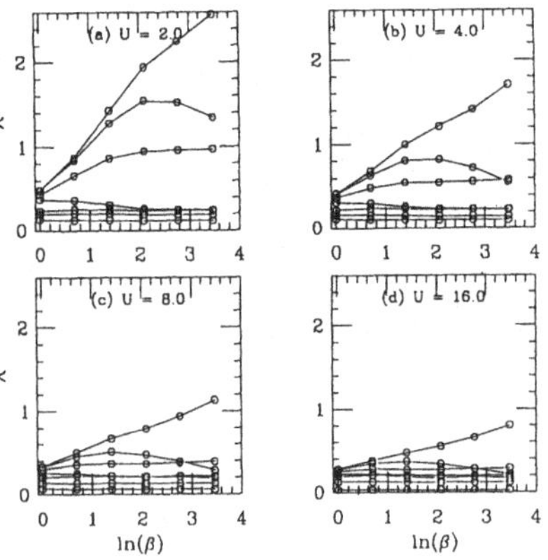

FIG. 4. All eigenvalues vs logarithm of the temperatures on a type-I (4×2) lattice with $N_e=4$, $t_x=1.0$, $t_y=0.50$, and $t_2=0$ for (a) $U=2$, (b) $U=4$, (c) $U=8$, and (d) $U=16$. There are two identical eigenvalues in this case.

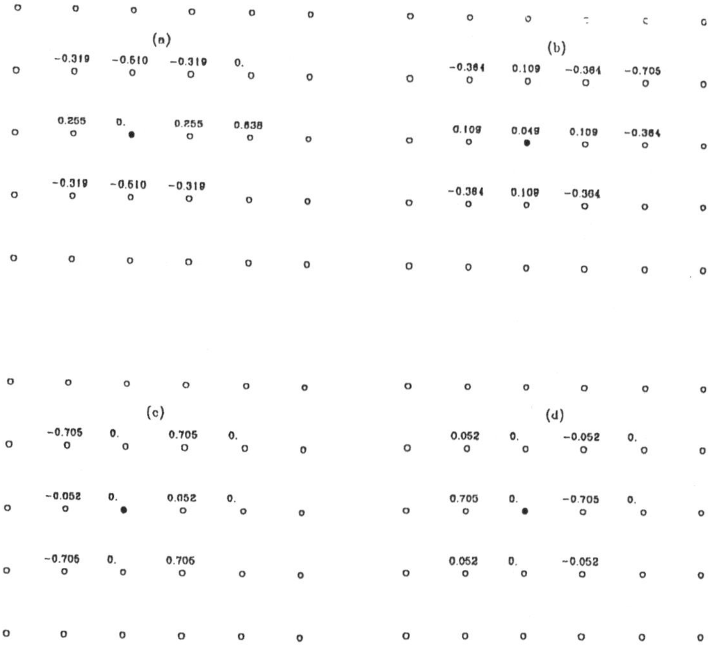

FIG. 5. Pairing structure of (a) the largest eigenvalue, (b) the second-largest eigenvalue, (c) the third-largest eigenvalue, and (d) the sixth-largest eigenvalue on a type-I (4×2) lattice with $N_e=4$, $t_x=1.0$, $t_y=0.50$, and $t_2=0$ for $U=16$, $\beta=32$.

of the first- and second-largest eigenvalues shown in Figs. 2(a) and 2(b) have mixed *s*- and *d*-wave symmetry. The eigenvectors of the third- and sixth-largest eigenvalue shown in Figs. 2(c) and 2(d) are *p*-wave-like. Note that the amplitude at the origin is always very small, due to the large on-site repulsion and that one also gets appreciable amplitudes beyond nearest neighbors. For the noninteracting case, the amplitudes at all sites are found to be identical for the eigenvector with the largest eigenvalue.

Case (ii)

$$N_e=4,\quad t_x=1.0,\quad t_y=0.50,\quad t_2=0\ .$$

This case differs from case (i) since the largest eigenvalue does not saturate as the temperature goes to zero. This is due to the degeneracy at the Fermi energy when $U=0$. Such unphysical degeneracy comes from the finiteness of the small cluster and can be avoided if one turns on the next-nearest-neighbor hopping t_2. For $t_2\neq0$, we find that the largest eigenvalue saturates as $T\to0$ (we do not show the results here). Here again all eigenvalues are suppressed by U, as one can see from Fig. 4. The pairing structure corresponding to the largest eigenvalue Fig. 5(a) again has mixed *s*-*d* symmetry while the structure of the second-largest eigenvalue is an ex-

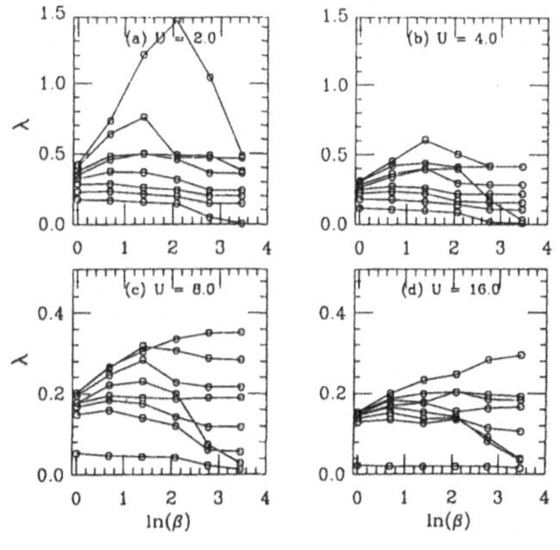

FIG. 6. All eigenvalues vs logarithm of the temperatures on a type-I (4×2) lattice with $N_e=6$, $t_x=1.0$, $t_y=0.25$, and $t_2=0$ for (a) $U=2$, (b) $U=4$, (c) $U=8$, and (d) $U=16$. Note the different scales used for (a), (b), and (c), (d).

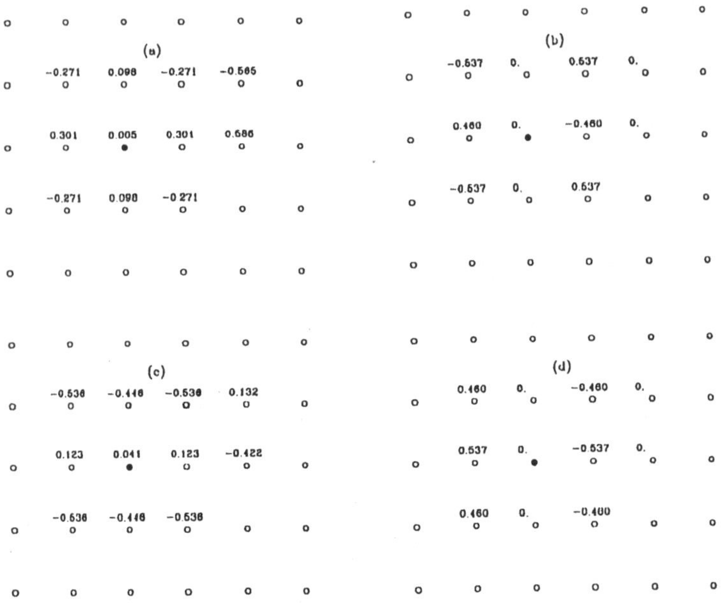

FIG. 7. Pairing structures of (a) the largest eigenvalue, (b) the second-largest eigenvalue, (c) the third-largest eigenvalue, and (d) the seventh-largest eigenvalue on a type-I (4×2) lattice with $N_e=6$, $t_x=1.0$, $t_y=0.25$, and $t_2=0$ for $U=16$, $\beta=32$.

tended s wave [Fig. 5(b)]. The structure of the eigenvectors of the third- and sixth-largest eigenvalues are also shown.

Case (iii)

$$N_e=6,\quad t_x=1.0,\quad t_y=0.25,\quad t_2=0\ .$$

Here the band filling is $\rho=0.75$, and the results for the eigenvalues and eigenvectors are shown in Figs. 6 and 7. For small U, some of the eigenvalues show a peak and do not vary with temperature monotonically. The eigenvector of the largest eigenvalue looks like an extended s wave and that of the second-largest eigenvalue has p_x-wave symmetry. d-wave pairing is now the third-largest eigenvalue. Note that the eigenvalues are decreasing functions of U as before.

The final case we will consider corresponds to a type-II lattice (tilted eight-site):

Case (iv)

$$N_e=4,\quad t_x=t_y=1.0,\quad t_2=0.125\ .$$

As previously discussed, we have set $t_2\neq0$ in order to remove the degeneracy at the Fermi surface when $U=0$. The structures of the pairing eigenvectors shown in Fig. 9 for $U=16$ and $\beta=32$ are simpler than in the other three cases. The eigenvalues show peaks at intermediate temperatures, and the largest eigenvalue drops as the temper-

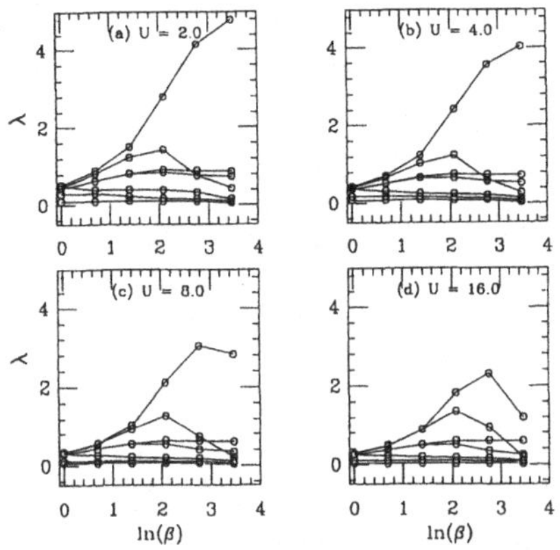

FIG. 8. All eigenvalues vs logarithm of the temperatures for the type-II (tilted square) lattice with $N_e=4$, $t_x=t_y=1.0$, and $t_2=0.125$ for (a) $U=2$, (b) $U=4$, (c) $U=8$, and (d) $U=16$. There are two identical eigenvalues in this case.

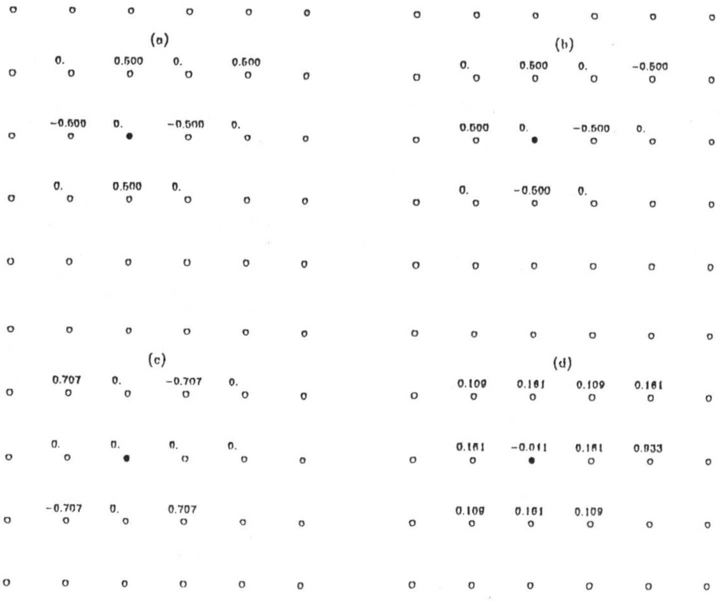

FIG. 9. Pairing structures of (a) the largest eigenvalue, (b) the second- and third-largest eigenvalues, (c) the fourth-largest eigenvalue, and (d) the fifth-largest eigenvalue for the type-II (tilted square) lattice with $N_e=4$, $t_x=t_y=1.0$, and $t_2=0.125$ for $U=16, \beta=32$.

ature approaches zero (Fig. 8). As seen from Fig. 9, the eigenvector of the largest eigenvalue is a d wave, and the eigenvectors of the second- and third-largest eigenvalues are p waves involving only nearest-neighbor sites. The eigenvector of the fourth-largest eigenvalue is d wave involving next-nearest-neighbor sites only, and the rest are extended s waves. Just as in the previous three cases, the largest eigenvalue decreases as U increases. These results clearly show that the Coulomb repulsion U suppresses the pairing susceptibilities of the most stable pairing structure regardless of what kind of symmetry it has.

As a further check on our procedure, we have performed the calculation for negative values of U and always found strong enhancement of the susceptibility corresponding to the largest eigenvalue compared with the noninteracting case, as expected. An example is shown in Fig. 10, for case (i), and the structure of the pairing eigenvectors is shown in Fig. 11.

IV. PERTURBATION THEORY

We have also investigated the various pairing susceptibilities using perturbation theory. As discussed in Ref. 14, perturbation theory can provide useful insight at higher temperatures and moderate U values.

In particular, here we are interested in determining whether the leading contributions of the interaction to the pair susceptibility enhance or reduce it. Consider, for example, the d-wave pair susceptibility

$$P_d=\int_0^\beta d\tau\langle \Delta_d(\tau)\Delta_d^\dagger(0)\rangle \tag{10}$$

with Δ_d given by Eq. (6) of Sec. II. Fourier transforming the site-operator expression gives

$$\Delta_d=\sum_p g(p)C_{p\uparrow}C_{-p\downarrow} \tag{11}$$

with the form factor

$$g(p)=\cos p_x-\cos p_y \ . \tag{12}$$

For a noninteracting system

$$P_d^{(0)}=\frac{T}{N}\sum_{p,n}\frac{g^2(p)}{\omega_n^2+\varepsilon_p^2}$$

$$=\frac{1}{N}\sum_p\frac{\tanh(\beta\varepsilon_p/2)}{2\varepsilon_p}(\cos^2 p_x+\cos^2 p_y)\ . \tag{13}$$

If $t_2=\mu=0$ one finds

$$P_d^{(0)}=\frac{1}{4\pi^2 t}\left[\ln^2\left[\frac{2t}{T}\right]+2\ln\left[\frac{16\gamma}{\pi}\right]\ln\left[\frac{2t}{T}\right]+c\right] \tag{14}$$

with $c=2.13$. Here, the $\ln^2(2t/T)$ term arises from the overlap of the Van Hove singularity in the density of

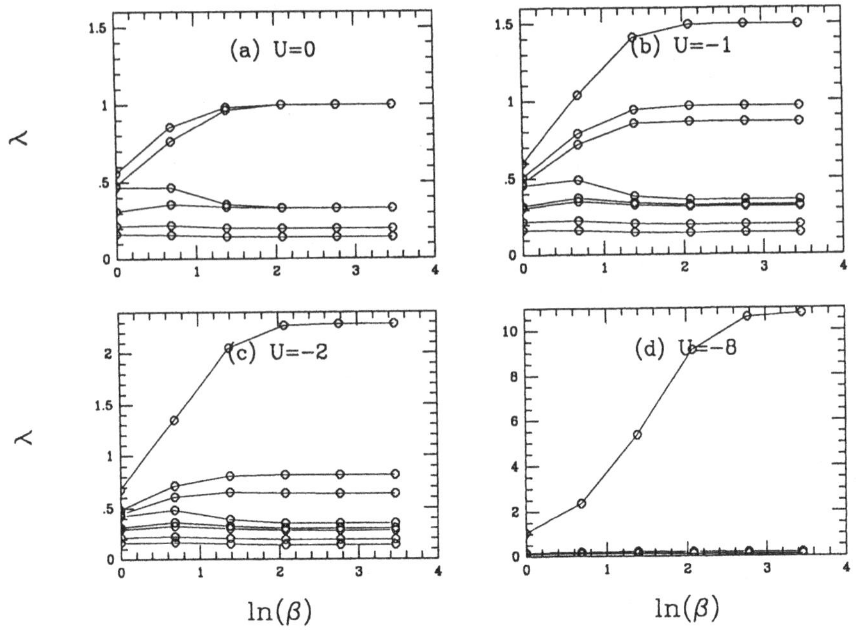

FIG. 10. All eigenvalues vs logarithm of the temperature on a type-I (4×2) lattice with $N_e=4$, $t_x=1.0$, $t_y=0.25$, and $t_2=0$ for (a) $U=0$, (b) $U=-1$, (c) $U=-2$, and (d) $U=-8$.

states with the usual BCS $\ln(2t/T)$ term. More generally,[14] if $t_2\neq0$ the perfect nesting of the Fermi surface is destroyed, but if $\mu=-4t_2$ the leading behavior of $P_d^{(0)}$ remains $\ln^2(2t/T)$ with the coefficient in Eq. (13) multiplied by

$$\left[1-\left[\frac{2t_2}{t}\right]^2\right]^{-1/2}.$$

As μ moves away from $4t_2$, the leading behavior of $P_d^{(0)}$ changes to the usual $N(\mu)\ln(2t/T)$ form, where $N(\mu)$ is the density of states at the Fermi surface.

The leading contributions from the interaction are shown in Fig. 12. The vertex corresponds to the form factor $g(p)$. To linear order in U, Fig. 12(a) gives

$$P_d^{(1)}=-U\left[\frac{T}{N}\right]^2\sum_{p,n}\sum_{p',n'}\frac{g(p)}{\omega_n^2+\epsilon_p^2}\frac{g(p')}{\omega_{n'}^2+\epsilon_{p'}^2}\ . \qquad (15)$$

For $g(p)=\cos p_x-\cos p_y$, the sum vanishes

$$\sum_p\frac{g(p)}{\omega_n^2+\epsilon_p^2}=0 \qquad (16)$$

so that for the d wave, the pair field is not reduced by the linear U term. The second-order terms, Figs. 12(b) and 12(c), correspond to the leading contributions from the interaction and self-energy, respectively. The latter self-energy part is twice the contribution of the diagram shown in Fig. 12(c), since either propagator can be dressed. From Eq. (16) it follows that the second-order ladder contribution vanishes.

The interaction Fig. 12(b) gives a positive contribution and enhances the pairing susceptibility. It is just the leading term in the RPA spin-fluctuation-mediated Berk-Schrieffer[15] interaction. However, the self-energy terms give a negative contribution and suppress the pair susceptibility. We have numerically evaluated the diagrams in Figs. 12(b) and 12(c) for a range of β. Figure 13 shows $(P_b+2P_c)/U^2$ versus β for various values of the chemical potential μ. Here P_b is the contribution from the diagram in Fig. 12(b), and $2P_c$ is twice the contribution of the diagram in Fig. 12(c) divided. Now to the leading order in the interaction we have

$$P_d=P_d^{(0)}+(P_b+2P_c)\ , \qquad (17)$$

and from Fig. 13 we see that the effect of U is to *reduce* the d-wave pairing susceptibility for a range of temperatures. As μ decreases and the system moves away from the region of large spin-density wave fluctuations, the temperature at which (P_b+2P_c) becomes positive in-

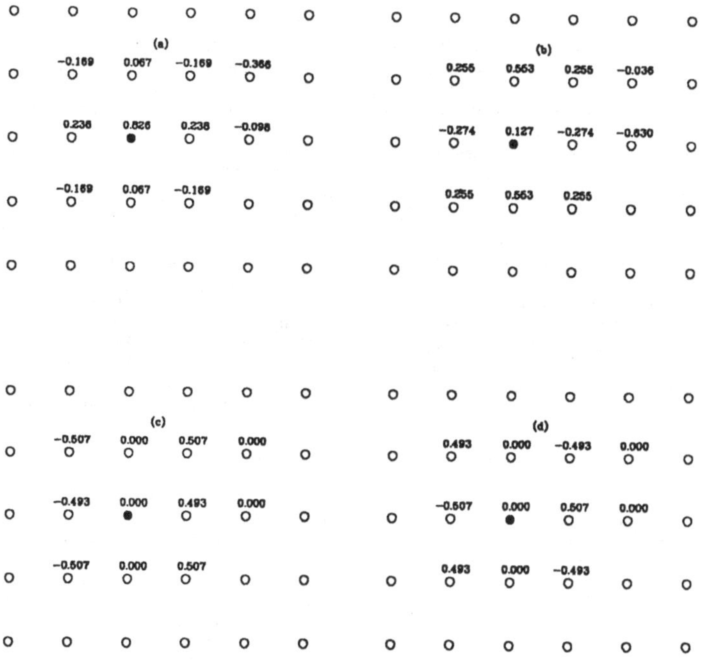

FIG. 11. Pairing structures of (a) the largest eigenvalue, (b) the second-largest eigenvalue, (c) the third-largest eigenvalues, and (d) the fourth-largest eigenvalue for the case of Fig. 10(b) ($U=-1$) for $\beta=32$.

creases. However, as this happens the size of the spin-fluctuation contribution decreases and the pairing interaction becomes weak.[16] Naturally, for a given value of U, as the temperature decreases, higher-order terms become important, and perturbation theory fails. Nevertheless, the fact that the leading-order term suppresses the pairing over a wide region of temperature is quite different from the case of the electron-phonon interaction or the behavior of χ for a Heisenberg ferromagnet as J is increased.

V. DISCUSSION

We have presented numerical results for pairing susceptibilities on eight-site Hubbard clusters. The results for such small systems are naturally very sensitive to the geometry, boundary conditions, and relative size of the hopping parameters. Nevertheless, in studying several different cases we have consistently found suppression of all pairing susceptibilities by the Hubbard repulsion. Our calculation took into account all possible pair wave func-

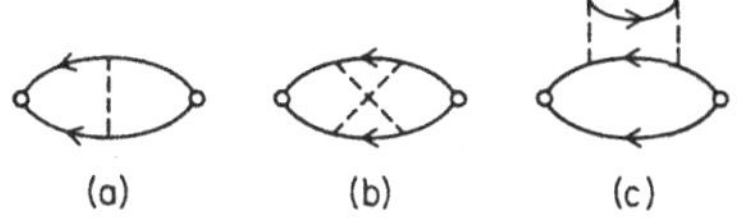

FIG. 12. Perturbation theory graphs for the d-wave pairing susceptibility (a) lowest order, (b) second-order interaction, (c) second-order self-energy. The vertices on each end of a graph correspond to the form factor $g(p)$.

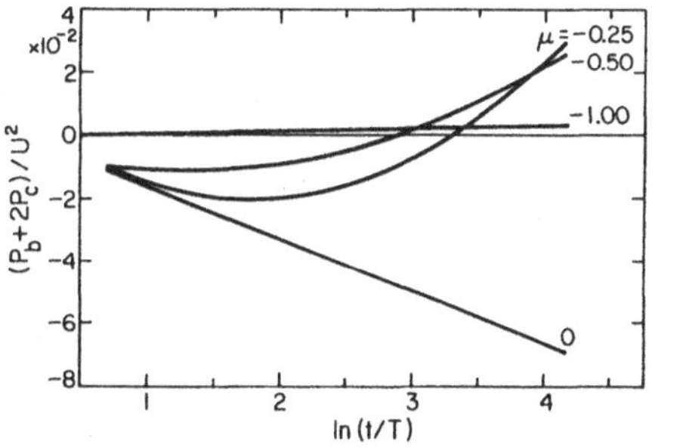

FIG. 13. The leading correction to the pairing susceptibility obtained from the graphs shown in Figs. 10(b) and 10(c), $(P_b+2P_c)/v^2$ vs $\ln(t/T)$ for various values of the chemical potential μ. Here μ is measured in units where $t=1$.

tions that fit onto the clusters considered.

Our results are in contradiction with expectations that in the presence of U the resulting superexchange interaction would induce superconductivity. In the presence of U, the electron pairs associated with the largest pairing susceptibility tend to rearrange themselves so that the weight of the pair wave function at the origin is small. This rearrangement of the wave function to avoid double occupancy evidently causes suppression of the pairing susceptibility in these small systems, and we expect this to carry over to large systems. Emery has suggested that this should occur in a one-band model but not in a two-band model.[17]

Our perturbation theory results also give some insight into why RPA-like calculations fail. The lowest-order self-energy diagram is larger and of opposite sign than the corresponding one for the effective interaction, yielding a net suppression of pairing. In third order we have found vertex corrections which also suppress pairing.

Our results here, together with recent Monte Carlo results,[12] suggests that the 2D Hubbard model does not exhibit superconductivity. This does not, however, necessarily rule out spin fluctuations as playing a role in high-T_c superconductivity, for example, in a two-band model[17,18] or with phonons also playing an essential role.[19] We can also not rule our superconductivity in the 2D Hubbard model at very low temperatures involving very extended pair states, although this would probably not be relevant for high T_c.

ACKNOWLEDGMENTS

This work was supported by the National Science Foundation under Grant Nos. DMR-85-17756, (H.Q.L., J.E.H.) and DMR-86-15454 (D.J.S.). Computations were performed at the Cray X-MP of the San Diego Supercomputer Center. J.H. is grateful to Cray Corporation and AT&T Bell Laboratories for financial support.

---

*Present address: Department of Physics, 510A Brookhaven National Laboratory, Upton, N.Y. 11973.

[1] P. W. Anderson, Science **235**, 1196 (1987).
[2] J. E. Hirsch, Phys. Rev. Lett. **54**, 1317 (1985).
[3] D. J. Scalapino, E. Loh, and J. E. Hirsch, Phys. Rev. B **34**, 8190 (1986).
[4] J. Miyake, S. Schmitt-Rink, and C. Varma, Phys. Rev. B **34**, 6554 (1986).
[5] G. Baskaran, Z. Zou, and P. W. Anderson, Solid State Commun. **63**, 973 (1987).
[6] A. Ruckenstein, P. Hirschfield, and J. Appel, Phys. Rev. B **36**, 857 (1987).
[7] C. Gros, R. Joynt, and T. M. Rice (unpublished).
[8] M. Cyrot, Solid State Commun. **62**, 821 (1987).
[9] H. J. Schulz (unpublished).
[10] F. J. Ohkawa, J. Phys. Soc. Jpn. (to be published).
[11] N. E. Bickers, D. J. Scalapino, and R. T. Scalettar (unpublished).
[12] J. E. Hirsch and H. Q. Lin, Phys. Rev. B (to be published).
[13] G. Kotliar (unpublished).
[14] J. E. Hirsch and D. J. Scalapino, Phys. Rev. Lett. **56**, 2732 (1986).
[15] N. F. Berk and J. R. Schrieffer, Phys. Rev. Lett. **17**, 433 (1986).
[16] See the results for the λ_d given in Ref. 11.
[17] V. J. Emery, Phys. Rev. Lett. **58**, 2794 (1987).
[18] J. E. Hirsch, Phys. Rev. Lett. **59**, 228 (1987).
[19] J. E. Hirsch, Phys. Rev. B **35**, 8726 (1987).